Fifth Edition

Fundamental Concepts in the Design of Experiments

Charles R. Hicks

Purdue University

Kenneth V. Turner, Jr.

Anderson University

New York • Oxford
OXFORD UNIVERSITY PRESS
1999

Oxford University Press

Oxford   New York
Athens   Auckland   Bangkok   Bogotá   Buenos Aires   Calcutta
Cape Town   Chennai   Dar es Salaam   Delhi   Florence   Hong Kong   Istanbul
Karachi   Kuala Lumpur   Madrid   Melbourne   Mexico City   Mumbai
Nairobi   Paris   São Paulo   Singapore   Taipei   Tokyo   Toronto   Warsaw

and associated companies in
Berlin   Ibadan

Published by Oxford University Press, Inc.
198 Madison Avenue, New York, New York 10016
http://www.oup-usa.org

Library of Congress Cataloging-in-Publication Data

Hicks, Charles Robert, 1920–
    Fundamental concepts in the design of experiments / Charles R. Hicks,
Kenneth V. Turner, Jr. — 5th ed.
        p.   cm.
    Includes bibliographical references and index.
    ISBN 0-19-512273-9 (cloth)
    1. Experimental design.   I. Turner, Kenneth V.   II. Title.
QA279.H53   1999
001.4'34—dc21   98-27756
                                                                    CIP

Printing (last digit): 9 8 7 6 5 4 3 2 1

Printed in the United States of America
on acid-free paper

CONTENTS

PREFACE

As stated in all previous editions, the primary purpose of this book is to present the fundamental concepts in the design of experiments using simple numerical problems, many from actual research work. These problems are selected to emphasize the basic philosophy of design. Another purpose is to present a logical sequence of designs that fit into a consistent outline; for every type of experiment, the distinctions among the experiment, the design, and the analyses are emphasized. Since this theme of experiment–design–analysis is to be highlighted throughout the text, the first chapter presents and explains this theme. The second chapter reviews statistical inference, and the remaining chapters follow a general outline, with an experiment–design–analysis summary at the end of each chapter.

The book is written for anyone engaged in experimental work who has a good background in statistical inference. It will be most profitable reading to those with a background in statistical methods including the analysis of variance.

I am pleased to have Dr. Kenneth V. Turner Jr. of Anderson University as coauthor of this fifth edition. Both Dr. Turner and I have worked as consultants for McElrath and Associates, Inc., and both of us have taught many courses using earlier editions of this text. He is largely responsible for this latest revision and brings a wealth of knowledge and experience to this effort, especially in the areas of new computer programs to handle the analysis phase of experiment, design, and analysis.

Charles R. Hicks

It has been a pleasure to contribute to the fifth edition of this book. The previous edition is a very good book. I hope that I have, in some small way, made this edition even better.

Users of the previous edition will find more emphasis, this time, on the use of computer outputs as part of the analyses. The discussion of formulas and their uses has not been omitted, however, for we found, when we used previous editions for in-plant training courses, that students often rely on hand calculations as an aid to understanding.

Advances in technology make possible many graphical procedures and techniques that were cumbersome, if not impossible, only a few years ago. As a result, greater emphasis has been given to graphical procedures such as residual plots and normal quantile plots.

The reader will also find a stronger emphasis on the use of orthogonal tables in discussions of 2^f and 3^f factorial experiments, fractional factorials, confounding, and aliases. This makes the consideration of Taguchi's use of orthogonal arrays and linear graphs seem quite natural.

Chapter 14 has been revised substantially, and the relationship between Taguchi's basic designs and Plackett–Burman designs has been noted. An introduction to outer arrays and signal-to-noise ratios has been added.

I wish to thank P. Duane Saylor, who worked as a consultant for McElrath and Associates, for giving me my first introduction to SAS. The SAS programs included in the text are a direct result of his tireless efforts as he taught courses based on previous editions of this book. Grateful acknowledgment is also given to Bill B. Morgan and Keith J. Sanborn of Delco Electronics for sharing their insights and encouragement, and for sharing some of the new problems with us.

Special thanks to Jacek Dmochowski, State University of New York at Buffalo; Ray Eberts, Purdue University; Tom Kuczek, Purdue University; Joseph Naus, Rutgers University; and William I. Notz, Ohio State University, for having reviewed the fourth edition and for providing comments that led to many of the revisions incorporated in this fifth edition.

We are indebted to Bill Zobrist and the staff at Oxford University Press for their assistance in making this edition a reality. We are also indebted to our families for bearing with us throughout this project.

Kenneth V. Turner Jr.

Chapter 1

THE EXPERIMENT, THE DESIGN, AND THE ANALYSIS

1.1 INTRODUCTION TO EXPERIMENTAL DESIGN

One of the most abused words in the English language, "research" means many things to many people. The high school student who visits the library to gather information on the Crusades is doing research. A man who checks the monthly sales of his company for the year 1995 is doing research. Many such studies make use of bibliothecal or statistical tools or similar research tools but are a far cry from a scientist's conception of the term. Let us narrow the concept a bit and define *research* as "a systematic quest for undiscovered truth." (See Leedy [18].[1])

In this definition we note that the quest must be "systematic," using sound scientific procedures—not a haphazard, seat-of-the-pants approach—to solving a problem. The truth sought should be something that is not already known, which implies a thorough literature search to ascertain that the answer is not available in previous studies.

True Experiments and Experimental Units

Not all studies are necessarily research, nor is all research experimental. A *true experiment* may be defined as a study in which certain independent variables are manipulated, their effect on one or more dependent variables is determined, and the levels (values) of these independent variables are assigned at random to the experimental units in the study. Such experiments allow the researcher to develop a statistical model and estimate its validity.

An *experimental unit* must be carefully defined. It may be a part, a person, a class, a lot of parts, a month, or some other unit depending on the problem at hand. The experimental units make up a *universe* of all such units, present or future, and a *population* is defined as measurements that might be taken on all experimental units in the universe. There may be several populations defined on the same experimental units. For example, if the universe is made up of all persons in a given school, we may conceive of at least two populations: one made up of the heights of all these persons and another of the weights of these same persons.

---

[1]Numbers in brackets refer to the references at the end of the book.

Quasi-Experiments

Manipulation and randomization are essential for a true experiment from which one may be able to infer cause and effect. It is not always physically possible to assign some variables to the experimental units at random, but it may be possible to run a quasi-experiment in which groups of units are assigned to various levels of the independent variable at random. For example, it may not be possible to study the effects of two methods of in-plant instruction on worker output by assigning workers at random to the two methods. Instead, it may be necessary to decide by a random flip of a coin which two of four intact classes will receive method I and which two will receive method II. In such quasi-experiments the experimenter is obligated to demonstrate that the four classes were similar at the start of the study. Since they can be dissimilar in many different ways, this is no easy task. Randomizing all individuals to the four groups will assure a high probability of group equality on many variables.

A similar situation arises when different batches (or lots) of material are used with each of several processing methods. Suppose four lots of 24 wafers are randomly assigned to four different metal evaporation systems used to transfer metal from a source to the surface of a wafer by means of a high-voltage beam and the metal thickness at a particular wafer position is measured. Since all information about the wafers in a lot is included in the information about the system used to process that lot, we say that *lot is confounded with system*. Any differences among the four sets of thicknesses may be due to differences among lots.

Ex-post-facto Research

There are other types of research that are not experimental. One of the most common is *ex-post-facto* research. The values of the variables have been determined by circumstances beyond the control of the experimenter, the variables have already acted, and the research measures only what has occurred. An example of some note would be the studies showing a high incidence of cancer in people who smoke heavily. There is no manipulation here: researchers did not assign X cigarettes per day to some people and none to another group, and then wait to see whether cancer developed. Instead they studied what they found from people's historical records and attempted to make an inference in each case. However, such inferences are dangerous, since the researchers cannot claim to have proved that smoking causes cancer unless they are able to rule out other competing hypotheses as to what may have caused cancer in the people studied.

In this category are survey research, which is also ex post facto, historical research, and developmental research. All these investigations may use statistical methods to analyze data collected but none is really experimental in nature.

■ **Example 1.1**

Illustrations of true experiments and ex-post-facto experiments can often be found in the techniques of regression and correlation. In one type of regression problem, levels of the independent variable X are set, a level of X is assigned at random to each experimental unit in the study, observations are made on the dependent variable Y, and an equation for predicting Y from X over the range of specified X's is sought. Such studies are true experiments.

A correlation problem may involve a sample of experimental units for which two variables (X and Y) have acted, the resulting (x, y) pairs are plotted to look for a relationship, and a correlation coefficient is calculated to measure the strength of the relationship. Such studies are ex-post-facto studies.

Assuming that true experiments are our basic concern, what steps should be taken to make a meaningful study?

To attack this problem one may easily recognize three important phases of every project: the experimental or planning stage, the design stage, and the analysis phase. The theme of this book might be called experiment, design, and analysis.

1.2 THE EXPERIMENT

The experiment includes a statement of the problem to be solved. This sounds rather obvious, but in practice it often takes quite a while to get general agreement as to the statement of a problem. It is important to bring out all points of view to establish just what the experiment is intended to do. A careful statement of the problem goes a long way toward its solution.

Too often problems are tossed about in very general terms—a sort of "idea burping" (see [18]). As an example, note how these two statements of a problem differ:

- What can Salk vaccine do for polio?
- Is there a difference in the percentage of first, second, and third grade children in the United States who contract polio within a year after being inoculated with Salk vaccine and those who are not inoculated?

Much care must be taken to spell out the problem in terms that are well understood and point the way in which the research might be conducted.

Response Variables

The statement of the problem must include reference to at least one characteristic of an experimental unit on which information is to be obtained. Such characteristics are called *response* (or *dependent*) *variables*. The response variable may be *qualitative*, such as whether or not an individual contracts polio, or *quantitative*, such as yield in grams of penicillin per kilogram of corn steep.

The outcomes of a true experiment occur at random, so a response variable is a *random variable*. Knowledge of such a variable is essential because the shape of its distribution often dictates what statistical tests can be used in the subsequent data analysis.

In addition to a statement of the problem that includes reference to a response variable, one must ask the following questions about the measurement method. What instruments are necessary to measure the variable? How accurately can the variable be measured? For a detailed discussion of this latter issue, see Chapter 9 of Ostle et al. [26].

Independent Variables

Many controllable experimental variables, called *independent variables* or *factors*, may contribute to the value of the response variable. Factor values are called *levels* of the factor. *Qualitative factors* have levels that vary by category rather than numerical degree. Vendor, machine, material, and furnace are examples of qualitative factors. *Quantitative factors*, on the other hand, have levels that are counts or measurements. Temperature in degrees Fahrenheit, oxygen level in parts per million, and glue bath depth in mils are examples of quantitative factors. The experimenter must decide which factors will be held constant, to be manipulated at certain specified levels, and which will be averaged out by a process of randomization.

Factors whose levels are set at specified values are called *fixed effects*, and those whose levels are randomly selected from all possible levels are called *random effects*. Are levels of some factors to be set at values specified by the experimenter (such as temperature at 70, 90, and 110°F) and those of other factors to be chosen at random from all possible levels? The answer will determine the nature of the information one can obtain from the experimental results.

When at least two factors are being considered, one must also ask: How are the factor levels to be combined? Each combination of levels from two or more factors is called a *treatment combination*. If at least one value of the response is observed at each treatment combination, the experiment is called a (complete) *factorial experiment*. An experiment in which the levels of one factor are chosen within the levels of another factor is a *nested experiment*.

■ **Example 1.2 (A Factorial Experiment)**

Consider an experiment in which the factor A is to be considered at three levels and a second factor B is to be considered at two levels. Three observations are to be made at each treatment combination. If the $3 \times 2 = 6$ treatment combinations can be set in the following way, a factorial experiment will be possible. Notice that both levels of factor B can be used with the three levels of factor A.

Factor B	Factor A		
	1	2	3
1	——— ——— ———	——— ——— ———	——— ——— ———
2	——— ——— ———	——— ——— ———	——— ——— ———

■ **Example 1.3 (A Nested Experiment)**

Sometimes one chooses three suppliers and randomly selects two batches of material from each for chemical analysis, letting A denote the supplier and B denote the batch. Such an experiment has the nested arrangement, as illustrated below. There are three "nests" (the suppliers) with two "eggs" (the batches) in each nest. This involves six sets of data, like the factorial arrangement in Example 1.2, but now the factors are in a decidedly different arrangement. Unlike the two levels of factor B in the factorial arrangement, the two levels of factor B (batch) associated with level 1 of factor A (supplier) cannot be used with levels 2 and 3 of factor A.

	Supplier, A				
1		2		3	
Batch, B		Batch, B		Batch, B	
1	2	3	4	5	6

1.3 THE DESIGN

Many times an experiment is agreed upon, data are collected, and conclusions are drawn, but little or no consideration is given to *how* the data were collected. In such cases, experimental error (noise) may cause the investigator to draw invalid conclusions; or, the data may not reveal an important result. To avoid such problems, the investigator needs an experimental design for obtaining data that provide objective results with a minimum expenditure of time and resources. The remainder of this book is devoted to the design phase of a project and the analysis of data obtained from a designed experiment. Of the two, design and analysis, the design of the experiment is of primary importance and will require the most effort.

How Many Observations Are Needed?

One of the first questions we face when designing an experiment is: How many observations are to be taken? Considerations of how large a difference is to be detected, how much variation is present, and what size risks can be tolerated are all important in answering this question. If such information is lacking, one is best advised to take as large a sample as possible. In practice this sample size is often quite arbitrary. However, the availability of tables and statistical software often make it possible to determine the sample size in a much more objective fashion (see Section 2.5).

Order of Experimentation: Handling the Independent Variables

Also of prime importance is the order in which the observations are obtained. Once a decision has been made to control certain variables at specified levels, there are always a number of other variables that cannot be controlled. To average out the effects of these uncontrolled variables, it is necessary to randomize the order of experimentation.

■ **Example 1.4**

An experimenter wishes to compare the average current flow through two types of computer. If 5 machines of each type are to be tested, in what order are all 10 to be tested?

If the 5 of type I are tested, followed by the 5 of type II, and any general "drift" in line voltage occurs during the testing, it may appear that the current flow is greater on the first 5 (type I) than on the second 5 (type II); yet the real cause is the "drift" in line voltage. This problem can be minimized by randomizing the order of testing. First, number the computers, using 1, 2, 3, 4, and 5 for those of type I and 6, 7, 8, 9, and 10 for those of type II. Then determine the testing order, either by means of a random number table, by drawing numbered slips of paper from a box, or by using statistical software. When the latter method was used by the authors, the following random testing order was obtained: 2, 10, 3, 4, 9, 8, 1, 7, 5, 6.

A random order for testing allows any time trends to average out. It is desirable to have the average current flow in the type I and type II computers equal, if the computer types do not differ in this respect. Randomization will help accomplish this. Randomization will also permit the experimenter to proceed as if the errors of measurement were independent, a common assumption in most statistical analyses.

What is meant by random order? Is the whole experiment to be completely randomized, with each observation made only after consulting a table of random numbers, or tossing dice, or flipping a coin, as in Figure 1.1? Or, are two temperature baths to be prepared, with one randomization made only within one particular temperature and another randomization made at the other temperature, as in Figure 1.2? In other words, what is the randomization procedure, and how are the units arranged for testing? Once this step has been agreed upon, the experimenter should keep a watchful eye on the experiment to see that it is actually conducted in the order prescribed.

Independent variables may be purposefully introduced into the study because they are of interest, because they represent restrictions on randomization, or, indeed, because they are somewhat extraneous and not of interest to the experimenter. When deciding how such variables should be handled, one should seek to maximize the systematic variation of factors of interest and to minimize error (unexplained) variation, including so-called errors of measurement, by controlling the systematic variation of the factors that are not of interest. Kerlinger [15] refers to this as the *max-min-con principle* of experimental design.

Temperature	Device Type		
	1	2	3
1	(10)	(18)	(6)
	(17)	(16)	(2)
	(11)	(5)	(1)
2	(12)	(8)	(13)
	(4)	(7)	(15)
	(9)	(3)	(14)

Figure 1.1   Data layout: a completely randomized experiment.

Temperature	Device Type		
	1	2	3
1	(6)	(5)	(4)
2	(1)	(3)	(2)

Figure 1.2   Data layout: randomization within temperature level.

To successfully apply the max-min-con principle in the design of an experiment, one should note that there are basically three ways to handle independent variables:

1. Rigidly controlled—the factors remain fixed throughout the experiment
2. Manipulated or set at levels of interest
3. Randomization

When factors remain fixed, any inferences drawn from the experiment are valid only for the fixed conditions. For example, one might decide to study only children from the first three grades of a 12-grade school system. Or, of 28 molding machines, only machines *A, B,* and *C* can be freed up long enough to investigate a quality problem related to the molded product.

Variables that are manipulated or set at levels of interest are those whose effects on the response variable are to be studied. They may be either qualitative or quantitative, and their levels may be either fixed or random. Every effort should be made to include extreme levels, to provide the opportunity to maximize any effects that may be present.

The order of experimentation is randomized to average out the effects of variables that cannot be controlled. Such averaging does not remove those effects completely, however; they still increase the variation in the observed data.

The Model Description

Having agreed upon the experiment and the randomization procedure, we can now set up a mathematical model to describe the experiment. This model will show the response variable as a function of all factors to be studied and any restrictions imposed on the experiment as a result of the method of randomization. A typical model description for the experimental layout in Figure 1.2 is

$$Y_{ij} = \mu + \beta_i + \tau_j + \varepsilon_{(ij)}$$

with Y_{ij} the measured variable, μ a common effect in all observations, β_i the temperature effect (where $i = 1, 2$), τ_j the device effect (where $j = 1, 2, 3$), and $\varepsilon_{(ij)}$ the random error in the experiment.

The Research Hypothesis

Since the objective of the research project is to shed light on a stated problem, one next expresses the problem in terms of a testable hypothesis or hypotheses. A research hypothesis is a formal statement of what the experimenter expects to find in the data. For example: "Inoculation with Salk vaccine will lower the incidence of polio." Or: "The four treatments will produce different average yields."

Statisticians usually state their hypotheses in null form, since these are the only easily testable statements. For example: "There will be no difference in the percentage of youngsters contracting polio between those inoculated and those not inoculated." Or: "The treatment means will have the same average yield." A null hypothesis usually can be expressed in terms of the mathematical model set up in the last step of the design phase.

1.4 THE ANALYSIS

The final step, analysis, begins with data collection as described in connection with the design. Notice that data collection begins *after* the nature of an experimental design has been developed. Make sure that easy-to-use, foolproof forms have been prepared. If the experiment is complicated, a pilot run may be used to check details and make sure that no critical details have been overlooked. Once collected, the data should be prepared for manual or computer analysis.

The preparation of graphical displays, inspection of those displays, and computation of appropriate numerical summaries should follow the data collection. In many instances, important conclusions can be made from the graphical displays. In others, values of the test statistic required for a formal hypothesis test will be necessary.

Once graphical displays and test statistics have been prepared, decisions must be made. A meeting of all those involved in the design, conduct, and analysis of the experiment is important at this stage. The decisions should be expressed in terms that are meaningful both to the experimenter and to those scheduled to receive a report. Such reports should not be couched in statistical jargon such as "the third-order $A \times B \times E$ interaction is significant at the 1% level." Instead, use graphical or tabular formats that will be clearly understood not only by the experimenter but by those who are to be "sold" by the experiment. The actual statistical tests may be included in an appendix.

An investigative process has built-in self-correction. As depicted in Figure 1.3, it is an iterative process that moves us closer to the "truth" at each stage. Statistics enters such processes at two points: (1) the design of the experiment and (2) the analysis and interpretation of the experimental results. When certain hypotheses appear to be tenable, these results should be used as "feedback" to design a better experiment.

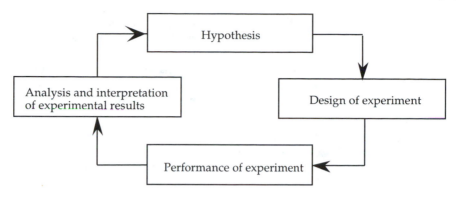

Figure 1.3   Structure of an investigative process.

1.5 EXAMPLES

Examples 1.5 and 1.6 are but two illustrations of how projects are organized into three phases: experiment, design, and analysis.

■ **Example 1.5 (Salk Vaccine Experiment)**

The experiment conducted in the United States in 1954 concerning the use of Salk vaccine in the control of polio has become a classic in its employment of good design principles. A more detailed description of the study can be found in *Statistics: A Guide to the Unknown* (see [34]).

Before the statement of the problem was agreed upon, there was much discussion as to what universe was to be sampled. Polio is a disease affecting predominantly children in the early grades. The designers of the experiment decided that the experimental unit should be a child and that the universe should be the set of all children in the United States in grades 1, 2, and 3. Since a child contracts polio or does not contract polio, the population of concern consists of 0's and 1's, with 0 assigned to each child in the universe who does not contract polio and 1 assigned to each child who does contract the disease.

Once the universe and population of interest had been determined, the researchers agreed on the following problem statement: Is there a difference in the percentage of first, second, and third grade children in the United States who contract polio within a year after being inoculated with Salk vaccine and those who are not inoculated? This statement allows the dependent variable to be either the percentage of students in the study who contract the disease or the number who contract it. The basic variable here is a qualitative one—a binomial variable Y—as children fall into just two categories: contracting polio or not contracting polio.

Factors that might affect the likelihood that a given child would contract polio included socioeconomic status, grade level, doctor making the diagnosis, geographic area of residence, and whether the child had been inoculated. Researchers agreed that the main factor of interest should be each child's status as inoculated or not inoculated, which is a fixed effect, and that all children in the study should be from the first three elementary grades without distinguishing between grade level. They decided that the other factors were noise (or nuisance) variables, to be handled in the design procedures.

Since there is only one qualitative, fixed factor of interest—inoculated or not inoculated—there is no basic concern about combining factors. However, because three grade levels were to be included in the experiment, the researchers decided to stratify their sample by grades; that is, they would use approximately equal sample sizes of inoculated and noninoculated children from each grade level.

Much planning went into the design phase. The researchers noted that the overall percentage of children contracting polio was quite small: about 8 or 10 cases in 10,000 children. To get percentages large enough to discriminate between those who were inoculated and those who were not, therefore, it was necessary to take a fairly large sample. The factor of interest is a classification variable x for which $x = 1$ if the child is inoculated and $x = 2$ if the child is not inoculated. So, the researchers decided to use an approximately equal number of children at each level of x. Those assigned to level 2 would serve as a control group whose members received a placebo (a salt solution).

At first it was proposed that children in grades 1 and 3 receive the vaccine and those in grade 2 be given the placebo treatment. This plan was later abandoned because polio is a contagious disease, and if children in a given grade get polio, their classmates might contract it as well. Another early idea was to assign children of parents who would consent to the inoculation to the "experimental" class, with children of parents who did not consent in the "control" category. This too was abandoned because parents of a higher socioeconomic status were thought to be more likely to consent than parents lower in socioeconomic status, and polio is a socially related disease—it affects more children from the upper echelons of society than from the lower!

The final decision called for a complete randomization countrywide involving about 1 million randomly selected children from grades 1 to 3 in areas of the United States where polio had been quite prevalent in the past. More than 400,000 sets of parents consented to have their children participate in the study. This produced a pool of approximately 750,000 children who could participate.

Realizing that some children in the pool identified probably would not participate when the actual experiment was started and knowing that the incidence of polio was low, the researchers decided to start by ascertaining whether the sample sizes would be adequate. They were concerned that after time and money had been invested in the experiment, they might come to an incorrect conclusion: that the vaccine reduced the incidence of polio among children in the early grades when in fact it had no such effect (a type I error), or that the vaccine was ineffective in reducing that incidence when, in fact, it was effective (a type II error). Using the fact that only about 0.09% of children contracted polio, they were able to determine that samples of at least 200,000 subjects and 200,000 controls would be adequate to make the likelihood of either error quite small.

The researchers also decided that the children who were able to participate at the time of the experiment should be divided into two groups at random, with group 1 getting the Salk vaccine and group 2 getting the placebo. The randomization would produce two groups that were approximately the same in terms of the noise factors. To handle other concerns, they decided to run the experiment as a "double-blind" experiment. The children would not know whether they were receiving the vaccine or the placebo, and the doctors who later diagnosed the children would not know which of their patients had received the vaccine and which had not.

A mathematical model could be written as

$$Y_{ij} = \mu + \tau_j + \varepsilon_{ij}$$

where Y_{ij} is equal to 0 or 1: 0 if no polio is diagnosed and 1 if polio is found. The subscript i represents the child, a number from 1 to approximately 200,000. The subscript j represents the treatment received by the ith child: 1 if treated with Salk vaccine and 2 if not so treated. Thus, μ is a constant or general average of the 0's and 1's for the whole population, τ_j is the effect of the jth treatment, and ε_{ij} is a random error associated with the ith child receiving treatment j.

Letting n_j denote the number of children who receive treatment j, the proportion of children contracting polio who received treatment j is

$$P_j = \frac{\sum_{i=1}^{n_j} Y_{ij}}{n_j} \tag{1.1}$$

and the proportion contracting polio in the total sample of $n = n_1 + n_2$ children is

$$P = \frac{\sum_{i=1}^{n_1} Y_{i1} + \sum_{i=1}^{n_2} Y_{i2}}{n_1 + n_2} \tag{1.2}$$

Use of P_j indicates that the proportion contracting polio who receive treatment j is a random variable *before* the experiment is conducted. The proportion observed *after* conducting the experiment is denoted p_j.

Letting θ_j denote the true proportion in the jth population who contract polio, the statistical hypothesis to be tested is

$$H_0 : \theta_1 = \theta_2$$

with alternative

$$H_1 : \theta_1 < \theta_2$$

The form of the alternative hypothesis indicates that we are interested in showing that the true proportion contracting polio will be less for treatment 1 (vaccine) than for treatment 2 (placebo).

Approximately 0.09% of all children contract polio, and (200,000)(0.0009) = 180 is sufficiently large, so the sample proportions have approximate normal distributions. Thus,

$$Z = \frac{P_1 - P_2}{\sqrt{P(1 - P)[(1/n_1) + (1/n_2)]}} \tag{1.3}$$

is an appropriate test statistic. When the null hypothesis ($H_0 : \theta_1 = \theta_2$) is true, Z has an approximate normal distribution with mean 0 and variance 1.

The results of the experiment are summarized in Table 1.1. When equations (1.1), (1.2), and (1.3) are used with those results, an observed value of

$$z \approx \frac{\left(28 \times 10^{-5}\right) - \left(71 \times 10^{-5}\right)}{\sqrt{\left[49 \times 10^{-5}\right]\left[1 - \left(49 \times 10^{-5}\right)\right]\left[(1/200, 745) + (1/201, 229)\right]}} \approx -6.14$$

is obtained. The probability of observing a value of Z as extreme as -6.14 when the null hypothesis is true is $P(Z \leq -6.14) \approx 4.126 \times 10^{-10}$. Thus, the likelihood of observing a difference of this magnitude when the Salk vaccine has no effect is less than one in a billion. This gives a very, very strong indication that the researchers should reject H_0 and conclude that the vaccine is effective. The wisdom in that decision has been strongly substantiated by the almost complete eradication of polio in the United States.

TABLE 1.1
Polio Experiment Results

Treatment	Sample Size	Number Contracting Polio	Proportion Contracting Polio
Salk vaccine	$n_1 = 200,745$	56	$p_1 \approx 28 \times 10^{-5}$
Placebo	$n_2 = 201,229$	142	$p_2 \approx 71 \times 10^{-5}$
Totals	$n = 401,974$	198	$p \approx 49 \times 10^{-5}$

■ Example 1.6 (A Factorial Experiment)

This example [21] shows the three phases of the design of an experiment. It is not assumed that the reader is familiar with the design principles or analysis techniques in this problem. The remainder of the book is devoted to a discussion of many such principles and techniques, including those used in this problem.

An experiment was to be designed to study the effect of several factors on the power requirements for cutting metal with ceramic tools. The metal was cut on a lathe, and the vertical component of a dynamometer reading was recorded. Since this component is proportional to the horsepower requirements in making the cut, it was taken as the measured variable, Y. The vertical component is measured in millimeters of deflection on a recording instrument. Some of the factors that might affect this deflection are tool types, angle of tool edge bevel, type of cut, depth of cut, feed rate, and spindle speed. After much discussion, it was agreed to hold three factors constant: depth of cut, at 0.100 inch; feed rate, at 0.012 in./min; and spindle speed, at 1000 rpm. These levels were felt to represent typical operating conditions. The main objective of the study was to determine the effect of the other three factors (tool type, angle of edge bevel, and type of cut) on the power requirements. Since only two ceramic tool types were available, this factor was considered at two levels. The angle of tool edge bevel was also set at two levels, 15° and 30°, representing the extremes for normal operation. The type of cut was either continuous or interrupted—again, two levels.

There are therefore two fixed levels for each of three factors, or eight experimental conditions (2^3), which may be set and may affect the power requirements or vertical deflection (Y) on the dynamometer. This is called a 2^3 factorial experiment, since both levels of each of the three factors are to be combined with both levels of all other factors. The levels of two factors (type of tool and type of cut) are qualitative, whereas the angle of edge bevel (15° and 30°) is a quantitative factor.

The question of design for this experiment involves the number of tests to be made under each of the eight experimental conditions. After some preliminary discussion of expected variability under the same set of conditions and the costs of wrong decisions, it was decided to take four observations under each of the eight conditions, making a total of 32 runs. The order in which the 32 units were to be put in a lathe and cut was to be completely randomized.

To completely randomize the 32 readings, the experimenter decided on the order of experimentation from the results of three coin tossings. A penny was used to represent the tool type T: heads for tool type 1, tails for type 2; a nickel represented the angle of bevel B: heads for 15°, tails for 30°; and a dime represented the type of cut C: heads for continuous, tails for interrupted. Thus, if the first set of tosses came up HTH, the first experimental setup to be used would be tool type 1 with a 30° angle of edge bevel and a continuous cut. The coin flipping was to continue until the order of all 32 runs had been decided upon.

The data layout given in Figure 1.4 shows each of the experimental conditions. The symbols (1), (2), . . . , (32) represent the orders obtained from the coin tosses. That is, (1), (2), (3), (4), and (5) indicate that the first five tosses of the coins came up HTH, HHT, TTH, TTH, and THT, respectively. In this layout note that the same set of conditions may be repeated (e.g., runs 3 and 4) before all eight conditions are run once. The only restriction on complete randomization here is that once four repeated measures have occurred in the same cell, no more will be run using those same conditions.

	Tool Type, T			
	1		2	
	Bevel Angle, B		Bevel Angle, B	
Type of Cut, C	15°	30°	15°	30°
Continuous	(8)	(1)	(11)	(3)
	(16)	(12)	(15)	(4)
	(20)	(27)	(21)	(19)
	(23)	(28)	(22)	(26)
Interrupted	(2)	(7)	(5)	(6)
	(24)	(9)	(14)	(13)
	(31)	(10)	(18)	(29)
	(32)	(17)	(25)	(30)

Figure 1.4   Data layout of power consumption for ceramic tools.

This experiment is a 2^3 factorial experiment with four observations per cell, run in a completely randomized manner. Complete randomization ensures the averaging out of any effects that might be correlated with the time of the experiment. If the lathe-spindle speed should vary, being faster at first and slower near the end of the experiment, and all the type 1 tools were run through first, followed by all the type 2 tools, this extraneous effect of lathe-spindle speed might appear as a difference between tool types.

The mathematical model for this experiment and design would be

$$Y_{ijkm} = \mu + T_i + B_j + TB_{ij} + C_k + TC_{ik} + BC_{jk} + TBC_{ijk} + \varepsilon_{m(ijk)}$$

with Y_{ijkm} the measured variable, μ a common effect in all observations (the true mean of the population from which all the data came), T_i the tool type effect (where $i = 1, 2$), B_j the angle of bevel (where $j = 1, 2$) with level 1 the 15° angle, C_k the type of cut (where $k = 1, 2$) with level 1 the continuous cut, and $\varepsilon_{m(ijk)}$ the random error in the experiment (where $m = 1, 2, 3, 4$). The other terms stand for interactions[2] between the main factors T, B, and C.

Analysis of this experiment begins with the collection of 32 items of data in the order indicated in Figure 1.4. The results in millimeter deflection are given in Table 1.2. Using Minitab[3] with the given model and fixed effects produces the ANOVA summary in Table 1.3.

The experimental design lets us test each null hypothesis (no type of tool effect, no bevel effect, no type of cut effect, no interaction between tool type and bevel angle, no interaction between tool type and type of cut, no interaction between bevel angle and type of cut, and no three-way interaction) by comparing the appropriate mean square and the error mean square of 2.227. The proper test statistic is the F statistic (Statistical Table D) with 1 and 24 degrees of freedom. In a test at the $100\alpha\%$ level, a null hypothesis can be rejected when the p value, $P(F_{(1,24)} \geq F_{observed})$, is less than or equal to α.

TABLE 1.2
Data for Power Requirement Example

	Tool Type, T			
	1		2	
	Bevel Angle, B		Bevel Angle, B	
Type of Cut, C	15°	30°	15°	30°
Continuous	29.0	28.5	28.0	29.5
	26.5	28.5	28.5	32.0
	30.5	30.0	28.0	29.0
	27.0	32.5	25.0	28.0
Interrupted	28.0	27.0	24.5	27.5
	25.0	29.0	25.0	28.0
	26.5	27.5	28.0	27.0
	26.5	27.5	26.0	26.0

---

[2]The meaning of an interaction between two terms is considered near the end of this example.
[3]Minitab is a registered trademark of Minitab Inc.

When considering each p value in Table 1.3, we find that the hypothesis of no angle of bevel effect can be rejected for any $\alpha \geq 0.006$ and that of no type of cut effect can be rejected for any $\alpha \geq 0.001$. None of the other hypotheses can be rejected at any reasonable significance level, and it is concluded that only angle of bevel and type of cut affect power consumption as measured by the vertical deflection on the dynamometer. Tool type appears to have little effect on the vertical deflection, and all interactions are negligible.

Calculations on the original data of Table 1.2 show the average vertical deflections given in Table 1.4. These averages seem to bear out the conclusions that bevel affects vertical deflection, with a 30° bevel requiring more power than the 15° bevel (note the difference in the sample means of 1.6 mm) and that type of cut affects vertical deflection, with a continuous cut averaging 2.0 mm more deflection than an interrupted cut. The difference of 0.6 mm in average deflection due to tool type is not significant for any $\alpha < 0.272$.

The graph of the sample means for the four B–C treatment combinations (Figure 1.5) illustrates the meaning of "no significant interaction." A brief examination of that graph indicates that the vertical deflection increases when the degree of bevel increases. The line for the continuous cut is above the line for the interrupted cut, indicating that the continuous cut requires more power. That the lines are nearly parallel is characteristic of no interaction between two factors. Stated differently, an increase in the degree of bevel produced about the same average increase in vertical deflection regardless of which type of cut was made, indicating the presence of no interaction between those factors.

The experiment here was a three-factor experiment with two levels for each factor. The design was a completely randomized design, and the analysis was a three-way ANOVA with four observations per cell. From the results of this experiment, the experimenter not only found that two factors (angle of bevel and type of cut) affect

TABLE 1.3
Minitab ANOVA for Power Requirement Example

Analysis of Variance (Balanced Designs)

Factor	Type	Levels	Values	
Tool	fixed	2	1	2
Bevel	fixed	2	1	2
Cut	fixed	2	1	2

Analysis of Variance for Vertical Deflection

Source	DF	SS	MS	F	P
Tool	1	2.820	2.820	1.27	0.272
Bevel	1	20.320	20.320	9.13	0.006
Tool*Bevel	1	0.195	0.195	0.09	0.770
Cut	1	31.008	31.008	13.93	0.001
Tool*Cut	1	0.008	0.008	0.00	0.953
Bevel*Cut	1	0.945	0.945	0.42	0.521
Tool*Bevel*Cut	1	0.195	0.195	0.09	0.770
Error	24	53.437	2.227		
Total	31	108.930			

TABLE 1.4
Sample Average Vertical Deflections

Tool type, T	1	28.1
	2	27.5
Bevel angle, B	15°	27.0
	30°	28.6
Type of cut, C	Continuous	28.8
	Interrupted	26.8

the power requirements, but also determined that within the range of the experiment it makes little difference which ceramic tool type is used and that there are no significant interactions among the three factors.

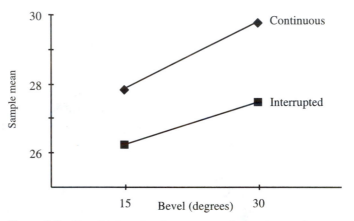

Figure 1.5 $B \times C$ interaction for power requirement example.

1.6 SUMMARY IN OUTLINE

I. *EXPERIMENT*
   A. Statement of problem
   B. Choice of response (dependent) variable
   C. Selection of factors (independent variables) to be varied
   D. Choice of levels of these factors
      1. *Quantitative or qualitative*
      2. *Fixed or random*

II. *DESIGN*
   A. Number of observations to be taken
   B. Order of experimentation
   C. Method of randomization to be used
   D. Mathematical model to describe the experiment

        E. Hypotheses to be tested

III. *ANALYSIS*
        A. Data collection and processing
        B. Computation of test statistics and preparation of graphics
        C. Interpretation of results for the experimenter

1.7 FURTHER READING

Box, G. E. P., W. G. Hunter, and J. S. Hunter. *Statistics for Experimenters: An Introduction to Design, Data Analysis, and Model Building.* New York: John Wiley & Sons, 1978, Chapter 1.

Gunter, B., "Fundamental Issues in Experimental Design," *Quality Progress*, June 1996, pp. 105–113.

Gunter, B., "Second-Class Citizens and Experimental Design," *Quality Progress,* October 1996, pp. 127–129.

Gunter, B., "Statistically Designed Experiments. Part 1: Quality Improvement, the Strategy of Experimentation, and the Road to Hell," *Quality Progress,* December 1989, pp. 63–64.

Gunter, B., "Statistically Designed Experiments. Part 2: The Universal Structure Underlying Experimentation," *Quality Progress,* February 1990, pp. 87–89.

Knowlton, J., and R. Keppinger, "The Experimentation Process," *Quality Progress,* February 1993, pp. 43–47.

Mason, R. L., R. F. Gunst, and J. L. Hess. *Statistical Design and Analysis of Experiments with Applications to Engineering and Science.* New York: John Wiley & Sons, 1989, Chapters 1 and 2.

Ostle, B., K. V. Turner Jr., C. R. Hicks, and G. W. McElrath. *Engineering Statistics: The Industrial Experience.* Belmont, CA: Duxbury Press, 1996, Chapter 1.

Snee, R. D., L. B. Hare, and J. R. Trout. *Experiments in Industry: Design, Analysis, and Interpretation of Results.* Milwaukee, WI: American Society for Quality, 1985.

PROBLEMS

1.1 In your own area of interest, write out the statement of a problem that you believe is researchable.

1.2 For the problem stated in Problem 1.1, name the experimental unit and universe.

1.3 For the problem and the universe stated in Problems 1.1 and 1.2, respectively, list the population(s) and response variable(s) of interest. Tell how you propose to measure each response variable.

1.4 For the problem stated in Problem 1.1, list the factors (independent variables) that might affect your response (dependent variable) and explain how you would handle each factor.

1.5 Prepare a data layout for collecting data on your problem. Is the layout one for a possible factorial experiment, nested experiment, or some other type of experiment?

1.6 From your experience, name some experimental unit and name at least two populations that might be generated on a universe of your experimental units.

1.7 Briefly discuss the importance of experimental design and analysis of data, and comment on their significance to the experimenter.

Chapter 2

REVIEW OF STATISTICAL INFERENCE

2.1 INTRODUCTION

In any experiment the experimenter is attempting to draw certain inferences or to make a decision about some hypothesis or "hunch" concerning the situation being studied. Life consists of a series of decision-making situations. As you taste your morning cup of coffee, almost unconsciously you decide that it is better than, the same as, or worse than the coffee you have been drinking in the past. If, based on your idea of a "standard" cup of coffee, the new cup is sufficiently "bad," you may pour it out and make a new batch. If you believe it superior to your "standard," you may try to determine whether the brand is different from the one you are used to or the brewing is different, and so forth. In any event, it is the extreme differences that may cause you to decide to take action. Otherwise, you behave as if nothing has changed. In such a trivial case, you run the risk that the decision based on so small an amount of data is wrong.

In deciding whether to carry an umbrella on a given morning, you "collect" certain data: you tap the barometer, look at the sky, read the newspaper forecast, listen to the radio, and so on. After quickly assimilating all available data—including such predictions as "a 30% probability of rain today"—you make a decision, somehow compromising between the inconvenience of carrying the umbrella and the possibility of having to spend money to have your clothes cleaned.

What Is Statistics?

Most everyday decisions are made in the light of uncertainty. *Statistics* may be defined as a tool for decision making in the light of uncertainty. Uncertainty does not imply no knowledge, but only that the exact outcome is not completely predictable. If 10 coins are tossed, one knows that the number of heads will be some integer between 0 and 10. However, each specific integer has a certain chance of occurring, and various results may be predicted in terms of their chance or probability of occurrence. When the observed results of an experiment could have occurred only 5 times in 100 by chance alone, most experimenters consider that this is a rare event and will state that the results are statistically significant at the 5% significance level. In such cases the hypothesis being tested is usually rejected as untenable. When statistical methods are used in experimentation, one can assess the magnitude of the risks taken in making a particular decision.

Statistical Inference

The process of inferring something about a population from a sample drawn from that population is called *statistical inference*. As noted in Chapter 1, a *population* is defined as all possible values of a response variable Y that might be obtained for all experimental units in the universe under consideration. The response Y may represent tensile strength, weight, score, reaction time, proportion of nonconforming units in a lot, or any other criterion being used to evaluate the experimental results. Characteristics of the population associated with this random variable are called *parameters*. Statistical inference may be divided into two parts: (1) estimation of a parameter and (2) testing a hypothesis about a parameter.

The Mean and Variance of a Random Variable

The *mean* or *expected value* of the random variable Y is designated as $E[Y] = \mu$. If the probability function defining the random variable is known,

$$E[Y] = \sum_i y_i p(y_i)$$

when Y is a discrete random variable with discrete probability function $p(y_i) = P(Y = y_i)$, and

$$E[Y] = \int_{-\infty}^{\infty} y f(y) dy$$

when Y is a continuous random variable with probability density function $f(y)$.

The long-range average of squared deviations from the mean of a random variable is called the *variance* of that variable. Designating the variance of Y by $\mathrm{Var}(Y) = \sigma_Y^2$,

$$\mathrm{Var}(Y) = E[(Y - \mu_Y)^2]$$

In the discrete case,

$$\mathrm{Var}(Y) = \sum_i (y_i - \mu_Y)^2 p(y_i)$$

and

$$\mathrm{Var}(Y) = \int_{-\infty}^{\infty} (y - \mu_Y)^2 f(y) dy$$

in the continuous case. The positive square root of the variance of Y is called the *standard deviation* of Y, denoted $\mathrm{SD}(Y) = \sigma_Y$.

Properties of Expected Values

A few basic theorems involving expected values that will be useful later in this book include

- $E[k] = k$, for k any constant
- $E[kY] = kE[Y]$
- $E[Y_1 + Y_2] = E[Y_1] + E[Y_2]$
- $\text{Var}(Y) = E[(Y - \mu_Y)^2] = E[Y^2] - \mu_Y^2$

Sample Statistics

Most statistical theory is based on the assumption that samples used are random samples. If n experimental units are selected in a manner that is equivalent to removing a "handful" of units from a "well-mixed" universe of experimental units, the resulting collection is called a random sample of n experimental units. The collection of n values of the response variable associated with the random sample of experimental units is called a *random sample of size n* from the corresponding population. Quantities computed from those sample values are called *sample statistics* or, simply, *statistics*. Examples include the *sample mean*

$$\bar{Y} = \frac{\sum_{i=1}^{n} Y_i}{n}$$

and the sample variance

$$S^2 = \frac{\sum_{i=1}^{n}(Y_i - \bar{Y})^2}{n - 1}$$

Before sampling occurs $Y_1, \ldots, Y_n$ are random variables, so $\bar{Y}$ and S^2 are random variables. Once sampling has occurred and the values $y_1, \ldots, y_n$ are obtained,

$$\bar{y} = \frac{\sum_{i=1}^{n} y_i}{n}$$

and

$$s^2 = \frac{\sum_{i=1}^{n}(y_i - \bar{y})^2}{n - 1}$$

are the observed values of $\bar{Y}$ and S^2, respectively. The positive square root of S^2, denoted S, is called the *sample standard deviation* and its observed value is denoted s.

2.2 ESTIMATION

The objective of statistical estimation is to use a statistic for a sample drawn from a population to estimate a parameter of that population. Estimates of two types are usually needed, point estimates and interval estimates.

Point Estimators and Point Estimates

A random variable such as $\bar{Y}$ or S^2, when used to obtain estimates of a population parameter, is called a *point estimator* of that parameter, and an observed value of such a statistic is called a *point estimate* of the parameter. Since they are random variables, the commonly used point estimators have probability distributions with means and variances. Statisticians usually use point estimators that are unbiased statistics.

Unbiased Statistics

A point estimator W of a population parameter θ is an *unbiased statistic* if $E[W] = \theta$. That is, the expected or average value of W taken over an infinite number of similar samples equals the population parameter being estimated. Since $E[kY] = kE[Y]$ and $E[Y_1 + Y_2] = E[Y_1] + E[Y_2]$ can be used to prove that

$$E[\bar{Y}] = E[Y] = \mu_Y \qquad \text{(or simply, } \mu)$$

$\bar{Y}$ is an unbiased estimator of μ and $\bar{y}$ is an unbiased estimate. Likewise,

$$E[S^2] = \text{Var}(Y) = \sigma_Y^2 \qquad \text{(or simply, } \sigma^2)$$

so S^2 is an unbiased estimator of σ^2 and s^2 is an unbiased estimate. On the other hand, the sample standard deviation S is not an unbiased estimator of σ, since $E[S] \neq \sigma$. For a normal population, $E[S] = c_4\sigma$ for c_4 a positive constant based on the sample size. This somewhat subtle point is proved in Burr (see [9], pp. 101–104).

Sum of Squares and Mean Square

For $Y_1, \ldots, Y_n$ a random sample from the distribution of a random variable Y with mean μ and standard deviation σ, the *sum of squares* is the statistic

$$SS_Y = \sum_{i=1}^{n}(Y_i - \bar{Y})^2$$

It is actually the sum of the squares of the deviations of the sample values from the mean of the sample. Before sampling occurs, SS_Y is a random variable. For $y_1, \ldots, y_n$ the observed sample values, the calculated constant

$$\sum_{i=1}^{n}(y_i - \bar{y})^2$$

is also denoted SS_Y. Whether the random variable or an observed value of the variable is under consideration is usually evident from the context of the discussion.

Since $SS_Y = (n-1)(S_Y^2)$, $E[(n-1)S_Y^2] = (n-1)E[S_Y^2]$, and $E[S_Y^2] = \sigma_Y^2$, we find that

$$E[SS_Y] = (n - 1)\sigma_Y^2$$

where $n - 1$ represents the degrees of freedom (*df*) associated with SS_Y.

Suppose the sum of squares SS_W is associated with a random variable W. Division of SS_W by its degrees of freedom indicates an averaging of the squared deviations from the mean. The resulting statistic, denoted MS_W is called a *mean square*. In simple cases like SS_Y considered in this section, $df = n-1$, $MS_Y = SS_Y/(n-1) = S_Y^2$, and $E[MS_Y] = \text{Var}(Y)$.

Interval Estimators and Interval Estimates

A *100(1 −α)% confidence interval estimator* of a population parameter is a random interval that is asserted to include the parameter in question with probability $1 − \alpha$ for some $0 < \alpha < 1$. The interval with end points determined from observed sample values is called a $100(1 − \alpha)\%$ *confidence interval* (*estimate*) for that parameter. For example, a 95% confidence interval for the mean μ of a normal distribution is given by

$$\bar{y} \pm 1.96 \frac{\sigma}{\sqrt{n}}$$

where $P(Z \leq -1.96) = 0.025$, $P(Z \geq 1.96) = 0.025$, n is the sample size, and σ is the population standard deviation. If 100 sample means based on n observations each are computed, and 100 confidence intervals are set up using the preceding formula, we expect that about 95 of the 100 intervals will include μ. If only one interval is set up based on one sample of n observations, as is usually the case, we can state that we have 95% confidence that the interval includes μ. Before sampling, the probability is 0.95 that we will obtain a sample for which the corresponding confidence interval will include μ. After sampling, the calculated interval either contains μ or does not contain μ.

In general, if Y has a normal distribution with mean μ and standard deviation σ, with $\bar{y}$ the mean of the observed values $y_1, \ldots, y_n$ for a random sample from the distribution of Y and $P(Z \leq Z_{1-\alpha/2}) = 1 - \alpha/2$ with $Z_{1-\alpha/2}$ obtained from Statistical Table A, the interval with end points

$$\bar{y} \pm Z_{1-\alpha/2} \frac{\sigma}{\sqrt{n}} \tag{2.1}$$

is a $100(1 - \alpha)\%$ confidence interval for μ. If σ is unknown and the population is normal, Student's t distribution (Statistical Table B) is used, and a $100(1 - \alpha)\%$ confidence interval is given by

$$\bar{y} \pm t_{1-\alpha/2} \frac{s}{\sqrt{n}} \tag{2.2}$$

where t has $n - 1$ degrees of freedom, $P(t \leq t_{1-\alpha/2}) = 1 - \alpha/2$, and s is the observed sample standard deviation. When Y is continuous, nonnormal and n is (say) at least 30, either (2.1) or (2.2) can be used to obtain an approximate $100(1 - \alpha)\%$ confidence interval for μ.

2.3 TESTS OF HYPOTHESES

A *statistical hypothesis* is an assertion about the sampled population that is to be tested. It usually consists of assigning a value to one or more parameters of the population. For example, it may be hypothesized that the average number of miles per gallon obtained with a certain carburetor is 19.5. This is expressed as $H_0 : \mu = 19.5$ mpg. The basis for the assignment of this value to μ usually rests on past experience with similar carburetors. Or, one might hypothesize that the variance in weight of filled vials for the week is 40 grams squared, or $H_0 : \sigma^2 = 40g^2$. When such hypotheses are to be tested, the other parameters of the population are either assumed or estimated from data taken on a random sample of experimental units from the universe. Assumptions about the population and its parameters are taken as true and, although not formally tested, must have a strong relationship with reality. Checks of the reasonableness of the assumptions often use graphical techniques such as histograms, box plots, and normal quantile plots.

What Is a Test of a Hypothesis?

A *test of a hypothesis* is simply a rule by which one determines whether a hypothesis is rejected or not rejected. Such a rule is usually based on sample statistics, called *test statistics* when they are used to test hypotheses. For example, the rule might be to reject $H_0 : \mu = 19.5$ mpg if a sample of 25 carburetors averaged 18.0 mpg or less when tested. In this case, $\bar{Y}$ is the test statistic.

Critical Regions

The *critical region* of a test statistic consists of all values of the test statistic for which the decision is made to reject the null hypothesis H_0. In the preceding example, the critical region for the test statistic $\bar{Y}$ is $\{\bar{y} | \bar{y} \leq 18.0 \text{ mpg}\}$.

Type I and Type II Errors

Since hypothesis testing is based on observed sample statistics computed on n observations, the decision is always subject to error. If a null hypothesis that is really true is rejected, a *type I error* has been committed, and its magnitude is denoted as α. For an example of a type I error, suppose that $H_0 : \mu = 19.5$ mpg is really true, the critical region for the test statistic $\bar{Y}$ is $\{\bar{y} | \bar{y} \leq 18.0 \text{ mpg}\}$, and the sample of 25 values attains a value of $\bar{y} = 17.8$ mpg; in such a case, a true null hypothesis is rejected.

If the null hypothesis is not true but is not rejected because the sample provides insufficient evidence for such rejection, a *type II error* has been made. In the previous example, if the true average gasoline mileage is $\mu = 19.2$ mpg, a sample of 25 values that attains a value of $\bar{y} = 18.9$ mpg will fail to cause the null hypothesis to be rejected, and a type II error will be committed. Also note that in this example a type II error occurs when $\bar{y}$ is any number greater than 18.0 and the true average gasoline mileage is some value $\mu < 19.5$. The magnitude of that error is denoted $\beta(\mu)$.

These α and β error probabilities are often referred to as the risks of making incorrect decisions. One of the objectives in hypothesis testing is to design a test for which α is small and β is known to be small for at least one value of the parameter at which the experimenter would like rejection to occur. In most such test procedures α is set at some predetermined level, and the decision rule is then formulated to minimize the other risk, β, at some reasonable value of the parameter being tested.

The Classical Approach to Hypothesis Testing

In many cases, the critical region of the test statistic, based on the accepted magnitude of α, is determined before sampling occurs. Then the value of the test statistic is calculated for the values of the obtained sample, and the null hypothesis is rejected when that value is in the critical region. Such an approach is referred to as the *classical approach to hypothesis testing*. A series of steps applicable to hypotheses and test statistics of most types can be taken when the classical approach is used. The following simple example will help clarify these steps and illustrate the procedure.

■ **Example 2.1**

Suppose the average number of miles per gallon obtained by a certain carburetor is to be at least 19.5 mpg. Further suppose that a customer of the supplier of those carburetors has reason to believe that the average is less than 19.5 mpg. The following parallel steps illustrate one way for the customer to test the hypothesis.

Steps in Hypothesis Testing	**Example**
1. Set up the hypothesis and its alternative.	1. $H_0 : \mu = 19.5$ mpg $H_1 : \mu < 19.5$ mpg
2. Set the significance level, α, of the test and the sample size n.	2. $\alpha = 0.05, n = 25$
3. Choose a test statistic to test H_0, noting any assumptions necessary when applying this statistic.	3. Test statistic: Standardized $\overline{Y}$, $Z = (\overline{Y} - \mu)/(\sigma/\sqrt{n})$ (Assume σ is known to be 2 mpg.)
4. Determine the sampling distribution of this test statistic when H_0 is true.	4. $Z = (\overline{Y} - 19.5)/(2/\sqrt{25}) =$ $5(\overline{Y} - 19.5)/2$ has a standard normal distribution.
5. Set up a critical region on this test statistic where the probability of rejecting H_0 is α when H_0 is true.	5. Since $P(Z \leq -1.645) = 0.05$, the critical region is $\{z \mid z \leq -1.645\}$. That is, reject H_0 if $z \leq -1.645$.
6. Choose a random sample of n observations, compute the test statistic, and make a decision on H_0.	6. If $n = 25$ and $\overline{y} = 18.9$ mpg, $z = \dfrac{5(18.9 - 19.5)}{2} = -1.5$ As $-1.5 > -1.645$, do not reject H_0.

Without consideration of the magnitude of a type II error, H_0 should not be accepted as true. For that reason, the decision in step 6 is *do not reject H_0*.

In Example 2.1 a one-sided or one-tailed test was used. This is dictated by the alternative hypothesis, since we wish to reject H_0 only when low values of $\overline{Y}$ are observed. When Y is normally distributed, the null hypothesis is true, and $\sigma = 2$; $\overline{Y}$ is normally distributed with mean 19.5 and standard deviation $\sigma/\sqrt{n} = 2/\sqrt{25} = 0.4$. The decision rule is expressing the experimenter's belief that a sample mean that is at least 1.645 standard deviations below 19.5 [i.e., $<$ than $19.5 - 1.645(0.4) = 18.842$ mpg] deviates far enough from the hypothesized mean to be due to something other than chance. Since $P(\overline{Y} \leq 18.842) \approx 0.05$, the significance level is $\alpha = 0.05$. The size of the significance level is a choice of the experimenter and is often set, by tradition, to values such as 0.05 or 0.01. It should reflect the seriousness of rejecting many carburetors when they are really satisfactory, or when the actual mean of the universe of the population is 19.5 mpg or better.

In general, if Y has a normal distribution with mean μ and standard deviation σ; $\bar{y}$ is the mean of the observed values $y_1, \ldots, y_n$ for a random sample from the distribution of Y; and $P(Z \leq Z_{1-\alpha/2}) = 1 - \alpha/2$ with $Z_{1-\alpha/2}$ obtained from Statistical Table A, a test of $H_0 : \mu = \mu_0$ versus $H_1 : \mu \neq \mu_0$ [or $\mu > \mu_0$ or $\mu < \mu_0$] can be conducted at the $100\alpha\%$ significance level by using

$$Z = \frac{\bar{Y} - \mu_0}{\sigma/\sqrt{n}} \tag{2.3}$$

as the test statistic and rejecting H_0 when $z = (\bar{y} - \mu_0)/(\sigma/\sqrt{n})$ and $|z| \geq Z_{1-\alpha/2}$ [or $z \geq Z_{1-\alpha}$ or $z \leq Z_\alpha$]. If σ is unknown and the population is normal, Student's t distribution (Statistical Table B) is used with the test statistic

$$t = \frac{\bar{Y} - \mu_0}{S/\sqrt{n}} \tag{2.4}$$

where t has $n - 1$ degrees of freedom, $P(t \leq t_{1-\alpha/2}) = 1 - \alpha/2$, S is the sample standard deviation, and H_0 is rejected when $t = (\bar{y} - \mu_0)/(s/\sqrt{n})$ and $|t| \geq t_{1-\alpha/2}$ [or $t \geq t_{1-\alpha}$ or $t \leq t_\alpha$]. When Y is nonnormal and n is (say) at least 30, either Equation (2.3) or (2.4) can be used to test at an approximate $100\alpha\%$ significance level.

The procedure outlined in Example 2.1 may be used to test many different hypotheses. The nature of the problem will indicate what test statistic is to be used, and proper tables (or computer software) can be found to set up the required critical region. Well known are tests such as those on two means with various assumptions about the corresponding variances, one variance, and two variances. These tests are reviewed in Sections 2.6 to 2.8.

The p Value of a Test of Hypothesis

In Example 2.1, $Z = (\bar{Y} - 19.5)/(2/\sqrt{25}) = (\bar{Y} - 19.5)/0.4$ has a standard normal distribution when Y is normally distributed with $\sigma = 2$ and the null hypothesis is true. After sampling, $\bar{y} = 18.9$ mpg and $z = -1.5$ were observed. The probability of observing a value of Z at least as extreme as -1.5 when $H_0 : \mu = 19.5$ mpg is true is $P(Z \leq -1.5) = 0.0668$ from Statistical Table A. This probability, called the p value of the test, indicates that the sample evidence allows one to reject the null hypothesis for any $\alpha \geq 0.0668$.

In general, the p value of a test is the probability of obtaining a value of the test statistic that is at least as extreme as the calculated value when the null hypothesis is true. It is the smallest significance level at which the null hypothesis can be rejected. Thus, if the agreed-upon value of α is less than the observed p value (as is the case in Example 2.1), the null hypothesis is not rejected.

2.4 THE OPERATING CHARACTERISTIC CURVE

In Example 2.1 no mention was made of the type II error. The magnitude of such an error, denoted as β, is the probability of failing to reject the original hypothesis H_0 when it is not true or when some alternative hypothesis H_1 is true. Now H_1 in that example states that the true average gasoline mileage is less than 19.5 mpg, and there are many possible means that satisfy this alternative hypothesis. Thus, β is a function of μ for $\mu < 19.5$. To see how

β varies with μ, let us consider several possible values of μ and sketch the distributions of $\bar{Y}$ that these values would generate (we are assuming that Y is normally distributed and $\sigma = 2$). Also note that the critical region for $\bar{Y}$ was established as $\{\bar{y}|\bar{y} \leq 18.842 \text{ mpg}\}$.

From Figure 2.1 it can be seen that as one considers μ farther and farther to the left of 19.5, the β error decreases. The size of any β error can be determined for any given μ from Figure 2.1 and a normal distribution table. For example, if μ is assumed to be 19.0, the standardized value of the critical mean 18.842 with respect to this assumed mean is

$$z = \frac{18.842 - 19.000}{2/\sqrt{25}} = \frac{-0.158}{0.400} = -0.395$$

indicating that the rejection point is approximately 0.4 standard deviation below 19.0. For a standard normal distribution (see Statistical Table A), $P(Z > -0.395) \approx P(Z > -0.40) = 1 - 0.3446 = 0.6554$. Thus, the β error for $\mu = 19.0$ is $\beta(19.0) \approx 0.6554 \approx 0.66$, and approximately 66% of the samples would fail to reject the null hypothesis that the average is 19.5 mpg when it has really dropped to 19.0.

If $\beta(\mu) = P[Z > (18.842 - \mu)/0.4|\mu < 19.5]$ is plotted for various values of μ, the resulting curve is called the *operating characteristic* (or OC) *curve* for this test. Using Microsoft Excel[1] (or Statistical Table A) a summary like that in Table 2.1 is easily

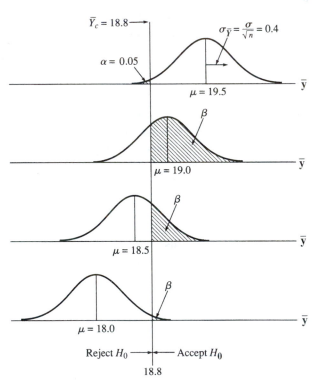

Figure 2.1 The effect of the mean on the β error.

---

[1]Microsoft Excel, referred to as Excel hereafter, is a registered trademark of Microsoft Corporation.

obtained. If one considers the mean μ at 18.0 as in the bottom sketch of Figure 2.1, $z = (18.842 - 18.0)/0.4 = 2.105$ and the β error is only $\beta(18.0) = P(Z > 2.105) = 1 - P(Z \le 2.105) = 1 - 0.9824 = 0.0176$.

The pairs $(\mu, \beta(\mu))$ determined in Table 2.1 can be plotted to obtain an OC curve for our example. However, some people prefer to plot $1 - \beta(\mu)$ versus μ. This curve is called the *power curve* of the test because $1 - \beta(\mu)$ is called the *power of the test* against the alternative μ. In this example, $1 - \beta(\mu) = P[Z > (18.842 - \mu)/0.4]$, selected values of which are summarized in the third column of Table 2.1. Since $1 - \beta(18.0) = 0.9824$, we expect approximately 98 samples in every 100 samples obtained when $\mu = 18.0$ to result in the rejection of $H_0 : \mu = 19.5$ mpg.

Using the data in Table 2.1 to plot the OC and power curves for our test produces Figure 2.2. Note that when μ is quite a distance from the hypothesized value of 19.5, the test does a fairly good job of detecting this shift; that is, if $\mu = 18.5$, the probability of rejecting H_0 is 0.8037, which is fairly high. On the other hand, if μ has shifted only slightly from 19.5,

TABLE 2.1
Data for Operating Characteristic Curve of Example 2.1

μ	$z = (18.842 - \mu)/0.4$	$1 - \beta(\mu)$	$\beta(\mu)$
18.0	2.105	0.9824	0.0176
18.1	1.855	0.9682	0.0318
18.2	1.605	0.9458	0.0542
18.3	1.355	0.9123	0.0877
18.4	1.105	0.8654	0.1346
18.5	0.855	0.8037	0.1963
18.6	0.605	0.7274	0.2726
18.7	0.355	0.6387	0.3613
18.8	0.105	0.5418	0.4582
18.9	−0.145	0.4424	0.5576
19.0	−0.395	0.3464	0.6536
19.1	−0.645	0.2595	0.7405
19.2	−0.895	0.1854	0.8146
19.3	−1.145	0.1261	0.8739
19.4	−1.395	0.0815	0.9185
19.5	−1.645	0.0500*	0.9500

*Equals α.

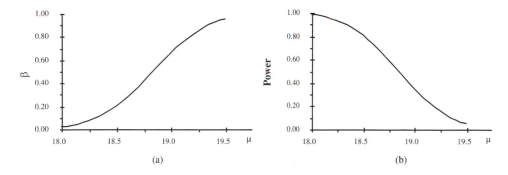

Figure 2.2   (a) Operating characteristic curve and (b) power curve.

say to 19.2, the probability of detection is only 0.1854. The power of the test to detect a shift in μ to 19.2 may be increased by increasing the sample size or increasing the α error.

2.5 HOW LARGE A SAMPLE?

Statisticians are often asked how large a sample to take from a population for making a test. To answer this question, the experimenter must be able to answer each of the following questions, which we answer for the test considered in Example 2.1:

1. How large a shift in a parameter do you wish to detect? [from 19.5 to 19.0]
2. How much variability is present in the population? [based on past experience, $\sigma = 2$ mpg]
3. What size risks are you willing to take? [$\alpha = 0.05$ and $\beta(19.0) = 0.10$]

If numerical values can be at least estimated in answering these questions, a good estimate of the sample size is possible.

To determine an appropriate sample size, set up two sampling distributions of $\bar{Y}$, one when H_0 is true (in this case, $\mu = 19.5$), and the other when the alternative to be detected is true (in this case, $\mu = 19.0$). Both situations are depicted in Figure 2.3. Indicate by $\bar{Y}_c$ a value between the two population means that will become a critical point, rejecting H_0 for observed values of $\bar{Y}$ below it and failing to reject H_0 for $\bar{Y}$ values above it. Indicate the α and β risks on the diagram. Set up two simultaneous equations, standardizing $\bar{Y}_c$ first with respect to a μ of 19.5 (α equation) and second with respect to a μ of 19.0 (β equation). The equations for this case are presented and solved as follows.

$$\alpha \text{ equation} : \frac{\bar{Y}_c - 19.5}{2/\sqrt{n}} = -1.645 \Rightarrow \bar{Y}_c = 19.5 - 1.645 \left(\frac{2}{\sqrt{n}}\right) \text{(based on } \alpha = 0.05)$$

$$\beta \text{ equation} : \frac{\bar{Y}_c - 19.0}{2/\sqrt{n}} = +1.282 \Rightarrow \bar{Y}_c = 19.0 + 1.282 \left(\frac{2}{\sqrt{n}}\right) \text{(based on } \beta = 0.10)$$

So, $19.5 - 1.645 \left(\frac{2}{\sqrt{n}}\right) = \bar{Y}_c = 19.0 + 1.282 \left(\frac{2}{\sqrt{n}}\right) \Rightarrow 0.5 = 2.927 \left(\frac{2}{\sqrt{n}}\right).$

Therefore, $\sqrt{n} = 5.854/0.5 = 11.708$ and $n = (11.708)^2 \approx 137.1$ or 138.

Keeping $\alpha = 0.05$ gives

$$\bar{Y}_c = 19.5 - 1.645 \left(\frac{2}{\sqrt{138}}\right) \approx 19.22 \text{ mpg}$$

The decision rule is then: Choose a random sample of 138 carburetors, and if the mean mpg of these is less than 19.22, reject H_0; otherwise, accept H_0.

OC curves that may be used to determine n for tests of hypotheses of several types are included in Ostle et al. [26, pp. 549–552]. Odeh [24] contains excellent tables that can be used for the same purpose. Hahn and Meeker [11] contains curves and tables to determine n for estimation of parameters.

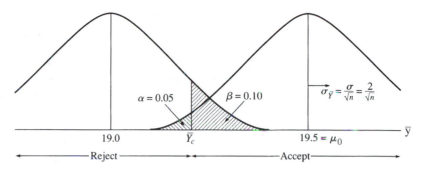

Figure 2.3 Determining sample size.

2.6 APPLICATION TO TESTS ON VARIANCES

To review how the concepts of confidence limits and hypothesis testing apply to variance tests, consider the following. These procedures are based on an assumption of normality and are strongly affected by moderate departures from normality. In the two-sample case with small sample sizes (say less than 15), the power of the test is so low when the larger variance exceeds the smaller by less than a factor of 4 that the test may be of little value.

I. Tests on a single variance

$$H_0 : \sigma^2 = \sigma_0^2$$

$$H_0 : \sigma^2 \neq \sigma_0^2 \left[\text{or } \sigma^2 > \sigma_0^2;\ \text{or } \sigma^2 < \sigma_0^2\right]$$

Test statistic: $W = (n-1)S^2/\sigma_0^2$, which follows a chi-square distribution with $\nu = n-1$ degrees of freedom if the sampled population is normally distributed, μ is unknown, and H_0 is true.

Decision rule: (Classical approach)
Reject H_0 if $w \geq W_{1-\alpha/2}$ or $w \leq W_{\alpha/2}$ [or if $w \geq W_{1-\alpha}$; or if $w \leq W_{\alpha}$].
(p-value approach) Reject H_0 if $\alpha \geq \min\{2P(W \leq w), 2P(W \geq w)\}$ [or if $\alpha \geq P(W \geq w)$; or if $\alpha \geq P(W \leq w)$].

100(1 - a)% Confidence Interval for σ^2: $[(n-1)s^2/\chi_{1-\alpha/2}^2,\ (n-1)s^2/\chi_{\alpha/2}^2]$, using the chi-square table (Statistical Table C) with $\nu = n-1$ degrees of freedom and assuming $Y \sim N(\mu, \sigma^2)$.

■ Example 2.2

Suppose we wish to determine whether a process standard deviation is greater than a specified value of 5 oz and prior experience indicates that the associated population is normally distributed. If a random sample of 10 weights is found to have an average of 35.2 oz and a standard deviation of 7 oz, what can we conclude about the standard deviation of this process?

We hypothesize $H_0 : \sigma^2 = 25$ oz^2 and take $H_1 : \sigma^2 > 25$ oz^2 as the alternative. When the process variance is 25 oz^2, $W = 9S^2/25$ has a chi-square distribution with 9 degrees of freedom. From Statistical Table C, the 95th percentile of W is 16.919. Using W as the test statistic and $\alpha = 0.05$, one should reject H_0 if $w \geq 16.919$.

From the sample data, $s^2 = 7^2 = 49$, so $w = 9(49/25) = 17.64$. We reject the null hypothesis and conclude that our process standard deviation is greater than 5 oz.

We might wish to set 90% confidence limits on σ^2 (or σ) based on our observed variance of 49. Since the 5th and 95th percentiles of a chi-square distribution with 9 degrees of freedom are 3.325 and 16.919, respectively, a 90% confidence interval for σ^2 is given as:

$$\frac{9(49)}{16.919} \leq \sigma^2 \leq \frac{9(49)}{3.325}$$

or

$$26.065 \leq \sigma^2 \leq 132.632$$

Taking square roots and rounding to the nearest tenth, $5.1 < \sigma < 11.5$, and we feel 90% confident that the process standard deviation is between 5.1 and 11.5 oz. We note that this band does not include the hypothesized value of 5 oz. We also note that this is a two-sided confidence band, whereas the hypothesis test is one-sided. Deleting the upper confidence limit, we are 95% confident that $\sigma > 5.1$ oz. By making the α for confidence limits twice the α in the test of H_0, we succeed in rendering the results comparable.

II. Tests on Two Variances: Independent Samples

$$H_0 : \sigma_1^2 = \sigma_2^2$$

$$H_1 : \sigma_1^2 \neq \sigma_2^2 [\text{or } \sigma_1^2 > \sigma_2^2]$$

Test statistic: $F = S_1^2/S_2^2$, which follows an F distribution with numerator degrees of freedom $v_1 = n_2 - 1$ and denominator degrees of freedom $v_2 = n_1 - 1$ if the sampled populations are normally distributed, the samples are independent, and H_0 is true. The F distribution appears as Statistical Table D.

Decision rule: (Classical approach) Reject H_0 if $f \geq F_{1-\alpha/2}$ or $f \leq F_{\alpha/2}$
[or if $f \geq F_{1-\alpha}$].
(p-value approach) Reject H_0 if $\alpha \geq \min\{2P(F \leq f), 2P(F \geq f)\}$
[or if $\alpha \geq P(F \geq f)$].

100(1−α)% Confidence Interval for σ_1^2/σ_2^2 $: [(s_1^2/s_2^2) \times F_{\alpha/2}, \ (s_1^2/s_2^2) \times F_{1-\alpha/2}]$, using the F table (Statistical Table D) with numerator degrees of freedom $\nu_1 = n_2 - 1$ and denominator degrees of freedom $\nu_2 = n_1 - 1$, and assuming that independent random samples were obtained from normally distributed populations.

■ **Example 2.3**

As another example of a test of a hypothesis, consider testing whether the variances of two normal populations are equal, based on independent random samples of sizes $n_1 = 8$ and $n_2 = 10$. If random sampling produces variances of $s_1^2 = 156$ and $s_2^2 = 100$, should we reject $H_0 : \sigma_1^2 = \sigma_2^2$ in favor of $H_1 : \sigma_1^2 \neq \sigma_2^2$?

When the population variances are equal, $F = S_1^2/S_2^2$ has an F distribution with numerator degrees of freedom $\nu_1 = 7$ and denominator degrees of freedom $\nu_2 = 9$. From Statistical Table D, we find that the 5th and 95th percentiles of F are 0.272 and 3.29, respectively. Using F as the test statistic and $\alpha = 0.10$, one should reject H_0 if $f \geq 0.272$ or $f \geq 3.29$.

From the sample data, $s_1^2 = 156$ and $s_2^2 = 100$, so $f = 156/100 = 1.56$. Hence the null hypothesis is not rejected. There is insufficient evidence to conclude that the population variances differ.

To obtain a 90% confidence interval estimate of σ_1^2/σ_2^2, we use the critical values of F with $\nu_1 = 9$ and $\nu_2 = 7$. Since $s_1^2/s_2^2 = 1.56$, $F_{0.05} = 0.304$, and $F_{0.95} = 3.68$, we obtain the interval $[(1.56)(0.304), (1.56)(3.68)] \approx [0.47, 5.74]$. This interval contains 1, confirming our earlier decision not to reject equality of variances.

2.7 APPLICATION TO TESTS ON MEANS

I. Tests on a Single Mean

$$H_0 : \mu = \mu_0$$

$$H_1 : \mu \neq \mu_0 [\text{or } \mu > \mu_0; \text{ or } \mu < \mu_0]$$

Test procedure when:

A. σ is known.

Test statistic: $Z = (\bar{Y} - \mu_0)/(\sigma/\sqrt{n})$, which follows a standard normal distribution when the sampled population is normally distributed and H_0 is true. (Rule of thumb: When the sampled population is not normally distributed, use only if $n \geq 25$.)

Decision rule: (Classical approach) Reject H_0 if $z \geq Z_{1-\alpha/2}$ or $z \leq Z_{\alpha/2}$
[or if $z \geq Z_{1-\alpha}$; or if $z \leq Z_{\alpha}$].
(p-value approach)   Reject H_0 if $\alpha \geq 2P(Z \geq |z|)$
[or if $\alpha \geq P(Z \geq z)$; or if $\alpha \geq P(Z \leq z)$].

Note: Example 2.1 illustrates this test.

100(1 - a)% Confidence Interval for μ: $\bar{y} \pm Z_{1-\alpha/2}\sigma/\sqrt{n}$, using the standard normal table (Statistical Table A) and assuming $Y \sim N(\mu, \sigma^2)$. If Y is nonnormal, use only if you have a large sample size, say $n \geq 25$.

B. *σ is unknown.*

Test statistic: $t = (\bar{Y} - \mu_0)/(S/\sqrt{n})$, which follows a t distribution with $n - 1$ degrees of freedom when the sampled population is normally distributed and H_0 is true. This procedure can also be used when Y is not normally distributed but n is large (say, $n \geq 30$).

Decision rule: (Classical approach) Reject H_0 if $|t| \geq t_{1-\alpha/2}$
[or if $t \geq t_{1-\alpha}$; or if $t \leq -t_{1-\alpha}$]
(p-value approach) Reject H_0 if $\alpha \geq 2P(t \geq |t_{observed}|)$.
[or $\alpha \geq P(t \geq t_{obs})$; or $\alpha \geq P(t \leq t_{obs})$]

100(1−a)% Confidence Interval for μ: $\bar{y} \pm t_{1-\alpha/2}s/\sqrt{n}$, using Student's t distribution (Statistical Table B) with $n - 1$ degrees of freedom and assuming $Y \sim N(\mu, \sigma^2)$. If Y is nonnormal, use only if you have a large sample size, say $n \geq 30$.

■ **Example 2.4**

Consider a sample of six cylinder blocks whose cope hardness (Y) values are 70, 75, 60, 75, 65, and 80. Is there evidence here that the average cope hardness has changed from its specified value of 75?

We hypothesize $H_0 : \mu = 75$ and take $H_1 : \mu \neq 75$ as the alternative. When the true average is 75 and Y is normally distributed, $t = (\bar{Y} - 75)/(S/\sqrt{6})$ has a t distribution with 5 degrees of freedom. From Statistical Table B, the 97.5th percentile of t is 2.57. Using t as the test statistic and $\alpha = 0.05$, one should reject H_0 if $|t_{obs}| \geq 2.57$.

From the given sample of 6, $\bar{y} = 70.8$ and $s = 7.4$ (to the nearest 0.1). So,

$$t_{obs} = \frac{70.8 - 75}{7.4/\sqrt{6}} \approx -1.39$$

Since $|t_{obs}| = 1.39 < 2.57$, there is insufficient evidence to reject H_0.

We might have used our data to set 95% confidence limits on μ. Since $t_{0.975} = 2.57$, those limits are $\bar{y} \pm 2.57(s/\sqrt{6}) = 70.8 \pm 2.57(7.4/\sqrt{6}) \approx 70.8 \pm 7.8$ or from 63.0 to 78.6. This interval includes the hypothesized value of 75, which agrees with our decision not to reject H_0.

II. Tests on Two Means

$$H_0 : \mu_1 = \mu_2$$

$$H_1 : \mu_1 \neq \mu_2 [\text{or } \mu_1 > \mu_2]$$

Test procedure when:

A. *Samples are independent and variances (σ_1^2 and σ_2^2) are known.*

Test statistic: $Z = \dfrac{\bar{Y}_1 - \bar{Y}_2}{\sqrt{(\sigma_1^2/n_1) + (\sigma_2^2/n_2)}}$, which follows a standard normal distribution when the sampled populations are normally distributed and H_0 is true. (Rule of thumb: When the sampled populations are not normally distributed, use only if $n_1 \geq 25$ and $n_2 \geq 25$.)

Decision rule: (Classical approach) Reject H_0 if $|z| \geq Z_{1-\alpha/2}$ [or if $z \geq Z_{1-\alpha}$].
(p-value approach) Reject H_0 if $\alpha \geq 2P(Z \geq |z|)$ [or of $\alpha \geq P(Z \geq z)$].

$100(1 - \alpha)\%$ Confidence Interval for $\mu_1 - \mu_2$:

$(\bar{y}_1 - \bar{y}_2) \pm Z_{1-\alpha/2}\sqrt{(\sigma_1^2/n_1) + (\sigma_2^2/n^2)}$, using the standard normal table (Statistical Table A) and assuming the sampled populations are normally distributed. If either distribution is nonnormal, use only if you have large sample sizes, say $n_1 \geq 25$ and $n_2 \geq 25$.

B. *Samples are independent and variances (σ_1^2 and σ_2^2) are unknown but equal.*

Test statistic: $t = (\bar{Y}_1 - \bar{Y}_2)/\sqrt{\dfrac{(n_1-1)s_1^2+(n_2-1)s_2^2}{n_1+n_2-2}\left(\dfrac{1}{n_1} + \dfrac{1}{n_2}\right)}$, which has a t distribution with $n_1 + n_2 - 2$ degrees of freedom when the sampled populations are normally distributed and H_0 is true. This procedure can also be used when either distribution is not normally distributed but sample sizes are large (say, $n_1 \geq 30$ and $n_2 \geq 30$).

Decision rule: (Classical approach) Reject H_0 if $|t| \geq t_{1-\alpha/2}$ [or if $t \geq t_{1-\alpha}$].
(p-value approach) Reject H_0 if $\alpha \geq 2P(t \geq |t_{\text{observed}}|)$. [or $\alpha \geq P(t \geq t_{\text{observed}})$]

$100(1 - \alpha)\%$ Confidence Interval for $\mu_1 - \mu_2$:

$(\bar{y}_1 - \bar{y}_2) \pm t_{1-(\alpha/2)}\sqrt{\dfrac{(n_1-1)s_1^2+(n_2-1)s_2^2}{n_1+n_2-2}\left(\dfrac{1}{n_1} + \dfrac{1}{n_2}\right)}$, using a t distribution with $n_1 + n_2 - 2$ degrees of freedom and assuming that the sampled populations are normally distributed. If either distribution is nonnormal, use only if you have large sample sizes, say $n_1 \geq 30$ and $n_2 \geq 30$.

C. *Samples are independent and variances (σ_1^2 and σ_2^2) are unknown and unequal.*

If a preliminary F test (or graphical analysis) shows the variances to be unequal, no test on means may be necessary as the two samples are so heterogeneous with respect to their variances. If, however, a test on means is desired, some tests are given in statistics books. This situation is often referred to as the Behrens–Fisher problem. One method is to take as the test statistic

$$t' = \frac{\bar{Y}_1 - \bar{Y}_2}{\sqrt{(S_1^2/n_1) + (S_2^2/n_2)}}$$

which has an approximate t distribution with degrees of freedom

$$df = \frac{[(s_1^2/n_1) + (s_2^2/n_2)]^2}{\dfrac{(s_1^2/n_1)^2}{n_1 + 1} + \dfrac{(s_2^2/n_2)^2}{n_2 + 1}} - 2 \tag{2.5}$$

when the sampled populations are normally distributed and H_0 is true. Notice that *df* is calculated *after* the data collection. Most computer software systems include an option that uses a formula similar to this one to calculate the degrees of freedom. [If either distribution is nonnormal, use only if you have large sample sizes.]

100(1 − α)% Confidence Interval for $\mu_1 - \mu_2$:

$(\bar{y}_1 - \bar{y}_2) \pm t_{1-\alpha/2}\sqrt{(s_1^2/n_1) + (s_2^2/n_2)}$, using a t distribution with degrees of freedom calculated according to Equation (2.5) and assuming that the sampled populations are normally distributed. If either distribution is nonnormal, use only if you have large sample sizes, say $n_1 \geq 30$ and $n_2 \geq 30$.

D. Samples are dependent or correlated.

Here one often has the same sample "before and after" some treatment has been applied. The usual procedure is to take differences between the first and second observation on the same experimental unit and test the hypothesis that the mean difference μ_D is zero. This reduces the problem to a test on a single mean for which a test of $H_0 : \mu_D = 0$ versus $H_1 : \mu_D \neq 0$ [or $\mu_D > 0$] is appropriate.

Choose a random sample of n experimental units. Take a pair of measurements, Y_1 and Y_2, on each unit in the sample. Let $D = Y_1 - Y_2$, $\bar{D}$ denote the mean of the sample of n differences, and S_D denote the standard deviation of that sample.

Test statistic: $t = \bar{D}/(S_D/\sqrt{n})$, which follows a t distribution with $n - 1$ degrees of freedom when the population of differences is normally distributed and H_0 is true. This procedure can also be used when D is not normally distributed but n is large (say, $n \geq 30$).

Decision rule: (Classical approach) Reject H_0 if $|t| \geq t_{1-\alpha/2}$ [or if $t \geq t_{1-\alpha}$].
(*p*-value approach) Reject H_0 if $\alpha \geq 2P(t \geq |t_{observed}|)$.
[or if $\alpha \geq P(t \geq t_{observed})$]

100(1 − α)% Confidence Interval for μ: $\bar{d} \pm t_{1-\alpha/2}s_D/\sqrt{n}$, using Student's t distribution (Statistical Table B) and assuming $D \sim N(\mu_D, \sigma_D^2)$. If D is nonnormal, use only if you have a large sample size, say $n \geq 30$.

■ **Example 2.5**

Measurements were made at each of two plants on the tensile strength of samples of the same type of steel. The data that follow are in thousands of pounds per square inch (psi). Do these data provide sufficient evidence to reject the hypothesis that the steels made in the two plants at the time of data collection have the same average tensile strength? Assume the data are random samples from normal populations.

Plant	Tensile Strength (psi $\times 10^3$)						
1	207	195	209	218	196	217	
2	219	228	222	197	225	184	218	211

Summary statistics (see Table 2.2) are: $n_1 = 6$ $\bar{y}_1 = 207$ $s_1^2 = 98.00$
$n_2 = 8$ $\bar{y}_2 = 213$ $s_2^2 \approx 230.29$

To test $H_0 : \sigma_1^2 = \sigma_2^2$ versus $H_1 : \sigma_1^2 \neq \sigma_2^2$ with $\alpha = 0.05$ when the sampled populations are normally distributed, we use the test statistic $F = S_1^2/S_2^2$ with degrees of freedom $\nu_1 = 5$ and $\nu_2 = 7$. From Statistical Table D, $F_{0.025} = 0.146$ and $F_{0.975} = 5.29$, so H_0 can be rejected when $f \leq 0.146$ or $f \geq 5.29$. Since $f = s_1^2/s_2^2 \approx 98/230.29 \approx 0.43$ falls between these two values, we have insufficient evidence to reject equality of variances.

Excel results of a test of $H_0 : \mu_1 = \mu_2$ (or $H_0 : \mu_1 - \mu_2 = 0$) versus $H_1 : \mu_1 \neq \mu_2$ assuming normality and equal variances are summarized in Table 2.2. The quantity labeled "Pooled Variance" is

$$s_{\text{pooled}}^2 = \frac{(n_1 - 1)s_1^2 + (n_2 - 1)s_2^2}{n_1 + n_2 - 2} = \frac{5(98) + 7(230.29)}{6 + 8 - 2} \approx 175.17$$

TABLE 2.2
Excel Output for a Test on Means using the Data in Example 2.5

	A	B	C	D	E
1	Plant 1	Plant 2	t-test: Two Sample Assuming Equal Variances	Plant 1	Plant 2
2	207	219	Mean	207	213
3	195	228	Variance	98	230.29
4	209	222	Observations	6	8
5	218	197	Pooled Variance	175.17	
6	196	225	Hypothesized Mean Difference	0	
7	217	184	df	12	
8		218	t Stat	-0.84	
9		211	P(T<=t) one-tail	0.21	
10			t Critical one-tail	1.78	
11			P(T<=t) two-tail	0.42	
12			t Critical two-tail	2.18	

and the quantity labeled "t Stat" is

$$t_{observed} = \frac{\bar{y}_1 - \bar{y}_2}{\sqrt{s^2_{pooled} \times [(1/n_1) + (1/n_2)]}} = \frac{207 - 213}{\sqrt{175.17[(1/6) + (1/8)]}} \approx -0.84$$

The p-value of the test (denoted P(T<=t) two-tail in Table 2.2) is $2P(t \geq 0.84) \approx 0.42$, so $H_0 : \mu_1 = \mu_2$ cannot be rejected for any reasonable value of α.

■ **Example 2.6**

Tabulated below are reflection light-box readings before and after dichromating the interior of a metal cone for each of eight cones. Do these data provide sufficient evidence to conclude that dichromating decreases the average light-box reading? Assume the differences are a random sample from a normal population.

				Cone				
	1	2	3	4	5	6	7	8
Before	6.5	6.0	7.0	6.8	6.5	6.8	6.2	6.5
After	4.4	4.2	5.0	5.0	4.8	4.6	5.2	4.9
Difference	2.1	1.8	2.0	1.8	1.7	2.2	1.0	1.6

The appropriate test is $H_0 : \mu_{before} = \mu_{after}$ versus $H_1 : \mu_{before} > \mu_{after}$. Since two readings are taken on each cone, we may have dependent samples. So, we use the differences (before − after) to test $H_0 : \mu_D = 0$ versus $H_1 : \mu_D > 0$. Excel results of this test are summarized in Table 2.3. The quantity labeled t Stat is

$$t_{observed} = \frac{\bar{d}}{s_D/\sqrt{8}} \approx \frac{1.7750}{0.3732/\sqrt{8}} \approx 13.45$$

where $\bar{d}$ and s_D are obtained using the 8 differences.

The p value of the test (denoted P(T<=t) one-tail in Table 2.3) is $P(t \geq 13.45) \approx 0.000001$. Thus, $H_0 : \mu_{before} = \mu_{after}$ can be rejected for any reasonable value of α. We conclude that the true average light-box reading before dichromating exceeds the true average light box reading after dichromating.

TABLE 2.3
Excel Output for a Test on Means using the Data in Example 2.6

	A	B	C	D	E
1	Before	After	t-test: Paired Two Sample for Means	Before	After
2	6.5	4.4	Mean	6.5375	4.7625
3	6.0	4.2	Variance	0.1084	0.1141

4	7.0	5.0	Observations	8	8
5	6.8	5.0	Pearson Correlation	0.37	
6	6.5	4.8	Hypothesized Mean Difference	0	
7	6.8	4.6	df	7	
8	6.2	5.2	t Stat	13.45	
9	6.2	4.9	P(T<=t) one-tail	0.000001	
10			t Critical one-tail	1.89	
11			P(T<=t) two-tail	0.000003	
12			t Critical two-tail	2.36	

2.8 ASSESSING NORMALITY

Statistical procedures used throughout this text are based on assumptions about the distributions of the sampled populations. For example, many procedures assume that the population is normally distributed. When we say that a procedure is *robust* to that assumption, we mean that the population must be severely nonnormal before the statistical results will be adversely affected. For example, the t procedures reviewed in Section 2.7 are quite robust to nonnormality. For those cases, exploratory procedures that help us identify outliers and extreme departures from normality, such as severe skewedness, are quite useful. On the other hand, inference procedures for variances are adversely affected by moderate departures from normality.

This section presents some exploratory techniques that can be used to assess a normality assumption. These subjective techniques should be combined with more objective methods when one is deciding whether to proceed with an analysis based on normality or to use some other procedure. Two such objective methods are the Shapiro–Wilk W test and the Lilliefors test.

Sample Quantiles

Let q denote a number between 0 and 1. The *100qth sample quantile*, denoted y_q, is an estimate of the *100qth quantile of Y*, denoted Y_q and defined by $P(Y \leq Y_q) = q$. To determine y_q for a sample of size n:

- Order the values $y_1, y_2, \ldots, y_n$ to obtain $y_{(1)} \leq y_{(2)} \leq \cdots \leq y_{(n)}$.
- Let $nq = w + d$ with w an integer and $0 \leq d < 1$. That is, w is the integer part of the product and d is the decimal part.
- The 100qth sample quantile is $y_q = y_{(w)} + d \times (y_{(w+1)} - y_{(w)})$.

If w is 0 or n, the $100q$th quantile does not exist. If $100q$ is an integer, y_q is called the *100qth percentile*. The quantiles $q_1 = y_{0.25}$, $m = y_{0.50}$, and $q_3 = y_{0.75}$ are called the *lower quartile*, *median*, and *upper quartile*, respectively. The *interquartile range* is IQR $= q_3 - q_1$.

■ **Example 2.7**

Measurements of the pull strength (in grams) required to break the wire in each of 12 integrated circuits were 11.0, 9.3, 11.3, 10.3, 11.0 9.8, 10.0, 8.0, 11.3, 5.8, 9.0, and 10.8. The ordered values are 5.8, 8.0, 9.0, 9.3, 9.8, 10.0, 10.3, 10.8, 11.0, 11.0, 11.3, and 11.3. Since $0.25(12+1) = 3.25, 0.50(12+1) = 6.50$, and $0.75(12+1) = 9.75$, the quartiles are:

$$q_1 = y_{(3)} + 0.25(y_{(4)} - y_{(3)}) = 9.0 + 0.25(9.3 - 9.0) = 9.075$$

$$m = y_{(6)} + 0.50(y_{(7)} - y_{(6)}) = 10.0 + 0.50(10.3 - 10.0) = 10.15$$

$$q_3 = y_{(9)} + 0.75(y_{(10)} - y_{(9)}) = 11.0 + 0.75(11.0 - 11.0) = 11.0$$

The interquartile range is IQR $= q_3 - q_1 = 11.0 - 9.075 = 1.925$.

The definition of a sample quantile presented here is the default definition used by SAS.[2] Other definitions exist, and indeed, SAS includes four other definitions as options (see [31], p. 579). The definition given here is adequate for our purposes.

Box Plots and a Check for Outliers

Tukey [33] developed the graphical display known as a *box plot* as part of his exploratory data analysis procedures. The original procedure requires the calculation of sample hinges, but we shall consider a modification that uses sample quartiles. Box plots obtained from some statistical software packages may differ slightly because of differences in the ways of defining sample quantiles.

The box plot procedure reviewed in this section includes a simple check for potential outliers. Construction of such graphics requires that we compute the three sample quartiles, interquartile range, and four additional measures of location called *fences*. The quantities LOF $= q_1 - 3($IQR$)$ and LIF $= q_1 - 1.5($IQR$)$ are the *lower outer fence* and *lower inner fence*, respectively; UIF $= q_3 + 1.5($IQR$)$ and UOF $= q_3 + 3($IQR$)$ are the *upper inner fence* and *upper outer fence*, respectively.

A generic box plot is depicted in Figure 2.4. The box, which gives a visual impression of the spread and location of the middle 50% of the data, is formed by constructing a rectangle with length the interquartile range (IQR) that is parallel to the measurement axis, begins at the lower quartile, and ends at the upper quartile. The vertical segment through the interior of the box is located at the sample median. The horizontal segments from the extremes of the box extend to the values nearest, but between, the inner fences. These values (referred to as *adjacent values*) are indicated by small vertical bars at the ends of each segment. Any value between an inner fence and an outer fence is indicated by an asterisk (*) on the display

---

[2]SAS is a registered trademark for the Statistical Analysis System of SAS Institute, Inc.

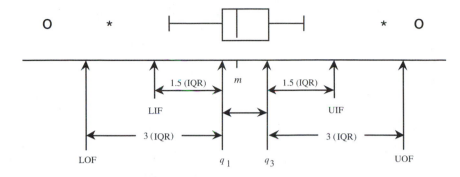

Figure 2.4   A generic box plot with legend.

and is called a *mild* (or *suspect*) *outlier*. Values outside an outer fence are indicated by open circles and are called (*strong*) *outliers*.

■ Example 2.8

Consider the ordered pull strength data (5.8, 8.0, 9.0, 9.3, 9.8, 10.0, 10.3, 10.8, 11.0, 11.0, 11.3, and 11.3) first considered in Example 2.7. In that example we found that $q_1 = 0.9075$, $m = 10.15$, $q_3 = 11.0$, and IQR $= 1.925$. Since $1.5(IQR) = 1.5(1.925) = 2.8875$ and $3(IQR) = 3(1.925) = 5.775$, the upper fences are UIF $= 11.0 + 2.8875 = 13.8875$, and UOF $= 11.0 + 5.775 = 16.775$. No value is beyond the upper inner fence, and the upper adjacent value is 11.3. The lower fences are LOF $= 9.075 - 5.775 = 3.300$ and LIF $= 9.075 - 2.8875 = 6.1875$. The lower adjacent value is 8.0 and 5.8, which is between the two lower fences, is a mild outlier. A box plot follows.

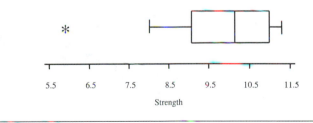

When a mild or strong outlier is observed (as in Example 2.8), the experimental unit associated with that value may have been measured incorrectly, the value may have been recorded incorrectly, or the experimental unit may be atypical. On the other hand, the observed value may be giving valid information about the sampled population not provided by the other values. The experimenter is faced with the task of investigating such values and determining the message they are sending.

   The width of a box plot usually is arbitrary and has no meaning. However, when several box plots are included in the same graphic (called *grouped box plots*) and the sample sizes differ, the width is sometimes made proportional to the sample size. Grouped box plots are useful when two or more populations are sampled, since they provide a graphical means of comparing variability, centering, and other characteristics.

Normal Quantile Plots

A plot of the quantiles of a sample versus the corresponding quantiles of the standard normal distribution is called a *normal quantile plot*. If the sampled population is that of a normally distributed random variable Y with mean μ and standard deviation σ, the relationship between Y and the standard normal Z is given by $Y = \mu + Z\sigma$. So, a normal quantile plot for a sample from such a population should fall near a line with intercept and slope approximately equal to μ and σ, respectively.

If $y_{(i)}$ denotes the ith ordered value of a sample of size n and $q_i = i/(n+1)$, then $y_{(i)}$ is the $100q_i$th quantile of the sample. A normal quantile plot is obtained by graphing the pairs $(z_i, y_{(i)})$, where z_i denotes the $100q_i$th quantile of the standard normal distribution. That is, $P(Z \le z_i) = q_i$. A reasonably linear plot supports an assumption of normality.

Shapiro–Wilk Test for Normality

If possible, graphical displays such as box plots and normal quantile plots should be supplemented with more objective statistical procedures when an assumption of normality is being investigated. The Shapiro–Wilk test, described by Shapiro [29] in the third volume of the *How to* series available through the American Society for Quality, is one such procedure. This test is appropriate when one is testing for normality of a population distribution for which a sample of independent observations is available. Even when observations are not independent, the test can be used as an approximate test for normality to supplement the graphical procedures. Rather than go into details for hand calculation, we will consider results obtained from statistical computing packages. The interested reader will find formulas and tables required for hand calculations in [29].

■ **Example 2.9**

Consider again the ordered pull strength data (5.8, 8.0, 9.0, 9.3, 9.8, 10.0, 10.3, 10.8, 11.0, 11.0, 11.3, and 11.3) in Example 2.7. Since the sample size is $n = 12$, let $q_i = i/13$ for $i = 1, 2, \ldots, 12$. Using Excel to calculate values of q_i and z_i, where $P(Z \le z_i) = q_i$, the following summary is obtained.

i	$y_{(i)}$	q_i	z_i
1	5.8	0.0769	−1.4261
2	8.0	0.1538	−1.0201
3	9.0	0.2308	−0.7363
4	9.3	0.3077	−0.5024
5	9.8	0.3846	−0.2934
6	10.0	0.4615	−0.0966
7	10.3	0.5385	0.0966
8	10.8	0.6154	0.2934
9	11.0	0.6923	0.5024

10	11.0	0.7692	0.7363
11	11.3	0.8462	1.0201
12	11.3	0.9231	1.4261

Using ChartWizard in Excel to plot the $(z_i, y_{(i)})$ pairs and the drawing tool to add a line through the main body of the points produces Figure 2.5. The curvature in the plot calls into question the assumption of normality. Much of that curvature is due to the suspect outlier. If that outlier is due to a special cause, the extreme value should be removed and the investigation should be repeated on the remaining 11 data values.

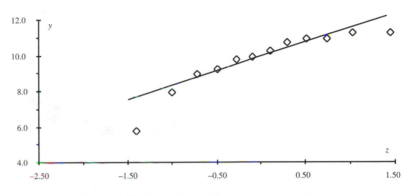

Figure 2.5   Normal quantile plot for strength data.

The Shapiro–Wilk test is one of the most powerful tests of normality when the sample size is less than 50. It is included in JMP, SAS, and SPSS procedures. Minitab includes the Ryan–Joiner test, which is similar to the Shapiro–Wilk test, as an option. Other possibilities are the Anderson–Darling, Kolmogorov–Smirnov, and Lilliefors tests. The UNIVARIATE procedure in SAS reports the results of the Shapiro–Wilk test for small samples, but uses the Kolmogorov–Smirnov test when the sample size exceeds 50. The Lilliefors test is included as an option in SPSS and Minitab, and a graphical version of that test is included with the normal quantile plots produced by recent versions of JMP. The Anderson–Darling normality test is, by default, included with Minitab probability plots.

■ **Example 2.10**

JMP outputs for the strength data in Example 2.8 are shown in Figure 2.6. The p value of the Shapiro–Wilk test (denoted Prob<W) indicates that normality can be rejected for any $\alpha \geq 0.0341$. When this information is combined with the suspect outlier in the box plot and curvature in the normal quantile plot, an assumption of normality seems unwise. Before that decision is finalized, an effort should be made to determine whether outside forces such as measurement error or a defective integrated circuit caused the extreme value. If such a cause is identified, one should delete the suspect outlier and reanalyze the values remaining.

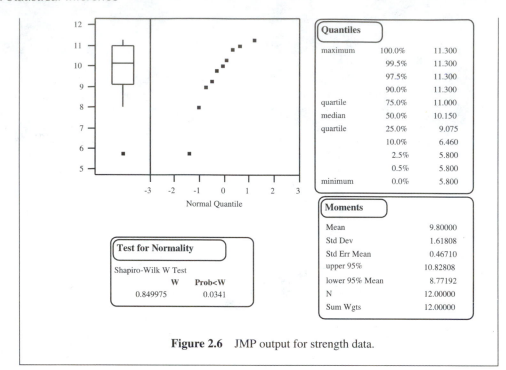

Figure 2.6   JMP output for strength data.

2.9 APPLICATION TO TESTS ON PROPORTIONS

I. Tests on a Single Proportion p

$$H_0 : p = p_0$$

$$H_1 : p \neq p_0 \text{ [or } p > p_0; \text{ or } p < p_0]$$

Test statistic: $Z = [(Y/n) - p_0]/\sqrt{p_0(1 - p_0)/n}$, where Y is the number of successes in the sample of n experimental units and n is large enough to make Z approximately standard normal when H_0 is true. As a rule of thumb, the sample size should satisfy $np_0 \geq 5$ and $n(1 - p_0) \geq 5$.

Decision rule: (Classical approach) Reject H_0 if $|z| \geq Z_{1-\alpha/2}$.
[or if $z \geq Z_{1-\alpha}$; or if $z \leq Z_\alpha$]
(p-value approach) Reject H_0 if $\alpha \geq 2P(Z \geq |z|)$
[or if $\alpha \geq P(Z \geq z)$; or if $\alpha \geq P(Z \leq z)$]

100(1 − α)% Confidence Interval for p: $(y/n) \pm Z_{1-\alpha/2}\sqrt{[y/n][1 - (y/n)]/n}$, where n is large enough that Y has an approximate normal distribution.

■ **Example 2.11**

In a class of 25 students, 10 are left-handed. Is this an excessive number of left-handed students, if the national average is 20%? Let $\alpha = 0.05$.

Assuming independence, the appropriate test is $H_0 : p = 0.20$ versus $H_1 : p > 0.20$. Both $25(0.20) = 5$ and $25(0.80) = 20$ are at least 5, so $z = [(10/25) - 0.20]/\sqrt{(0.20)(0.80)/25} = 2.5$ is the observed value of an approximate standard normal random variable. Thus, $P(Z \geq 2.5) = 0.0062$ is the approximate p value of the test. Since H_0 can be rejected for any $\alpha \geq 0.0062$, we conclude that the class does have an excessive number of left-handers.

II. Tests on Two Independent Proportions

$$H_0 : p_1 = p_2$$

$$H_1 : p_1 \neq p_2 \text{ [or } p_1 > p_2]$$

Test statistic: $Z = \left[\left(\frac{Y_1}{n_1}\right) - \left(\frac{Y_2}{n_2}\right)\right] / \sqrt{\left(\frac{Y_1+Y_2}{n_1+n_2}\right)\left(1 - \frac{Y_1+Y_2}{n_1+n_2}\right)\left(\frac{1}{n_1} + \frac{1}{n_2}\right)}$, where Y_1 is the number of successes in a random sample of n_1 experimental units, Y_2 is the number of successes in a random sample of n_2 experimental units, the two samples are independent, and sample sizes are large enough to make Z approximately standard normal when H_0 is true.

Decision rule: (Classical approach) Reject H_0 if $|z| \geq Z_{1-\alpha/2}$.
[or if $z \geq Z_{1-\alpha}$]
(p-value approach) Reject H_0 if $\alpha \geq 2P(Z \geq |z|)$.
[or if $\alpha \geq P(Z \geq z)$]

$100(1 - \alpha)\%$ Confidence Interval for $p_1 - p_2$:

$[(y_1/n_1) - (y_2/n_2)] \pm Z_{1-\alpha/2}\sqrt{\frac{(y_1/n_1)(1-y_1/n_1)}{n_1} + \frac{(y_2/n_2)(1-y_2/n_2)}{n_2}}$, where n_1 and n_2 are large enough for Y_1 and Y_2 to have approximate normal distributions.

■ **Example 2.12**

The number of nonconforming units found in a sample of 100 drawn from a production process was 12 on Monday. On Tuesday, a sample of 200 from the same process showed 16 defectives. Has the process improved? Use $\alpha = 0.05$.

Letting p_1 denote the true proportion of nonconforming items produced on Monday and p_2 denote the true proportion of nonconforming items produced on Tuesday, we test $H_0 : p_1 = p_2$ versus $H_1 : p_1 > p_2$. From the given information, $y_1 = 12, n_1 = 100, y_2 = 16$, and $n_2 = 200$; so

$$z = \frac{(12/100) - (16/200)}{\sqrt{(28/300)(272/300)[(1/100) + (1/200)]}} \approx 1.12$$

Assuming approximate normality, $P(Z \geq 1.12) = 0.1314$ is the p value of the test. Since this p value exceeds $\alpha = 0.05$, one cannot say that the process has improved significantly.

2.10 ANALYSIS OF EXPERIMENTS WITH SAS

The calculations necessary to summarize a relatively small amount of data can be tedious and time-consuming. To simplify these calculations for studies and experiments of various types, a preprogrammed package of statistical techniques is very useful. We present a brief introduction to SAS (Statistical Analysis System) to give an idea of the power of this package and the ease with which results are obtained. Our intent is to present illustrative examples of the use of SAS to solve some of the text examples. More detailed discussions of the capabilities of SAS are found in the software documentation [31].

The first step in using SAS is to create a data set, a series of statements that name the data set (the **DATA** statement), describe the arrangement of the data lines (the **INPUT** statement), and signal the beginning of the data itself (the **CARDS** statement). SAS statements always end with a semicolon (;).

The **DATA** statement must include the word "DATA" followed by a space, which is followed by a name for the data set. The name chosen must be a one-word name of eight or fewer characters starting with an alphabetic. The **INPUT** statement must include the word "INPUT" followed by a space followed by the names of the variables in the order in which their values appear on the data lines. Variable names are separated by spaces. The **CARDS** statement, which signals that data follow, is followed by lines of data. A semicolon (;) is placed on the first line that follows the last line of data.

Once the data have been entered in a SAS command file, statements of procedures to be performed on the data can be added. Each procedure statement begins with the keyword **PROC** followed by the name of the procedure, which can be followed by any of the available options for that procedure. Remember to end the statement with a semicolon.

Examples in this section will be used to illustrate the use of three SAS procedures. The first, **UNIVARIATE**, can be used with a sample taken from a single population to obtain summary statistics, the p value of the test that the mean of the sampled population is 0, a box plot for the sample data, and the p value of a test for normality. **TTEST** can be used to test for equality of the means or variances when independent samples from two populations are available. **MEANS** gives the p value of a test on a single mean and can be used to test for equality of means when paired data are involved.

The UNIVARIATE Procedure

Measurements of tensile strength (in psi) for 50 steel bars produced the data in Table 2.4. The output of the SAS command file in Table 2.5 will include the mean, variance, and other statistics for that data. Notice that PSI and COUNT are the names of the variables representing the tensile strength in psi and the frequency of that value. As SAS reads the data lines, the first value encountered is read as a tensile strength and the next value is read as the count for that strength. Once the two values on a data line have been read, SAS moves to the next line in the command file.

The procedure line in Table 2.5 instructs SAS to use the **UNIVARIATE** procedure. One of the options, **PLOT**, in the **UNIVARIATE** procedure was included to have SAS include a stem-and-leaf display, box plot, and normal quantile plot in the output. Other available options are **FREQ** (to have a frequency table printed), **NORMAL** (to have the results of a Shapiro–Wilk test for normality included), and **DEF=** *value*. The latter option is used to select one of four methods (other than the default method) to calculate percentiles.

The SAS output for our command file is shown in Figure 2.7. From that output we find $\bar{y} = 18{,}472.9$ and $s = 6.45945 \approx 6.5$. The value denoted USS (read "uncorrected sum of squares") is the sum of the squares of the 50 individual values. The value denoted CSS (read "corrected sum of squares"), which we refer to as SS_Y, is the sum of the squares of the 50 deviations from the sample mean.

The **UNIVARIATE** procedure includes a test for normality as one of its options. If the keyword **NORMAL** is included in the procedure statement, the p value of such a test will be included in the output. For samples of size 50 or less, the p value of a Shapiro–Wilk test (denoted PROB<W) is reported; the p value of a Kolmogorov–Smirnov test (denoted PROB>D) is reported when n exceeds 50.

TABLE 2.4
Steel Bar Data

Tensile Strength	Frequency
18,461	2
18,466	12
18,471	15
18,476	10
18,481	8
18,486	3
Total	50

TABLE 2.5
SAS Command File for Steel Bar Study

```
DATA TENSILE;              names the data set
  INPUT PSI COUNT;         names the variables in order
CARDS;                     signals that data follow
  18461  2                 data begin (two entries per line)
  18466 12
  18471 15
  18476 10
  18481  8
  18486  3
;                          signals the end of data
PROC UNIVARIATE PLOT;      identifies the procedure and an option
```

```
                                         SAS
VARIABLE=PSI
              MOMENTS                                        QUANTILES(DEF=4)                        EXTREMES
N                  50  SUM WGTS          50     100%  MAX   18486   99%     18486       LOWEST   HIGHEST
MEAN          18472.9  SUM           923645      75%  Q3    18476   95%     18486        18461     18466
STD DEV       6.45945  VARIANCE     41.7245      50%  MED   18471   90%     18481        18466     18471
SKEWNESS     0.305914  KURTOSIS   -0.659197      25%  Q1    18466   10%     18466        18471     18476
USS         1.706E+10  CSS           2044.5       0%  MIN   18461    5%   18463.7        18476     18481
CV          0.0349672  STD MEAN    0.913504                          1%     18461        18481     18486
T:MEAN=0        20222  PROB>|T|      0.0001    RANGE           25
SGN RANK        637.5  PROB>|S|      0.0001    Q3--Q1          10
NUM                50                            MODE       18471

Stem  Leaf                 #
18E3  000                  3
18E3
18E3
18E3  00000000             8
18E3
18E3  0000000000          10
18E3
18E3
18E3  000000000000000     15
18E3
18E3  000000000000        12
18E3
18E3
18E3  00                   2
```

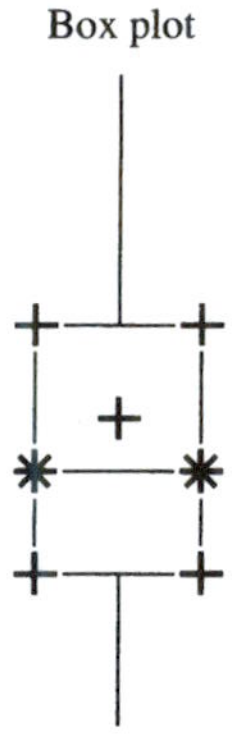

Box plot

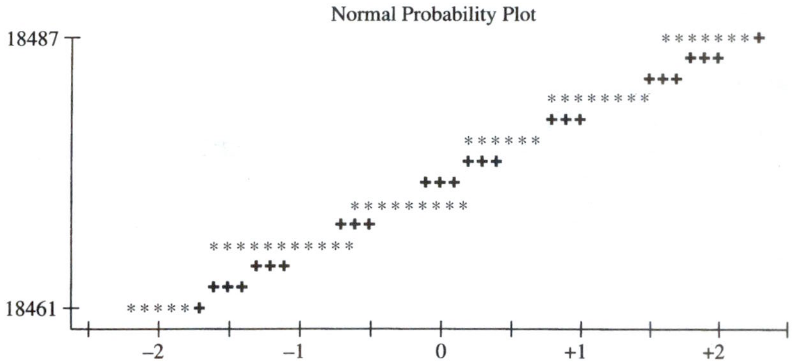

Normal Probability Plot
18487
18461
−2
−1
0
+1
+2

Figure 2.7 SAS output for steel bar study.

If the procedure line in Table 2.5 is changed to

`PROC UNIVARIATE PLOT NORMAL;`

SAS outputs the information in Figure 2.7 with

`W:NORMAL   0.917173    PROB<W   0.0017`

included. Upon combining this information with the visual information from the stem-and-leaf display and the box plot, we realize that an assumption of normality should not be made.

The TTEST Procedure

The **TTEST** procedure can be used to test for equality of both variances and means of two populations when independent random samples from normally distributed populations are available. This procedure requires that a grouping variable be named in the **INPUT** statement and that a **CLASS** statement be included to identify the grouping variable.

The tensile strength data considered in Example 2.5 are summarized in Table 2.6. We will use **TTEST** output for the command file in Table 2.7 to answer the question: Is there sufficient evidence to reject the hypothesis that the steels made in the two plants have the same mean tensile strength?

We begin the command file in Table 2.7 with an **OPTIONS** statement. That statement restricts the SAS output to 80 characters, thereby making the output fit on standard-size paper.

Two variables, PLANT and PSI, are identified in the **INPUT** statement. The first is a grouping variable that identifies the plant from which the steel sample was obtained. The second identifies the response variable.

TABLE 2.6
Steel Plant Data From Example 2.5

Plant	psi ($\times 10^3$)							
1	207	195	209	218	196	217		
2	219	228	222	197	225	184	218	211

TABLE 2.7
SAS Command File for Steel Plant Study

`OPTIONS LINESIZE=80;`	*restricts the output to 80 characters per line*
`DATA STEEL;`	*names the data set*
`  INPUT PLANT PSI @@;`	*names the variables and the input format*
`CARDS;`	*signals that data follow*
`  1   207   1   195   1   209   1   218   1   196   1   217   2   219`	
`  2   228   2   222   2   197   2   225   2   184   2   218   2   211`	
`;`	*signals the end of data*
`PROC TTEST;`	*identifies the procedure to be used*
`  CLASS PLANT;`	*identifies the grouping variable*
`  VAR PSI;`	*identifies the response variable*

The symbols @@ in the input statement tell the computer that more than one set of variables will be entered on a line instead of a single set as before. This allows us to type a value of PLANT, then type a space, then type the corresponding value of PSI, then type a space, and so forth until all ordered pairs have been typed. Once all pairs have been entered, remember to type a semicolon on the next line to signal the end of the data.

The last three lines of Table 2.7 identify the procedure to be used, the grouping variable, and the response variable. Since only one response variable was input, inclusion of the statement **VAR** *name* with the **TTEST** procedure is optional. However, if another variable (say length) had been included in the input statement and data set, that statement would be required.

Output for the command file in Table 2.7 is given in Figure 2.8. The p value of a two-tailed test for equality of variances (PROB > F' = 0.3640) supports an assumption of equal variances. This agrees with the result in Example 2.5. The p value of a two-tailed test for equality of means (assuming equal variances) is approximately 0.42, as before, so there is insufficient evidence to reject equality of means.

The MEANS Procedure

The **MEANS** procedure can be used to test a hypothesis on a population mean when the variance is unknown. If requested, the value of the test statistic and the p value for a two-tailed alternative will be included in the output. By default, this procedure tests H_0: $\mu = 0$ versus H_a: $\mu \neq 0$. Thus, to test H_0: $\mu_Y = \mu_0$ when μ_0 is not zero, one should define a new variable, say W with $W = Y - \mu_0$, and test H_0 : $\mu_w = 0$.

In Example 2.6, data for a typical "before-after" experiment performed on the same items were used to test H_0 : $\mu_{\text{before}} = \mu_{\text{after}}$ versus H_1 : $\mu_{\text{before}} > \mu_{\text{after}}$. The usual procedure is to use the difference between "before" and "after" and test H_0 : $\mu_D = 0$ versus H_1 : $\mu_D > 0$. For convenience, the data in Example 2.6 are summarized in Table 2.8.

```
                          TTEST PROCEDURE
VARIABLE: PSI
PLANT       N      MEAN      STD DEV    STD ERROR    MINIMUM     MAXIMUM
  1         6    207.0000    9.89949    4.04145     195.0000    218.0000
  2         8    213.0000   15.17517    5.36523     184.0000    228.0000
VARIANCES        T     DF   PROB > | T |
UNEQUAL       -0.8932  11.9    0.3895
EQUAL         -0.8394  12.0    0.4176
FOR H0: VARIANCES ARE EQUAL,  F' = 2.35 WITH 7 AND 5 DF PROB > F' = 0.3640
```

Figure 2.8   SAS Output for the steel plant study.

TABLE 2.8
Light—Box Readings from Example 2.6

| | | | | Cone | | | | |
	1	2	3	4	5	6	7	8
Before	6.5	6.0	7.0	6.8	6.5	6.8	6.2	6.5
After	4.4	4.2	5.0	5.0	4.8	4.6	5.2	4.9

A SAS command file for our test is summarized in Table 2.9. The **INPUT** statement identifies **BEFORE** and **AFTER** as the variables involved. The difference variable (DIFF) is defined by adding the statement **DIFF=BEFORE—AFTER**; immediately after the input statement. The title line has the information enclosed in single quotes typed with the outputs for any procedures that follow.

A new procedure, **PRINT**, is invoked after the title line in Table 2.9. This will have SAS print a listing of the values of the three variables.

The statement requesting the use of the **MEANS** procedure contains a listing of the options to be included. The options **T** and **PRT** cause SAS to print the value of the test statistic and the p value for $H_0 : \mu_D = 0$ versus $H_1 : \mu_D \neq 0$. By choosing **MEAN**, **VAR**, and **STDERR**, we can have $\bar{d}$, s_D, and $s_D/\sqrt{n}$, respectively, printed. Output is restricted to two decimal places by including the option **MAXDEC=2**.

The last line of the command file specifies the variable on which we want the test made. Without that line, results of $H_0 : \mu = 0$ versus $H_1 : \mu \neq 0$ would be included for each of **BEFORE**, **AFTER**, and **DIFF**.

Outputs for the command file in Table 2.9 are included in Figure 2.9. We observe that the t test value of 13.45 was also computed in Example 2.6.

TABLE 2.9
SAS Command File for Light—Box Study

```
OPTIONS LINESIZE=80;
DATA LIGHTBOX;
   INPUT BEFORE AFTER @@;
   DIFF=BEFORE—AFTER;
CARDS;
   6.5   4.4   6.0   4.2   7.0   5.0   6.8   5.0
   6.5   4.8   6.8   4.6   6.2   5.2   6.5   4.9
;
TITLE1 'TESTING LIGHT READINGS';
PROC PRINT;
PROC MEANS MEAN VAR STDERR T PRT MAXDEC=2;
   VAR DIFF;
```

TESTING LIGHT READINGS

OBS	BEFORE	AFTER	DIFF
1	6.5	4.4	2.1
2	6.0	4.2	1.8
3	7.0	5.0	2.0
4	6.8	5.0	1.8
5	6.5	4.8	1.7
6	6.8	4.6	2.2
7	6.2	5.2	1.0
8	6.5	4.9	1.6

TESTING LIGHT READINGS

VARIABLE	MEAN	VARIANCE	STD ERROR OF MEAN	T	PR> |T|
DIFF	1.78	0.14	0.13	13.45	0.0001

Figure 2.9   SAS output for the light-box study.

2.11 FURTHER READING

Gunter, B., "Q–Q Plots," *Quality Progress,* February 1994, pp. 81–86.

Nelson, W. *How to Analyze Data with Simple Plots.* Milwaukee, WI: American Society for Quality, 1986. This is Volume 1 in the ASQ's *How to* series.

PROBLEMS

2.1 For the following data on tensile strength, determine the sample mean and variance.

Tensile Strength (psi)	Frequency
18,461	2
18,466	12
18,471	15
18,476	10
18,481	8
18,486	3
Total	50

2.2 Using the results in Problem 2.1 and assuming that $\sigma^2 = 40$ psi², is there sufficient evidence to reject the hypothesis that the mean tensile strength of the population sampled is 18,470 psi? Use $\alpha = 0.05$.

2.3 Plot the operating characteristic curve for the test in Problem 2.2.

2.4 Determine how large a sample would be needed to detect an increase of 10 psi in the mean tensile strength of Problem 2.2 if $\alpha = 0.05$, $\beta = 0.02$, and $\sigma^2 = 40$ psi².

2.5 Repeat Problem 2.4 for the detection of a shift of 10 psi in the mean in either direction.

2.6 For a sample mean of 124 g based on 16 observations from a normal population whose variance is 25 g², set up 90% confidence limits on the mean of the population.

2.7 For a sample variance of 62 based on 12 observations of a normally distributed random variable, test the hypothesis that the population variance is 40. Use a one-sided test and a 1% significance level.

2.8 For Problem 2.7 set up 90% confidence limits (two-sided) on σ^2.

2.9 Two samples, one from each of the weekly productions of two machines, produced the following statistics:

$$n_1 = 8 \qquad \bar{y}_1 = 42.2\text{g} \qquad s_1^2 = 10\text{g}^2$$

$$n_2 = 15 \qquad \bar{y}_2 = 44.5\text{g} \qquad s_2^2 = 18\text{g}^2$$

Test these results for a significant difference in variances. Assume normality.

2.10 Test the results in Problem 2.9 for a significant difference in means. Assume normality.

2.11 A group of 10 students was pretested before instruction and posttested after 6 weeks of instruction with the following achievement scores:

Student	Before	After	Difference = d
1	14	17	3
2	12	16	4
3	20	21	1
4	8	10	2
5	11	10	−1
6	15	14	−1
7	17	20	3
8	18	22	4
9	9	14	5
10	7	12	5

Is there evidence of an improvement in achievement over this 6-week period? Justify using an appropriate statistical test.

2.12 To test the hypothesis that the nonconforming proportion p of a process is 0.20, a sample of 100 pieces was drawn at random.
   a. Letting $\alpha = 0.05$, set up a critical region for the number of observed defectives to test this hypothesis against a two-sided alternative.
   b. If the process now shifts to a 0.10 fraction nonconforming, find the probability of a type II error for the alternative in part a.
   c. Without repeating the work, would you expect the probability of a type II error to be the same if the process shifted to 0.30? Explain.

2.13 Data on a random variable Y were 12, 8, 14, 20, 26, 26, 20, 21, 18, 24, 30, 21, 18, 16, 10, and 20. Assuming this is a random sample, Y is normally distributed, and $\sigma = 7$, test each of the following. Let $\alpha = 0.05$.
   a. $H_0 : \mu = 12$ versus $H_1 : \mu > 12$
   b. $H_0 : \mu = 16$ versus $H_1 : \mu \neq 16$
   c. $H_0 : \mu = 18$ versus $H_1 : \mu > 18$

2.14 For the test in Problem 2.13a, evaluate the power of the test to reject H_0 when μ is 13, 14, 15, 18, 20, 22, 23, and 24. Plot the power curve.

2.15 a. Repeat Problem 2.14 with $\alpha = 0.01$.
   b. Plot the power curve in Problems 2.14 and in 2.15a on the same graph and compare.

2.16 In a study of the current flow (in amperes) through a cereal-forming machine, values of 8.2, 8.3, 8.2, 8.6, 8.0, 8.4, 8.5, and 7.3 were obtained. Assuming that these data represent a random sample from a normally distributed population, test $H_0 : \sigma = 0.35$ A versus $H_1 : \sigma > 0.35$ A.

2.17 For the data in Problem 2.16, test the hypothesis that the true average current flow is 7.0 A.

2.18 Random samples of a puffed cereal were taken from a day's production from two different "guns." The moisture content was as follows.

Gun	Moisture (%)						
I	3.6	3.8	3.6	3.3	3.7	3.4	
II	3.7	3.9	4.2	4.9	3.6	3.5	4.0

Test the hypothesis of equal variances and that of equal means. Use any assumptions you believe to be appropriate.

2.19 Pretest data for experimental and control groups on course content in a special vocational-industrial course indicated:

$$\text{Experimental:} \quad \bar{y}_1 = 9.333 \quad s_1 = 4.945 \quad n_1 = 12$$

$$\text{Control:} \quad \bar{y}_2 = 8.375 \quad s_2 = 1.187 \quad n_2 = 8$$

Test the hypothesis of equal variances and that of equal means. Comment on the results.

2.20 The WISC performance scores of 10 boys and 8 girls were recorded as follows:

Boys:	101	86	72	129	99	118	104	125	90	107
Girls:	97	107	94	101	90	108	108	86		

Test whether there is a significant difference ($\alpha = 0.10$) in the means of boys and girls. State all assumptions.

2.21 To evaluate the relative merits of two prosthetic devices designed to facilitate manual dexterity, an occupational therapist assigned 21 patients with identical handicaps to wear one or the other of the two devices while performing a certain task. Eleven patients wore device A and 10 wore device B. The researcher recorded the time each patient required to perform a certain task, with the following results:

$$\bar{y}_A = 65 \text{ seconds} \qquad s_A^2 = 81 \text{ seconds}^2$$

$$\bar{y}_B = 75 \text{ seconds} \qquad s_B^2 = 64 \text{ seconds}^2$$

Do these data provide sufficient evidence to indicate that device A is more effective than device B?

2.22 An anthropologist believes that individuals in two populations will show double occipital hair whorls in the same proportion. To see whether there is any reason to doubt this hypothesis, the anthropologist takes independent random samples from each of the two populations and determines the number in each sample with this characteristic. On the basis of the following results, is there reason to doubt this hypothesis?

Population	n	Number with Characteristic
1	100	23
2	120	32

2.23 A psychologist randomly selected 10 couples from among the residents of an urban area and asked them to complete a questionnaire designed to measure their level of satisfaction with the community in which they lived. Do the following results indicate that the husbands are better satisfied with the community than the wives? State all assumptions and justify the procedure used.

	Couple									
	1	2	3	4	5	6	7	8	9	10
Wife	33	57	32	54	52	34	60	40	59	39
Husband	44	60	55	68	40	48	57	49	47	52

2.24 If two samples randomly selected from two independent normal populations give

$$n_1 = 9 \qquad \bar{y}_1 = 16.0 \qquad s_1^2 = 5.0$$

$$n_2 = 4 \qquad \bar{y}_2 = 12.0 \qquad s_2^2 = 2.0$$

is there enough evidence to claim that the mean of population 1 is greater than the mean of population 2?

2.25 The standard medical treatment for a certain ailment has proved to be about 85% effective. One hundred patients were given an alternative treatment, and 78 showed marked improvement. Is the new treatment actually inferior, or could the difference noted be reasonably attributed to chance?

2.26 Use the sample of differences in Problem 2.11 and check for normality.

2.27 Consider Problem 2.13. Using an appropriate procedure, investigate the wisdom of assuming normality in the sampled population.

2.28 The following data on the coded amount of residual binder in sand used to cast manganese steel railway components are given in control chart format. Preliminary analyses indicate that (1) the data were obtained from a stable process and (2) the sampled population has a nonnormal distribution.
a. Test the hypothesis that the coded average amount of residual binder is 0.19585.
b. Determine the p value of the preceding test.
c. Determine a 99% confidence interval for the coded average amount of residual binder in the sand used to cast such components.

Sample	Observations					Sample	Observations				
	1	2	3	4	5		1	2	3	4	5
1	0.1776	0.1830	0.2632	0.1995	0.1954	14	0.2171	0.1812	0.2072	0.1731	0.2556
2	0.2142	0.2039	0.2343	0.2247	0.1656	15	0.1644	0.1923	0.2200	0.1672	0.2892
3	0.2428	0.1701	0.2022	0.1623	0.1735	16	0.1840	0.1762	0.1713	0.1698	0.1786
4	0.1607	0.1969	0.1465	0.1443	0.1582	17	0.1891	0.2238	0.1852	0.1709	0.1977
5	0.1616	0.1628	0.1509	0.2202	0.2303	18	0.2041	0.1644	0.1880	0.1801	0.2245
6	0.1684	0.1719	0.1623	0.1755	0.2015	19	0.1993	0.1726	0.1666	0.1621	0.1882
7	0.1860	0.1842	0.1990	0.1770	0.1814	20	0.2109	0.1827	0.2369	0.1959	0.2059

8	0.2021	0.2261	0.1992	0.2080	0.2718	21	0.1879	0.2097	0.2680	0.1905	0.2232
9	0.2042	0.2762	0.1683	0.2414	0.2386	22	0.2912	0.1881	0.1659	0.1682	0.2086
10	0.2072	0.2618	0.1948	0.2351	0.1771	23	0.1736	0.1616	0.1940	0.1442	0.2297
11	0.1678	0.2260	0.2212	0.2241	0.2282	24	0.2021	0.2186	0.1968	0.1765	0.1579
12	0.1661	0.2114	0.2036	0.2244	0.1855	25	0.1883	0.1986	0.1585	0.1719	0.1688
13	0.2311	0.1952	0.1795	0.2380	0.2171						

2.29 Consider Problem 2.28.
   a. Prepare a normal quantile plot and comment.
   b. Prepare a box plot and comment. Include a check for outliers.

2.30 Calculate summary statistics (mean, median, variance, standard deviation, range, quartiles, etc.) for the data in Problem 2.28. Calculate the proportion of values within 1, 2, and 3 standard deviations of the mean and compare with that for a normal distribution. Comment.

2.31 When each student in a random sample of students from a particular university was asked to tap a pencil with each hand as fast as possible for 15 seconds, the following data were obtained.

Person	Hand Used		Person	Hand Used	
	Dominant	**Nondominant**		**Dominant**	**Nondominant**
1	94	86	13	100	95
2	78	76	14	99	95
3	86	82	15	77	66
4	106	99	16	82	68
5	110	102	17	87	81
6	43	36	18	80	74
7	87	86	19	90	78
8	80	70	20	65	62
9	56	59	21	61	59
10	79	63	22	82	78
11	55	55	23	90	81
12	109	94	24	94	91

   a. Do these data indicate that for students at this university, the true average number of taps made for 15 seconds with the dominant hand is greater than that for the nondominant hand? Justify your claim with an appropriate test procedure.
   b. Find a lower 95% confidence limit for the difference between the average number of taps a person makes in 15 seconds with the dominant hand and that for the nondominant hand by preparing a 90% confidence interval and removing the right end point. Interpret in practical terms.
   c. Using a horizontal scale for the number of taps recorded and a vertical scale for the hand used (with 1 = "dominant" and 2 = "nondominant"), construct box plots for the given data. Comment on the graphical information obtained and explain why it may not convey the same information as the preceding hypothesis test.

Chapter 3

SINGLE-FACTOR EXPERIMENTS WITH NO RESTRICTIONS ON RANDOMIZATION

3.1 INTRODUCTION

This chapter begins our consideration of single-factor experiments. To ensure that a given design will be completely randomized, Chapter 3 places no restrictions on the randomization. Many of the techniques of analysis for a completely randomized single-factor experiment can be applied with little alteration to more complex experiments.

For example, the single factor could be steel manufacturers, where the main interest of an analyst centers on the effect of several different manufacturers on the hardness of steel purchased from these vendors. It could be temperature, if the experimenter is concerned about the effect of this variable on penicillin yield. In cases such as these, where only one factor is varied, the experiment is referred to as a *single-factor experiment*. The symbol τ_j will be used to indicate the effect of the jth level of the factor, suggesting that the general factor may be thought of as a "treatment" effect.

The Completely Randomized, Single-Factor Experiment

If the order of experimentation applied to the several levels of the factor is completely random, so that any material to which the treatments might be applied is considered to be approximately homogeneous, the design is called a *completely randomized design*. The number of observations for each level of the factor will be determined from cost considerations and the power of the test. The model then becomes

$$Y_{ij} = \mu + \tau_j + \varepsilon_{ij} \tag{3.1}$$

with Y_{ij} the ith observation ($i = 1, 2, \ldots, n_j$) on the jth treatment ($j = 1, 2, \ldots, k$); μ a common effect for the whole experiment; τ_j the effect of the jth treatment; and ε_{ij} the random error present in the ith observation on the jth treatment. The error term ε_{ij}, usually considered to be a normally and independently distributed (NID) random effect with mean zero and variance the same for all treatments (levels), is expressed as follows:

The ε_{ij}s are NID$(0, \sigma_\varepsilon^2)$, where σ_ε^2 is the common variance within all treatments.

The common effect μ is always a fixed parameter and $\tau_j = \mu_j - \mu$, where μ_j is the true mean of the jth population. The types of inference that can be made depend on the assumptions about the treatment effects.

The Fixed Model

Suppose a manufacturer of clothing is concerned about the "wearability" of four competing fabrics (A, B, C, and D). A completely randomized, single-factor experiment is to be designed to enable the manufacturer to choose the fabric that wears the least over a given period of time. Since the experimenter's interest is restricted to the fabrics under consideration, the four treatments (fabrics) are referred to as *fixed effects*. The resulting model is called a *fixed* model.

When the treatment effect in model Equation (3.1) is fixed, the following assumptions are added to those already mentioned:

- $\tau_1, \tau_2, \ldots, \tau_j, \ldots, \tau_k$ are considered to be fixed parameters
- $\sum_{j=1}^{k} \tau_j = 0$, implying that $\mu = (1/k) \sum_{j=1}^{k} \mu_j$

The analysis usually includes a one-way analysis of variance (ANOVA) test of $H_0 : \tau_1 = \cdots = \tau_k = 0$ versus H_1: "At least one of the treatment effects is not 0." This test is discussed in detail in Section 3.2. If H_0 is true, no treatment effects exist, the mean of the jth population is μ, and each observation Y_{ij} is made up of its population mean μ and a random error ε_{ij}. If graphical and inferential procedures indicate that H_0 should be rejected, a number of tests (see Section 3.3) may follow.

■ Example 3.1

In the manufacture of clothing a wear-testing machine is used to measure the resistance to abrasion of different fabrics. The dependent variable Y gives the loss of weight of the material in grams after a specified number of cycles. The problem is to determine whether there is a difference in the average weight loss among four competing fabrics. Here there is but one factor of interest: the fabric at levels A, B, C, and D. These are qualitative and fixed levels of this single factor. It is agreed to test four samples of each fabric and completely randomize the order of testing of the 16 samples. The mathematical model for this example then is

$$Y_{ij} = \mu + \tau_j + \varepsilon_{ij}$$

with $i = 1, 2, 3, 4$ and $j = 1, 2, 3, 4$, since there are four samples for each of four treatments (fabrics). The resulting data are shown in Table 3.1.

Excel results of a test of $H_0 : \tau_1 = \tau_2 = \tau_3 = \tau_4 = 0$ are summarized in Table 3.2. Assuming that the model assumptions have been adequately satisfied, the test statistic has an F distribution with 3 and 12 degrees of freedom. Since the observed value of that statistic is $f = 0.1734/0.0203 = 8.53$ and $P(F_{3,12} \geq 8.53) = 0.0026$, H_0 can be rejected for any $\alpha \geq 0.0026$. Thus, we reject the null hypothesis and claim that there is a considerable difference in average wear resistance among the four fabrics. [**Note:** The value $F \ crit = 5.95$ in Table 3.2 indicates that H_0 can be rejected at the 1% significance level if the value of the test statistic exceeds 5.95.]

TABLE 3.1
Fabric Wear Resistance Data

		Fabric		
	A	*B*	*C*	*D*
	1.93	2.55	2.40	2.33
	2.38	2.72	2.68	2.40
	2.20	2.75	2.31	2.28
	2.25	2.70	2.28	2.25

TABLE 3.2
Excel Output for the One-Way Analysis-of-Variance Test in Example 3.1

	A	**B**	**C**	**D**	**E**	**F**	**G**
1		Fabric					
2		A	B	C	D		
3		1.93	2.55	2.40	2.33		
4		2.38	2.72	2.68	2.40		
5		2.20	2.75	2.31	2.28		
6		2.25	2.70	2.28	2.25		
7	Anova: Single Factor						
8	SUMMARY						
9	*Groups*	*Count*	*Sum*	*Average*	*Variance*		
10	A	4	8.76	2.190	0.035800		
11	B	4	10.72	2.680	0.007933		
12	C	4	9.67	2.418	0.033225		
13	D	4	9.26	2.315	0.004300		
14							
15	ANOVA						
16	*Source*	*SS*	*df*	*MS*	*F*	*P-value*	*F crit*
17	Between	0.5201	3	0.1734	8.53	0.0026	5.95
18	Within	0.2438	12	0.0203			
19	Total	0.7639	15				

The Random Model

In Example 3.1, the levels of the factor were considered to be fixed, since only four fabrics were available and a decision was desired on the effect of these four fabrics only. When the levels of the factor are chosen at random from a large number of possible levels, the model is called a *random model*. In a random model, the experimenter is not usually interested in testing the hypothesis of no treatment effect, but in estimating the proportion of variability in the dependent variable that is attributable to differences among the levels of the factor. This is addressed in detail in Section 3.6.

For a random model, the following assumptions are added to those stated with model Equation (3.1):

- $\tau_1, \tau_2, \ldots, \tau_j, \ldots, \tau_k$ are considered to be random variables.
- the τ_j's are NID$(0, \sigma_\tau^2)$.

For this model, a test of $H_0 : \sigma_\tau^2 = 0$, which is equivalent to a test of the hypothesis that the population means are equal, is conducted in the same way as the test illustrated in Example 3.1.

3.2 ANALYSIS OF VARIANCE RATIONALE

To review the basis for the F test in a one-way analysis of variance, k populations, each representing one level of treatment, can be considered with observations as shown in Table 3.3. Since each observation could be returned to the population and measured, an infinite number of observations could be taken on each population. So, the average of Y_{i1} is $E(Y_{i1}) = \mu_1, E(Y_{i2}) = \mu_2$, and so on. The parameter μ represents an overall model parameter. The treatment effect in the model is $\tau_j = \mu_j - \mu$, the random error is $\varepsilon_{ij} = Y_{ij} - \mu_j$, and the model is either

$$Y_{ij} = \mu + \tau_j + \varepsilon_{ij}$$

or

TABLE 3.3
Population Layout for One-Way ANOVA

	Treatment					
	1	2	$\cdots$	j	$\cdots$	k
	Y_{11}	Y_{12}		Y_{1j}		Y_{1k}
	Y_{21}	Y_{22}		Y_{2j}		Y_{2k}
	Y_{31}	Y_{32}		—		—
	—	—		—		—
	Y_{i1}	Y_{i2}		Y_{ij}		Y_{ik}
Population means	μ_1	μ_2	$\cdots$	μ_j	$\cdots$	μ_k

$$Y_{ij} \equiv \mu + (\mu_j - \mu) + (Y_{ij} - \mu_j)$$

This last expression is seen to be an identity true for all values of Y_{ij}. Expressed in another way,

$$Y_{ij} - \mu \equiv (\mu_j - \mu) + (Y_{ij} - \mu_j) \tag{3.2}$$

Since these means are unknown, random samples are drawn from each population and estimates can be made of the treatment means and the grand mean. If n_j observations are taken for each treatment where the numbers need not be equal, a sample layout would be as shown in Table 3.4.

Here the "dot notation" indicates a summing over all observations in the sample. The $T_{.j}$ represents the total of the observations taken under treatment j, n_j represents the number of observations taken for treatment j, and $\bar{Y}_{.j}$ is the sample mean for treatment j. Also note that the grand total of all observations taken is

$$T_{..} = \sum_{j=1}^{k} \sum_{i=1}^{n_j} Y_{ij} = \sum_{j=1}^{k} T_{.j}$$

the total number of observations is

$$N = \sum_{j=1}^{k} n_j$$

and the mean of all N observations is

$$\bar{Y}_{..} = \sum_{j=1}^{k} n_j \bar{Y}_{.j} / N = \frac{T_{..}}{N}$$

TABLE 3.4
Sample Layout for One-Way ANOVA

	Treatment						
	1	**2**	$\cdots$	**j**	$\cdots$	**k**	
	Y_{11}	Y_{12}	$\cdots$	Y_{1j}	$\cdots$	Y_{1k}	
	Y_{21}	Y_{22}	$\cdots$	Y_{2j}	$\cdots$	Y_{2k}	
	$\vdots$	$\vdots$		$\vdots$		$\vdots$	
	Y_{i1}	Y_{i2}	$\cdots$	Y_{ij}	$\cdots$	Y_{ik}	
	$\vdots$	$\vdots$		$\vdots$		$\vdots$	
	$Y_{n_1 1}$	$\vdots$		$Y_{n_j j}$		$\vdots$	
		$Y_{n_2 2}$				$Y_{n_k k}$	Overall totals
Column totals	$T_{.1}$	$T_{.2}$	$\cdots$	$T_{.j}$	$\cdots$	$T_{.k}$	$T_{..}$
Number	n_1	n_2	$\cdots$	n_j	$\cdots$	n_k	N
Means	$\bar{Y}_{.1}$	$\bar{Y}_{.2}$	$\cdots$	$\bar{Y}_{.j}$	$\cdots$	$\bar{Y}_{.k}$	$\bar{Y}_{..}$

If these sample statistics are substituted for their corresponding population parameters in Equation (3.2), we get a sample equation (also an identity) of the form

$$Y_{ij} - \bar{Y}_{..} \equiv \left(\bar{Y}_{.j} - \bar{Y}_{..} \right) + \left(Y_{ij} - \bar{Y}_{.j} \right) \tag{3.3}$$

This equation states that the deviation of an observation from the grand mean can be broken into two parts: the deviation of the observation from its own treatment mean plus the deviation of the treatment mean from the grand mean.

If both sides of Equation (3.3) are squared and then added over both i and j, we have

$$\sum_{j=1}^{k} \sum_{i=1}^{n_j} \left(Y_{ij} - \bar{Y}_{..} \right)^2 = \sum_{j=1}^{k} \sum_{i=1}^{n_j} \left(\bar{Y}_{.j} - \bar{Y}_{..} \right)^2 + \sum_{j=1}^{k} \sum_{i=1}^{n_j} \left(Y_{ij} - \bar{Y}_{.j} \right)^2$$

$$+ 2 \sum_{j=1}^{k} \sum_{i=1}^{n_j} \left(\bar{Y}_{.j} - \bar{Y}_{..} \right) \left(Y_{ij} - \bar{Y}_{.j} \right)$$

Examining the last expression on the right, we find that

$$2 \sum_{j=1}^{k} \sum_{i=1}^{n_j} \left(\bar{Y}_{.j} - \bar{Y}_{..} \right) \left(Y_{ij} - \bar{Y}_{.j} \right) = 2 \sum_{j=1}^{k} \left(\bar{Y}_{.j} - \bar{Y}_{..} \right) \left[\sum_{i=1}^{n_j} \left(Y_{ij} - \bar{Y}_{.j} \right) \right]$$

The term in brackets is seen to equal zero, since the sum of the deviation about the mean within a given treatment equals zero. Hence

$$\sum_{j=1}^{k} \sum_{i=1}^{n_j} \left(Y_{ij} - \bar{Y}_{..} \right)^2 = \sum_{j=1}^{k} \sum_{i=1}^{n_j} \left(\bar{Y}_{.j} - \bar{Y}_{..} \right)^2 + \sum_{j=1}^{k} \sum_{i=1}^{n_j} \left(Y_{ij} - \bar{Y}_{.j} \right)^2 \tag{3.4}$$

Equation (3.4) may be referred to as *the fundamental equation of analysis of variance*. It expresses the idea that the sum of the squares of the deviations from the grand mean is equal to the sum of the squares of the deviations between treatment means and the grand mean plus the sum of the squares of the deviations within treatments. That is,

$$SS_{total} = SS_{between} + SS_{within}$$

This is also expressed as

$$SS_{total} = SS_{treatment} + SS_{error}$$

In Chapter 2 an unbiased estimate of a population variance was determined by dividing the sum of squares $\sum_{i=1}^{n}(Y_i - \bar{Y})^2$ by the corresponding degrees of freedom, $n - 1$. If the model assumptions are satisfied and the hypothesis being tested in a one-way analysis of variance is true (i.e., $\tau_j = 0$ for all j, or there is no treatment effect), then each of the k treatment means equals μ and the model contains only the population mean μ and random error ε_{ij}. Then any one of the three terms in Equation (3.4) may be used to obtain an unbiased estimate of the common population variance.

The Error Mean Square

Within the jth treatment, $\sum_{i=1}^{n_j}(Y_{ij} - \bar{Y}_{.j})^2$ divided by $n_j - 1$ df would yield an unbiased estimate of the variance within the jth treatment. If the variances within the k treatments are really all alike, their estimates may be pooled to give $\mathrm{SS}_{\mathrm{error}} = \sum_{j=1}^{k} \sum_{i=1}^{n_j}(Y_{ij} - \bar{Y}_{.j})^2$ with $\sum_{j=1}^{k}(n_j - 1) = N - k$ degrees of freedom. Dividing $\mathrm{SS}_{\mathrm{error}}$ by its degrees of freedom, we obtain the *error mean square*

$$\mathrm{MS}_{\mathrm{error}} = \frac{\sum_{j=1}^{k} \sum_{i=1}^{n_j}(Y_{ij} - \bar{Y}_{.j})^2}{N - k} = \frac{\mathrm{SS}_{\mathrm{error}}}{N - k} \tag{3.5}$$

Before sampling, $\mathrm{MS}_{\mathrm{error}}$ is an unbiased estimator of the common population variance, regardless of whether the hypothesis of no treatment effect is true or false.

The Treatment Mean Square

Another unbiased estimate of the common population variance can be made by using the variance of the k sample means. Dividing $\mathrm{SS}_{\mathrm{treatment}}$ by its degrees of freedom, we obtain the *treatment mean square*

$$\mathrm{MS}_{\mathrm{treatment}} = \frac{\sum_{j=1}^{k} n_j(\bar{Y}_{.j} - \bar{Y}_{..})^2}{k - 1} = \frac{\mathrm{SS}_{\mathrm{treatment}}}{k - 1} \tag{3.6}$$

When the k population means are equal, $\mathrm{MS}_{\mathrm{treatment}}$ is an unbiased estimator of σ_ε^2. To aid understanding of this statement, consider the case where $n_1 = n_2 = \cdots = n_k = n$. Let

$$S_{\bar{Y}}^2 = \frac{\sum_{j=1}^{k}(\bar{Y}_{.j} - \bar{Y}_{..})^2}{k - 1}$$

denote the variance of the sample of k sample means. *If there are no real differences in the k population means,*

$$E[S_{\bar{Y}}^2] = \frac{\sigma_\varepsilon^2}{n}$$

and

$$E[nS_{\bar{Y}}^2] = \sigma_\varepsilon^2$$

In this case, $nS_{\bar{Y}}^2 = \mathrm{MS}_{\mathrm{treatment}}$.

We have noted that $\mathrm{MS}_{\mathrm{treatment}}$ is an unbiased estimator of σ_ε^2 when the population means are equal (i.e., the treatment effect is 0). In Section 3.6, the average value of $\mathrm{MS}_{\mathrm{treatment}}$ is shown to exceed σ_ε^2 when at least two population means differ.

The F Ratio

The sums of squares in Equation (3.4) are not independent, since the total sum of squares is the sum of the treatment sum of squares and the error sum of squares. *When the hypothesis of no treatment effect is true,* however, one can show that the treatment mean square and the error mean square:

- are unbiased estimators of σ_ε^2
- have independent chi-square distributions with $k - 1$ and $N - k$ degrees of freedom, respectively

If two such independent unbiased estimates of the same variance are compared, their ratio can be shown to be distributed as F with $\nu_1 = k - 1$ and $\nu_2 = N - k$. That is,

$$F = \frac{\text{MS}_{\text{treatment}}}{\text{MS}_{\text{error}}} \sim F_{(k-1, N-k)} \tag{3.7}$$

when the model assumptions are satisfied and the population means are equal.

The One-Way ANOVA

Suppose the model assumptions of this section are satisfied and a test of

$$H_0 : \tau_1 = \cdots = \tau_k = 0$$

versus

$$H_1 : \tau_j \neq 0 \text{ for some } j \in \{1, 2, \ldots, k\}$$

is to be conducted. We use $F = \text{MS}_{\text{treatment}}/\text{MS}_{\text{error}}$ as the test statistic. Since the expected value of $\text{MS}_{\text{treatment}}$ equals that of MS_{error} when H_0 is true, but exceeds that of MS_{error} when H_0 is false, values of F that are much larger than 1 support rejection of H_0. Thus, the test is an upper-tailed test. Letting f denote the observed value of F, the p value of the test is $P(F_{(k-1, N-k)} \geq f)$. If this p value is less than or equal to α, we have strong evidence that there is a real difference in treatment means $(\mu_1, \mu_2, \ldots, \mu_k)$ and that H_0 should be rejected. Alternatively, H_0 can be rejected if $f \geq F_{1-a}$ where α is the area above F_{1-a}.

The preceding F test and Equation (3.4) are usually summarized in an *analysis of variance* (or ANOVA) table such as Table 3.5, where SS and MS denote "sum of squares" and "mean square," respectively. The quantities in the SS column represent the fundamental equation $\text{SS}_{\text{total}} = \text{SS}_{\text{treatment}} + \text{SS}_{\text{error}}$.

The actual computing of sums of squares needed for the ANOVA summary is much easier if the terms are first expanded and rewritten in terms of treatment totals. Even though computer programs such as Excel, Minitab, and SAS make such hand calculations unnecessary, we have found that many students report a better understanding after having calculated at least one ANOVA summary by hand. In such cases, the following formulas may be used.

TABLE 3.5
One-Factor ANOVA Test for Equality of $k \geq 2$ Means

Source	df	SS	MS	F	p value
Treatment	$\nu_1 = k - 1$	$\text{SS}_{\text{treatment}}$	$\text{MS}_{\text{treatment}}$	$f = \frac{\text{MS}_{\text{treatment}}}{\text{MS}_{\text{error}}}$	$P(F_{(\nu_1, \nu_2)} \geq f)$
Error	$\nu_2 = N - k$	SS_{error}	MS_{error}		
Totals	$N - 1$	SS_{total}			

$$\text{SS}_{\text{total}} = \sum_{j=1}^{k}\sum_{i=1}^{n_j}\left(Y_{ij} - \bar{Y}_{..}\right)^2 = \sum_{j=1}^{k}\sum_{i=1}^{n_j} Y_{ij}^2 - \frac{T_{..}^2}{N} \tag{3.8}$$

$$\text{SS}_{\text{treatment}} = \sum_{j=1}^{k} n_j\left(\bar{Y}_{.j} - \bar{Y}_{..}\right)^2 = \sum_{j=1}^{k}\frac{T_{.j}^2}{n_j} - \frac{T_{..}^2}{N} \tag{3.9}$$

$$\text{SS}_{\text{error}} = \sum_{j=1}^{k}\sum_{i=1}^{n_j}\left(Y_{ij} - \bar{Y}_{.j}\right)^2 = \text{SS}_{\text{total}} - \text{SS}_{\text{treatment}} \tag{3.10}$$

We should also note that the value of S^2, the variance of the entire set of N observations, is easily obtained with most calculators and that $\text{SS}_{\text{total}} = (N - 1)S^2$.

Example 3.2

Table 3.6 shows treatment totals $T_{.j}$'s and the sums of squares of the Y_{ij}'s for each treatment, along with the results added across all treatments for the fabric wear resistance data first considered in Example 3.1. The sums of squares can be computed quite easily from this table.

TABLE 3.6
Fabric Wear Resistance Data

	Fabric				
	A	*B*	*C*	*D*	
	1.93	2.55	2.40	2.33	
	2.38	2.72	2.68	2.40	
	2.20	2.75	2.31	2.28	
	2.25	2.70	2.28	2.25	
$T_{.j}$:	8.76	10.72	9.67	9.26	$T_{..} = 38.41$
n_j :	4	4	4	4	$N = 16$
$\sum_{i=1}^{n_j} Y_{ij}^2$:	19.2918	28.7534	23.4769	21.4498	$\sum_{j=1}^{k}\sum_{i=1}^{n_j} Y_{ij}^2 = 92.9719$

The expanded total sum of squares formula [Equation (3.8)] states, "Square each observation, add over all observations, and subtract the correction term." The correction term is the grand total squared and divided by the total number of observations. Thus,

$$\text{SS}_{\text{total}} = \sum_{j=1}^{k}\sum_{i=1}^{n_j} Y_{ij}^2 - \frac{T_{..}^2}{N} = 92.9719 - \frac{38.41^2}{16} = 0.7639$$

The sum of squares between treatments [Equation (3.9)] is found by totaling n_j observations for each treatment, squaring the total, dividing by the number of observations, adding for all treatments, and then subtracting the correction term. So,

$$\text{SS}_{\text{treatment}} = \frac{\sum_{j=1}^{k} T_{.j}^2}{n_j} - \frac{T_{..}^2}{N}$$

$$= \frac{8.76^2}{4} + \frac{10.72^2}{4} + \frac{9.67^2}{4} + \frac{9.26^2}{4} - \frac{38.41^2}{16} = 0.5201$$

The sum of squares for error is then determined by subtraction [Equation (3.10)]. That is,

$$\text{SS}_{\text{error}} = \text{SS}_{\text{total}} - \text{SS}_{\text{treatment}} = 0.7639 - 0.5201 = 0.2438$$

These results are displayed as in Table 3.2, and the F test is run on $H_0 : \tau_j = 0$ as discussed in Example 3.1.

The computation of the sums of squares can be simplified by coding the data. For example, one could code the data in Table 3.2 by subtracting a convenient constant such as 2.00 and then multiplying all y's by 100 to eliminate the decimals. The subtraction of the constant will not affect the sums of squares, but multiplying by 100 will multiply all SS terms by $(100)^2$ or 10,000. However, since the F statistic is the ratio of two mean squares, the value of F will be unaffected by this coding scheme.

3.3 AFTER ANOVA—WHAT?

After concluding, as in the fabric wear resistance study, that there is a difference in treatment means, we must ask more questions such as:

- Which treatment is best?
- Does the mean wear resistance of fabric A differ from that of C?
- Does the mean of A and B together differ from that of C and D?

To answer such questions, one must consider the design of the experiment and the manner in which the experiment was actually conducted. One must also have reliable evidence that the model assumptions are reasonable.

The decision tree of Figure 3.1 may be helpful in outlining some tests after analysis of variance. Tests for a fixed model with a qualitative factor are considered in Section 3.4, and estimation of the components of variance for a random model is considered in Section 3.6. These procedures also are based on the assumptions of the original model. Methods of checking the model assumptions are considered in Section 3.7.

3.4 TESTS ON MEANS

In considering qualitative levels of treatments as in Example 3.1, any comparison of means will depend on whether the means to be compared were selected before the experiment was run or after the results were examined. If the means were decided on before the experiment was run, orthogonal contrasts may be used to make the appropriate comparisons. A multiple

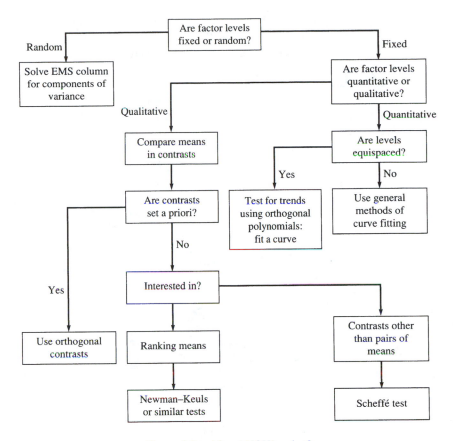

Figure 3.1 After ANOVA, what?

comparison test may be used when the means are not selected until after the ANOVA results have been examined.

3.4.1 Orthogonal Contrasts

If the means to be compared are selected prior to the running of an experiment, such comparisons can usually be set without disturbing the risk α of the original ANOVA. This means that the contrasts must be chosen with care, and the number of such contrasts should not exceed the number of degrees of freedom between the treatment means. The method usually used here is called the *method of orthogonal contrasts*. The following discussion develops the method using the fabric wear resistance data.

What Is a Contrast?

In comparing fabrics 1 and 2 it would seem quite logical to examine the difference in their means or their totals, since all treatment totals are based on the same sample size. One such comparison might be $T_{.1} - T_{.2}$. If, on the other hand, one wished to compare the first and second treatment with the third, the third treatment total should be weighted by a factor of 2, since it is based on only four observations and the treatment totals of 1 and 2 together represent eight observations. Such a comparison would then be $T_{.1} + T_{.2} - 2T_{.3}$. The sum of the coefficients of the treatment totals for $T_{.1} - T_{.2}$ is $(+1) + (-1) = 0$ and for $T_{.1} + T_{.2} - 2T_{.3}$ is $(+1) + (+1) + (-2) = 0$. That is, for both these comparisons, the sum of the coefficients is zero. Hence, a *contrast C_m in the treatment totals* is defined as follows when sample sizes are equal:

$$C_m = c_{1m} T_{.1} + c_{2m} T_{.2} + \cdots + c_{km} T_{.m}$$

where

$$c_{1m} + c_{2m} + \cdots + c_{km} = 0$$

That is, when sample sizes are equal,

$$C_m = \sum_{j=1}^{k} c_{jm} T_{.j}$$

with

$$\sum_{j=1}^{k} c_{jm} = 0$$

is a contrast in the treatment totals.

Meaning of Orthogonal Contrasts

For experiments that entail several contrasts, it is highly desirable to have the contrasts independent of each other. When the contrasts are independent—one having no projection on any of the others—they are said to be orthogonal contrasts, and independent tests of hypotheses can be made by comparing the mean square of each such contrast with the mean square of the error term in the experiment. Each contrast carries one degree of freedom, so the sum of squares for a contrast and the mean square are equal.

It may be recalled that when two straight lines, say

$$c_{11} x + c_{21} y = b_1$$
$$c_{12} x + c_{22} y = b_2$$

are perpendicular, or orthogonal, to each other, the slope of one line is the negative reciprocal of the slope of the other line. Writing the preceding equations in slope–intercept form gives

$$y = (-c_{11}/c_{21})x + (b_1/c_{21})$$
$$y = (-c_{12}/c_{22})x + (b_2/c_{22})$$

The requirement for orthogonality is expressed as follows:

$$(-c_{11}/c_{21}) = -(-c_{22}/c_{12}) \Leftrightarrow c_{11} c_{12} = -c_{21} c_{22} \Leftrightarrow c_{11} c_{12} + c_{21} c_{22} = 0$$

Thus, the lines are orthogonal if and only if the sum of the products of the corresponding coefficients (on x and y) is zero. This idea may be extended to a general definition for contrasts.

When sample sizes are equal, the contrasts $C_m = c_{1m}T_{.1} + \cdots + c_{km}T_{.k}$ and $C_q = c_{1q}T_{.1} + \cdots + c_{kq}T_{.k}$ are *orthogonal contrasts*, provided $c_{1m}c_{1q} + \cdots + c_{km}c_{kq} = 0$.

■ **Example 3.3**

Returning to the fabric wear resistance study, three orthogonal contrasts may be set up since there are 3 degrees of freedom between treatments. One such set of three is:

$$
\begin{aligned}
C_1 &= T_{.1} & - T_{.4} \\
C_2 &= & T_{.2} - T_{.3} \\
C_3 &= T_{.1} - T_{.2} - T_{.3} + T_{.4}
\end{aligned}
$$

C_1 is a contrast to compare the mean of the first treatment with the fourth, C_2 compares the second treatment with the third, and C_3 compares the average of treatments 1 and 4 with the average of 2 and 3.

The coefficients of the $T_{.j}$'s for the three contrasts are given in Table 3.7. Since the sum of the coefficients in each row is 0 and sample sizes are equal, we know that C_1, C_2, and C_3 are contrasts. Also, the sum of products of corresponding coefficients of each pair of contrasts is zero, so any two of the three contrasts are orthogonal.

TABLE 3.7
Orthogonal Coefficients: Equal Sample Sizes

	$T_{.1}$	$T_{.2}$	$T_{.3}$	$T_{.4}$
C_1	+1	0	0	−1
C_2	0	+1	−1	0
C_3	+1	−1	−1	+1

Sums of Squares for Contrasts

If $C_m = c_1 T_{.1} + \cdots + c_k T_{.k}$ is a contrast in the treatment totals and each sample consists of n observations, the sum of squares for C_m is

$$
\text{SS}_{C_m} = \frac{C_m^2}{n \sum_{j=1}^{k} c_{jm}^2} \tag{3.11}
$$

■ **Example 3.4**

We know that $C_1 = T_{.1} - T_{.4}$, $C_2 = T_{.2} - T_{.3}$, and $C_3 = T_{.1} - T_{.2} - T_{.3} + T_{.4}$ are orthogonal contrasts for a factor with $k = 4$ levels when sample sizes are equal. From

Table 3.6, $T_{.1} = 8.76$, $T_{.2} = 10.72$, $T_{.3} = 9.67$, and $T_{.4} = 9.26$. So, the observed values of these contrasts are

$$c_1 = 1(8.76) + 0(10.72) + 0(9.67) - 1(9.26) = -0.50$$

$$c_2 = 0(8.76) + 1(10.72) - 1(9.67) + 0(9.26) = 1.05$$

$$c_3 = 1(8.76) - 1(10.72) - 1(9.67) + 1(9.26) = -2.37$$

and the observed sums of squares are

$$SS_{C_1} = \frac{c_1^2}{4[1^2 + 0^2 + 0^2 + (-1)^2]} = \frac{(-0.50)^2}{8} = 0.0312$$

$$SS_{C_2} = \frac{c_2^2}{4[0^2 + 1^2 + (-1)^2 + 0^2]} = \frac{(1.05)^2}{8} = 0.1378$$

$$SS_{C_3} = \frac{c_3^2}{4[1^2 + (-1)^2 + (-1)^2 + 1^2]} = \frac{(-2.37)^2}{16} = 0.3511$$

The sum of these three sums of squares is $0.0312 + 0.1378 + 0.3511 = 0.5201 = SS_{treatment}$. This additivity occurs because the contrasts are orthogonal and the treatment sum of squares has 3 degrees of freedom.

Hypothesis Tests Using Contrasts

For samples of size n, the expected value and variance of $T_{.j}$ are $E[T_{.j}] = n\mu_j$ and $Var(T_{.j}) = n\sigma^2$, respectively. We leave it to the reader to use these facts and the basic properties of expected values to show that the contrast $C_m = c_{1_m}T_{.1} + \cdots + c_{k_m}T_{.k}$ has expected value

$$E[C_m] = n(c_{1m}\mu_1 + \cdots + c_{km}\mu_k)$$

and variance

$$Var(C_m) = (c_{1m}^2 + \cdots + c_{km}^2)(n\sigma^2)$$

Further, if the sampled populations are normally distributed, C_m is normally distributed. Thus, under the usual assumptions of normality, independence, and common variance

$$Z = \frac{C_m - n(c_{1m}\mu_1 + \cdots + c_{km}\mu_k)}{\sqrt{(c_{1m}^2 + \cdots + c_{km}^2)(n\sigma^2)}}$$

has a standard normal distribution. However σ^2 is unknown, so we substitute MSE for σ^2. The resulting statistic

$$T = \frac{C_m - n(c_{1m}\mu_1 + \cdots + c_{km}\mu_k)}{\sqrt{n(c_{1m}^2 + \cdots + c_{km}^2)(\text{MSE})}} \tag{3.12}$$

is distributed as t with $N - k$ degrees of freedom.

The statistic in Equation (3.12) can be used to conduct a test of $H_0 : c_{1m}\mu_1 + \cdots + c_{km}\mu_k = 0$. When H_0 is true,

$$T = \frac{C_m}{\sqrt{n(c_{1m}^2 + \cdots + c_{km}^2)(\text{MSE})}} \sim t_{(N-k)} \qquad (3.13)$$

Letting T_{obs} denote the observed value of T, H_0 is rejected in favor of $H_1 : c_{1m}\mu_1 + \cdots + c_{km}\mu_k \neq 0$ when $2P(t_{(N-k)} \geq |T_{\text{obs}}|) \leq \alpha$.

The preceding test is often conducted using an F statistic rather than the t statistic in Equation (3.13). Squaring T, we obtain

$$F = T^2 = \frac{C_m^2}{n(c_{1m}^2 + \cdots + c_{km}^2)(\text{MSE})} \sim F_{(1,N-k)} \qquad (3.14)$$

Letting f denote the observed value of F, we reject H_0 in favor of $H_1 : c_{1m}\mu_1 + \cdots + c_{km}\mu_k \neq 0$ when $P(F_{(1,N-k)} \geq f) \leq \alpha$.

The F test described with Equation (3.14) is easily included in an ANOVA table if we combine Equations (3.11) and (3.14). Noting that the sum of squares of a contrast has 1 degree of freedom, we find that the F statistic in Equation (3.14) can be expressed as

$$F = \frac{\text{SS}_{C_m}}{\text{MSE}} = \frac{\text{MS}_{C_m}}{\text{MSE}} \qquad (3.15)$$

which is distributed as F with degrees of freedom 1 and $N - k$ when H_0 is true.

■ Example 3.5

Again consider the wear resistance data of Table 3.1. Suppose we wish to test the hypothesis $H_{01} : \mu_1 - \mu_4 = 0$ versus $H_{11} : \mu_1 - \mu_4 \neq 0$. Since the observed sum of squares for $C_1 = T_{.1} - T_{.4}$ is 0.0312 (from Example 3.4) and the observed value of MSE is 0.0203 (from Table 3.2), F as described in Equation (3.15) has observed value $f = 0.0312/0.0203 \approx 1.54$. The p value of this test is $P(F_{(1,12)} \geq 1.54) \approx 0.238$, so we do not reject H_{01}.

The preceding test is summarized in Table 3.8. Two other tests based on contrasts are included in that table. The contrasts $C_2 = T_{.2} - T_{.3}$ and $C_3 = T_{.1} - T_{.2} - T_{.3} + T_{.4}$ are used to test $H_{02} : \mu_2 - \mu_3 = 0$ versus $H_{12} : \mu_2 - \mu_3 \neq 0$ and $H_{03} : (\mu_1 + \mu_4) - (\mu_2 + \mu_3) = 0$ versus the alternative $H_{13} : (\mu_1 + \mu_4) - (\mu_2 + \mu_3) \neq 0$, respectively. H_{02} can be rejected for any $\alpha \geq 0.023$ and H_{03} can be rejected for any $\alpha \geq 0.001$. We conclude that the mean wear resistance of fabric 2 (or B) differs from that of fabric 3 (or C). We also conclude that the mean wear resistance of fabrics 1 and 4 (A and D) differs from that of fabrics 2 and 3 (B and C).

TABLE 3.8
ANOVA Summary for Fabric Wear Study

Source	df	SS	MS	F	p value
Treatment	3	0.5201	0.1734	8.53	0.003
Contrast 1	1	0.0312	0.0312	1.54	0.238
Contrast 2	1	0.1378	0.1378	6.76	0.023
Contrast 3	1	0.3511	0.3511	17.30	0.001
Error	12	0.2438	0.0203		
Totals	15	0.7639			

Contrasts Expressed in Terms of Sample Means

Some practitioners prefer to express contrasts in the sample means instead of totals. For samples of size n, such a contrast is of the form

$$C_q = c_{1q}\bar{Y}_{.1} + \cdots + c_{kq}\bar{Y}_{.k}$$

with $c_{1q} + \cdots + c_{kq} = 0$. If the same coefficients are used with the treatment totals, giving

$$C_r = c_{1q}T_{.1} + \cdots + c_{kq}T_{.k},$$

the observed value of C_r will exceed that of C_q by a factor of n. The sums of squares will be equal, since

$$\begin{aligned}
SS_{C_r} &= \frac{(c_{1q}T_{.1} + \cdots + c_{kq}T_{.k})^2}{n(c_{1q}^2 + \cdots + c_{kq}^2)} \\
&= \frac{n(c_{1q}\bar{Y}_{.1} + \cdots + c_{kq}\bar{Y}_{.k})^2}{c_{1q}^2 + \cdots + c_{kq}^2} = SS_{C_q}
\end{aligned}$$

JMP can be used to analyze a number of models, including that for a one-way analysis of variance. Contrasts and their corresponding tests can be included in the output. For the fabric wear data of Table 3.1, JMP produces a summary that includes the output in Figure 3.2. JMP uses contrasts (printed in a column format) in the sample means and codes the contrasts so that the sum of the absolute values of the coefficients is 2. Notice also that the associated tests are t tests instead of the F tests considered in Example 3.5.

What if Sample Sizes Differ?

Since the method of orthogonal contrasts is used quite often in experimental design work, definitions and formulas for the case of unequal numbers of observations per treatment will be useful. Suppose $T_{.1}$, $T_{.2}$, $\ldots$, and $T_{.k}$ are based on samples of sizes $n_1, n_2, \ldots$, and n_k, respectively. Then:

- $C_m = c_{1m}T_{.1} + \cdots + c_{km}T_{.k}$ is a *contrast in the treatment totals*, provided $n_1 c_{1m} + \cdots + n_k c_{km} = 0$.

Contrast			
1	1	0	0.5
2	0	1	-0.5
3	0	-1	-0.5
4	-1	0	0.5
Estimate	-0.125	0.2625	-0.296
Std Error	0.1008	0.1008	0.0713
t Ratio	-1.240	2.6046	-4.157
Prob>|t|	0.2386	0.0230	0.0013
SS	0.0313	0.1378	0.3511
Sum of Squares	0.5201		
Numerator DF	3		
F Ratio	8.5344		
Prob > F	0.0026		

Figure 3.2   JMP contrasts for the fabric wear study.

- The contrasts $C_m = c_{1m}T_{.1} + \cdots + c_{km}T_{.k}$ and $C_q = c_{1q}T_{.1} + \ldots + c_{kq}T_{.k}$ are *orthogonal contrasts* if $n_1 c_{1m} c_{1q} + \cdots + n_k c_{km} c_{kq} = 0$.
- The *sum of squares* for the contrast $C_m = c_{1m}T_{.1} + \cdots + c_{km}T_{.k}$ is given by:

$$\text{SS}_{C_m} = \frac{C_m^2}{n_1 c_{1m}^2 + \cdots + n_k c_{km}^2}$$

3.4.2 Multiple Comparison Procedures

If the designers of an experiment postpone selection of the comparisons to be made until after the data have been examined, comparisons may still be made, but the α level is altered because such decisions are not taken at random but are based on observed results. Several methods have been introduced to handle such situations [20], but only the Student–Newman–Keuls (SNK) range test [16] and Scheffé's test [28] are described here for balanced data.

The Student–Newman–Keuls Range Test

Sometimes an analysis of variance test leads to rejection of the hypothesis that the effect of a qualitative treatment is zero, and we then wonder which population means differ. The SNK range test (Table 3.9) is often used to make such determinations.

■ **Example 3.6**

To see how the Student–Newman–Keuls range test works, consider the data of Example 3.1 as given in Table 3.6. Following the steps in Table 3.9:

TABLE 3.9
Steps for the Student–Newman–Keuls Range Test

1. Arrange the k sample means in order from low to high.
2. Enter the ANOVA table and take the error mean square and error df.
3. Obtain the standard error of the mean for each treatment

$$s_{\bar{Y}_{.j}} = \sqrt{\frac{\text{error mean square}}{\text{number of observations in } \bar{y}_{.j}}}$$

where the error mean square is the one used as the denominator in the F test on the population means.

4. Enter a Studentized range table (Statistical Table E.1 or E.2) of significant ranges at the α level desired. Using n_2 as the error degrees of freedom and $p = 2, 3, \ldots, k$, list the $k - 1$ tabled ranges.
5. Multiply these ranges by $s_{\bar{Y}_{.j}}$ to obtain $k - 1$ least significant ranges.
6. Test the observed ranges between means, beginning with the largest versus smallest, which is compared with the least significant range for $p = k$; then test largest versus second smallest with the least significant range for $p = k-1$; and so on. Continue the comparisons for second largest versus smallest, and so forth, until all $k(k - 1)/2$ possible pairs have been tested. The sole exception to this rule is that no difference between two means can be declared significant if both these means are contained in a subset with a nonsignificant range.

---

1. $k = 4$ means are: 2.19   2.32   2.42   2.68
   for treatments: A D C B
2. From Table 3.2 the error mean square is 0.0203; the error sum of squares has 12 df.
3. The standard error of a mean is $s_{\bar{Y}_{.j}} = \sqrt{0.0203/4} = 0.0712$.
4. From Statistical Table E.1 with $\alpha = 0.05$ and $n_2 = 12$, we find

p:	2	3	4
Ranges:	3.08	3.77	4.20

5. We multiply each range in step 4 by the standard error of 0.0712 to find the least significant ranges (LSR):

p:	2	3	4
LSR:	0.22	0.27	0.30

6. Beginning with the largest versus the smallest, we make the following conclusions at the 5% significance level:

$\bar{y}_B - \bar{y}_A = 2.68 - 2.19 = 0.49 > 0.30$; so conclude that $\mu_B > \mu_A$

$\bar{y}_B - \bar{y}_D = 2.68 - 2.32 = 0.36 > 0.27$; so conclude that $\mu_B > \mu_D$

$\bar{y}_B - \bar{y}_C = 2.68 - 2.42 = 0.26 > 0.22$; so conclude that $\mu_B > \mu_C$

$\bar{y}_C - \bar{y}_A = 2.42 - 2.19 = 0.23 < 0.27$; so μ_A and μ_C may be equal

$\bar{y}_C - \bar{y}_D = 2.42 - 2.32 = 0.10 < 0.22$; so μ_C and μ_D may be equal

$\bar{y}_D - \bar{y}_A = 2.32 - 2.19 = 0.13 < 0.22$; so μ_A and μ_D may be equal

The results in step 6 indicate that the effect of B differs significantly from those of A, D, and C, but the effects of A, D, and C do not differ significantly from each other. These results are shown in Figure 3.3, where the true means of the populations associated with any sample means underscored by the same line are not significantly different.

Figure 3.3 Newman–Keuls test means for Example 3.6.

In practice, several configurations of Figure 3.3 are encountered. For five means, several outcomes may be seen, as in Figure 3.4.

If one seeks a minimum response as the "best" treatment in Figure 3.4a, B looks best, but it is no better than A. We would recommend B, however, since its mean is significantly less than that of either D, C, or E (no common line), whereas the mean of A is not significantly less than that of D or C (have a line in common).

In Figure 3.4b, B or A means are less than C or E means, whereas D is not. So, we would choose either B or A as best.

Scheffé's Test

Since the Newman–Keuls test is restricted to comparing pairs of means and it is often desirable to examine other contrasts that represent combinations of treatments, many experimenters prefer a test devised by Scheffé [96]. Although Scheffé's method uses the concept of contrasts presented earlier, the contrasts need not be orthogonal. In fact, any and

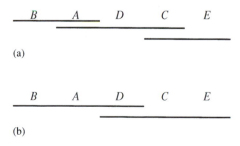

Figure 3.4 Some possible means.

all conceivable contrasts may be tested for significance. Since comparing means in pairs is a special case of contrasts, Scheffé's scheme is more general than the Newman–Keuls procedure. However, since the Scheffé method must be valid for such a large set of possible contrasts, it requires larger observed differences to be significant than some of the other schemes.

The method is summarized in Table 3.10. The steps outlined are taken after the data have been compiled and the error mean square has become available. The chosen value of α is the overall significance level. That is, α is the probability that at least one type I error will occur. Also, at least one contrast will be declared significant if and only if the ANOVA test is significant at the $100\alpha\%$ level.

■ **Example 3.7**

To see how Scheffé's test works, consider the fabric wear data first considered in Example 3.1. From the ANOVA summary (Table 3.2), we find that the error sum of squares has 12 degrees of freedom and the observed error mean square is 0.0203. Using Table 3.6 with fabrics A, B, C, and D coded as fabrics 1, 2, 3, and 4, we find $T_{.1} = 8.76$, $T_{.2} = 10.72$, $T_{.3} = 9.67$, and $T_{.4} = 9.26$. Then we proceed through the steps in Table 3.10:

1. Consider the contrasts in the following summary.

Contrast	Observed value
$C_1 = T_{.1} - T_{.2}$	$c_1 = 8.76 - 10.72 = -1.96$
$C_2 = 3T_{.1} - T_{.2} - T_{.3} - T_{.4}$	$c_2 = 3(8.76) - 10.72 - 9.67 - 9.26 = -3.37$

(**Note:** These contrasts are not orthogonal. The first compares the mean of treatment A with that of B and the second compares the mean of treatment A with that of the other three treatments.)

2. If $\alpha = 0.05$, since $k - 1 = 3$, $N - k = 12$, and $P(F_{(3,12)} \geq 3.49) = 0.05$, let $f = 3.49$.

3. $A = \sqrt{3(3.49)} = 3.24$

4. $s_{C_1} = \sqrt{0.0203[4(1)^2 + 4(-1)^2 + 4(0)^2 + 4(0)^2]} = 0.40$ and $As_{c_1} = 3.24(0.40) = 1.30$

   $s_{C_2} = \sqrt{0.0203[4(3)^2 + 4(-1)^2 + 4(-1)^2 + 4(-1)^2]} = 0.99$ and $As_{C_2} = 3.24(0.99) = 3.21$

5. Since $|c_1| = 1.96 > 1.30$, conclude that the means of fabrics A and B differ. Since $|c_2| = 3.37 > 3.21$, conclude that the mean of fabric A differs from that of the other three fabrics.

TABLE 3.10
Steps for Scheffé's Test

---

1. Set up all contrasts of interest to the experimenter and compute their numeric value.
2. Determine the number f for which $P(F_{(k-1,N-k)} \geq f) = \alpha$.
3. Compute $A = \sqrt{(k-1)f}$, using f from step 2.
4. Compute the standard error of each contrast to be tested. For the contrast $C_m = c_{1m}T_{.1} + \cdots + c_{km}T_{.k}$, this standard error is given by

$$s_{C_m} = \sqrt{(\text{observed error square})(n_1 c_{1m}^2 + \cdots + n_k c_{km}^2)}$$

5. Let c_m denote the observed value of C_m. Reject the hypothesis that the true contrast among means is zero if $|c_m| > A s_{C_m}$.

Using Computer Packages to Compare Pairs of Means

Many statistical computing packages for personal computers include at least one multiple comparison procedure that can be used to conduct tests like those in Example 3.6. For example, SPSS[1] includes the Student–Newman–Keuls test as an option in the **One-Way ANOVA** procedure. Minitab and JMP include the Tukey–Kramer HSD test (see [26], pp. 448–451] instead of the SNK procedure.

SPSS output for the Student–Newman–Keuls test based on the fabric wear data is summarized in Figure 3.5. The three asterisks indicate that the average wear for fabric 2 (or *B*) is significantly different from that of each of the other fabrics. The "Subset 1" summary indicates that the averages for fabrics 1, 3, and 4 (or *A*, *C*, and *D*) are not significantly different and the "Subset 2" summary indicates that the true average wear for fabric 2 differs significantly from the averages of the other fabrics.

3.5 CONFIDENCE LIMITS ON MEANS

After an analysis of variance it is often desirable to set confidence limits on a treatment mean. The $100(1 - \alpha)\%$ confidence limits on μ_j are given by

$$\bar{y}_{.j} \pm t_{1-\alpha/2} \sqrt{\frac{\text{mean square in denominator of } F \text{ ratio used to test treatment mean}}{n_j}} \qquad (3.16)$$

where "mean square in denominator of F ratio used to test treatment mean" is the error mean square in a one-way ANOVA but may be a different mean square in more complex analyses. The degrees of freedom of the t statistic are the degrees of freedom that correspond to the mean square in the denominator of the F ratio used to test the treatment mean.

---

[1]SPSS is a registered trademark of SPSS Inc.

```
                        - - - - - ONEWAY - - - - -
      Variable  WEAR
   By Variable  FABRIC
Multiple Range Tests: Student-Newman-Keuls test with significance
   level .05

The difference between two means is significant if
   MEAN(J)-MEAN(I) >= .1008 * RANGE * SQRT(1/N(I) + 1/N(J))
   with the following value(s) for RANGE:

Step      2      3      4
RANGE   3.08   3.77   4.20

   (*) Indicates significant differences which are shown in the
      lower triangle

                      G G G G
                      r r r r
                      p p p p

                      1 4 3 2
   Mean      Fabric

   2.1900   Grp 1
   2.3150   Grp 4
   2.4175   Grp 3
   2.6800   Grp 2   * * *

   Homogeneous Subsets (highest and lowest means are not
      significantly different)
Subset 1
Group      Grp 1      Grp 4      Grp3
Mean       2.1900     2.3150     2.4175
-------------------------------------
Subset 2
Group      Grp 2
Mean          2.6800
--------------------
```

Figure 3.5   SPSS output for the fabric wear study.

■ **Example 3.8**

From the ANOVA summary (Table 3.2) for the fabric wear data first considered in Example 3.1, we can find the sample mean for fabric B ($\bar{y}_{.2} = 2.68$), and we see that the error sum of squares has 12 degrees of freedom and the observed error mean square is 0.0203. When these values and $t_{0.975} = 2.18$ are substituted into Equation (3.16), the 95% confidence limits on the mean wear resistance for fabric B are found to be $2.68 \pm 2.18\sqrt{0.0203/4}$ or 2.68 ± 0.16 or 2.52 to 2.84.

3.6 COMPONENTS OF VARIANCE

In Example 3.1 the levels of the factor were considered to be fixed, since only four fabrics were available and a decision was desired on the effect of these four fabrics only. If, however, the levels of the factor are random (such as operators, days, or samples, where the levels in the experiment might have been chosen at random from a large number of possible levels), the model is called a *random model*, and inferences are to be extended to all levels of the population (of which the observed four levels are random samples). In a random model, the experimenter is not usually interested in testing hypotheses, setting confidence limits, or making contrasts in means, but in estimating components of variance. How much of the variance in the experiment might be attributable to true differences in treatment means, and how much might be due to random error about these means?

In the next part of this section we show that, for a one-factor random model with n observations per treatment level, the expected value of the error mean square is

$$E(\mathrm{MS_{error}}) = \sigma_\varepsilon^2 \qquad (3.17)$$

and that for the treatment mean square is

$$E(\mathrm{MS_{treatment}}) = \sigma_\varepsilon^2 + n\sigma_\tau^2 \qquad (3.18)$$

The following example illustrates how these results can be used to estimate how much of the variance in the experiment can be attributed to treatment differences and how much can be attributed to random error.

■ **Example 3.9**

A company supplies a customer with several hundred batches of a raw material every year. The customer is interested in a high yield from the raw material in terms of percent usable chemical. Usually, three sample determinations of yield are made from each batch in order to control the quality of the incoming material. Some variation occurs between determinations on a given batch, but the customer suspects that there may be significant batch-to-batch variation as well.

To check this, five batches are randomly selected from several available batches and three yield determinations are run per batch. Thus, the 15 yield determinations are completely randomized. The mathematical model is again

$$Y_{ij} = \mu + \tau_j + \varepsilon_{ij}$$

except that in this experiment the k levels of the treatment (batches) are chosen at random, rather than being fixed levels. The data are shown in Table 3.11.

Next the customer enters the data in a Minitab worksheet and obtains the ANOVA summary in Figure 3.6. Notice that the F test indicates a highly significant difference among batches ($p = 0.000$).

TABLE 3.11
Chemical Yield by Batch Data

		Batch		
1	2	3	4	5
74	68	75	72	79
76	71	77	74	81
75	72	77	73	79

Since the batches used in the experiment are but a random sample of batches, we may be interested in how much of the variance in the experiment might be attributed to batch differences and how much to random error.

From Equation (3.17), the expected mean square for error is found to be σ_ε^2. Minitab indicates this in Figure 3.6 by placing (2) in the error row of the expected mean square column, since the error term is listed as source 2 among the sources of variation. The observed value of $\mathrm{MS_{error}}$ is 1.800, so $\sigma_\varepsilon^2 \approx 1.800$.

Using Equation (3.18) with $n = 3$, we find that $\sigma_\varepsilon^2 + 3\sigma_\tau^2$ is the expected mean square for batches. Minitab indicates this by placing $(2) + 3.0000(1)$ in the batch row of the expected mean square column in Figure 3.6. The observed value of $\mathrm{MS_{treatment}}$ is 36.933, so $\sigma_\varepsilon^2 + 3\sigma_\tau^2 \approx 36.933$.

Letting $s_\varepsilon^2 = 1.800$ and $s_\varepsilon^2 + 3s_\tau^2 = 36.933$, where s^2 indicates that these are estimates of the values of corresponding σ^2, we find

$$s_\tau^2 = \frac{36.933 - 1.800}{3} = 11.711$$

In Figure 3.6, s_ε^2 and s_τ^2 are listed in the error and batch rows, respectively, of the variance component column.

The total variance can be estimated as

$$\sigma_{total}^2 \approx s_{total}^2 = s_\tau^2 + s_\varepsilon^2 = 11.711 + 1.800 = 13.511$$

```
Analysis of Variance (Balanced Designs)

Factor       Type        Levels      Values
Batch        random           5           1     2     3     4     5

Analysis of Variance for Yield

Source       DF               SS          MS           F            P
Batch         4          147.733      36.933        20.52        0.000
Error        10           18.000       1.800
Total        14          165.733

Source              Variance        Error        Expected Mean Square
                    component        term        (using unrestricted model)
1 Batch               11.711           2          (2)  +  3.0000 (1)
2 Error                1.800                       (2)
```

Figure 3.6   Minitab output for chemical yield study.

and the standard deviation as $s_{\text{total}} = \sqrt{13.511} = 3.68$. One might therefore expect all the data in such a small experiment ($N = 15$) to fall within four standard deviations or within $4(3.68) = 14.72$. The actual range is 13 (high, 81; low, 68). This breakdown of total variance into its two components then seems quite reasonable.

It is useful to express the components of variance as percentages of the total variance. In this case, 11.711/13.511 or 86.7% of the total variance is attributable to batch and 13.3% is attributable to errors within batches.

Some readers may wonder why the total sum of squares is never divided by its degrees of freedom to estimate the total variance. If the methods of the next section are used to determine the expected mean square for the total, it is found to be a biased estimate of the total variance when $H_0 : \sigma_\tau^2 = 0$ is false. In fact, a derivation will show that the expected mean square for the total in a one-way ANOVA is equal to

$$E[\text{MS}_{\text{total}}] = \frac{n(k-1)}{N-1}\sigma_\tau^2 + \sigma_\varepsilon^2$$

Since the coefficient of σ_τ^2 is always less than one, this expected mean square will be less than $\sigma_{\text{total}}^2 = \sigma_\tau^2 + \sigma_\varepsilon^2$.

Expected Mean Square Derivation

For a completely randomized, single-factor experiment with samples of size n at each of the k treatment levels,

$$Y_{ij} = \mu + \tau_j + \varepsilon_{ij}$$

with

$$i = 1, \ldots, n \qquad j = 1, \ldots, k \tag{3.19}$$

Assuming that the model assumptions discussed in Section 3.1 are satisfied, we can derive the expected values of the error and treatment mean squares. Before doing so, recall that

$$\text{SS}_{\text{error}} = \sum_{j=1}^{k}\sum_{i=1}^{n}(Y_{ij} - \bar{Y}_{.j})^2 \tag{3.20}$$

and

$$\text{SS}_{\text{treatment}} = \sum_{j=1}^{k} n(\bar{Y}_{.j} - \bar{Y}_{..})^2 \tag{3.21}$$

for the sample layout in Table 3.4 when $n_1 = \cdots = n_k = n$.

We begin by noting that

$$\bar{Y}_{.j} = \frac{\sum_{i=1}^{n} Y_{ij}}{n}$$

$$= \frac{\sum_{i=1}^{n}(\mu + \tau_j + \varepsilon_{ij})}{n} \quad \text{from model Equation (3.29)}$$

$$= \frac{n\mu}{n} + \frac{n\tau_j}{n} + \frac{\sum_{i=1}^{n} \varepsilon_{ij}}{n}$$

$$= \mu + \tau_j + \frac{\sum_{i=1}^{n} \varepsilon_{ij}}{n} \tag{3.22}$$

Also,

$$\bar{Y}_{.j} = \frac{\sum_{j=1}^{k} \sum_{i=1}^{n} Y_{ij}}{nk}$$

$$= \frac{\sum_{j=1}^{k} \sum_{i=1}^{n}(\mu + \tau_j + \varepsilon_{ij})}{nk} \quad \text{from model Equation (3.19)}$$

$$= \frac{nk\mu}{nk} + \frac{n\sum_{j=1}^{k} \tau_j}{nk} + \frac{\sum_{i=1}^{k} \sum_{i=1}^{n} \varepsilon_{ij}}{nk}$$

$$= \mu + \frac{\sum_{j=1}^{k} \tau_j}{k} + \frac{\sum_{i=1}^{k} \sum_{i=1}^{n} \varepsilon_{ij}}{nk} \tag{3.23}$$

Subtracting Equation (3.23) from (3.22) gives

$$\bar{Y}_{.j} - \bar{Y}_{..} = \left(\tau_j - \frac{\sum_{j=1}^{k} \tau_j}{k} \right) + \left(\frac{\sum_{i=1}^{n} \varepsilon_{ij}}{n} - \frac{\sum_{j=1}^{k} \sum_{i=1}^{n} \varepsilon_{ij}}{nk} \right)$$

Squaring gives

$$(\bar{Y}_{.j} - \bar{Y}_{..})^2 = \left(\tau_j - \frac{\sum_{j=1}^{k} \tau_j}{k} \right)^2 + \frac{1}{n^2} \left(\sum_{i=1}^{n} \varepsilon_{ij} - \frac{\sum_{j=1}^{k} \sum_{i=1}^{n} \varepsilon_{ij}}{k} \right)^2$$

$$+ \frac{2}{n} \left(\tau_j - \frac{\sum_{j=1}^{k} \tau_j}{k} \right) \left(\sum_{i=1}^{n} \varepsilon_{ij} - \frac{\sum_{j=1}^{k} \sum_{i=1}^{n} \varepsilon_{ij}}{k} \right)$$

Multiplying by n and summing over j gives

$$\text{SS}_{\text{treatment}} = \sum_{j=1}^{k} n(\bar{Y}_{.j} - \bar{Y}_{..})^2 \quad \text{from Equation (3.21)}$$

$$= n \sum_{j=1}^{k} \left(\tau_j - \frac{\sum_{j=1}^{k} \tau_j}{k} \right)^2 + \frac{1}{n} \sum_{j=1}^{k} \left(\sum_{i=1}^{n} \varepsilon_{ij} - \frac{\sum_{j=1}^{k} \varepsilon_{ij}}{k} \right)^2$$

$$+ 2 \sum_{j=1}^{k} \left(\tau_j - \frac{\sum_{j=1}^{k} \tau_j}{k} \right) \left(\sum_{i=1}^{n} \varepsilon_{ij} - \frac{\sum_{j=1}^{k} \sum_{i=1}^{n} \varepsilon_{ij}}{k} \right) \tag{3.24}$$

Since τ_j and ε_{ij} are independent random variables for the random model, $\tau_j - \sum_{j=1}^{k} \tau_j/k$ is a constant for the fixed model, and $E(\sum_{i=1}^{n} \varepsilon_{ij} - \sum_{j=1}^{k}\sum_{i=1}^{n}\varepsilon_{ij}/k) = 0$, the expected value of the third term in Equation (3.24) equals zero. Thus, for either model

$$E[SS_{treatment}] = E\left[n\sum_{j=1}^{k}\left(\tau_j - \frac{\sum_{j=1}^{k}\tau_j}{k}\right)^2\right]$$

$$+ E\left[\frac{1}{n}\sum_{j=1}^{k}\left(\sum_{i=1}^{n}\varepsilon_{ij} - \frac{\sum_{j=1}^{k}\sum_{i=1}^{n}\varepsilon_{ij}}{k}\right)^2\right] \qquad (3.25)$$

But,

$$E\left[\frac{1}{n}\sum_{j=1}^{k}\left(\sum_{i=1}^{n}\varepsilon_{ij} - \frac{\sum_{j=1}^{k}\sum_{i=1}^{n}\varepsilon_{ij}}{k}\right)^2\right] = E\left[\frac{1}{n}\sum_{j=1}^{k}(U_j - \bar{U})^2\right] \text{ with } U_j = \sum_{i=1}^{n}\varepsilon_{ij}$$

$$= E\left[\frac{1}{n}(k-1)S_U^2\right]$$

$$= \frac{1}{n}(k-1)(n\sigma_\varepsilon^2)$$

$$= (k-1)\sigma_\varepsilon^2$$

Substituting this result into Equation (3.25), we find that for either model

$$E[SS_{treatment}] = E\left[n\sum_{j=1}^{k}\left(\tau_j - \frac{\sum_{j=1}^{k}\tau_j}{k}\right)^2\right] + (k-1)\sigma_\varepsilon^2 \qquad (3.26)$$

Equation (3.26) provides us with the information needed to prove the following theorem.

THEOREM 3.1   The expected value of the treatment mean square for a one-factor, completely randomized experiment with equal ns is:

(a) $E[MS_{treatment}] = \sigma_\varepsilon^2 + n\phi_\tau$ with $\phi_\tau = \sum_{j=1}^{k}\tau_j^2/(k-1)$ for the mixed model.

(b) $E[MS_{treatment}] = \sigma_\varepsilon^2 + n\sigma_\tau^2$ for the random model.

Proof: If the treatment levels are fixed, the summation in the first term of Equation (3.26) is a constant with $\sum_{j=1}^{k}\tau_j = 0$. It follows that $E[MS_{treatment}] = E[SS_{treatment}/(k-1)] = [n\sum_{j=1}^{k}\tau_j^2 + (k-1)\sigma_\varepsilon^2]/(k-1) = n\sum_{j=1}^{k}\tau_j^2/(k-1) + \sigma_\varepsilon^2$.

If treatment levels are random, $SS_\tau = \sum_{j=1}^{k}(\tau_j - \sum_{j=1}^{k}\tau_j/k)^2$ is the sum of squares for a random sample of size k from a population with variance σ_τ^2. Substituting in Equation (3.26), we write

$$E[\text{MS}_{\text{treatment}}] = E\left(\frac{\text{SS}_{\text{treatment}}}{k-1}\right)$$

$$= E\left[\frac{n(\text{SS}_\tau)}{(k-1)}\right] + \frac{(k-1)\sigma_\varepsilon^2}{k-1}$$

$$= \frac{n(k-1)\sigma_\tau^2}{(k-1)} + \sigma_\varepsilon^2$$

$$= n\sigma_\tau^2 + \sigma_\varepsilon^2.$$

Subtracting Equation (3.22) from model equation (3.19) gives

$$Y_{ij} - \bar{Y}_{.j} = \varepsilon_{ij} - \frac{\sum_{i=1}^n \varepsilon_{ij}}{n}$$

Squaring and summing over both i and j gives

$$\text{SS}_{\text{error}} = \sum_{j=1}^k \sum_{i=1}^n (Y_{ij} - \bar{Y}_{.j})^2 = \sum_{j=1}^k \sum_{i=1}^n \left(\varepsilon_{ij} - \frac{\sum_{i=1}^n \varepsilon_{ij}}{n}\right) \qquad (3.27)$$

which is used in the proof of the following theorem.

THEOREM 3.2   For both the fixed and random models, the expected value of the error mean square for a one-factor, completely randomized experiment is:

$$E[\text{MS}_{\text{error}}] = \sigma_\varepsilon^2$$

Proof: For equal n's

$$E[\text{SS}_{\text{error}}] = E\left[\sum_{j=1}^k \sum_{i=1}^n \left(\varepsilon_{ij} - \frac{\sum_{i=1}^n \varepsilon_{ij}}{n}\right)\right] \quad \text{from Equation (3.27)}$$

$$= \sum_{j=1}^k E\left[\sum_{i=1}^n \left(\varepsilon_{ij} - \frac{\sum_{i=1}^n \varepsilon_{ij}}{n}\right)\right]$$

$$= \sum_{j=1}^k (n-1)\sigma_\varepsilon^2$$

$$= k(n-1)\sigma_\varepsilon^2$$

$$= (nk - k)\sigma_\varepsilon^2$$

Thus, $E[\text{MS}_{\text{error}}] = E[\text{SS}_{\text{error}}/(nk-k)] = (nk-k)\sigma_\varepsilon^2/(nk-k) = \sigma_\varepsilon^2.$

TABLE 3.12
Modified One-Factor ANOVA Summary: Random Model with Equal *ns*

Source	df	SS	MS	EMS
Treatment	$\nu_1 = k - 1$	$SS_{treatment}$	$MS_{treatment}$	$\sigma_\varepsilon^2 + n\sigma_\tau^2$
Error	$\nu_2 = nk - k$	SS_{error}	MS_{error}	σ_ε^2
Totals	$nk - 1$	SS_{total}		

The expected mean squares (EMSs) in Theorems 3.1 and 3.2 are often placed in an ANOVA summary as illustrated in Table 3.12. In later chapters, such a table will help us determine the appropriate comparisons or components of variance.

3.7 CHECKING THE MODEL

The ANOVA techniques considered in this chapter are theoretically based on independence, random samples, normal distributions, and equal population variances. In practice, we do not expect the model assumptions to be satisfied exactly. For the procedures to yield reliable results, however, those assumptions must be reasonably satisfied. Fortunately, simple plots of the data and residuals often reveal discrepancies between the proposed model and reality.

The residuals in an experiment are what remains after the estimated effects in the model have been subtracted from the values of the response variable. Thus, the residual for the ith observation at the jth treatment level of a single-factor experiment is $Y_{ij} - \hat{Y}_{ij}$, where $\hat{Y}_{ij}$ is the value of Y_{ij} predicted from the model. Since $\tau_j = \mu_j - \mu$, $Y_{ij} = \mu + \tau_j + \varepsilon_{ij} = \mu_j + \varepsilon_{ij}$. To estimate μ_j we use the mean of the jth sample. So, $\hat{Y}_{ij} = \bar{Y}_{.j}$ and the *residual of Y_{ij}* is

$$E_{ij} = Y_{ij} - \bar{Y}_{.j} \tag{3.28}$$

In this single-factor case, the residual of Y_{ij} is simply the error estimate.

■ **Example 3.10**

For convenience, the fabric wear data originally considered in Table 3.1 are included in Table 3.13. Using Equation (3.28) with $y_{11} = 1.93$ and $\bar{y}_{.1} = 2.19$, $e_{11} = 1.93 - 2.19 = -0.26$ is the observed value of the residual for the first observation on fabric 1 (or A). Proceeding in like manner, we obtain the residuals summarized in Table 3.13.

TABLE 3.13
Fabric Wear Data and Residuals

	Fabric							
	$A = 1$		$B = 2$		$C = 3$		$D = 4$	
i	y_{i1}	e_{i1}	y_{i2}	e_{i2}	y_{i3}	e_{i3}	y_{i4}	e_{i4}
1	1.93	−0.26	2.55	−0.13	2.40	−0.0175	2.33	0.015
2	2.38	0.19	2.72	0.04	2.68	0.2625	2.40	0.085
3	2.20	0.01	2.75	0.07	2.31	−0.1075	2.28	−0.035
4	2.25	0.06	2.70	0.02	2.28	−0.1375	2.25	−0.065
Means	2.19	0.00	2.68	0.00	2.4175	0.0000	2.315	0.000

Assessing Normality

Several studies have shown that lack of normality in the response variable does not seriously affect the analysis of the means for a fixed model when the number of observations per treatment is the same for all treatments. However, nonnormality can have a serious effect on estimates of components of variance for the random model.

The residuals are normally distributed when the model assumptions are met. Thus, exploratory studies indicating that the sample of residuals may have come from a nonnormal population would give one reason to believe that the normality assumption is invalid.

Figure 3.7 contains a box plot and stem-and-leaf display of the residuals in Table 3.13. No outliers are indicated and, for such a small sample size, the graphics are reasonably symmetric. These provide no strong evidence of nonnormality.

A normal quantile plot and Shapiro–Wilk test summary have also been included in Figure 3.7. The normal quantile plot includes the Lilliefors graph as a test for normality. If

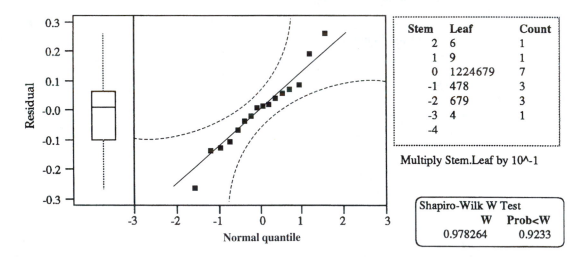

Figure 3.7   JMP graphics for the residuals in Table 3.13.

any points fall outside the Lilliefors bounds (indicated by the dashed curves), the sampled population should be considered nonnormal. Otherwise, the sample could have come from a normally distributed population. In this case, the reasonably linear plot in the residuals, the Lilliefors procedure, and the large p value of 0.9233 for the Shapiro-Wilk test all support an assumption of normality.

Assessing Equality of Variances

To check for homogeneous variances within treatments, several tests have been suggested. One quick check is to examine the ranges of the observations within each treatment. If the average range $\bar{R}$ is multiplied by a factor D_4, which is found in most quality control texts, and all ranges are less than $D_4 \bar{R}$, it is quite safe to assume homogeneous variances. Values of D_4 for a few selected sample sizes are given in Table 3.14.

■ **Example 3.11**

The ranges of the fabric wear data in Table 3.13 are 0.45, 0.20, 0.40, and 0.15. Here $\bar{R} = 0.30$. Since $n = 4$, $D_4 = 2.282$ and $D_4 \bar{R} \approx 0.68$. All four ranges are well below 0.68, so homogeneity of variance may be reasonably assured.

We could get some idea of the homogeneity of the variances of the residuals around their own treatment means from a plot of each treatment separately, as seen in Figure 3.8. This "look test" can be only a rough one with small samples, but it does seem reasonable in light of the result in Example 3.11.

TABLE 3.14
D_4 **Values for Sample Size** n

If $n =$	2	3	4	5	6	7	8	9	10
$D_4 =$	3.267	2.575	2.282	2.115	2.004	1.924	1.864	1.816	1.777

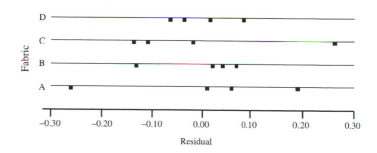

Figure 3.8   Dot diagram of residuals by fabrics A, B, C, and D.

It should be noted that the quick test illustrated in Example 3.11 can also be used with the residuals. When the model assumptions are met, the variances of the residuals within treatments are homogeneous. Evidence to the contrary leads one to suspect that the assumption of homogeneous variances within treatments is invalid. We should also note that ANOVA results are particularly sensitive to unequal variances when sample sizes differ substantially. Thus, if at all possible, equal sample sizes should be used.

Assessing Independence

It is important that data for a one-factor, completely randomized experiment be obtained in a completely random manner. Failure to do so may invalidate the model assumption that the error terms are independently distributed. Lack of independence seriously affects the inferences, so every effort must be made to experiment under conditions that minimize the possibility of correlated errors. We use randomization in an attempt to cover all contingencies.

An autocorrelation coefficient, which indicates how the residuals are correlated with themselves, is often used to investigate the independence assumption. To compute this, we correlate the observed residuals (in time series order) with the same residuals moved one or more (say k) positions from the originals. If $k = 1$, the time series $e_1, e_2, \ldots, e_{n-1}$ is used with the series $e_2, e_3, \ldots, e_n$. The *lag 1 autocorrelation coefficient* for the pairs (e_1, e_2), $(e_2, e_3), \ldots, (e_{n-1}, e_n)$, which is approximately equal to the sample correlation coefficient for those pairs, is

$$r_1 = \frac{\sum_{i=1}^{n-1} e_i e_{i+1}}{\sum_{i=1}^{n} e_i^2} \tag{3.29}$$

Please notice that this sample statistic is meaningful *only* for time series data.

When the errors in the model equation are normally and independently distributed, the sampling distribution of the lag 1 autocorrelation coefficient associated with a sample of size n is approximately normal with mean 0 and standard deviation $1/\sqrt{n}$. Thus, independence of the errors should be questioned when the absolute value of r_1 exceeds (say) $1.96/\sqrt{n}$.

■ **Example 3.12**

The time series of the residuals for the fabric wear study is $-0.26, 0.19, 0.01, \ldots,$ $-0.04, -0.07$. Applying Equation (3.29) to this series gives

$$r_1 = \frac{(-0.26)(0.19) + (0.19)(0.01) + \cdots + (-0.035)(-0.065)}{(-0.26)^2 + (0.19)^2 + \cdots + (-0.035)^2 + (-0.065)^2} \approx -0.31$$

The absolute value of r_1 is less than $1.96/\sqrt{n} = 1.96/\sqrt{16} = 0.49$, so the assumption of independence among the model errors seems reasonable.

The procedure illustrated in Example 3.12 is essentially the Durbin–Watson test reported in many computer printouts. If applied to the time-ordered residuals, the observed value of the Durbin–Watson statistic is

$$DW = \frac{\sum_{i=2}^{n}(e_i - e_{i-1})^2}{\sum_{i=1}^{n} e_i^2} \tag{3.30}$$

As a rule of thumb, values of DW in excess of 1.7 support an assumption of independence. If DW is numerically less than 1, the validity of that assumption should be questioned. This test is meaningful only for time-series data and does not detect other types of departures from independence.

■ **Example 3.13**

Applying Equation (3.30) to the time series of residuals in Example 3.12, we find

$$DW = \frac{[0.19 - (-0.21)]^2 + [0.01 - 0.19]^2 + \cdots + [(-0.065) - (-0.035)]^2}{(-0.26)^2 + (0.19)^2 + \cdots + (-0.035)^2 + (-0.065)^2} \approx 2.3254$$

This value exceeds 1.7, supporting an assumption of independent errors.
[**Note:** A more formal conclusion is possible by consulting a table of Durbin–Watson test bounds. For example, when $n = 16$ and $\alpha = 0.05$, values of DW in excess of 1.37 support independence and those less than 1.10 do not.]

3.7.1 Plot the Residuals Against Other Variables

In addition to the considerations just mentioned, plots of the residuals versus other variables should be examined for patterns that are unexplained by the model. One should look for trends, curvature, extreme points (or outliers, which might indicate nonnormality), wedge-shaped patterns (which may indicate heterogeneity of variances), and peculiar characteristics. Unexplained, systematic patterns often indicate inadequacies in the model.

As noted in Section 3.1, we commonly assume that errors are NID$(0, \sigma_\varepsilon^2)$. When these assumptions are met, the residuals:

- are normally distributed
- have equal variances
- are *not* independent

The lack of independence among the residuals is a result of a number of constraints that they must satisfy. As seen in Table 3.13, the residuals must sum to zero at each treatment level. This lack of independence does not, however, seriously affect the graphical procedure illustrated in this section.

A plot of the residuals versus the values predicted by the model is a good beginning. Figure 3.9 is such a plot for the fabric wear study. Other than the possibility of nonhomogeneous variances, no particular characteristic stands out. Earlier considerations led us to conclude that homogeneity of variances is a reasonable assumption.

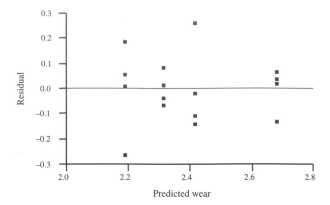

Figure 3.9   Residual plot for Table 3.13 data.

If the experiment has been run in a completely randomized order, we do not expect any pattern to emerge when the residuals are plotted against the time that they were recorded. If we take the time order of the 16 residuals in Table 3.13 to be $-0.26, 0.19, 0.01, \ldots$, -0.065, we obtain the graph of Figure 3.10a. Here the pattern looks quite like a random walk. If, on the other hand, some definite pattern is seen, it is a good indication of a deficiency in the model.

If $e_1, e_2, \ldots, e_n$ represents the residuals *in time series order*, a lack of independence in the model errors may be indicated by a pattern in the plot of the pairs $(e_1, e_2), \ldots, (e_{n-1}, e_n)$. Figure 3.10b gives such a plot for the time-ordered residuals of the fabric wear data (see the preceding paragraph). Again, no obvious pattern is seen. This finding, combined with the results in Examples 3.12 and 3.13, allows us to conclude that the independence assumption seems reasonable.

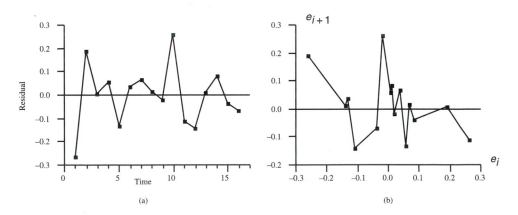

(a)   (b)

Figure 3.10   Plots of residuals for fabric wear study: (a) residuals versus time and (b) consecutive residuals.

What If the Model Is Inadequate?

Suppose procedures such as those considered above indicate that all is not well with the model or the model assumptions. What steps should we take next?

One obvious possibility is to consider a different model. When it is impossible or impractical to randomize completely, for example, the restriction on randomization should be included in the way the model is formulated. Failure to do so may result in an incorrect inference, or an important difference may go undetected. Designs are discussed throughout remaining chapters that address this issue.

In some instances a data transformation will adequately correct model inadequacies. A transformation to correct nonhomogeneous variances or nonnormality may correct both, since these model deficiencies often occur together. Of course the transformed data should be checked for model inadequacies before proceeding.

At times the means of the treatments of measured variables and the variances within the treatments are related in some way. For example, a Poisson variable has its mean and variance equal. It can be shown mathematically that a necessary and sufficient condition for normality of a variable is that means and variances of samples drawn from that population are independent of each other. If it is known or observed that the sample means and sample variances are not independent but are related to each other, it may be possible to "break" this correlation by a suitable transformation of the original variable Y. Table 3.15 suggests several transformations to try for certain specified relationships.

The transformations y^3, y^2, $\sqrt{y}$, $\ln(y)$, $-1/\sqrt{y}$, and $-1/y$ are among those often used, where y denotes an observed value of the response variable. The first two shorten the tail of a left-skewed distribution and the last four shorten the tail of a right-skewed distribution. Using -1 in the numerators of the last two transformations preserves the order of the y values.

The $\ln(y)$ and $\sqrt{y}$ transformations are often used to correct nonnormality. If any of the sample values are negative, a positive constant of sufficient size may be added to each sample value before the transformation is applied.

All statistical computing packages illustrated in this chapter allow the data analyst to transform data and then carry out analyses, both exploratory and formal, on the transformed data.

3.8 SAS PROGRAMS FOR ANOVA AND TESTS AFTER ANOVA

SAS contains three procedures that can be used to analyze data for a completely randomized, one-factor experiment: **ANOVA**, **GLM**, and **NESTED**. The **ANOVA** procedure can be used

TABLE 3.15
Transformations in ANOVA

If	Is Proportional to	Transform y_{ij} to
s_j^2	$\bar{y}_{.j}$	$\sqrt{y_{ij}}$ (Poisson case)
s_j^2	$\bar{y}_{.j}(1 - \bar{y}_{.j})$	arcsin $\sqrt{y_{ij}}$ (binomial case)
s_j	$\bar{y}_{.j}$	$\ln(y_{ij})$ or $\ln(y_{ij} + 1)$
s_j	$\bar{y}_{.j}^2$	$1/y_{ij}$

with a variety of balanced (equal sample sizes at all treatment combinations) experimental designs but may provide invalid results with unbalanced data. **GLM** permits the use of both balanced and unbalanced data and will generate contrasts. **NESTED** can be used to evaluate the components of variance for the random model.

The ANOVA Procedure

ANOVA can be used to test $H_0: \mu_1 = \cdots = \mu_k$ when sample sizes are equal. A class variable that identifies the independent variable and a response variable must be included in the **INPUT** statement of the command file. If levels of the independent variable will be entered as alphabetics, the **INPUT** statement must contain the variable name followed by a space followed by a dollar ($) symbol. A **CLASS** statement to identify the independent variable is required, as is a **MODEL** statement to identify the variables and the nature of the assumed model. To cause SAS to return an ANOVA summary, we let X and Y denote the treatment and response variables, respectively, and place the statements

```
PROC ANOVA;
CLASS X;
MODEL Y=X;
```

after the data in the appropriate SAS command file. Notice that the parameter μ and an error term are not included in the **MODEL** statement.

If the statement **MEANS X/SNK;** is placed after the model statement, SAS will conduct a Student–Newman–Keuls analysis on the pairs of treatment means. The slash (/) in the statement indicates that a listing of options will follow. If the statement **MEANS X;** is used instead, only the k treatment means will be output.

A SAS command file for a one-way analysis of variance followed by a Student–Newman–Keuls analysis on the fabric wear data of Example 3.1 is given in Table 3.16. The outputs of that command file are given in Figure 3.11.

TABLE 3.16
SAS Command File for Fabric Wear Study

`OPTIONS LINESIZE=80;`	*restricts width of output*
`DATA RAW;`	*names the data set*
`  INPUT FABRIC $ WEARLOSS @@;`	*names the variables in order*
`CARDS;`	*signals that data follow*
`  A 1.93 A 2.38 A 2.20 A 2.25`	*data begin*
`  B 2.55 B 2.72 B 2.75 B 2.70`	
`  C 2.40 C 2.68 C 2.31 C 2.28`	
`  D 2.33 D 2.40 D 2.28 D 2.25`	
`;`	*signals the end of data*
`PROC ANOVA;`	*identifies the procedure*
`CLASS FABRIC;`	*identifies the treatment variable*
`MODEL WEARLOSS=FABRIC;`	*identifies the model*
`MEAN FABRIC/SNK;`	*requests a Student–Newman–Keuls analysis on pairs of means*

```
                        ANALYSIS OF VARIANCE PROCEDURE
DEPENDENT VARIABLE: WEARLOSS

                                  SUM OF          MEAN
SOURCE                    DF      SQUARES        SQUARE    F VALUE    PR > F
MODEL                      3   0.52011875    0.17337292       8.53    0.0026
ERROR                     12   0.24377500    0.02031458
CORRECTED TOTAL           15   0.76389375

                   R-SQUARE            C.V.     ROOT MSE      WEARLOSS MEAN
                   0.680878          5.9372   0.14252924        2.40062500

SOURCE                    DF      ANOVA SS      F VALUE     PR > F
FABRIC                     3   0.52011875         8.53     0.0026

                        ANALYSIS OF VARIANCE PROCEDURE

STUDENT-NEWMAN-KEULS TEST FOR VARIABLE:WEARLOSS
NOTE: THIS TEST CONTROLS THE TYPE I EXPERIMENTWISE ERROR RATE
      UNDER THE COMPLETE NULL HYPOTHESIS BUT NOT UNDER PARTIAL
      NULL HYPOTHESES

                  ALPHA=0.05   DF=12   MSE=.0203146

NUMBER OF MEANS          2              3            4
CRITICAL RANGE     0.219578       0.268881     0.299213

   MEANS WITH THE SAME LETTER ARE NOT SIGNIFICANTLY DIFFERENT

          SNK GROUPING           MEAN          N          FABRIC
                    A          2.6800          4          B
                    B          2.4175          4          C
                    B
                    B          2.3150          4          D
                    B
                    B          2.1900          4          A
```

Figure 3.11   SAS output for the fabric wear study using the ANOVA procedure.

The SAS output of Figure 3.11 includes the sample value of the *coefficient of determination* (R-SQUARE). That sample statistic (usually denoted R^2) measures how much variation in the response variable is associated with the model. R^2, which can range from 0 to 1, is the ratio of the sum of squares for the model divided by the (corrected) total sum of squares. In general, the larger the R^2 value, the better the model fits the data.

The SAS output for the fabric wear study also includes the sample value of the *coefficient of variation* (C.V.). That value is 100 times the standard deviation of the N response values divided by the mean of those values.

The GLM (General Linear Model) Procedure

The **GLM** procedure can be used for many different analyses, including simple and multiple regression, analyses of variance (especially for unbalanced data), analysis of covariance, response-surface models, weighted regression, polynomial regression, partial correlation, and contrasts (a customized hypothesis test). Many options and statements are available, and we will introduce a few, as they are needed.

As with the **ANOVA** procedure, **CLASS** and **MODEL** statements are required with the **GLM** procedure, with many options available with each. One option, **CONTRAST**, can be used to generate contrasts such as those considered in Section 3.4.1. That option is not available in the **ANOVA** procedure. To cause SAS to return a one-way ANOVA summary, we let X and Y denote the treatment and response variables, respectively, and place the statements

```
PROC GLM;
CLASS X;
MODEL Y=X;
```

after the data in the appropriate SAS command file. If contrasts are to be included with the analysis, **CONTRAST** statements must come after the **MODEL** statement. As with **ANOVA**, placing the statement **MEANS X/SNK;** after the **MODEL** statement will result in the inclusion of a Student–Newman–Keuls analysis in the outputs.

There is no limit to the number of contrasts that can be requested. The analyst is responsibile for determining the appropriate coefficients and checking orthogonality.

The general form of a **CONTRAST** statement is:

$$\text{CONTRAST 'contrast label' } X \ c_1 \ c_2 \ \ldots \ c_k;$$

A label of 20 or fewer characters, placed in single quotes, must follow the word CONTRAST. The label itself is followed by a space, which is followed by the effect. The effect, also, is followed by a space, and then the contrast coefficients are entered, with a space before each one. Like all SAS statements, the **CONTRAST** statement must end with a semicolon (;).

A SAS command file for a one-way analysis of variance on the fabric wear data of Example 3.1 is given in Table 3.17. The outputs of that command file are given in Figure 3.12. **PROC GLM** is used so that the contrasts considered in Example 3.5 can be included. The analyses give the same F tests on the orthogonal contrasts as those included with Example 3.5.

TABLE 3.17
SAS Command File for Fabric Wear Study: PROC GLM

```
OPTIONS LINESIZE=80;              restricts width of output
DATA RAW;                          names the data set
  INPUT FABRIC WEARLOSS @@;        names the variables in order
CARDS;                             signals that data follow
  1 1.93 1 2.38 1 2.20 1 2.25      data begin
  2 2.55 2 2.92 2 2.75 2 2.70
  3 2.40 3 2.68 3 2.31 3 2.28
  4 2.33 4 2.40 4 2.28 4 2.25
;                                  signals the end of data
PROC GLM;                          identifies the procedure
CLASS FABRIC;                      identifies the treatment variable
MODEL WEARLOSS=FABRIC;             identifies the model
CONTRAST '1 VS 4' FABRIC 1 0 0 −1; contrasts begin
CONTRAST '2 VS 3' FABRIC 0 1 −1 0;
CONTRAST '1&4 VS 2&3' FABRIC 1 −1 −1 1;
```

GENERAL LINEAR MODELS PROCEDURE

DEPENDENT VARIABLE: WEARLOSS

SOURCE	DF	SUM OF SQUARES	MEAN SQUARE	F VALUE	PR > F
MODEL	3	0.52011875	0.17337292	8.53	0.0026
ERROR	12	0.24377500	0.02031458		
CORRECTED TOTAL	15	0.76389375			

	R-SQUARE	C.V.	ROOT MSE	WEARLOSS MEAN
	0.680878	5.9372	0.14252924	2.40062500

SOURCE	DF	TYPE I SS	MEAN SQUARE	F VALUE	PR > F
FABRIC	3	0.52011875	0.17337292	8.53	0.0026

SOURCE	DF	TYPE III SS	MEAN SQUARE	F VALUE	PR > F
FABRIC	3	0.52011875	0.17337292	8.53	0.0026

GENERAL LINEAR MODELS PROCEDURE

DEPENDENT VARIABLE: WEARLOSS

CONTRAST	DF	CONTRAST SS	MEAN SQUARE	F VALUE	PR > F
1 VS 4	1	0.03125000	0.03125000	1.54	0.2386
2 VS 3	1	0.13781250	0.13781250	6.78	0.0230
1&4 VS 2&3	1	0.35105625	0.35105625	17.28	0.0013

Figure 3.12   SAS output for the fabric wear study using the **GLM** procedure.

In Section 3.7 we considered the importance of checking the model, which often requires careful consideration of residuals. Changing the **MODEL** statement in the **GLM** procedure to

$$\text{MODEL } Y=X/P;$$

causes the observed values to be output, along with their corresponding predicted values and residuals. This also has the lag 1 autocorrelation coefficient and the Durbin–Watson statistic output, as illustrated in the next example. For those statistics to be meaningful, the data listing in the command file must be in time series order.

■ **Example 3.14**

The effect of three different tube types on cathode warm-up time in seconds was to be studied. The three tube types are A, B, and C, and eight observations are to be made (in random order) on each type of tube. The data are reported in the SAS command file of Table 3.18 and the outputs are included in Figure 3.13. The F test (p value = 0.0797) indicates that there may be a difference among the population means. Further analyses should follow.

If the data were entered in time series order, the small value of the lag 1 auto-correlation coefficient ($r_1 = -0.06872537$) and the large value of the Durbin–Watson statistic (DW = 2.12076765) support an independence assumption. Otherwise, these statistics have no meaningful interpretation.

TABLE 3.18
SAS Command File for Example 3.14

`OPTIONS LINESIZE=80;`	*restricts width of output*
`DATA RAW;`	*names the data set*
`  INPUT TUBE $ WARMTIME @@;`	*names the variables in order*
`CARDS;`	*signals that data follow*
`  A 19 A 23 A 26 A 18 A 20 A 20`	*data begin*
`  A 18 A 35 B 20 B 20 B 32 B 27`	
`  B 40 B 24 B 22 B 18 C 16 C 15`	
`  C 18 C 26 C 19 C 17 C 19 C 18`	
`;`	*signals the end of data*
`PROC GLM;`	*identifies the procedure*
`CLASS TUBE;`	*identifies the treatment variable*
`MODEL WARMTIME=TUBE/P;`	*identifies the model and requests the output of predicted values with residuals*

GENERAL LINEAR MODELS PROCEDURE

DEPENDENT VARIABLE: WARMTIME

SOURCE	DF	SUM OF SQUARES	MEAN SQUARE	F VALUE	PR > F
MODEL	2	190.08333333	95.04166667	2.86	0.0797
ERROR	21	697.75000000	33.22619048		
CORRECTED TOTAL	23	887.83333333			

R-SQUARE	C.V.	ROOT MSE	WEARLOSS MEAN
0.214098	26.1021	5.76421638	22.08333333

SOURCE	DF	TYPE I SS	MEAN SQUARE	F VALUE	PR > F
TUBE	2	190.08333333	95.04166667	2.86	0.0797

SOURCE	DF	TYPE III SS	MEAN SQUARE	F VALUE	PR > F
TUBE	2	190.08333333	95.04166667	2.86	0.0797

OBSERVATION	OBSERVED VALUE	PREDICTED VALUE	RESIDUAL
1	19.00000000	22.37500000	-3.37500000
2	23.00000000	22.37500000	0.62500000
3	26.00000000	22.37500000	3.62500000
4	18.00000000	22.37500000	-4.37500000
5	20.00000000	22.37500000	-2.37500000
6	20.00000000	22.37500000	-2.37500000
7	18.00000000	22.37500000	-4.37500000
8	35.00000000	22.37500000	12.62500000
9	20.00000000	25.37500000	-5.37500000
10	20.00000000	25.37500000	-5.37500000
11	32.00000000	25.37500000	6.62500000
12	27.00000000	25.37500000	1.62500000
13	40.00000000	25.37500000	14.62500000
14	24.00000000	25.37500000	-1.37500000
15	22.00000000	25.37500000	-3.37500000
16	18.00000000	25.37500000	-7.37500000
17	16.00000000	18.50000000	-2.50000000
18	15.00000000	18.50000000	-3.50000000
19	18.00000000	18.50000000	-0.50000000

20	26.00000000	18.50000000	7.50000000
21	19.00000000	18.50000000	0.50000000
22	17.00000000	18.50000000	−1.50000000
23	19.00000000	18.50000000	0.50000000
24	18.00000000	18.50000000	−0.50000000

SUM OF RESIDUALS	0.00000000
SUM OF SQUARED RESIDUALS	697.75000000
SUM OF SQUARED RESIDUALS − ERROR SS	−0.00000000
FIRST ORDER CORRELATION	−0.06872537
DURBIN-WATSON D	2.12076765

Figure 3.13   SAS output for Example 3.14.

The NESTED Procedure

The **NESTED** procedure can be used to evaluate the components of variance of a nested or hierarchical random model. It requires a **CLASS** statement and will perform an analysis of variance for all numeric values included in the **CLASS** statement unless a **VAR** statement is used to identify the variables to be analyzed.

The data must be entered or presorted by the order in which they are given in the **CLASS** statement. The second effect is assumed to be nested within the first effect, the third effect is assumed to be nested within the second effect, and so on.

We use the chemical yield data analyzed in Example 3.9 to illustrate the **NESTED** procedure. The data, originally presented in Table 3.11, are found in the SAS command file of Table 3.19. The resulting outputs are summarized in Figure 3.14. Except for rounding, the variance components are the same as those found in Example 3.9. Notice that almost 87% of the total variance is attributable to variability among the batches.

```
                    COMPONENTS OF VARIANCE
                 CHEMICAL YIELD BY BATCH DATA
                      FROM TABLE 3.11

              COEFFICIENTS OF EXPECTED MEAN SQUARES
```

	SOURCE	BATCH	ERROR
	BATCH	3	1
	ERROR	0	1

```
                  ANALYSIS OF VARIANCE YIELD
```

VARIANCE SOURCE	D.F.	SUM OF SQUARES	MEAN SQUARES	VARIANCE COMPONENT	PERCENT
TOTAL	14	165.73333	11.838095	13.511111	100
BATCH	4	147.73333	36.933333	11.711111	86.6776
ERROR	10	18	1.8	1.8	13.3224

MEAN	74.8666667
STANDARD DEVIATION	1.34164079
COEFFICIENT OF VARIATION	1.79204023

Figure 3.14   SAS output for the chemical yield study.

TABLE 3.19
SAS Command File for Chemical Yield Study

`OPTIONS LINESIZE=80;`	*restricts width of output*
`DATA CHEMICAL;`	*names the data set*
`INPUT BATCH YIELD @@;`	*names the variables in order*
`CARDS;`	*signals that data follow*
`  1 74 1 76 1 75 2 68 2 71`	*data begin*
`  2 72 3 73 3 77 3 77 4 72`	
`  4 74 4 73 5 79 5 81 5 79`	
`;`	*signals the end of data*
`TITLE 'COMPONENTS OF VARIANCE';`	*places caption on first row of output*
`TITLE2 'CHEMICAL YIELD BY BATCH DATA';`	
`TITLE3 'FROM TABLE 3.11';`	
`PROC NESTED;`	*identifies the procedure*
`CLASSES BATCH;`	*identifies the hierarchy*
`VAR YIELD;`	*identifies the variables to be analyzed*

3.9 SUMMARY

In this chapter consideration has been given to

Experiment	Design	Analysis
I. Single factor	Completely randomized	One-way ANOVA

In applying the ANOVA techniques, certain assumptions should be kept in mind:

1. The process is in control; that is, it is repeatable.
2. The population distribution being sampled is normal.
3. The variance of the errors within all k levels of the factor is homogeneous.

If the process is not repeatable, inferences are applicable to the populations in existence at the time of the study, but not to any future populations. In such cases, we have what are known as *enumerative studies*. On the other hand, when the process is repeatable, inferences are applicable to both the populations in existence at the time of the study and those that may exist in the near future. Studies performed under such circumstances are known as *analytical studies*.

3.10 FURTHER READING

Anscombe, F. J., and J. W. Tukey, "The Examination and Analysis of Residuals," *Technometrics*, Vol. 5, No. 2, May 1963, pp. 141–160.

Box, G. E. P., W. G. Hunter, and J. S. Hunter, *Statistics for Experimenters: An Introduction to Design, Data Analysis, and Model Building.* New York: John Wiley & Sons, 1978, Chapter 6.

Inman, R. L., *A Data-Based Approach to Statistics.* Belmont, CA: Duxbury Press, 1994, Chapter 15.

Neter, J., and W. Wasserman, *Applied Linear Statistical Models.* Homewood, IL: Richard D. Irwin, 1974, Chapters 13–16.

Ostle, B., K. V. Turner Jr., C. R. Hicks, and G. W. McElrath, *Engineering Statistics: The Industrial Experience.* Belmont, CA: Duxbury Press, 1996, Chapter 12.

Wooding, W. M., "The Computation and Use of Residuals in the Analysis of Experimental Data," *Journal of Quality Technology* Vol. 1, No. 3, July 1969, pp. 175–188.

PROBLEMS

3.1 Assuming a completely randomized design, do a one-way analysis of variance on the following data to familiarize yourself with the technique.

	Factor A Level				
	1	**2**	**3**	**4**	**5**
Measurement	8	4	1	4	10
	6	−2	2	6	8
	7	0	0	5	7
	5	−2	−1	5	4
	8	3	−3	4	9

3.2 The cathode warm-up time in seconds was determined for three different tube types using eight observations on each type of tube. The order of experimentation was completely randomized. The results were as tabulated.

	Tube Type					
	A		**B**		**C**	
Warm-up time	19	20	20	40	16	19
(seconds)	23	20	20	24	15	17
	26	18	32	22	18	19
	18	35	27	18	26	18

    a. Do an analysis of variance on these data and test the hypothesis that the three tube types require the same average warm-up time.

    b. Using residual plots and other procedures as appropriate, assess the model.

3.3 For Problem 3.2, set up orthogonal contrasts between the tube types and test your contrasts for significance.

3.4 Set up 95% confidence limits for the average warm-up time for tube type *C* in Problem 3.2.

3.5 Use the Newman–Keuls range method to test for differences between tube types.

3.6 The following data give the pressure in a torsion spring for several settings of the angle between the legs of the spring in a free position.

	Angle of Legs of Spring (degrees)				
	67	**71**	**75**	**79**	**83**
Pressure (psi)	83	84	86	89	90
	85	85	87	90	92
		85	87	90	
		86	87	91	
		86	88		
		87	88		
			88		
			88		
			88		
			89		
			90		

Assuming a completely randomized design, complete a one-way analysis of variance for this experiment and state your conclusion concerning the effect of angle on the pressure in the spring.

3.7 Set up orthogonal contrasts for the angles in Problem 3.6.

3.8 Show that the expanded forms for the sums of squares in Equations (3.8) to (3.10) are correct.

3.9 Assume that the levels of factor A in Problem 3.1 were chosen at random and determine the proportion of variance attributable to differences in level means and the proportion due to error.

3.10 Verify the results in Figure 3.6 based on the data of Table 3.11.

3.11 Set up two or more contrasts for the data of Problem 3.2 and test their significance by the Scheffé method.

3.12 Since the Scheffé method is not restricted to equal sample sizes, set up several contrasts for Problem 3.6 and test for significance by the Scheffé method.

3.13 It is suspected that the environmental temperature at which batteries are activated affects their activated life. Thirty homogeneous batteries were tested, six at each of five temperatures, and the following data were obtained.

	Temperature (°C)				
	0	**25**	**50**	**75**	**100**
Activated life	55	60	70	72	65
(seconds)	55	61	72	72	66

57	60	73	72	60
54	60	68	70	64
54	60	77	68	65
56	60	77	69	65

a. Analyze and interpret the data.
b. Using residual plots and other procedures as appropriate, assess the model.

3.14 A highway research engineer wishes to determine the effect of four types of subgrade soil on the moisture content in the topsoil. He takes five samples of each type of subgrade soil and the total sum of squares is computed as 280, whereas the sum of squares among the four types of subgrade soil is 120.
a. Set up an analysis of variance table for these results.
b. Set up a mathematical model to describe this problem, define each term in the model, and state the assumptions made on each term.
c. Set up a test of the hypothesis that the four types of subgrade soil have the same effect on moisture content in the topsoil.
d. Set up a set of orthogonal contrasts for this problem.
e. Explain briefly how to set up a test on means after the analysis of variance for these data.
f. Set up an expression for 90% confidence limits on the mean of type 2 subgrade soil. Insert all numerical values that are known.

3.15 Data collected on the effect of four fixed types of television tube coating on the conductivity of the tubes are given as

Coating			
I	II	III	IV
56	64	45	42
55	61	46	39
62	50	45	45
59	55	39	43
60	56	43	41

Do an analysis of variance on these data and test the hypothesis that the four coatings yield the same average conductivity.

3.16 For Problem 3.15, set up orthogonal contrasts between tube coatings and test them for significance.

3.17 Use a Newman–Keuls range test on the data of Problem 3.15 and discuss any significant results that you find.

3.18 Using the data of Problem 3.15, set up contrasts and tests using Scheffé's method to answer the following questions: Do the means of coating I and II differ? the means of II and IV? the mean of I and II versus the mean of III and IV?

3.19 Four bonding machines (A, B, C, D) are used in a certain plant to bond circuit wires onto a board. To determine which bonder does the best job, samples of 10 are taken at random from each machine, and the strength needed to break the bond (in grams) is determined. The data are as follows.

	Bonder		
A	**B**	**C**	**D**
204	197	264	248
181	223	226	138
201	206	228	273
203	232	249	220
214	213	246	186
262	207	255	304
246	259	186	330
230	223	237	268
256	195	236	295
288	197	240	276

a. Analyze these data and make your recommendations to management.
b. Using residual plots and other procedures as appropriate, assess the model.

3.20 To determine the effect of several teaching methods on student achievement, 30 students were assigned to five treatment groups with 6 students per group. The treatments given for the semester are as follows:

Treatment	Description
1	Current textbook
2	Textbook A with teacher
3	Textbook A with machine
4	Textbook B with teacher
5	Textbook B with machine

At the end of a semester of instruction, achievement scores were recorded and the following statistics were developed.

	Treatment						
	1	**2**	**3**	**4**	**5**		
Totals	120	600	720	240	420		
Source		df	SS	MS	F	$F_{0.95}$	
Between treatments			340				
Error							
Total			465				

a. Complete the ANOVA table.
b. Write the mathematical model assumed here and state the hypothesis to be tested.
c. Test the hypothesis and state your conclusion.

3.21 Set up one set of orthogonal contrasts that might seem reasonable from the treatment descriptions in Problem 3.20.

3.22 Test whether the mean achievement under treatment 1 of Problem 3.20 differs significantly from the mean achievement under treatment 5 in a Newman–Keuls sense for Problem 3.20.

3.23 Determine the standard error of the contrast $4T_{.1} - T_{.2} - T_{.3} - T_{.4} - T_{.5}$ based on Problem 3.20 data.

3.24 Three fertilizers are tried on 27 plots of land in a random fashion such that each fertilizer is applied to nine plots. The total yield for each fertilizer type is given by

	Type		
	1	2	3
$T_{.j}$	240	320	180

and the ANOVA table is

Source	df	SS	MS
Between fertilizers	2	1096	548
Error	24	1440	60
Total	26	2536	

a. Set up one set of orthogonal contrasts that might be used on these data.
b. For your first contrast, determine the sum of squares due to this contrast.
c. For your second contrast, find its standard error.
d. If a farmer wished to compare types 1 and 2 and also the average of 1 and 3 versus 2, which method would you recommend and why?

3.25 Birth weights in pounds of five Poland China pigs were recorded from each of six randomly chosen litters. The ANOVA printout showed:

Source	df	SS	MS
Between litters	5	6.0	1.20
Within litters	24	9.6	0.40

Determine what percentage of the total variance in this experiment can be attributed to litter-to-litter differences.

3.26 To determine whether there is a difference in leakage between the capacitors of three vendors (A, B, C), six samples were randomly drawn from each vendor and the following leakage readings (in milliamperes) were found:

A	B	C
7.3	10.7	10.5
8.0	10.2	10.1
8.1	10.2	10.8
8.5	10.7	11.6
8.4	9.9	11.4
7.5	11.0	10.8

a. Analyze these data and make recommendations.
b. Using residual plots and other procedures as appropriate, assess the model.

3.27 If vendor A in Problem 3.26 is the present supplier and vendors B and C are competing for the job, set up two reasonable orthogonal contrasts, test them, and discuss the results.

Chapter 4

SINGLE-FACTOR EXPERIMENTS: RANDOMIZED BLOCK AND LATIN SQUARE DESIGNS

4.1 INTRODUCTION

Consider the problem of determining whether different brands of tires exhibit different amounts of tread loss after 20,000 miles of driving. A fleet manager wishes to consider four brands that are available and decide which brand is likely to show the least amount of tread wear after 20,000 miles. The brands to be considered are *A, B, C,* and *D,* and although driving conditions might be simulated in the laboratory, the manager wants to try these four brands under actual driving conditions. The variable to be measured is the difference in maximum tread thickness on a tire between the time it is mounted on the wheel of a car and after it has completed 20,000 miles on that car. The measured variable Y_{ij} is this difference in thickness in mils (1 mil = 0.001 in.), and the only factor of interest is brands, say τ_j, where $j = 1, 2, 3,$ and 4.

Since the tires must be tried on cars, and since some measure of error is necessary, more than one tire of each brand must be used and a set of four of each brand would seem quite practical. This means 16 tires, four each of four different brands, and a reasonable experiment would involve at least four cars. Designating the cars as I, II, III, and IV, one might put brand A's four tires on car I, brand B's on car II, and so on, with a design as shown in Table 4.1.

TABLE 4.1
Design 1 for Tire Brand Test

	Car			
	I	**II**	**III**	**IV**
Brand	*A*	*B*	*C*	*D*
distribution	*A*	*B*	*C*	*D*
	A	*B*	*C*	*D*
	A	*B*	*C*	*D*

One look at this design shows its fallibility, since averages for brands are also averages for cars. If the cars travel over different terrains, using different drivers, any apparent brand differences are also car differences. This design is called *completely confounded,* since we cannot distinguish between brands and cars in the analysis.

A second attempt at design might be to try a completely randomized design, as given in Chapter 3. Assigning the 16 tires to the four cars in a completely random manner might give results as in Table 4.2, which shows the loss in thickness for each of the 16 tires. The purpose of complete randomization here is to average out any car differences that might affect the results. The model would be

$$Y_{ij} = \mu + \tau_j + \varepsilon_{ij}$$

with

$$j = 1, 2, 3, 4 \quad i = 1, 2, 3, 4$$

A Minitab ANOVA on these data gives the results in Figure 4.1.

The *p* value (0.115) in Figure 4.1 indicates that the hypothesis of equal average tread loss among the four brands can be rejected when $\alpha \geq 0.115$. Thus, if testing is at the 5% significance level, one would have insufficient evidence to reject that hypothesis.

TABLE 4.2
Design 2 for Tire Brand Test

	Car			
	I	**II**	**III**	**IV**
Brand distribution	C(12)	A(14)	C(10)	A(13)
and loss in thickness	A(17)	A(13)	D(11)	D(9)
(mils)	D(13)	B(14)	B(14)	B(8)
	D(11)	C(12)	B(13)	C(9)

One-Way Analysis of Variance

```
Analysis of Variance on Loss
Source      DF        SS        MS       F        p
Brand        3     30.69     10.23    2.44    0.115
Error       12     50.25      4.19
Total       15     80.94

                                   Individual 95% CIs for Mean
                                   Based on Pooled StDev
Level        N      Mean     StDev   ------+--------+--------+--------+
   1         4    14.250     1.893                  (--------*--------)
   2         4    12.250     2.872            (--------*--------)
   3         4    10.750     1.500   (--------*--------)
   4         4    11.000     1.633     (--------*--------)
                                   ------+--------+--------+--------+
Pooled StDev =     2.046
```

Figure 4.1 Minitab ANOVA for design 2.

Figure 4.1 also contains graphical displays of the confidence intervals for the population means. Notice that any two intervals overlap, indicating that the corresponding population means may be equal.

4.2 RANDOMIZED COMPLETE BLOCK DESIGN

A more careful examination of design 2 in Table 4.2 will reveal some glaring disadvantages of the completely randomized design in this problem. One thing to be noted is that brand A is never used on car III or brand B on car I. Also, any variation within brand A may reflect variation between cars I, II, and IV. Thus the random error may not be merely an experimental error but may include variation between cars. Since *the chief objective of experimental design is to reduce the experimental error,* a better design might be one in which car variation is removed from error variation. Although the completely randomized design averaged out the car effects, it did not eliminate the variance among cars. A design that requires that each brand be used once on each car is a *randomized complete block design,* given in Table 4.3.

In this design the order in which the four brands are placed on a car is random and each car gets one tire of each brand. In this way better comparisons can be made between brands, which are all driven over approximately the same terrain, and so on. This provides a more homogeneous environment in which to test the four brands. In general, these groupings for homogeneity are called *blocks* and randomization is now restricted within blocks. This design also allows the car (block) variation to be independently assessed and removed from the error term. The model for this design is

$$Y_{ij} = \mu + \beta_i + \tau_j + \varepsilon_{ij} \tag{4.1}$$

where β_i now represents the block effect (car effect) in the example above.

The analysis of this model is a two-way analysis of variance, since the block effect may now also be isolated. A slight rearrangement of the data in Table 4.3 gives Table 4.4.

Calling the number of treatments (brands) in general k and the number of blocks (cars) n, we can compute the total sum of squares as in Chapter 3:

$$SS_{total} = \sum_{j=1}^{k} \sum_{i=1}^{n} y_{ij}^2 - \frac{T_{..}^2}{N}$$

$$= 2409 - \frac{(193)^2}{16} = 80.94$$

TABLE 4.3
Design 3: Randomized Block Design for Tire Brand Test

	Car			
	I	II	III	IV
Brand distribution	$B(14)$	$D(11)$	$A(13)$	$C(9)$
and loss in thickness	$C(12)$	$C(12)$	$B(13)$	$D(9)$
(mils)	$A(17)$	$B(14)$	$D(11)$	$B(8)$
	$D(13)$	$A(14)$	$C(10)$	$A(13)$

TABLE 4.4
Randomized Block Design Data for Tire Brand Test

Car	Brand				$T_{i.}$
	A	B	C	D	
I	17	14	12	13	56
II	14	14	12	11	51
III	13	13	10	11	47
IV	13	8	9	9	39
$T_{.j}$	57	49	43	44	$T_{..} = 193$
$\sum_i y_{ij}^2$	823	625	469	492	$\sum_j \sum_i y_{ij}^2 = 2409$

The brand (treatment) sum of squares is computed as usual:

$$SS_{brands} = \sum_{j=1}^{k} \frac{T_{.j}^2}{n} - \frac{T_{..}^2}{N}$$

$$= \frac{(57)^2 + (49)^2 + (43)^2 + (44)^2}{4} - \frac{(193)^2}{16} = 30.69$$

Since the car (block) effect is similar to the brand effect but totaled across the rows of Table 4.4, the car sum of squares is computed exactly like the brand sum of squares, using row totals $T_{i.}$ instead of column totals. Since the number of treatments (brands) is in general k,

$$SS_{car} = \sum_{i=1}^{n} \frac{T_{i.}^2}{k} - \frac{T_{..}^2}{N}$$

$$= \frac{(56)^2 + (51)^2 + (47)^2 + (39)^2}{4} - \frac{(193)^2}{16} = 38.69$$

The error sum of squares is now the remainder after both brand and car sum of squares have been subtracted from the total sum of squares

$$SS_{error} = SS_{total} - SS_{brand} - SS_{car}$$

$$= 80.94 - 30.69 - 38.69 = 11.56$$

Table 4.5 is an ANOVA table for these data; the EMS values given are based on a random factor (car) and a fixed factor (brand). Derivation of such EMSs is considered in Chapter 6.

To test the hypothesis, $H_{01} : \mu_{.1} = \mu_{.2} = \mu_{.3} = \mu_{.4}$, the ratio is $F_{(3,9)} = MS_{brand}/MS_{error}$, the observed value of which is

$$f = \frac{10.2}{1.3} = 7.8$$

Since f is significantly larger than the corresponding critical F even at the 1% level (Statistical Table D), the hypothesis of equal brand means is rejected. Note that with a

TABLE 4.5
ANOVA for Randomized Block Design of Tire Brand Test

Source	df	SS	MS	EMS
Brands	3	30.69	10.2	$\sigma_\varepsilon^2 + 4\phi_\tau$
Cars	3	38.69	12.9	$\sigma_\varepsilon^2 + 4\sigma_\beta^2$
Error	9	11.56	1.3	σ_ε^2
Totals	15	80.94		

completely randomized design, this hypothesis would not be rejected. The randomized block design that allows for removal of the block (car) effect definitely reduced the error variance estimate from 4.2 to 1.3.

It is also possible, if desired, to test the hypothesis that the average tread loss is the same for all four cars. To do so, use $H_{02}: \mu_{1.} = \mu_{2.} = \mu_{3.} = \mu_{4.}$ and the ratio $F_{(3,9)} = MS_{car}/MS_{error}$. The observed value of this statistic is $f = 12.9/1.3 = 9.9$, which is also significant at the 1% level (Statistical Table D). Here this hypothesis is rejected and a car-to-car variation is detected.

The preceding calculations were included to aid understanding of an ANOVA summary for a randomized complete block design. Using Minitab with the same data produces the ANOVA of Figure 4.2. The sums of squares, mean squares, and F ratios are (except for rounding) the same as those obtained manually. The test for equality of average wear among brands has a p value of 0.007, indicating that rejection is possible for any $\alpha \geq 0.007$. [**Note:** This conclusion is valid when the model assumptions (see Section 4.3) are adequately satisfied.]

Even though an effect due to cars (blocks) has been isolated, the main objective is still to test brand differences. Thus it is still a single-factor experiment, the blocks representing only a restriction on complete randomization due to the environment in which the experiment was conducted. Other examples include testing differences in materials that are fed into several different machines, testing differences in fertilizers that must be spread on several different plots of ground, and testing the effect of different teaching methods on several pupils. In these examples the blocks are machines, plots, and pupils, respectively, and the levels of the factors of interest can be randomized within each block.

```
Analysis of Variance (Balanced Designs)

Factor        Type Levels Values
Car           random    4    1    2    3    4
Brand         fixed     4    1    2    3    4

Analysis of Variance for Loss

Source        DF          SS          MS          F          P
Car            3      38.688      12.896      10.04      0.003
Brand          3      30.687      10.229       7.96      0.007
Error          9      11.563       1.285
Total         15      80.937
```

Figure 4.2 Minitab ANOVA for randomized block design of tire brand test.

Data Table Formats

It is seldom possible to determine how data presented in tabular form were collected. Was the randomization complete over all N observations, or was the experiment run in blocks with randomization restricted to within the blocks? To help signify the design of the experiment, it is suggested that a completely randomized design be presented without horizontal or vertical lines, as in Table 4.2. When randomization has been restricted, either vertical or horizontal lines as shown in Tables 4.3 and 4.4, respectively, can be used to indicate this restriction on the randomization. As more complex designs are presented, this scheme may require some double lines for one restriction and single lines for a second restriction, and so on. It is hoped, however, that such a scheme will help the observer in noting just how the randomization has been restricted.

Paired Data as Randomized Blocks

Since blocking an experiment is a very useful procedure to reduce the experimental error, we present a second example to illustrate the situation in which the same experimental units (animals, people, parts, etc.) are measured before the experiment and then again at the conclusion of the experimental treatment. These before–after, pretest–posttest designs can be treated very well as randomized block designs where the experimental units are the blocks. Interest is primarily in the effect of the treatment, and the block effects can be removed to reduce the experimental error. Some refer to these designs as repeated-measures designs because data are repeated on the same units a second time. The following example will illustrate the procedure.

■ **Example 4.1**

A study on a physical strength measurement in pounds on seven subjects before and after a specified training period gave the results shown in Table 4.6. This problem could be handled by the methods of Chapter 2 using a t test on differences where $D =$ posttest measure–pretest measure. Here $\bar{d} = 15.71$, $s_D = 3.45$, and $t = 12.05$, which is highly significant (p value = 0.00002) with 6 degrees of freedom.

TABLE 4.6
Pretest and Posttest Strength Measures for Example 4.1

Subjects	Pretest	Posttest
1	100	115
2	110	125
3	90	105
4	110	130
5	125	140

6	130	140
7	105	125

If the study is considered as a single-factor experiment at two levels (pre- and post-) with seven blocks (subjects) making up a randomized block design, the ANOVA summary in the partial Excel worksheet of Figure 4.3 is obtained. Here the large value of the F statistic ($f = 145.2$) indicates a strong treatment effect. This value is equal to t^2 given by the method of differences, as $t = 12.05$ and $(12.05)^2 = 145.2$.

	A	B	C	D	E	F	G	H
1				Anova: Two-Factor Without Replication				
2								
3	Subject	Pretest	Posttest	SUMMARY	Count	Sum	Average	Variance
4	1	100	115	1	2	215	107.50	112.5000
5	2	110	125	2	2	235	117.50	112.5000
6	3	90	105	3	2	195	97.50	112.5000
7	4	110	130	4	2	240	120.00	200.0000
8	5	125	140	5	2	265	132.50	112.5000
9	7	130	140	6	2	270	135.00	50.0000
10	7	105	125	7	2	230	115.00	200.0000
11								
12				Pretest	7	770	110.00	191.6667
13				Posttest	7	880	125.71	161.9048
14								
15	ANOVA							
16	Source	SS	df	MS	F	P-value	F crit	
17	Subject	2085.71	6	347.62	58.4	0.00005	4.284	
18	Treatment	864.29	1	864.29	145.2	0.00002	5.987	
19	Error	35.71	6	5.95				
20	Total	2985.71	13					

Figure 4.3 Data and Excel outputs for the physical strength study.

4.3 ANOVA RATIONALE

Let μ_i and μ_j denote the true means of the populations associated with the ith block and the jth treatment, respectively. For a randomized complete block design, the model is

$$Y_{ij} = \mu + \beta_i + \tau_j + \varepsilon_{ij}; i = 1, \ldots, n \text{ and } j = 1, \ldots, k \qquad (4.2)$$

where $\beta_i = \mu_i - \mu$ denotes the true effect of the ith block, $\tau_j = \mu_j - \mu$ denotes the true effect of the jth treatment, and $\varepsilon_{ij} \sim NID(0, \sigma_\varepsilon^2)$. This is equivalent to

$$Y_{ij} = \mu + (\mu_i - \mu) + (\mu_j - \mu) + (Y_{ij} - \mu_i - \mu_j + \mu) \qquad (4.3)$$

The last term can be obtained by subtracting the treatment and block deviations from the overall deviation as follows:

$$(Y_{ij} - \mu) - (\mu_i - \mu) - (\mu_j - \mu) \equiv Y_{ij} - \mu_i - \mu_j + \mu$$

Best estimates of the parameters in Equation (4.3) give the sample model (after moving $\bar{Y}_{..}$ to the left of the equation):

$$Y_{ij} - \bar{Y}_{..} = (\bar{Y}_{i.} - \bar{Y}_{..}) + (\bar{Y}_{.j} - \bar{Y}_{..}) + (\bar{Y}_{ij} - \bar{Y}_{i.} - \bar{Y}_{.j} + \bar{Y}_{..})$$

Squaring both sides and summing over $i = 1, 2, \ldots, n$ and $j = 1, 2, \ldots, k$, we have

$$\sum_{i=1}^{n}\sum_{j=1}^{k}(Y_{ij} - \bar{Y}_{..})^2 = \sum_{i}^{n}\sum_{j}^{k}(\bar{Y}_{i.} - \bar{Y}_{..})^2$$

$$+ \sum_{i}^{n}\sum_{j}^{k}(\bar{Y}_{.j} - \bar{Y}_{..})^2 + \sum_{i}^{n}\sum_{j}^{k}(Y_{ij} - \bar{Y}_{i.} - \bar{Y}_{.j} + \bar{Y}_{..})^2$$

$$+ 3 \text{ cross products} \qquad (4.4)$$

A little algebraic work on the sums of the cross products will show that they all reduce to zero, and the remaining equation becomes the fundamental equation of a two-way ANOVA. The equation states that

$$SS_{total} = SS_{block} + SS_{treatment} + SS_{error}$$

Thus, there is one sum of squares for each variable term in the model [Equation (4.2)]. Each sum of squares has associated with it its degrees of freedom, and dividing any sum of squares by its degrees of freedom will yield an unbiased estimate of population variance σ_ε^2 if the hypotheses under test are true.

The breakdown of degrees of freedom here is

$$\overset{\text{total}}{(nk - 1)} = \overset{\text{blocks}}{(n - 1)} + \overset{\text{treatments}}{(k - 1)} + \overset{\text{error}}{(n - 1)(k - 1)}$$

The error degrees are derived from the remainder:

$$(nk - 1) - (n - 1) - (k - 1) = nk - n - k + 1 = (n - 1)(k - 1)$$

Statistical theory tells us that when U and V are independent chi-square random variables with degrees of freedom v_1 and v_2, respectively, $F = (U/v_1)/(V/v_2)$ has an F distribution with degrees of freedom v_1 and v_2. When the model assumptions for a randomized complete block experiment are satisfied and $H_{01} : \tau_1 = \tau_2 = \ldots = \tau_k = 0$ is true, the random variables $U_1 = \text{SS}_{\text{treatment}}/\sigma_\varepsilon^2$ and $V = \text{SS}_{\text{error}}/\sigma_\varepsilon^2$ are independent chi-square random variables with degrees of freedom $v_1 = k - 1$ and $v_2 = (n - 1)(k - 1)$, respectively [15, pp. 427–428]. Thus, $F_1 = \text{MS}_{\text{treatment}}/\text{MS}_{\text{error}}$ has an F distribution with degrees of freedom v_1 and v_2 when H_{01} is true. Since the expected value of $\text{MS}_{\text{treatment}}$ is greater than that of MS_{error} when H_{01} is false, large values of F_1 support rejection of H_{01}.

Similarly, when $H_{02} : \beta_1 = \beta_2 = \ldots = \beta_n = 0$ is true, the random variables $U_2 = \text{SS}_{\text{block}}/\sigma_\varepsilon^2$ and $V = \text{SS}_{\text{error}}/\sigma_\varepsilon^2$ are independent chi-square random variables with degrees of freedom $n - 1$ and $(n - 1)(k - 1)$, respectively. It follows that $F_2 = \text{MS}_{\text{block}}/\text{MS}_{\text{error}}$ has an F distribution with those degrees of freedom. When H_{02} is false, the expected value of MS_{block} is greater than that of MS_{error}. Thus, large values of F_2 support rejection of H_{02}.

The sums of squares formulas given in Equation (4.4) are usually expanded and rewritten to give formulas that are easier to apply. These are shown in Table 4.7. They are the formulas applied to the data of Table 4.4 where $nk = N$.

TABLE 4.7
ANOVA for Randomized Block Design

Source	df	SS	MS
Between blocks β_i	$n - 1$	$\displaystyle\sum_i^n \frac{T_{i.}^2}{k} - \frac{T_{..}^2}{nk}$	$\dfrac{\text{SS}_{\text{block}}}{n - 1}$
Between treatments τ_j	$k - 1$	$\displaystyle\sum_j^k \frac{T_{.j}^2}{n} - \frac{T_{..}^2}{nk}$	$\dfrac{\text{SS}_{\text{treatment}}}{k - 1}$
Error ε_{ij}	$(n - 1)(k - 1)$	$\displaystyle\sum_i^n \sum_j^k Y_{ij}^2 - \sum_{i=1}^n \frac{T_{i.}^2}{k}$ $- \displaystyle\sum_{j=1}^k \frac{T_{.j}^2}{n} + \frac{T_{..}^2}{nk}$	$\dfrac{\text{SS}_{\text{error}}}{(n - 1)(k - 1)}$
Totals	$nk - 1$	$\displaystyle\sum_i^n \sum_j^k Y_{ij}^2 - \frac{T_{..}^2}{nk}$	

4.4 MISSING VALUES

Occasionally in a randomized block design an observation is lost. A vial may break, an animal may die, or a tire may disintegrate, so that the data set lacks one or more observations. For a single-factor, completely randomized design a missing value presents no problem, since the analysis of variance can be run with unequal n_j's. But for a two-way analysis this incompleteness means a loss of orthogonality, since for some blocks the $\sum_j \tau_j$ no longer equals zero, and for some treatment the $\sum_i \beta_i$ no longer equals zero. When blocks and treatments are orthogonal, the block totals are added over all treatments, and vice versa. If one or more observations are missing, the usual procedure is to replace each missing value with one that makes the sum of the squares of the errors a minimum.

In the tire brand test example, suppose that the brand C tire on car III blew out and was ruined before completing the 20,000 miles. The resulting data, coded by subtracting 13 mils, appear in Table 4.8, where y is inserted in place of this missing value.

Now,

$$SS_{error} = SS_{total} - SS_{treatment} - SS_{block}$$

$$= \sum_i \sum_j Y_{ij}^2 - \sum_j \frac{T_{.j}^2}{n} - \sum_i \frac{T_{i.}^2}{k} + \frac{T_{..}^2}{nk}$$

For this example,

$$SS_{error} = 4^2 + 1^2 + \cdots + y^2 + \cdots + (-4)^2$$
$$- \frac{(5)^2 + (-3)^2 + (y-6)^2 + (-8)^2}{4}$$
$$- \frac{(4)^2 + (-1)^2 + (y-2)^2 + (-13)^2}{4} + \frac{(y-12)^2}{16}$$

To find the y value that will minimize this expression, it is differentiated with respect to y and set equal to zero. As all constant terms have their derivatives zero,

$$\frac{d(SS_{error})}{dy} = 2y - \frac{2(y-6)}{4} - \frac{2(y-2)}{4} + \frac{2(y-12)}{16} = 0$$

Solving gives

$$16y - 4y + 24 - 4y + 8 + y - 12 = 0$$
$$9y = -20$$
$$y = -\frac{20}{9} = -2.2 \text{ or } 10.8 \text{ mils}$$

If this value is now used in the y position, the resulting ANOVA table is as shown in Table 4.9. This is an approximate ANOVA, since the sums of squares are slightly biased. The resulting ANOVA is not too different from before, but the degrees of freedom for the error term are reduced by one, since there are only 15 actual observations, and y is determined from these 15 readings.

TABLE 4.8
Missing Value Example

Car	Brand				$T_{i.}$
	A	B	C	D	
I	4	1	-1	0	4
II	1	1	-1	-2	-1
III	0	0	y	-2	$y - 2$
IV	0	-5	-4	-4	-13
$T_{.j}$	5	-3	$y - 6$	-8	$y - 12 = T_{..}$

TABLE 4.9
ANOVA for Tire Brand Test Example Adjusted for a Missing Value (approximate)

Source	df	SS	MS
Brands τ_j	3	28.7	9.5
Cars β_i	3	38.3	12.7
Error ε_{ij}	8	11.2	1.4
Totals	14	78.2	

This procedure can be used on any reasonable number of missing values by differentiating the error sum of squares partially with respect to each such missing value and setting it equal to zero, giving as many equations as unknown values to be solved for these missing values.

4.4.1 Computer Analysis with Missing Values

The better statistical computing packages contain a general linear models (GLM) program that will calculate sums of squares and prepare an appropriate ANOVA summary for an unbalanced design or a data table with missing values. Figure 4.4 contains such Minitab outputs for the coded data in Table 4.8. The appropriate sums of squares, referred to as type III sums of squares in SAS, are those in the adjusted sums of squares (Adj SS) column. The error sum of squares is the same as that obtained manually. Notice, however, that the brand and total sums of squares differ from those in Table 4.9. Also notice that the sum of the three adjusted sums of squares differs from the total sum of squares.

As discussed in Chapter 3, consideration of the data and residuals should be a part of any statistical analysis. The Minitab outputs of Figure 4.4 include a warning that one of the observations may be an outlier. That message is based on the magnitude of a standardized residual (st.resid.). Since the mean of the residuals is 0, a *standardized residual* is the residual divided by its standard deviation. When the model assumptions are met, most residuals should have standardized values between -2 and 2. For this reason, Minitab prints R next to any standardized residual with absolute value greater than 2.

```
General Linear Model

Analysis of Variance for value

Source    DF      Seq SS     Adj SS     Adj MS        F          P
Car        3      38.233     38.278     12.759      9.10      0.006
Brand      3      26.944     26.944      8.981      6.40      0.016
Error      8      11.222     11.222      1.403
Total     14      76.400

Unusual Observations for value

Obs.    value        Fit   Stdev.Fit   Residual   St.Resid
 14   -5.00000   -3.11111     0.78959   -1.88889    -2.14R

R denotes an obs. with a large st. resid.
```

Figure 4.4 Minitab ANOVA for tire brand test with missing data.

4.5 LATIN SQUARES

The reader may have wondered about a possible position effect in the problem on testing tire brands. Experience shows that rear tires get different wear from front tires, and even different sides of the same car may show different amounts of tread wear. In the randomized block design, the four brands were randomized onto the four wheels of each car with no regard for position. The effect of position on wear could be balanced out by rotating the tires every 5000 miles, giving each brand 5000 miles on each wheel. However, if this is not feasible, the positions can impose another restriction on the randomization in such a way that each brand is not only used once on each car but also only once in each of the four possible positions: left front, left rear, right front, and right rear.

A design in which each treatment appears once and only once in each row (position) and once and only once in each column (cars) is called a *Latin square design*. Interest is still centered on one factor, treatments, but two restrictions are placed on the randomization. An example of one such 4 × 4 Latin square is shown in Table 4.10.

Such a design is possible only when the number of levels of both restrictions equals the number of treatment levels. In other words, it must be a square. It is not true that all randomization is lost in this design: the particular Latin square to be used on a given problem may be chosen at random from several possible Latin squares of the required size.

A 4 × 4 Latin square can be constructed beginning with the row A, B, C, D. Each succeeding row is obtained by taking the first letter of the preceding row and placing it last, which has the effect of shifting the other letters one position to the left, as in Table 4.11. Randomly assign one blocking factor to the rows and the other to the columns. Then, randomly assign levels of the row factor, column factor, and treatment to row positions, column positions, and letters, respectively. For the tire wear study, if positions 1, 2, 3, and 4 are assigned to rows 3, 2, 1, and 4 of Table 4.11, respectively; cars I, II, III, and IV are assigned to columns 1, 4, 3, and 2, respectively; brands *A, B, C,* and *D* are assigned to the letters *D, C, B,* and *A*, respectively; and the renamed rows and columns are then placed in numerical order, the result is a design we have already seen: Table 4.10.

The procedure for generating a 4 × 4 Latin square can be used for any size Latin square. For those wishing to avoid this chore, tables of Latin squares are found in references such as [8], [10], and [23].

The analysis of the data in a Latin square design is a simple extension of earlier analyses in which the data are now added in a third direction—positions. If the data of

TABLE 4.10
4 × 4 Latin Square Design

	Car			
Position	I	II	III	IV
1	*C*	*D*	*A*	*B*
2	*B*	*C*	*D*	*A*
3	*A*	*B*	*C*	*D*
4	*D*	*A*	*B*	*C*

TABLE 4.11
A Basic Latin Square

Row	Column			
	1	2	3	4
1	A	B	C	D
2	B	C	D	A
3	C	D	A	B
4	D	A	B	C

Table 4.4 were imposed on the Latin square of Table 4.10, the results could be those shown in Table 4.12.

Treatment totals for A, B, C, and D brands are 57, 49, 43, and 44 as before, where the model is now

$$Y_{ijk} = \mu + \beta_i + \tau_j + \gamma_k + \varepsilon_{ijk}$$

and γ_k represents the positions' effect. Since the only new totals are for positions, a position sum of squares can be computed as

$$SS_{position} = \frac{(44)^2 + (50)^2 + (50)^2 + (49)^2}{4} - \frac{(193)^2}{16} = 6.19$$

and

$$SS_{error} = SS_{total} - SS_{brand} - SS_{car} - SS_{position}$$
$$= 80.94 - 30.69 - 38.69 - 6.19 = 5.37$$

Figure 4.5 contains an ANOVA table for these data. Unlike the missing data ANOVA of Figure 4.4, corresponding sequential sums of squares (Seq SS) and adjusted sums of squares (Adj SS) are equal. Those sums of squares will be equal for any Latin square with no missing values.

Once again restriction placed on the randomization has further reduced the experimental error by identifying another possible source of variation, although the position effect is not

TABLE 4.12
Latin Square Design Data on Tire Wear

Position	Car				$T_{..k}$
	I	II	III	IV	
1	C 12	D 11	A 13	B 8	44
2	B 14	C 12	D 11	A 13	50
3	A 17	B 14	C 10	D 9	50
4	D 13	A 14	B 13	C 9	49
$T_{i..}$	56	51	47	39	193

General Linear Model

```
Analysis of Variance for Loss
Source      DF      Seq SS      Adj SS      Adj MS          F          P
Position     3      6.1875      6.1875      2.0625       2.30      0.177
Car          3     38.6875     38.6875     12.8958      14.40      0.004
Brand        3     30.6875     30.6875     10.2292      11.42      0.007
Error        6      5.3750      5.3750      0.8958
Total       15     80.9375
```

Figure 4.5 Minitab Latin square ANOVA for tire wear study.

significant at the 5% level. This further reduction of error variance is attained at the expense of degrees of freedom, since now the estimate of σ_ε^2 is based on only 6 df instead of 9 df as in the randomized block design. This means less precision in estimating this error variance. But the added restrictions should be made if the environmental conditions suggest them.

Should Insignificant Effects Be Pooled with the Error?

After discovering that position had no significant effect, some investigators might "pool" the position sum of squares with the error sum of squares and obtain a more precise estimate of σ_ε^2, namely, 1.3, as given in Table 4.5. However, there is a danger in "pooling," since it means "accepting" a hypothesis of no position effect, and the investigator has no idea about the possible error involved in "accepting" a hypothesis. Naturally, if the degrees of freedom on the error term are reduced much below the amount of Figure 4.5, there will have to be some pooling to get a reasonable yardstick for assessing other effects.

4.6 INTERPRETATIONS

After any ANOVA one usually wishes to investigate the single factor further for a more detailed interpretation of the effect of the treatments. Here, after the Latin square ANOVA of Figure 4.5, we are probably concerned about which brand to buy. Using a Newman–Keuls approach (see Table 3.9), the results are as follows:

1. Means: 10.75   11.00   12.25   14.25
   For brands: C   D   B   A
2. MSE = 0.8958 with 6 df.
3. $s_{\bar{y}_{.j}} = \sqrt{0.8958/4} = 0.47$
4. For $p =$ 2   3   4
   Tabled ranges: 3.46   4.34   4.90
5. LSRs: 1.63   2.04   2.30
6. Testing averages:

$$\bar{A} - \bar{C} = 3.50 > 2.30; \text{ so, conclude that } \mu_A > \mu_C$$

$$\bar{A} - \bar{D} = 3.25 > 2.04; \text{ so, conclude that } \mu_A > \mu_D$$

$$\bar{A} - \bar{B} = 2.00 > 1.63; \text{ so, conclude that } \mu_A > \mu_B$$
$$\bar{B} - \bar{C} = 1.50 < 2.04; \text{ so, } \mu_B, \mu_C, \text{ and } \mu_D \text{ may be equal}$$

Thus only A differs from the other brands in tread wear. A has the largest tread wear and so is the poorest brand to buy. One could recommend brand B, C, or D, whichever is the cheapest or whichever is "best" according to some criterion other than tread wear.

4.7 ASSESSING THE MODEL

As noted in Section 3.7, plots of the data and residuals enable us to investigate the model and check that the model assumptions are reasonably satisfied. The procedures discussed in that section can be applied to randomized block and Latin square designs. These will be illustrated for the Latin square design of the tire wear study.

We begin by having a computer program generate the predicted values, residuals, and standardized residuals given in Table 4.13. A *standardized residual* is the value of the residual divided by the standard error of the residuals. The standardized residuals have mean 0 and standard deviation 1, so those having absolute values in excess of (say) 2 may be associated with an outlier. One residual was unusually large for the randomized block design, but such is not the case when position is added as another blocking factor.

A plot of the observed values versus the predicted values allows us to visually evaluate the "fit" of the model. Such a plot (Figure 4.6a) reveals a good fit. A normal quantile plot of the residuals (Figure 4.6b) supports the assumption that the errors are normally distributed. The result of a Shapiro–Wilk test (p value $= 0.2065$) adds further support to that assumption.

Using the ranges for position (4, 3, 8, and 5), car (5, 3, 3, and 5), and brand (4, 6, 3, and 4), the average range is found to be $(4 + \ldots + 4)/12 = 53/12 \approx 4.417$. There are

TABLE 4.13
Predicted Values and Residuals for Tire Wear Study: Latin Square Design

Position	Car	Brand	Wear	Predicted	Residual	Standardized Residual
1	1	3	12	11.625	0.375	0.64700
1	2	4	11	10.625	0.375	0.64700
1	3	1	13	12.875	0.125	0.21567
1	4	2	8	8.875	-0.875	-1.50966
2	1	2	14	14.625	-0.625	-1.07833
2	2	3	12	11.875	0.125	0.21567
2	3	4	11	11.125	-0.125	-0.21567
2	4	1	13	12.375	0.625	1.07833
3	1	1	17	16.625	0.375	0.64700
3	2	2	14	13.375	0.625	1.07833
3	3	3	10	10.875	-0.875	-1.50966
3	4	4	9	9.125	-0.125	-0.21567
4	1	4	13	13.125	-0.125	-0.21567
4	2	1	14	15.125	-1.125	-1.94099
4	3	2	13	12.125	0.875	1.50966
4	4	3	9	8.625	0.375	0.64700

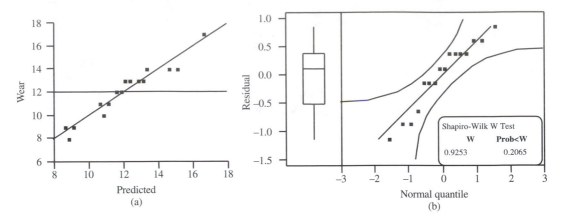

Figure 4.6   Latin square design JMP graphics for tire wear study: (a) observed versus predicted values and (b) normal quantile plot of the residuals.

four observations at each treatment level, and $D_4 = 2.282$ for samples of size 4. Since $D_4 \bar{R} = (2.282)(4.417) = 10.080$, our quick check indicates that an assumption of equal error variances is reasonable.

Some additional plots of the residuals are given in Figure 4.7. The plots provide no strong evidence against an assumption of equal error variances. Combining this information with that obtained from Figure 4.6 and observing that no unusual patterns are present in Figure 4.7, we have reason to believe that the basic assumptions of the Latin square model are reasonably satisfied.

The Adjusted Coefficient of Multiple Determination

Since a significant treatment (brand) effect was detected with the randomized complete block experiment and the position effect was not significant, we are faced with the problem of deciding which of the two models is better, based on the available data. One method used by many practitioners involves comparing the adjusted coefficients of multiple determination for both models. If N observations are made and c denotes the sum of the degrees of freedom associated with the sums of squares for treatments and blocks, the *adjusted coefficient of multiple determination* is

$$R^2_{adj} = \frac{(N-1)R^2 - c}{N - c - 1} \tag{4.5}$$

with

$$R^2 = \frac{\sum \text{treatment and block sums of squares}}{SS_{total}} \tag{4.6}$$

the *coefficient of multiple determination*. This statistic adjusts the proportion of unexplained variation upward, so $R^2_{adj} < R^2$. If there is little difference between the adjusted coefficients for the two models, the experimenter has evidence that supports using the simpler one. Such a decision should not be based solely on the magnitudes of the adjusted coefficients of

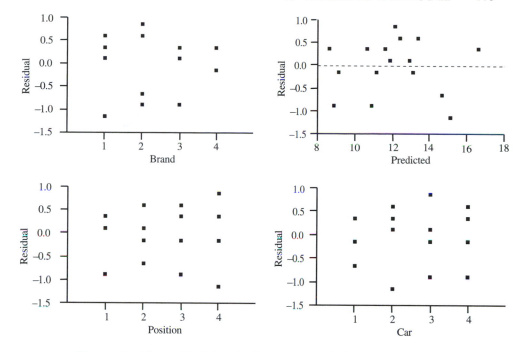

Figure 4.7   More residual plots for the tire wear study: Latin square design.

multiple determination. Results from a thorough analysis of the residuals for both models should also be considered.

■ **Example 4.2**

Consider the tire wear study and the ANOVA tables for the randomized block experiment and the Latin square experiment.

a. **Randomized block:** From Figure 4.2, $SS_{car} = 38.688$ and $SS_{brand} = 30.687$, each with 3 degrees of freedom. Also, $SS_{total} = 80.937$. Letting $c = 3 + 3 = 6$ and $N = 16$,

$$R^2 = \frac{SS_{car} + SS_{brand}}{SS_{total}} = \frac{38.688 + 30.687}{80.937} = 0.857$$

and

$$R^2_{adj} = \frac{(N-1)R^2 - c}{N - c - 1} = \frac{(16-1)(0.857) - 6}{16 - 6 - 1} = 0.762$$

b. **Latin square:** From Figure 4.5, $SS_{position} = 6.1875$, $SS_{car} = 38.6875$, and $SS_{brand} = 30.6875$, each with 3 degrees of freedom. Also, $SS_{total} = 80.9375$. Letting $c = 3 + 3 + 3 = 9$ and $N = 16$, it is left to the reader to show that $R^2 = 0.9336$ and $R^2_{adj} = 0.8340$.

> Since the adjusted coefficient of multiple determination for the Latin square design exceeds that for the randomized complete block design by 0.0720, we have reason to conclude that the Latin square design provides a better fit to the data. Residual analysis of the randomized block design reveals one extreme observation, whereas no extreme observations are present for the Latin square design. These facts seem to indicate that the model based on a Latin square design is better.

4.8 GRAECO–LATIN SQUARES

In some experiments still another restriction may be imposed on the randomization. The design may be a Graeco–Latin square such as Table 4.14 exhibits. This design is obtained by superimposing the Latin squares in Figure 4.8. Such a design might be useful for (say) a study in which the durability of four types of paint is to be assessed by means of a wear tester that can hold four specimens, where four holders are required and a holder can be placed in one of four tester positions.

In this design the third restriction is at levels α, β, γ, δ, and not only do these each appear once and only once in each row and each column, but they appear once and only once with each level of treatment A, B, C, or D. The model for this would be

$$Y_{ijkm} = \mu + \beta_i + \tau_j + \gamma_k + \omega_m + \varepsilon_{ijkm}$$

	Column					Column			
Row	**1**	**2**	**3**	**4**	**Row**	**1**	**2**	**3**	**4**
1	A	B	C	D	1	α	β	γ	δ
2	B	C	D	A	2	γ	δ	α	β
3	C	D	A	B	3	δ	γ	β	α
4	D	A	B	C	4	β	α	δ	γ
		(a)					(b)		

Figure 4.8   Two 4×4 Latin squares.

TABLE 4.14
Graeco–Latin Square Design

	Car			
Position	**I**	**II**	**III**	**IV**
1	$A\alpha$	$B\beta$	$C\gamma$	$D\delta$
2	$B\gamma$	$A\delta$	$D\alpha$	$C\beta$
3	$C\delta$	$D\gamma$	$A\beta$	$B\alpha$
4	$D\beta$	$C\alpha$	$B\delta$	$A\gamma$

where ω_m is the effect of the latest restriction with levels α, β, γ, and δ. An outline of the analysis appears in Table 4.15.

Such a design may not be very practical, since only 3 df is left for the error variance. Improvement may be possible by *replicating* the experiment. That is, one may run the experiment more than once (with a different randomization each time), adding "replicate" as another block effect.

TABLE 4.15
Graeco–Latin ANOVA Outline

Source	df
β_i	3
τ_j	3
γ_k	3
ω_m	3
ε_{ijkm}	3
Total	15

4.9 EXTENSIONS

In some situations it is not possible to place all treatments in every block of an experiment. This leads to a design known as an *incomplete block design*. Sometimes a whole row or column of a Latin square is missing and the resulting incomplete Latin square is called a *Youden square*. These two special designs are discussed in Chapter 16.

4.10 SAS PROGRAMS FOR RANDOMIZED BLOCKS AND LATIN SQUARES

The ANOVA procedure can be used for randomized complete block and Latin square designs if no values are missing from the data. The **GLM** procedure is more versatile, since it can also be used when values are missing. For illustration, **ANOVA** will be used to analyze the randomized complete block version of the tire wear study; **GLM** will be used with the Latin square version.

Randomized Complete Block Designs

Data collected for a randomized complete block design of the tire wear study, first presented in Table 4.4, are summarized in Table 4.16. A SAS command file for those data is given in Table 4.17. As you can see, the more complicated problem of two or more factors is handled quite easily by recognizing these variables in the **CLASS, MODEL,** and **MEANS** statement. Using CAR and BRAND as **CLASS** variables with WEAR the response variable, the data entry 1 A 17 indicates that the brand *A* tire on car 1 experienced 17 mils of wear. Even though the parameter μ and an error term are not included, notice the similarity of the **MODEL** statement to the mathematical model $Y_{ij} = \mu + \beta_i + \tau_j + \varepsilon_{ij}$.

The analysis is a two-way ANOVA with blocking on cars. Although we are primarily interested in brand differences, the added variable removes the car-to-car variability from the error term. This makes the test more sensitive if the reduction in the error sum of squares is sufficiently large to compensate for the loss of 3 degrees of freedom. From the output in Figure 4.9 we see that our reward for blocking is the detection of a significant brand effect.

Addition of the SNK analysis in the command file produced the analysis on pairs of means given in Figure 4.9. Since brand A shares no letter with the other three brands, we have reason to believe that the average wear is greatest for brand A and the other three averages may be equal.

Latin Square Designs

Data collected for a Latin square design of the tire wear study, first presented in Table 4.12, are summarized in Table 4.18. A SAS command file for those data is given in Table 4.19. Only slight modifications to the statements in Table 4.17 are needed to obtain this command file. The **INPUT**, **CLASS**, and **MODEL** statements contain the name of one

TABLE 4.16
Randomized Block Design Data for Tire Brand Test

Car	Brand			
	A	*B*	*C*	*D*
1	17	14	12	13
2	14	14	12	11
3	13	13	10	11
4	13	8	9	9

TABLE 4.17
SAS Command File for Tire Wear Study: Randomized Blocks

```
OPTIONS LINESIZE=80;                      restricts width of output
DATA RAW;                                    names the data set
INPUT CAR BRAND $ WEAR @@;          names the variables in order
CARDS;                                    signals that data follow
1 A 17 1 B 14 1 C 12 1 D 13                       data begin
2 A 14 2 B 14 2 C 12 2 D 11
3 A 13 3 B 13 3 C 10 3 D 11
4 A 13 4 B  8 4 C  9 4 C  9
;                                         signals the end of data
PROC ANOVA;                              identifies the procedure
CLASS CAR BRAND;                identifies the treatment variables
MODEL WEAR=CAR BRAND;                     identifies the model
MEANS BRAND/SNK;                requests a Student–Newman–Keuls
                                  analysis on pairs of brand means
```

```
                        ANALYSIS OF VARIANCE PROCEDURE
DEPENDENT VARIABLE: WEAR
                                   SUM OF           MEAN
SOURCE                    DF      SQUARES         SQUARE   F VALUE        PR > F
MODEL                      6  69.37500000   11.56250000      9.00        0.0022
ERROR                      9  11.56250000    1.28472222
CORRECTED TOTAL           15  80.93750000

                R-SQUARE          C.V.    ROOT MSE              WEAR MEAN
                0.857143        9.3965  1.13345588            12.06250000

SOURCE                    DF     ANOVA SS     F VALUE    PR > F
CAR                        3  38.68750000       10.04    0.0031
BRAND                      3  30.68750000        7.96    0.0067

                        ANALYSIS OF VARIANCE PROCEDURE

STUDENT-NEWMAN-KEULS TEST FOR VARIABLE: WEAR
NOTE: THIS TEST CONTROLS THE TYPE I EXPERIMENTWISE ERROR RATE
      UNDER THE COMPLETE NULL HYPOTHESIS BUT NOT UNDER PARTIAL
      NULL HYPOTHESES

                    ALPHA=0.05   DF=9   MSE=1.28472

              NUMBER OF MEANS         2        3        4
              CRITICAL RANGE    1.81314  2.23767  2.50205

      MEANS WITH THE SAME LETTER ARE NOT SIGNIFICANTLY DIFFERENT

                    SNK GROUPING        MEAN    N   BRAND
                              A      14.2500    4   A

                              B      12.2500    4   B
                              B
                              B      11.0000    4   D
                              B
                              B      10.7500    4   C
```

Figure 4.9   SAS output for the tire wear study: randomized complete block design.

more variable, position. Using POSITION, CAR, and BRAND as **CLASS** variables with WEAR the response variable, the data entry 1 1 C 12 indicates that the brand C tire in position 1 of car 1 experienced 12 mils of wear. Letting γ, β, and τ denote the position, car, and brand effects, respectively, notice the similarity of the **MODEL** statement to the mathematical model $Y_{ij} = \mu + \gamma_i + \beta_j + \tau_k + \varepsilon_{ijk}$.

TABLE 4.18
Latin Square Design Data for Tire Brand Test

Position	Car							
	1		**2**		**3**		**4**	
1	C	12	D	11	A	13	B	8
2	B	14	C	12	D	11	A	13
3	A	17	B	14	C	10	D	9
4	D	13	A	14	B	13	C	9

TABLE 4.19
SAS Command File for Tire Wear Study: Latin Square

```
OPTIONS LINESIZE=80;                          restricts width of output
DATA RAW;                                      names the data set
INPUT CAR BRAND POSITION $ WEAR @@;            names the variables in order
CARDS;                                         signals that data follow
1 A 3 17 1 B 2 14 1 C 1 12                     data begin
1 D 4 13 2 A 4 14 2 B 3 14
2 C 2 12 2 D 1 11 3 A 1 13
3 B 4 13 3 C 3 10 3 D 2 11
4 A 2 13 4 B 1  8 4 C 4  9
4 D 3  9
;                                              signals the end of data
PROC GLM;                                      identifies the procedure
CLASS CAR BRAND POSITION;                      identifies the treatment variables
MODEL WEAR=CAR POSITION BRAND/P;               identifies the model
MEANS BRAND/SNK;                               requests a Student–Newman–Keuls
                                               analysis on pairs of brand means
```

GENERAL LINEAR MODELS PROCEDURE

DEPENDENT VARIABLE: WEAR

SOURCE	DF	SUM OF SQUARES	MEAN SQUARE	F VALUE	PR > F
MODEL	9	75.56250000	8.39583333	9.37	0.0066
ERROR	6	5.37500000	0.89583333		
CORRECTED TOTAL	15	80.93750000			

	R-SQUARE	C.V.	ROOT MSE		WEAR MEAN
	0.933591	7.8465	0.94648472		12.06250000

SOURCE	DF	TYPE I SS	MEAN SQUARE	F VALUE	PR > F
CAR	3	38.68750000	12.89583333	14.40	0.0038
POSITION	3	6.18750000	2.06250000	2.30	0.1769
BRAND	3	30.68750000	10.22916667	11.42	0.0068

SOURCE	DF	TYPE III SS	MEAN SQUARE	F VALUE	PR > F
CAR	3	38.68750000	12.89583333	14.40	0.0038
POSITION	3	6.18750000	2.06250000	2.30	0.1769
BRAND	3	30.68750000	10.22916667	11.42	0.0068

OBSERVATION	OBSERVED VALUE	PREDICTED VALUE	RESIDUAL
1	17.00000000	16.62500000	0.37500000
2	14.00000000	14.62500000	−0.62500000
3	12.00000000	11.62500000	0.37500000
4	13.00000000	13.12500000	−0.12500000
5	14.00000000	15.12500000	−1.12500000
6	14.00000000	13.37500000	0.62500000
7	12.00000000	11.87500000	0.12500000
8	11.00000000	10.62500000	0.37500000
9	13.00000000	12.87500000	0.12500000
10	13.00000000	12.12500000	0.87500000
11	10.00000000	10.87500000	−0.87500000
12	11.00000000	11.12500000	−0.12500000
13	13.00000000	12.37500000	0.62500000
14	8.00000000	8.87500000	−0.87500000
15	9.00000000	8.62500000	0.37500000
16	9.00000000	9.12500000	−0.12500000

SUM OF RESIDUALS   0.00000000

```
SUM OF SQUARED RESIDUALS                    5.37500000
SUM OF SQUARED RESIDUALS - ERROR SS    −0.00000000
FIRST ORDER CORRELATION                    −0.45639535
DURBIN-WATSON D                              2.88372093
```
STUDENT-NEWMAN-KEULS TEST FOR VARIANCE:WEAR
NOTE: THIS TEST CONTROLS THE TYPE I EXPERIMENTWISE ERROR RATE UNDER THE
      COMPLETE NULL HYPOTHESIS BUT NOT UNDER PARTIAL NULL HYPOTHESES

            ALPHA=0.05   DF=6   MSE=0.895833

        NUMBER OF MEANS   2   3   4
        CRITICAL RANGE   1.63761 2.05357 2.31683
    MEANS WITH THE SAME LETTER ARE NOT SIGNIFICANTLY DIFFERENT
            SNK GROUPING   MEAN   N   BRAND
                        A   14.2500   4   A

                        B   12.2500   4   B
                        B
                        B   11.0000   4   D
                        B
                        B   10.7500   4   C
```

**Figure 4.10**    SAS output for the tire wear study: Latin square design.

The outputs from our command file are given in Figure 4.10. As before, a significant brand effect has been detected.

## 4.11  SUMMARY

| Experiment | Design | Analysis |
|---|---|---|
| I. Single factor | 1. Completely randomized $Y_{ij} = \mu + \tau_j + \varepsilon_{ij}$ | 1. One-way ANOVA |
| | 2. Complete randomized block $Y_{ij} = \mu + \tau_j + \beta_i + \varepsilon_{ij}$ | 2. Two-way ANOVA |
| | 3. Complete Latin square $Y_{ijk} = \mu + \beta_i + \tau_j + \gamma_k + \varepsilon_{ijk}$ | 3. Three-way ANOVA |
| | 4. Graeco–Latin square $Y_{ijkm} = \mu + \beta_i + \tau_j$ $+\gamma_k + \omega_m + \varepsilon_{ijkm}$ | 4. Four-way ANOVA |

It is emphasized once again that our interest in Chapters 3 and 4 has been centered on a single factor (treatments). The special designs simply represent restrictions on the randomization. In the next chapter, two or more factors are considered.

## 4.12  FURTHER READING

Anderson, V. L. and R. A. McClean, *Design of Experiments: A Realistic Approach*. New York: Marcel Dekker, 1974. See Chapters 5 and 8 for discussions of randomized complete block and Latin square designs.

Box, G. E. P., W. G. Hunter, and J. S. Hunter, *Statistics for Experimenters: An Introduction to Design, Data Analysis, and Model Building*. New York: John Wiley & Sons, 1978. See Chapters 4, 7, and
```

8 for discussions of randomization, blocking, randomized complete blocks, Latin squares, and Graeco–Latin squares.

Gunter, B., "Improve Experimental Sensitivity Through Blocking," *Quality Progress,* August 1996, pp. 124–129.

Mason, R. L., R. F. Gunst, and J. L. Hess, *Statistical Design and Analysis of Experiments with Applications to Engineering and Science.* New York: John Wiley & Sons, 1989. See Chapter 8 for a discussion of blocking, randomized complete block designs, and Latin square designs.

Ostle, B., K. V. Turner Jr., C. R. Hicks, and G. W. McElrath, *Engineering Statistics: The Industrial Experience.* Belmont, CA: Duxbury Press, 1996. See pages 456 through 464 for a discussion of randomized complete block experiments.

PROBLEMS

4.1 The effects of four types of graphite coater on light-box readings are to be studied. Since the readings might differ from day to day, observations are to be taken on each of the four types every day for three days. The order of testing of the four types on any given day can be randomized. The results are as follows:

| Day | **Graphite Coater Type** | | | |
	M	*A*	*K*	*L*
1	4.0	4.8	5.0	4.6
2	4.8	5.0	5.2	4.6
3	4.0	4.8	5.6	5.0

a. Analyze these data as a randomized block design and state your conclusions.
b. Using residual plots and other procedures as appropriate, assess the model.

4.2 Set up one set of orthogonal contrasts among coater types and analyze for Problem 4.1.

4.3 Use the Newman–Keuls range test to compare four coater-type means for Problem 4.1.

4.4 If the reading on type *K* in Problem 4.1 for the second day was missing, what missing value should be inserted and what is the analysis now?

4.5 Set up at least four contrasts in Problem 4.1 and use the Scheffé method to test the significance of these.

4.6 Researchers at Purdue University studied five electrode shapes *A, B, C, D,* and *E,* to determine metal removal rates. The removal was accomplished by an electric discharge between the electrode and the material being cut. For this experiment five holes were cut in five workpieces, and the order of electrodes was arranged so that only one electrode shape was used in the same position on each of the five workpieces. Thus, the design was a Latin square design, with workpieces (strips) and positions on the strip as restrictions on the randomization. Several variables were studied, one of which was the Rockwell hardness of metal where each hole was to be cut. The results were as follows:

Strip	Position				
	1	2	3	4	5
I	A(64)	B(61)	C(62)	D(62)	E(62)
II	B(62)	C(62)	D(63)	E(62)	A(63)
III	C(61)	D(62)	E(63)	A(63)	B(63)
IV	D(63)	E(64)	A(63)	B(63)	C(63)
V	E(62)	A(61)	B(63)	C(63)	D(62)

a. Analyze these data and test for an electrode effect, position effect, and strip effect on Rockwell hardness.
b. Using residual plots and other techniques as appropriate, assess the model.

4.7 The times in hours necessary to cut the holes in Problem 4.6 were recorded as follows:

Strip	Position				
	1	2	3	4	5
I	A(3.5)	B(2.1)	C(2.5)	D(3.5)	E(2.4)
II	E(2.6)	A(3.3)	B(2.1)	C(2.5)	D(2.7)
III	D(2.9)	E(2.6)	A(3.5)	B(2.7)	C(2.9)
IV	C(2.5)	D(2.9)	E(3.0)	A(3.3)	B(2.3)
V	B(2.1)	C(2.3)	D(3.7)	E(3.2)	A(3.5)

Analyze these data for the effect of electrodes, strips, and positions on time.

4.8 Analyze the electrode effect further in Problem 4.7 and make some statement as to which electrodes are best if the shortest cutting time is the most desirable factor.

4.9 For an $m \times m$ Latin square, prove that a missing value may be estimated by

$$Y_{ijk} = \frac{m(T'_{i..} + T'_{.j.} + T'_{..k}) - 2T'_{...}}{(m-1)(m-2)}$$

where the primes indicate totals for row, treatment, and column containing the missing value and $T'_{...}$ is the grand total of all actual observations.

4.10 A composite measure of screen quality was made on screens using four lacquer concentrations, four standing times, four acryloid concentrations (*A, B, C, D*), and four acetone concentrations (α, β, Γ, Δ). A Graeco–Latin square design was used with data recorded as follows:

	Lacquer Concentration			
Standing Time	**0.5**	**1**	**1.5**	**2**
30	$C\beta(16)$	$B\Gamma(12)$	$D\Delta(17)$	$A\alpha(11)$
20	$B\alpha(15)$	$C\Delta(14)$	$A\Gamma(15)$	$D\beta(14)$
10	$A\Delta(12)$	$D\alpha(6)$	$B\beta(14)$	$C\Gamma(13)$
5	$D\Gamma(9)$	$A\beta(9)$	$C\alpha(8)$	$B\Delta(9)$

Do a complete analysis of these data.

4.11 Explain why a three-level Graeco–Latin square is not a feasible design.

4.12 In a chemical plant, five experimental treatments are to be used on a basic raw material in an effort to increase the chemical yield. Since batches of raw material may differ, five batches are chosen at random. The order of the experiments may affect yield as well, as may the operator who performs the experiment. With order and operators imposing further restrictions on randomization, set up for this experiment a suitable design that will be as inexpensive as possible to run. Outline its analysis.

4.13 Three groups of students are to be tested for the percentage of high-level questions asked by each group. As questions can be on various types of material, six lessons are taught to each group and a record is made of the percentage of high-level questions asked by each group on all six lessons. Show a data layout for this situation and outline its ANOVA table.

4.14 Data from the results of Problem 4.13 were as follows:

	Group		
Lesson	*A*	*B*	*C*
1	13	18	7
2	16	25	17
3	28	24	14
4	26	13	15
5	27	16	12
6	23	19	9

Do an analysis of these data and state your conclusions.

4.15 For Problem 4.14 analyze further in an attempt to determine which group asks the highest percentage of high-level questions.

4.16 Thirty students who were pretested for science achievement were also posttested using the same test after 15 weeks of special instruction. A t test on the difference scores of those students was found to be 2.3. Consider this problem as a randomized block design and show a data layout and outline its ANOVA table with the proper degrees of freedom. Also indicate what F value would have to be exceeded to permit one to conclude that the training made a difference at the 5% significance level.

4.17 A student decides to investigate the accuracy of five scales in various local commercial places. She decides to account for the variations in her weight at different times of day and also on different days. She will collect data only on Saturdays, which she has free. The design will be a Latin square, so she chooses the five times of day as 9:00 A.M., 11:00 A.M., 1:00 P.M., 3:00 P.M., and 5:00 P.M. One possible Latin square design is as follows:

	Time				
Day	**9**	**11**	**1**	**3**	**5**
1	A	B	C	D	E
2	B	C	D	E	A
3	C	D	E	A	B
4	D	E	A	B	C
5	E	A	B	C	D

a. Give the model and write out the ANOVA table.
b. Suppose the student does a Scheffé test on the times of day and finds that the results at 3 P.M. and 5 P.M. are not significantly different. If she is going to repeat the experiment on some other scales, should she do a simpler randomized block design with five days as blocks and make observations at 3 P.M. and 5 P.M.? Give reasons for your decision.

4.18 Workers at an agricultural experiment station conducted an experiment to compare the effect of four different fertilizers on the yield of a cane crop. They divided four blocks of soil into four plots of equal size and shape and assigned the fertilizers to the plots at random such that each fertilizer was applied once in each block. Data were collected on yield in hundredweights per plot. The analysis gave a total sum of squares of 540 with fertilizer sum of squares at 210 and block sum of squares at 258. Set up an ANOVA table for these results.

4.19 From the data of Problem 4.18 state the hypothesis to be tested in terms of the mathematical model, run the test, and state your conclusions.

4.20 In Problem 4.18 if the observed mean yield of cane for fertilizer B was 41.25 hundredweight, what is its standard error?

4.21 In Problem 4.18 the original data show that all readings are divisible by 5. Assuming that the data are coded by subtracting 40 and dividing by 5, set up an expression for 95% confidence limits on the mean of fertilizer B in terms of the coded data.

4.22 Write a contrast to be used in comparing the mean of fertilizers A and B with the mean of fertilizer D and also show its standard error for Problem 4.18.

4.23 Set up three contrasts in which one contrast is orthogonal to the other two but these two are *not* orthogonal to each other.

4.24 a. What is the main advantage of a randomized block design over a completely randomized design?
   b. Under what conditions would the completely randomized design have been better?

4.25 Material is analyzed for weight in grams from three vendors—A, B, C—by three different inspectors—I, II, III—and using three different scales—1, 2, 3. The experiment is set up on a Latin square plan with results as follows:

Inspector	Scale		
	1	**2**	**3**
I	$A = 16$	$B = 10$	$C = 11$
II	$B = 15$	$C = 9$	$A = 14$
III	$C = 13$	$A = 11$	$B = 13$

Analyze the experiment to test for a difference in weight between vendors, between inspectors, and between scales.

4.26 A school district was concerned about its history curriculum. The board members found two new curricula that proponents said would improve students' achievement in history. To check these claims, three high schools (I, II, III) were chosen at random in the district and curricula A, B, and C (the present one) were used in each school. It was also decided to try these three curricula at three different grade levels (9, 10, 11). An experiment was designed in which each curriculum was tried once and only once in each school and once and only once at each grade level. The criterion used to measure achievement was the change in test scores from the beginning to the end of a semester. Devise a data layout for this experiment and outline its ANOVA.

4.27 The changes in test scores in Problem 4.26 are as follows:

Schools	Curricula	Grade	Score
I	A	9	28
I	C	10	31
I	B	11	30

II	B	9	31
II	A	10	30
II	C	11	32
III	C	9	30
III	B	10	30
III	A	11	29

Do a complete analysis of these data and state your recommendations.

4.28 To check on a possible laboratory bias in the reported ash content of coal, 10 samples of coal were split in half and then sent at random to each of two laboratories (L_1 and L_2). The laboratories reported the following ash content data:

Sample	Lab 1	Lab 2
1	5.47	5.13
2	5.31	5.46
3	5.46	5.54
4	5.55	5.54
5	5.93	6.00
6	5.97	5.99
7	6.32	6.43
8	6.02	6.13
9	5.87	5.87
10	5.58	5.60

Try two different methods to analyze these data and show that the results agree. What is your conclusion?

4.29 To study aluminum thickness, samples were taken from five machines. This cycle of sampling and testing was repeated three times. Results were as follows.

Replication	1	2	3	4	5
			Machine		
I	175	95	180	170	155
II	190	185	180	200	190
III	185	165	175	195	200

Analyze these data and state your conclusions.

4.30 Consider a randomized complete block experiment with one missing value y_{ij}.
a. Following the procedure illustrated in Section 4.4, show that

$$y_{ij} = \frac{nT'_{i.} + kT'_{.j} - T'_{..}}{(n-1)(k-1)}$$

where the primed totals $T'_{i.}$, $T'_{.j}$, and $T'_{..}$ use the totals indicated without the missing value.

b. Use the result in part a and find y_{33}, which is missing in Table 4.8.

4.31 A researcher wants to determine which of five types of plastic material has the least shrink when molded. Five molds, each with five cavities, are to be used with the plastic types under consideration. Only one part will be prepared for each treatment combination.

a. Prepare a data layout for this experiment, assuming that a Latin square design is planned.

b. Write a model equation for this experiment.

c. Using the method illustrated in Section 4.5, construct a 5×5 Latin square for this experiment.

4.32 A high-voltage beam is used to transfer metal from a source to the surface of a semiconductor wafer. In a study of the uniformity of the transferred metal, the metal thicknesses of nine wafers were measured at five fixed wafer positions. The resulting coded measurements follow.

Wafer	Position				
	1	2	3	4	5
1	6.17	6.17	5.91	6.08	5.91
2	6.08	6.00	6.00	6.08	5.83
3	6.17	6.17	6.08	6.17	6.00
4	6.08	6.08	6.00	6.00	5.83
5	6.17	6.17	6.08	6.25	6.00
6	6.08	6.00	5.91	5.91	5.91
7	6.08	6.08	5.91	6.00	5.91
8	6.00	6.00	5.83	6.00	5.83
9	6.17	6.08	5.91	6.08	5.91

a. State an appropriate model and all assumptions.

b. Using an appropriate ANOVA procedure, test for differences in position means. Include p values and comment.

c. If appropriate, conduct a Student–Newman–Keuls analysis on the position means. Interpret.

d. Using plots of residuals and other procedures as appropriate, assess the model.

Chapter 5

 FACTORIAL EXPERIMENTS

5.1 INTRODUCTION

In the preceding three chapters all the experiments involved a single factor and its effect on a measured variable. Several different designs were considered, but they all represented restrictions on the randomization in which interest was still centered on the effect of a single factor.

A One-Factor-at-a-Time Experiment

Suppose there are now two factors of interest to the experimenter: for example, the effect of both temperature and altitude on the current flow in an integrated circuit. One traditional method is to hold altitude constant and vary the temperature and then hold temperature constant and change the altitude, or, in general, hold all factors constant except one and take current flow readings for several levels of this one factor, then choose another factor to vary, holding all others constant, and so forth. To examine this type of experimentation, consider a very simple example in which temperature is to be set at 25 and 55°C only and altitudes of 0 K (K = 10, 000 feet) and 3 K (30,000 feet) are to be used. If one factor is to be varied at a time, the altitude may be set for sea level or 0 K and the temperature varied from 25°C to 55°C. Suppose the current flow changed from 210 mA to 240 mA with this temperature increase. Now there is no way to assess whether this 30 mA increase is real or due to chance. Unless there is available some previous estimate of the error variability, the experiment must be repeated to obtain an estimate of the error or chance variability within the experiment. If the experiment is now repeated and the readings are 205 and 230 mA as the temperature is varied from 25°C to 55°C, it seems obvious, without any formal statistical analysis, that there is a real increase in current flow, since for each repetition the increase is large compared with the variation in current flow within a given temperature. Graphically these results appear as in Figure 5.1.

Four experiments have now been run to determine the effect of temperature at 0 K altitude only. To check the effect of altitude, the temperature can be held at 25°C and the altitude varied to 3 K by adjustment of pressure in the laboratory. Assuming that the results obtained at 0 K are representative, we combine them with two observations of current flow, taken at 3 K, to produce the results shown in Figure 5.2.

From these experiments the temperature increase is seen to increase the current flow an average of

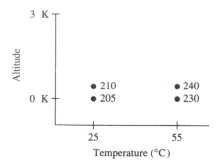

Figure 5.1   Temperature effect on current flow.

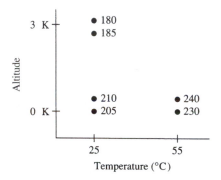

Figure 5.2   Temperature and altitude effect on current flow.

$$\frac{(240 - 210) + (230 - 205)}{2} = \frac{30 \text{ mA} + 25 \text{ mA}}{2} = 27.5 \text{ A}$$

and the increase in altitude decreases the current flow on the average of

$$\frac{|180 - 201| + |185 - 205|}{2} = \frac{30 \text{ mA} + 20 \text{ mA}}{2} = 25 \text{ mA}$$

This information is gained after six experiments have been performed, and no information is available on what would happen at a temperature of 55°C and an altitude of 3 K.

A Factorial Arrangement

An alternative experimental arrangement would be a factorial arrangement in which each temperature level is combined with each altitude and only four experiments are run. Results of four such experiments might appear as in Figure 5.3.

   With this experiment, one estimate of temperature effect on current flow is 240 − 210 mA = 30 mA at 0 K, and another estimate is 200 − 180 mA = 20 mA at 3 K. Hence two estimates can be made of temperature effect [average (30 + 20) / 2 = 25 mA] using all four observations, without necessarily repeating any observation at the same point. Using the

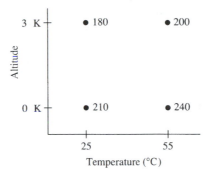

Figure 5.3   Factorial arrangement of temperature and altitude effect on current flow.

same four observations, two estimates of altitude effect can be determined: $180-210 = -30$ mA at 25°C, and $200 - 240 = -40$ mA at 55°C, an average decrease of 35 mA for a 3 K increase in altitude. Here, with just four observations instead of six, valid comparisons have been made on both temperature and altitude, and in addition some information has been obtained as to what happens at 55°C and altitude 3 K.

Interaction Between Factors

In the preceding example, if the current flow were 160 mA for 55°C and 3 K, the results could be plotted as in Figure 5.4. From Figure 5.4a note that as the temperature is increased from 25°C to 55°C at 0 K, the current flow increases by 30 mA, but at 3 K for the same temperature increase the current flow *decreases* by 20 mA. When a change in one factor produces a different change in the response variable at two levels of another factor, there is an *interaction* between the two factors. This is also observable in Figure 5.4b, where the two altitude lines are not parallel.

If the data of Figure 5.3 are plotted, we get the relations pictured in Figure 5.5. It is seen that the lines are much more nearly parallel, and no interaction is said to be present. An increase in temperature at 0 K produces about the same increase in current flow (30 mA) as at 3 K (20 mA). Now a word of warning is necessary. Lines can be made to look nearly parallel or quite diverse depending on the scale chosen; therefore, it is necessary to run statistical tests to determine whether the interaction is statistically significant. It is possible to test the significance of such interaction only if more than one observation is taken for each experimental condition. The graphic procedure shown here merely provides some insight into what interaction is and how it might be displayed for explaining the factorial experiment.

Advantages of Factorials

From this simple example, some of the advantages of a factorial experiment can be seen:

1. More efficiency is possible than with one-factor-at-a-time experiments (here four-sixths or two-thirds the amount of experimentation).

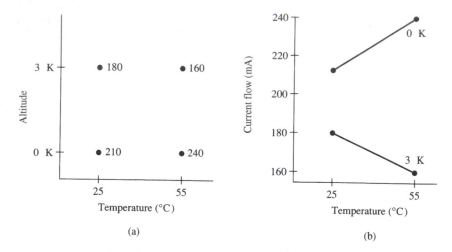

Figure 5.4   Interaction in a factorial experiment.

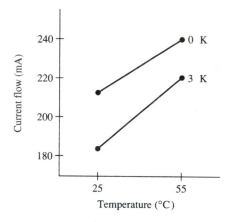

Figure 5.5   No-interaction temperature diagram in a temperature–altitude study.

2. All data are used in computing both effects. (Note that all four observations are used in determining the average effect of temperature and the average effect of altitude.)

3. Some information is gleaned on possible interaction between the two factors. (In the example, the increase in current flow of 20 mA at 3 K was about the same order of magnitude as the 30 mA increase at 0 K. If these increases had differed considerably, interaction might be said to be present.)

These advantages are even more pronounced as the number of levels of the two factors is increased. A *factorial experiment* is one in which all levels of a given factor are combined with all levels of every other factor in the experiment. Thus, if four temperatures are considered at three altitudes, a 4×3 factorial experiment would be run requiring 12 different experimental conditions.

In addition to the advantages of a factorial experiment, it may be that when the response variable Y is a yield and one is concerned with which combination of two independent variables X_1 and X_2 will produce a maximum yield, the factorial will give a combination near the maximum whereas the one-factor-at-a-time procedure will not do so. Figure 5.6 shows two cases in which the contours indicate the yields. In Figure 5.6a, X_2 is set and X_1 is varied until the maximum yield is found; then X_1 is held there while X_2 is varied and a maximum is found near the top of the mound, where the yield is around 60 units. But in Figure 5.6b the same procedure misses the maximum by a long way, giving about a 37-unit yield.

If, now, a factorial arrangement is used, one first notes the extremes of each independent variable and may also choose one or more points between these extremes. This can be done for the diagrams in Figure 5.6, as illustrated by Figure 5.7, where the maximum is reached in both cases. In some cases the maximum may not have been reached, but the factorial will indicate the direction to be followed in the next experiment to get closer to such a maximum. This concept is presented in Chapter 16.

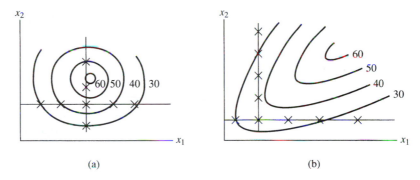

(a)   (b)

Figure 5.6   Yield contours for (a) a mound and (b) a rising ridge, created with the one-factor-at-a-time method.

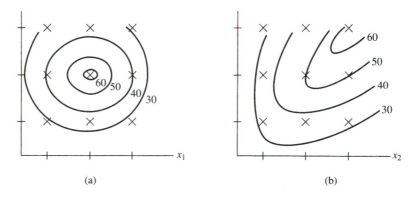

(a)   (b)

Figure 5.7   Yield contour for (a) a mound and (b) a rising ridge, created with a factorial method.

5.2 FACTORIAL EXPERIMENTS: AN EXAMPLE

Vacuum tubes are used in the best of audio equipment. To determine the effect of exhaust index (in seconds) and pump heater voltage (in volts) on the pressure inside a vacuum tube (in micrometers of mercury), three exhaust indexes and two voltages were chosen at fixed levels. It was decided to run two experiments at each of these six treatment conditions (three exhaust indexes × two voltages). The order for running the 12 experiments was completely randomized by labeling the six treatment conditions 1 through 6 and tossing a die to decide the order in which to run the experiments.

The Randomization Procedure

Table 5.1 shows the order of one such complete randomization for the 12 experiments. In such a procedure some treatment combinations may be run twice before others are run once, but this is the nature of complete randomization. In addition, numbers representing a combination that has already been run twice may come up on the die. In these cases the third time is ignored and the flipping continues until all treatment combinations have been run twice. This is, of course, a slight restriction to keep the number of observations per treatment constant. The resulting experiment may be labeled a 3 × 2 factorial experiment with two observations per treatment with all $N = 12$ observations run in a completely randomized order. Of course, this random order can be determined before the experiment is actually run, and it is important to supervise the experiment to make sure that it is actually run in the order prescribed. [**Note:** If six runs were randomized for each combination and then, separately, six more were randomized, the result would be a restricted randomization and the design could not be said to be completely randomized.]

The Mathematical Model

The mathematical model for this experiment can be written as

$$Y_{ijk} = \mu + \tau_{ij} + \varepsilon_{k(ij)} \tag{5.1}$$

TABLE 5.1
Data Layout for Vacuum Tube Pressure Experiment

Pump Heater Voltage (volts)	Exhaust Index (seconds)					
	60		90		150	
127	1		4 9	2		1 11	3		6 12
220	4		2 8	5		3 5	6		7 10

where

$i = 1, 2, 3$ for the three exhaust indexes

$j = 1, 2$ for the two voltages

$k = 1, 2$ for the two observations in each i, j treatment combination

$\tau_{ij} = $ the six treatment effects

$\varepsilon_{k(ij)} = $ the error within each of the six treatments

The notation $k(ij)$ indicates that the errors are unique to each i, j combination or are nested within each i, j.

The Data and Analysis

Following the prescribed randomization procedures, Table 5.2 shows the data that result in a between-treatments format after all results have been multiplied by 1000 to eliminate decimal points. In that table E_1, E_2, and E_3 denote exhaust indices of 60, 90, and 150 seconds, respectively; whereas V_1 and V_2 denote pump heater voltages of 127 and 220 volts, respectively. If the data of Table 5.2 were subjected to a one-way ANOVA, the results given in Table 5.3 would be obtained.

In Table 5.3 the treatment effects are really combinations of exhaust indexes and voltages, and if rearranged to show these main effects, the data might appear as in Table 5.4.

TABLE 5.2
Vacuum Tube Pressure as a One-Way Experiment

	E_1V_1	E_2V_1	E_3V_1	E_1V_2	E_2V_2	E_3V_2
	48	28	7	62	14	6
	48	33	15	54	10	9
$T_{ij.}$	106	61	22	116	24	15
$\sum_{k=1}^{2} y_{ijk}^2$	5668	1873	274	6760	296	117

$$T_{..} = 344$$

$$\sum_{i=1}^{3}\sum_{j=1}^{1}\sum_{k=1}^{2} y_{ijk}^2 = 14{,}988$$

TABLE 5.3
One-Way ANOVA for Vacuum Tube Pressure Experiment

Source	df	SS
Between treatments	5	4987.67
Within treatments	6	139.00
Totals	11	5126.67

TABLE 5.4
Two-Way Layout for Vacuum Tube Pressure Experiment

Pump Heater Voltage, V	Exhaust Index, E			
	60	90	150	$T_{.j.}$
127	48	28	7	189
	58	33	15	
220	62	14	9	155
	54	10	6	
$T_{i..}$	222	85	37	$T_{...} = 344$

If the sums of squares for the two main effects are computed from the marginal totals of Table 5.4, we find

$$SS_E = \frac{(222)^2 + (85)^2 + (37)^2}{4} - \frac{(344)^2}{12} = 4608.17$$

$$SS_V = \frac{(189)^2 + (155)^2}{6} - \frac{(344)^2}{12} = 96.33$$

These two add to 4704.50, which is less than the sum of squares between treatments of 4987.67. It is this difference between the treatment SS and the SS of the two main effects that shows the interaction between the two main effects. Here

$$SS_{E \times V \text{ interaction}} = SS_{\text{treatments}} - SS_E - SS_V$$

$$= 4987.67 - 4608.17 - 96.33 = 283.17$$

Since the treatments carried 5 degrees of freedom, the interaction has $5 - 2 - 1 = 2$ df. These results may now be displayed in the ANOVA of Table 5.5. In this final table, each main effect (E and V) and their interaction can be tested for significance by comparing each mean square with the error mean square. Exhaust index is seen to be highly significant. Voltage is not significant at the 5% significance level but the interaction is significant at the 5% level.

TABLE 5.5
Two-Way ANOVA for Vacuum Tube Pressure Experiment

Source	df	SS	MS	F	Probability
Between E's	2	4608.17	2304.08	99.5	0.001
Between V's	1	96.33	96.33	4.2	0.088
$E \times V$ interaction	2	283.17	141.58	6.1	0.036
Error (within treatments)	6	139.00	23.17		
Totals	11	5126.67			

The Two-Factor Factorial Model

The mathematical model of Equation (5.1) can now be expanded to read

$$Y_{ijk} = \mu + \overbrace{\tau_{ij}}^{} + \varepsilon_{k(ij)}$$

$$= \mu + \overbrace{E_i + V_j + EV_{ij}}^{} + \varepsilon_{k(ij)} \qquad (5.2)$$

where E_i represents the exhaust index effect in this problem, V_j the voltage effect, and EV_{ij} the interaction. Note that whereas in Chapter 4 randomization restrictions came from the error term, these effects come from the treatment term. In Equation (5.2) one tests the following hypotheses:

$$H_{0_1} : E_i = 0 \qquad \text{for all } i$$

$$H_{0_2} : V_j = 0 \qquad \text{for all } j$$

$$H_{0_3} : EV_{ij} = 0 \qquad \text{for all } i \text{ and } j$$

as both main effects are fixed effects.

Computer Outputs

ANOVA summaries such as those in Tables 5.3 and 5.5 are usually obtained using a statistical computing package. For example, the **Simple Factorial** procedure in SPSS produces the summary in Figure 5.8. The Explained, Residual, and Total rows contain the information given in the Between Treatments, Within Treatments, and Totals rows, respectively, of Table 5.3. The INDEX, VOLTAGE, INDEX*VOLTAGE, and Residual rows contain the information given in the Between E's, Between V's, $E \times V$ interaction, and Error (within treatments) rows, respectively, of Table 5.5.

For the mathematical model $Y_{ijk} = \mu + \tau_{ij} + \varepsilon_{k(ij)}$ of Equation (5.1), the total sum of squares is partitioned as depicted in Figure 5.9. The F test in the Explained row of Figure 5.8 is a test of $H_{01} : \tau_{ij} = 0$ for all i and j. If the usual assumptions of independence, equal variances, and normality are adequately satisfied, the p-value of that test (0.000) indicates that at least two of the six treatment effects differ.

For the mathematical model $Y_{ijk} = \mu + (E_i + V_j + EV_{ij}) + \varepsilon_{k(ij)}$ of Equation (5.2), the total sum of squares is partitioned as depicted in Figure 5.10. The F test in

```
              * * * ANALYSIS OF VARIANCE * * *

                      Sum of                Mean                   Sig
Source of Variation   Squares    DF        Square        F        of F

Main Effects          4704.500    3       1568.167     67.691     0.000
   INDEX              4608.167    2       2304.083     99.457     0.000
   VOLTAGE              96.333    1         96.333      4.158     0.088
INDEX*VOLTAGE          283.167    2        141.583      6.112     0.036

Explained             4987.667    5        997.533     43.059     0.000
Residual               139.000    6         23.167
Total                 5126.667   11        466.061
```

Figure 5.8   SPSS ANOVA summary for vacuum tube pressure experiment.

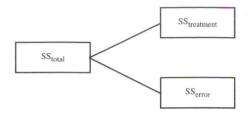

Figure 5.9 Partitioning associated with model equation (5.1).

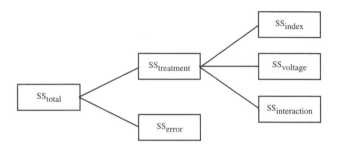

Figure 5.10 Partitioning associated with model equation (5.2).

the INDEX*VOLTAGE row of Figure 5.8 is a test of $H_{02} : EV_{ij} = 0$ for all i and j. Under the usual ANOVA assumptions, the p value of that test (0.036) indicates that H_{02} can be rejected for any $\alpha \geq 0.036$.

The test associated with the first model [equation (5.1)] and the test for an interaction effect in the second model [equation (5.2)] are appropriate regardless of the nature of the main effects (in this case, E and V). When a significant interaction is present, attention usually turns to the means of the populations associated with the treatment combinations. In such cases, the tests on the main effects should be ignored.

When the interaction effect is not significant, one proceeds to analyze the main effects. The tests on main effects as given in Figure 5.8 are appropriate when those effects are fixed. Tests on random main effects are considered in Chapter 6.

5.3 INTERPRETATIONS

Since the interaction is significant in the vacuum tube example, one should be very cautious in interpreting the main effects. A significant interaction here means that the effect of exhaust index on vacuum tube pressure at one voltage is different from its effect at the other voltage. This can be seen graphically by plotting the six treatment means as shown in Figure 5.11. Note that the lines in the figure are not parallel.

If one wishes to optimize the response variable, a reasonable procedure for interpreting such an interaction is to run a Newman–Keuls test on the six means. Had the main effects not interacted, one could treat each set of main effect means separately. Here testing main effects is not recommended because the results depend on how these main effects combine.

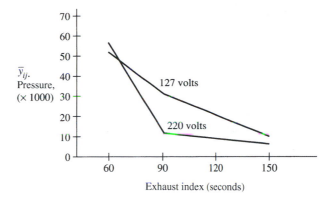

Figure 5.11   Plot of interaction for vacuum tube current flow example.

When the totals in Table 5.2 are used to determine the six means, Newman–Keuls gives the following values:

1. Means are   7.5   11   12   30.5   53   58
   for treatments: $E_3 V_2$ $E_3 V_1$ $E_2 V_2$ $E_2 V_1$ $E_1 V_1$ $E_1 V_2$
2. $s_\varepsilon^2 = 23.17$ with 6 df from Table 5.5.
3. $s_{\bar{y}_{ij.}} = \sqrt{23.17/2} = 3.40$.
4. For p: 2   3   4   5   6
   5% ranges: 3.46   4.34   4.90   5.31   5.63
5. LSR: 11.76,   14.76,   16.66,   18.05,   19.14
6. Checking means:

Comparison	Conclusion
$58 - 7.5 = 50.5 > 19.14$	$\mu_{12} > \mu_{32}$
$58 - 11 = 47 > 18.05$	$\mu_{12} > \mu_{31}$
$58 - 12 = 46 > 16.66$	$\mu_{12} > \mu_{22}$
$58 - 30.5 = 27.5 > 14.76$	$\mu_{12} > \mu_{21}$
$58 - 53 = 5 < 11.76$	μ_{12} and μ_{11} may be equal
$53 - 7.5 = 45.5 > 18.05$	$\mu_{11} > \mu_{32}$
$53 - 11 = 42 > 16.66$	$\mu_{11} > \mu_{31}$
$53 - 12 = 41 > 14.76$	$\mu_{11} > \mu_{22}$
$53 - 30.5 = 22.5 > 11.76$	μ_{11} and μ_{21} may be equal
$30.5 - 7.5 = 23 > 16.66$	$\mu_{21} > \mu_{32}$
$30.5 - 11 = 19.5 > 14.76$	$\mu_{21} > \mu_{31}$
$30.5 - 12 = 18.5 > 11.76$	$\mu_{21} > \mu_{22}$
$12 - 7.5 = 4.5 < 14.76$	$\mu_{22}, \mu_{31},$ and μ_{32} may be equal

Hence there are three groups of means:

$$\underline{7.5, \ 11, \ 12} \qquad \underline{30.5} \qquad \underline{53, \ 58}$$

If one is looking for the lowest pressure, any one of the three combinations in the first group will minimize the pressure in the vacuum tube. Thus one can recommend a 150-second exhaust index at either voltage or a 90-second exhaust index at 220 volts. From a practical point of view this really gives two choices: Pump at 127 volts for 150 seconds or at 220 volts for 90 seconds, whichever is cheaper.

5.4 THE MODEL AND ITS ASSESSMENT

The model for a completely randomized two-factor factorial experiment with n observations per treatment combination is of the form

$$Y_{ijk} = \mu + \tau_{ij} + \varepsilon_{k(ij)}$$

with

$$i = 1, 2, \ldots, a \qquad j = 1, 2, \ldots, b \qquad k = 1, 2, \ldots, n$$

where

$$\tau_{ij} = A_i + B_j + AB_{ij}$$

and

$$\varepsilon_{k(ij)} \sim \text{NID}(0, \sigma_\varepsilon^2)$$

In this case, A_i denotes the effect of the ith level of factor A, B_j denotes the effect of the jth level of factor B, AB_{ij} denotes the interaction effect of the ith level of A and the jth level of B, and $\varepsilon_{k(ij)}$ denotes the deviation of the kth value at the ijth treatment level from the population average at that level.

Again, consider the vacuum tube pressure experiment. The F tests summarized in Table 5.5 are based on the assumptions of this model. Such tests are valid only when the model assumptions are adequately satisfied. Thus, it seems prudent to use graphical and/or formal methods to investigate those assumptions.

A plot of the observed pressure versus that predicted by the model is given in Figure 5.12. The points fall near the line Pressure = Predicted, indicating a good fit. Further evidence of a good fit is provided by the large value of R^2 : 0.972887.

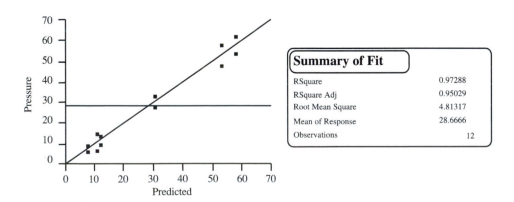

Figure 5.12   Using JMP to assess the fit for the vacuum tube pressure model.

The box plot, normal quantile plot, and summary for a Shapiro–Wilk test on the set of 12 residuals given in Figure 5.13 can be used to assess the normality assumption. These graphic aids show no strong evidence against that assumption.

Figure 5.14 gives a plot of the residuals versus treatment combinations. For such small sample sizes, there is no strong evidence for unequal variances. Further, in the data of Table 5.4 the ranges of the six treatment combinations are 10, 5, 8, 8, 4, and 3, which average

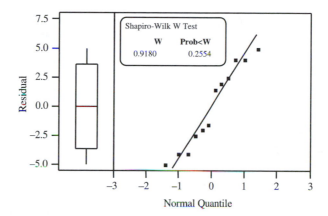

Figure 5.13   JMP box plot and normal quantile plot for the vacuum tube pressure residuals.

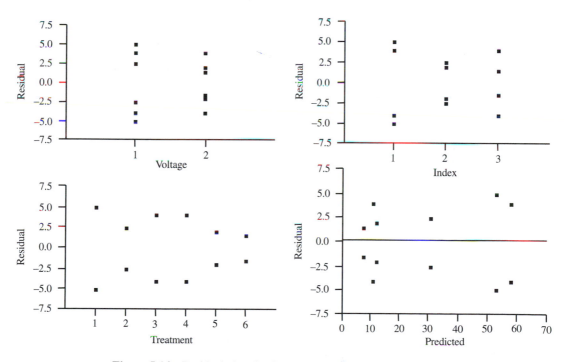

Figure 5.14   Residual plots for the vacuum tube pressure experiment.

6.333. From Table 3.14, $D_4 = 3.267$ for samples of size 2, so $D_4 \bar{R} = 20.680$ and all six ranges are well within this maximum range. These two observations support the assumption of equal variances.

It may be of interest to note that the error sum of squares is equal to the sum of the squares of the ranges divided by 2 when there are but two observations per treatment. Here $\sum R^2 = 278$ and $\sum R^2/2 = 139$, which equals the error sum of squares of Table 5.5.

The box plot in Figure 5.13 indicates no outliers among the residuals. Further, the standardized residuals (not included here) range from -1.1753 to 1.1753, indicating no unusually large residuals.

Table 5.1 indicated an experimental order for data collection. When the data are entered in that order, JMP returns an observed value of DW $= 2.232$ for the Durbin–Watson statistic, supporting the independence assumption. Recall that this statistic is valid only for time series data.

5.5 ANOVA RATIONALE

For a two-factor factorial experiment with n observations per cell, run as a completely randomized design, a general model would be

$$Y_{ijk} = \mu + A_i + B_j + AB_{ij} + \varepsilon_{k(ij)} \qquad (5.3)$$

where A and B represent the two factors, $i = 1, 2, \ldots, a$ levels of factor A, $j = 1, 2, \ldots, b$ levels of factor B, and $k = 1, 2, \ldots, n$ observations per cell. In terms of population means this becomes

$$Y_{ijk} - \mu \equiv (\mu_i - \mu) + (\mu_j - \mu) + (\mu_{ij} - \mu_i - \mu_j + \mu) + (Y_{ijk} - \mu_{ij}) \qquad (5.4)$$

where μ_{ij} represents the true mean of the i, j cell or treatment combination. Justification for the interaction term in the model comes from subtracting A and B main effects from the cell effect as follows:

$$(\mu_{ij} - \mu) - (\mu_i - \mu) - (\mu_j - \mu) = \mu_{ij} - \mu_i - \mu_j + \mu$$

If each mean is now replaced by its sample estimate, the resulting sample model is

$$Y_{ijk} - \bar{Y}_{...} = (\bar{Y}_{i..} - \bar{Y}_{...}) + (\bar{Y}_{.j.} - \bar{Y}_{...}) + (\bar{Y}_{ij.} - \bar{Y}_{i..} - \bar{Y}_{.j.} + \bar{Y}_{...}) + (Y_{ijk} - \bar{Y}_{ij.})$$

If this expression is now squared and summed over, i, j, and k, all cross products vanish, and we have

$$\sum_i^a \sum_j^b \sum_k^n (Y_{ijk} - \bar{Y}_{...})^2 = \sum_i^a \sum_j^b \sum_k^n (\bar{Y}_{i..} - \bar{Y}_{...})^2 + \sum_i^a \sum_j^b \sum_k^n (\bar{Y}_{.j.} - \bar{Y}_{...})^2$$

$$+ \sum_i^a \sum_j^b \sum_k^n (\bar{Y}_{ij.} - \bar{Y}_{i..} - \bar{Y}_{.j.} + \bar{Y}_{...})^2$$

$$+ \sum_i^a \sum_j^b \sum_k^n (Y_{ijk} - \bar{Y}_{ij.})^2$$

which again expresses the idea that the total sum of squares can be broken down into the sum of squares between means of factor A, plus the sum of squares between means of factor B, plus the sum of squares of $A \times B$ interaction, plus the error sum of squares (or within cell sum of squares). Each sum of squares is seen to be independent of the others. Hence, if any such sum of squares is divided by its associated degrees of freedom, and the model assumptions are true, the results are independently chi-square distributed, and F tests may be run.

The degree-of-freedom breakdown would be

$$(abn - 1) \equiv (a - 1) + (b - 1) + (a - 1)(b - 1) + ab(n - 1)$$

the interaction being cell df $= (ab - 1)$ minus the main effect df, $(a - 1)$ and $(b - 1)$, or $(ab - 1) - (a - 1) - (b - 1) = ab - a - b + 1 = (a - 1) \times (b - 1)$. Within each cell, the degrees of freedom are $n - 1$ and there are ab such cells, giving $ab(n - 1)$ df for error. An ANOVA table can now be set up expanding and simplifying the sum of squares expressions using totals (Table 5.6).

The formulas for sum of squares in Table 5.6 provide good computational formulas for a two-way ANOVA with replication. The error sum of squares might be rewritten as

$$\text{SS}_{\text{error}} = \sum_i^a \sum_j^b \left[\sum_k^n Y_{ijk}^2 - \frac{T_{ij.}^2}{n} \right]$$

to emphasize that the sum of squares within each of the $a \times b$ cells is being pooled or added for all such cells. This depends on the assumption that the variance within all cells came from populations with equal variance. The interaction sum of squares can also be rewritten as

$$\left(\sum_i^a \sum_j^b \frac{T_{ij.}^2}{n} - \frac{T_{...}^2}{nab} \right) - \left(\sum_i^a \frac{T_{i..}^2}{nb} - \frac{T_{...}^2}{nab} \right) - \left(\sum_j^b \frac{T_{.j.}^2}{na} - \frac{T_{...}^2}{nab} \right)$$

TABLE 5.6
General ANOVA for Two-Factor Factorial with n Replications per Cell

Source	df	SS	MS
Factor A	$a - 1$	$\sum_i^a \dfrac{T_{i..}^2}{nb} - \dfrac{T_{...}^2}{nab}$	Each SS divided by its df
Factor B	$b - 1$	$\sum_j^b \dfrac{T_{.j.}^2}{na} - \dfrac{T_{...}^2}{nab}$	
$A \times B$ interaction	$(a - 1)(b - 1)$	$\sum_i^a \sum_j^b \dfrac{T_{ij.}^2}{n} - \sum_i^a \dfrac{T_{i..}^2}{nb}$ $- \sum_j^b \dfrac{T_{.j.}^2}{na} + \dfrac{T_{...}^2}{nab}$	
Error $\varepsilon_{k(ij)}$	$ab(n - 1)$	$\sum_i^a \sum_j^b \sum_k^n Y_{ijk}^2 - \sum_i^a \sum_j^b \dfrac{T_{ij.}^2}{n}$	
Totals	$abn - 1$	$\sum_i^a \sum_j^b \sum_k^n Y_{ijk}^2 - \dfrac{T_{...}^2}{nab}$	

0

which shows again that interaction is calculated by subtracting the main effect sum of squares from the cell sum of squares.

■ **Example 5.1**

To extend the factorial idea a bit further, consider a problem with three factors. Such a problem was presented in Chapter 1 on the effect of tool type, angle of bevel, and type of cut on power consumption for ceramic tool cutting (see Example 1.6). Reference to this problem will point out the phases of experiment, design, and analysis as followed in Section 5.2.

It is a $2 \times 2 \times 2$ factorial experiment with four observations per cell run in a completely randomized manner. The mathematical model is

$$Y_{ijkm} = \mu + T_i + B_j + TB_{ij} + C_k + TC_{ik} + BC_{jk} + TBC_{ijk} + \varepsilon_{m(ijk)}$$

where TBC_{ijk} represents a three-way interaction.

The data for this example are given in Table 1.2 and the ANOVA table in Table 1.3. That this analysis is a simple extension of the methods used in Section 5.2 will be shown with the coded data from Table 1.2 (see Table 5.7).

Table 5.7 shows the total for each small cell and the sum of the squares of the cell observations. These results will be useful in doing the ANOVA. By this time we should be able to set up the steps in the analysis without recourse to formulas in dot notation.

Total sum of squares: Add the squares of all readings (the circled numbers) and subtract a correction term. The correction term is the grand total (-13) squared, divided by the number of observations (32). That is,

TABLE 5.7
Coded Ceramic Tool Data of Table 1.2, Code: $2(y - 28.0)$

		Tool Type				
		1		2		
		Bevel Angle		Bevel Angle		
Type of Cut	15°	30°		15°	30°	
Continuous	2	1		0	3	
	−3	1		1	8	
	5 (42)	4 (99)		0 (37)	2 (77)	25
	−2	9		−6	0	
	2	15		−5	13	
Interrupted	0	−2		−7	−1	
	−6	2		−6	0	
	−3 (54)	−1 (10)		0 (101)	−2 (21)	−38
	−3	−1		−4	−4	
	−12	−2		−17	−7	
Totals	−10	13		−22	+6	−13

$$SS_{total} = 441 - \frac{(-13)^2}{32} = 435.72$$

Tool type sum of squares: Add for each tool type. The totals are $+3$ and -16; square these and divide by the number of observations per type (16), add these results for both types, and subtract the correction term. Thus,

$$SS_{tool\ type} = \frac{3^2 + (-16)^2}{16} - \frac{(-13)^2}{32} = 11.28$$

Bevel angle sum of squares: Same procedure on the totals for each bevel angle, -32 and 19:

$$SS_{bevel\ angle} = \frac{(-32)^2 + (19)^2}{16} - \frac{(-13)^2}{32} = 81.28$$

Type of cut sum of squares: Same procedure, with cut totals of 25 and -38:

$$SS_{type\ of\ cut} = \frac{(25)^2 + (-38)^2}{16} - \frac{(-13)^2}{32} = 124.03$$

For the $T \times B$ interaction: Ignore type of cut and use cell totals for the $T \times B$ cells. These are $-10, 13, -22, +6$:

$$SS_{T \times B\ interaction} = \frac{(-10)^2 + (13)^2 + (-22)^2 + (6)^2}{8} - \frac{(-13)^2}{32} - 11.28 - 81.28$$
$$= 0.78$$

For $T \times C$ interaction: Ignore bevel angle and the cell totals become $17, -14, 8, -24$:

$$SS_{T \times C\ interaction} = \frac{(17)^2 + (-14)^2 + (8)^2 + (-24)^2}{8}$$
$$- \frac{(-13)^2}{32} - 11.28 - 124.03$$
$$= 0.03$$

For $B \times C$ interaction: Ignore tool type and the cell totals become $-3, 28, -29, -9$:

$$SS_{B \times C\ interaction} = \frac{(-3)^2 + (28)^2 + (-29)^2 + (-9)^2}{8}$$
$$- \frac{(-13)^2}{32} - 81.28 - 124.03$$
$$= 3.78$$

For the three-way interaction $T \times B \times C$: Consider the totals of the smallest cells, $2, 15, -5, 13, -12, -2, -17$, and -7. From this cell sum of squares subtract *not only* the main effect sum of squares *but also* the three two-way interaction sums of squares. Thus,

$SS_{T \times B \times C \text{ interaction}}$

$$= \frac{(2)^2 + (15)^2 + (-5)^2 + (13)^2 + (-12)^2 + (-2)^2 + (-17)^2 + (-7)^2}{4}$$

$$- \frac{(-13)^2}{32} - 11.28 - 81.28 - 124.03 - 0.78 - 0.03 - 3.78$$

$$= 0.79$$

Error sum of squares: By subtraction, we have

$$SS_{\text{error}} = 213.75$$

These results are displayed in Table 5.8. Although they appear to differ considerably from those in Table 1.3, in fact they give the same F test results. In Table 5.8, however, the data were coded, whereas the data of Table 1.3 are uncoded. The coding involved multiplication by 2, resulting in multiplication by 4 of the variance or mean square; thus if all mean squares in Table 5.8 are divided by 4, the results are the same as in Table 1.3 (e.g., on tool types $11.281/4 = 2.820$). It is worth noting that decoding is unnecessary for determining the F ratios. To find confidence limits on the original data or components of variance on the original data, however, it may be necessary to decode the results.

The results of this example were interpreted in the original presentation (Example 1.6). We present the material again to show that factorial experiments with three or more factors can easily be analyzed by simple extensions of the methods of this chapter.

TABLE 5.8
Minitab ANOVA for Ceramic Tool Problem

Source	DF	SS	MS	F	P
Tool	1	11.281	11.281	1.27	0.272
Bevel	1	81.281	81.281	9.13	0.006
Tool*Bevel	1	0.781	0.781	0.09	0.770
Cut	1	124.031	124.031	13.93	0.001
Tool*Cut	1	0.031	0.031	0.00	0.953
Bevel*Cut	1	3.781	3.781	0.42	0.521
Tool*Bevel*Cut	1	0.781	0.781	0.09	0.770
Error	24	213.750	8.906		
Total	31	435.719			

Significantly Small F Ratios

The F ratio for the interaction between total and cut is significantly small, since a p value of 0.953 is as rare as a p value of 0.047, which most practitioners consider significant. What can cause such low F ratios? Three possibilities are chance, improper randomization, and incorrect model. If every attempt has been made to perform the experiment in a completely randomized manner and a significantly small F ratio occurs, a model assumption may have been seriously violated. Investigation of that possibility by means of data plots, residual plots, and other checks relevant to this experiment is left to the reader (Problem 5.28).

5.6 ONE OBSERVATION PER TREATMENT

Since Example 5.1 contained several replications within a cell, it is well to examine a situation involving only one observation per cell. In this case $k = 1$, and the model is written as

$$Y_{ij} = \mu + A_i + B_j + AB_{ij} + \varepsilon_{ij}$$

A glance at the last two terms indicates that we cannot distinguish between the interaction and the error—they are hopelessly confounded. Then the only reasonable means of running one observation per cell is one in which past experience generally assures us that there is no interaction. In such a case, the model is written as

$$Y_{ij} = \mu + A_i + B_j + \varepsilon_{ij}$$

It may also be noted that this model looks very much like the model for a randomized block design for a single-factor experiment (Chapter 4). In Chapter 4 that model was written as

$$Y_{ij} = \mu + \tau_j + \beta_i + \varepsilon_{ij}$$

Even though the models do look alike and an analysis would be run in the same way, this latter is a single-factor experiment—treatments are the factor—and β represents a restriction on the randomization. In the factorial model, there are two factors of interest, A and B, and the design is completely randomized. It is, however, assumed in the randomized block situation that there is no interaction between treatments and blocks. This is often a more reasonable assumption for blocks and treatments, since blocks are often chosen at random. For a two-factor experiment an interaction between A and B may very well be present, and some external information must be available to justify the assumption that no such interaction exists. An experimenter who is not sure about interaction must take more than one observation per cell and test the hypotheses of no interaction.

As the number of factors increases, however, the presence of higher order interactions is much more unlikely, so it is fairly safe to assume no four-way, five-way, . . . , interactions. Even if these were present they would be difficult to explain in practical terms.

Analysis Involving One Observation per Treatment

The following set of steps might be used to analyze data from an experiment in which there are several factors but, because of the size of the experiment, only one observation can be afforded per treatment, namely $n = 1$.

1. Analyze the complete factorial including all main effects and all interactions in your model.

2. Examine the mean squares for all interactions and use all three-way and higher interactions as an error term on a second run of the analysis. Some three-way interactions may be left in the model if their mean squares look large compared with those to be eliminated. There is no F test here, for with $n = 1$ there is no error term. [**A word of caution:** If some interactions are to be retained in the model, be sure to include in the model all main effects and lower order interactions that appear in the interaction to be retained.]

3. After a second run, if the "pooled" error term from the first run has adequate degrees of freedom (say df ≥ 6), interpret the F tests as shown.

4. If the degrees of freedom in the error term are not adequate, repeat the analysis, placing all effects (even main effects, sometimes) and all interactions that are not significant at a high α level—say 0.25—in a new error term. Since this "pooling" assumes acceptance of null hypotheses, a large α (≥ 0.25) may help in reducing the β error when hypotheses are accepted.

5. Repeat this procedure until you feel you have an adequate model to explain the phenomena in the experiment. Base your "feeling" on a careful assessment of the model and assumptions.

5.7 SAS PROGRAMS FOR FACTORIAL EXPERIMENTS

Only slight modifications to the **INPUT, CLASS,** and **MODEL** statements used in Chapters 3 and 4 are needed to analyze any balanced factorial. This will be illustrated for the examples considered earlier in this chapter. The use of SAS to obtain interaction plots will also be illustrated.

Basically, analyses of data for a completely randomized two-factor factorial are handled in SAS like those for a randomized complete block experiment, with the exception that the interaction between the factors must be included in the model statement. To write a model statement to include an interaction between two factors, say A and B, one of two methods can be used. Letting RDG denote the response variable, the model statement can be expressed in the form

```
MODEL RDG=A B A*B;
```

or

```
MODEL RDG=A|B;
```

An asterisk between two factors indicates an interaction effect for the model. Use of a "pipe" (upright symbol) between the factors is equivalent to writing a model with all main effects and all possible interaction effects.

An interaction plot is useful when one is investigating the nature of a significant interaction. Here the **MEANS** procedure is used to calculate the cell means and output those means to a new data set. First, however, SAS requires that the cell data be prepared by means of the **SORT** procedure. A general form for these sorting procedures is as follows:

```
PROC SORT;
  BY A B;
PROC MEANS MEAN;
  VAR RDG;
  BY A B;
  OUTPUT OUT=CELL MEAN=CELLMEAN;
```

The third, fourth, and fifth lines of this program ask the **MEANS** procedure to calculate the average readings for the *A, B* treatment combinations. The phrase OUT=CELL in the **OUTPUT** statement tells the computer to store those averages in a new data set called CELL, and the phrase MEAN=CELLMEAN directs that the cell averages be treated as observations of a new variable called CELLMEAN.

Once the means have been calculated and stored, the **PLOT** procedure is used to plot the means versus levels of the factors. The following general form will cause the points (*A*, CELLMEAN) to be plotted for each level of *B*. The plotted points are identified by the first letter of the level of *B*.

```
PROC PLOT DATA=CELL;
PLOT CELLMEAN*A=B;
```

Only a scatterplot is prepared. To obtain a result like Figure 5.11, the points must be connected manually.

When one is interested in determining the optimum conditions and the analysis shows no significant interaction, a Student–Newman–Keuls procedure may be run for each factor. This can be accomplished by placing the statement

```
MEAN A B/SNK;
```

after the model statement.

If, on the other hand, the interaction is significant, it makes little sense to follow the procedure for main effects. Instead, a Student–Newman–Keuls procedure should be used with the cell means. One simple way to do this is to include a treatment combination variable with the input data and then follow the two-factor ANOVA with a one-factor ANOVA that includes the SNK procedure on the cell means.

The Vacuum Tube Pressure Experiment

The data for the vacuum tube pressure experiment are reproduced in Table 5.9. A SAS command file for a two-factor analysis of variance, SNK analysis of the cell means, and interaction plot is given in Table 5.10. The variable TC in the **INPUT** line identifies the treatment combination, which will be used with a one-factor ANOVA to run the SNK test. The **ANOVA** procedure is used for the two-factor factorial and then the one-factor model, because it is more efficient than **GLM**.

SAS outputs for the preceding command file are summarized in Figures 5.15 and 5.16. Notice that the results are the same as those shown in Sections 5.2 and 5.3. The SAS version used by the authors did not connect the points in the interaction plot—the connecting lines were added manually.

TABLE 5.9
Data for Vacuum Tube Pressure Experiment

Pump Heater Voltage	Exhaust Index		
	60	90	150
127	48	28	7
	58	33	15
220	62	14	9
	54	10	6

TABLE 5.10
SAS Command File for Vacuum Tube Pressure Experiment

`OPTIONS LINESIZE=80;`	*restricts width of output*
`DATA VACUUM;`	*names the data set*
`INPUT E V TC RDG @@;`	*names the variables in order*
`CARDS;`	*signals that data follow*
` 60 127 1 48   60 127 1 58`	*data begin*
` 60 220 2 62   60 220 2 54`	
` 90 127 3 28   90 127 3 33`	
` 90 220 4 14   90 220 4 10`	
`150 127 5  7 150 127 5 15`	
`150 220 6  9 150 220 6  6`	
`;`	*signals the end of data*
`PROC ANOVA;`	*identifies the procedure*
`  CLASS E V;`	*identifies the treatment variables*
`  MODEL VACUUM=E|V;`	*a two-factor factorial model*
`PROC ANOVA;`	*identifies another procedure*
`  CLASS TC;`	*identifies the treatment variable*
`  MODEL VACUUM=TC;`	*a one-factor model*
`  MEANS TC/SNK;`	*requests SNK test for tc's*
`PROC SORT;`	*sorting to be done for plots*
`  BY E V;`	*data sorted into treatment levels*
`PROC MEANS MEAN;`	*means to be calculated*
`  VAR VACUUM;`	*identifies the variable to be used*
`  BY E V;`	*means calculated for tc's*
`  OUTPUT OUT=CELL MEAN=CELLMEAN;`	*means output to new file and named*
`PROC PLOT DATA=CELL;`	*names procedure and data file*
`  PLOT CELLMEAN*E=V;`	*describes plot*

```
                    ANALYSIS OF VARIANCE PROCEDURE
                                SUM OF               MEAN
SOURCE               DF        SQUARES             SQUARE      F VALUE
MODEL                 5  4987.66666667      997.53333333        43.06
ERROR                 6   139.00000000       23.1666667        PR > F
CORRECTED TOTAL      11  5126.66666667                         0.0001

R-SQUARE           C.V.      ROOT MSE            RDG MEAN
0.972887        16.7902    4.81317636         28.66666667

SOURCE               DF      ANOVA SS     F VALUE     PR > F
E                     2  4608.16666667       99.46     0.0001
V                     1    96.33333333        4.16     0.0875
E*V                   2   283.16666667        6.11     0.0357
```

ANALYSIS OF VARIANCE PROCEDURE

DEPENDENT VARIABLE: VACUUM

SOURCE	DF	SUM OF SQUARES	MEAN SQUARE	F VALUE
MODEL	5	4987.66666667	997.53333333	43.06
ERROR	6	139.00000000	23.16666667	PR > F
CORRECTED TOTAL	11	5126.66666667		0.0001

R-SQUARE	C.V.	ROOT MSE	RDG MEAN
0.972887	16.7902	4.81317636	28.66666667

SOURCE	DF	ANOVA SS	F VALUE	PR > F
TC	5	4987.66666667	43.06	0.0001

Figure 5.15 SAS outputs for the vacuum tube pressure experiment.

STUDENT-NEWMAN-KEULS TEST FOR VARIABLE:VACUUM
NOTE: THIS TEST CONTROLS THE TYPE I EXPERIMENTWISE ERROR RATE UNDER THE
         COMPLETE NULL HYPOTHESIS BUT NOT UNDER PARTIAL NULL HYPOTHESES

ALPHA=0.05   DF=6   MSE=23.1667

NUMBER OF MEANS	2	3	4	5	6
CRITICAL RANGE	11.7772	14.7687	16.662	18.0545	19.155

MEANS WITH THE SAME LETTER ARE NOT SIGNIFICANTLY DIFFERENT

SNK GROUPING	MEAN	N	TC
A	58.000	2	2
A			
A	53.000	2	1
B	30.500	2	3
C	12.000	2	4
C			
C	11.000	2	5
C			
C	7.500	2	6

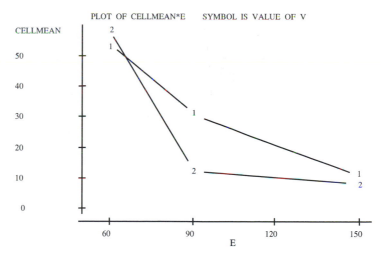

Figure 5.16 SAS outputs for the vacuum tube pressure experiment (continued).

The Ceramic Tool Experiment

The methods illustrated for the vacuum tube pressure experiment can be extended to any completely randomized factorial experiment with three or more factors and two or more observations per treatment combination. If, for example, the factors are *A, B,* and *C* and all interactions are to be included, the model equation can be expressed in the form

$$\text{MODEL RDG=A B A*B C A*C B*C A*B*C;}$$

or

$$\text{MODEL RDG=A|B|C;}$$

If some interactions are not to be included, the second form of the model statement cannot be used. If, for example, the three-way interaction is to be pooled with error instead of being included in the model, we write

$$\text{MODEL RDG=A B A*B C A*C B*C;}$$

To illustrate, the coded ceramic tool data in Table 5.7 were used to generate the SAS command file given in Table 5.11 and the resulting output (Figure 5.17). Except for rounding, the ANOVA results in Figure 5.17 are the same as those in Table 5.8.

TABLE 5.11
SAS Command File for Ceramic Tool Experiment

OPTIONS LINESIZE=80;								*restricts width of output*
DATA CERAMIC;								*names the data set*
INPUT T B C $ POWER @@;								*names the variables in order*
CARDS;								*signals that data follow*
1	15	CONT	2	1	15	CONT	−3	*data begin*
1	15	CONT	5	1	15	CONT	−2	
1	15	INTR	0	1	15	INTR	−6	
1	15	INTR	−3	1	15	INTR	−3	
1	30	CONT	1	1	30	CONT	1	
1	30	CONT	4	1	30	CONT	9	
1	30	INTR	−2	1	30	INTR	2	
1	30	INTR	−1	1	30	INTR	−1	
2	15	CONT	0	2	15	CONT	1	
2	15	CONT	0	2	15	CONT	−6	
2	15	INTR	−7	2	15	INTR	−6	
2	15	INTR	0	2	15	INTR	−4	
2	30	CONT	3	2	30	CONT	8	
2	30	CONT	2	2	30	CONT	0	
2	30	INTR	−1	2	30	INTR	0	
2	30	INTR	−2	2	30	INTR	−4	
;								*signals the end of data*
PROC ANOVA;								*identifies the procedure*
CLASS T B C;								*identifies the treatment variables*
MODEL POWER=T|B|C;								*a three-factor, full factorial model*

```
                          ANALYSIS OF VARIANCE PROCEDURE
         DEPENDENT VARIABLE: POWER
                                       SUM OF            MEAN
         SOURCE                 DF     SQUARES          SQUARE     F VALUE
         MODEL                   7  221.96875000    31.70982143       3.56
         ERROR                  24  213.75000000     8.90625000     PR > F
         CORRECTED TOTAL        31  435.71875000                    0.0091
         R-SQUARE             C.V.     ROOT MSE        RDG MEAN
         0.509431          734.6053  2.98433410     -0.40625000
         SOURCE                 DF     ANOVA SS       F VALUE      PR > F
         T                       1   11.28125000        1.27      0.2715
         B                       1   81.28125000        9.13      0.0059
         T*B                     1    0.78125000        0.09      0.7696
         C                       1  124.03125000       13.93      0.0010
         T*C                     1    0.03125000        0.00      0.9533
         B*C                     1    3.78125000        0.42      0.5209
         T*B*C                   1    0.78125000        0.09      0.7696
```

Figure 5.17   SAS ANOVA for the ceramic tool experiment.

5.8 SUMMARY

Experiment	Design	Analysis
I. Single factor		
	1. Completely randomized $Y_{ij} = \mu + \tau_j + \varepsilon_{ij}$	1. One-way ANOVA
	2. Complete randomized block $Y_{ij} = \mu + \tau_j + \beta_i + \varepsilon_{ij}$	2. Two-way ANOVA
	3. Complete Latin square $Y_{ijk} = \mu + \beta_i + \tau_j + \gamma_k + \varepsilon_{ijk}$	3. Three-way ANOVA
	4. Graeco-Latin square $Y_{ijkm} = \mu + \beta_i + \tau_j + \gamma_k + \omega_m + \varepsilon_{ijkm}$	4. Four-way ANOVA
II. Two or more factors Factorial (crossed)	1. General case for completely randomized $Y_{ijk} = \mu + A_i + B_j + AB_{ij} + \varepsilon_{k(ij)} \cdots$ for more factors	1. ANOVA with interactions

5.9 FURTHER READING

Gunter, B., "Statistically Designed Experiments. Part 3: Interaction," *Quality Progress,* April, 1990, pp. 74–75.

Gunter, B., "Statistically Designed Experiments. Part 4: Multivariate Optimization," *Quality Progress,* June 1990, pp. 68, 70. Factorial experimentation and the one-factor-at-a-time procedure are compared.

Gunter, B., "Statistically Designed Experiments. Part 5: Robust Process and Product Design and Related Matters," *Quality Progress,* August 1990, pp. 107–108. Ridges in the response surface are illustrated.

Nelson, L. S., "What Do Low *F* Ratios Tell You?" *Journal of Quality Technology,* October 1985, pp. 8–9.

Pendleton, O. J., "Influential Observations in the Analysis of Variance," *Communications in Statistics Part A: Theory and Methods,* Vol. 14, No. 3, 1985, pp. 551–565.

PROBLEMS

5.1 To determine the effect of two glass types and three phosphor types on the light output of a television tube, light output is measured by the current required in series with the tube to produce 30 foot-lamberts of light output. Thus the higher the current, in microamperes, the poorer the tube in light output. Three observations were taken under each of the six treatment conditions and the experiment was completely randomized. The following data were recorded.

Glass Type	Phosphor Type		
	A	*B*	*C*
1	280	300	270
	290	310	285
	285	295	290
2	230	260	220
	235	240	225
	240	235	230

a. Do an ANOVA on these data and test the effect of glass type, phosphor types, and interaction on the current flow.

b. Assess the model assumptions.

5.2 Plot the results of Problem 5.1 to show that your conclusions are reasonable.

5.3 For any significant effects in Problem 5.1 test further between the levels of the significant factors.

5.4 Based on the results in Problems 5.1–5.3, what glass type and phosphor type (or types) would you recommend if a low current is most desirable?

5.5 Adhesive force on gummed material was determined under three fixed humidity and three fixed temperature conditions. Four readings were made under each set of conditions. The experiment was completely randomized and the results set out in an ANOVA table as follows:

Source	df	SS	MS
Humidity		9.07	
Temperature		8.66	

$H \times T$ interaction	6.07
Error	
Total	52.30

Complete this table.

5.6 For the data in Problem 5.5 test all indicated hypotheses and state your conclusions.

5.7 Set up a mathematical model for the experiment in Problem 5.5 and indicate the hypotheses to be tested in terms of your model. State all assumptions.

5.8 The object of an experiment is to determine thrust forces in drilling at different speeds and feeds, and in different materials. Five speeds and three feeds are used, and two materials with two samples are tested under each set of conditions. The order of the experiment is completely randomized, and the levels of all factors are fixed. After 200 had been subtracted from all readings the following data on thrust forces were recorded.

		Speed				
Material	**Feed**	**100**	**220**	**475**	**715**	**870**
B_{10}	0.004	122	108	108	66	80
		110	85	60	50	60
	0.008	332	276	248	248	276
		330	310	295	275	310
	0.014	640	612	543	612	696
		500	500	450	610	610
V_{10}	0.004	192	136	122	108	136
		170	130	85	75	75
	0.008	386	333	318	472	499
		365	330	330	350	390
	0.014	810	779	810	893	1820
		725	670	750	890	890

Do a complete analysis of this experiment and state your conclusions.

5.9 Plot any results in Problem 5.8 that are significant.

5.10 Set up tests on means where suitable and draw conclusions from Problem 5.8.

5.11 In an experiment for testing rubber materials interest centered on the effect of the mix (A, B, or C), the laboratory involved (1, 2, 3, or 4), and the temperature (145, 155, 165°C) on the time in minutes to a rise above the minimum time of 2 inch-pounds. Assuming a completely randomized design, do an ANOVA on the following data.

Laboratory	Temperature (°C)								
	145			155			165		
	Mix			Mix			Mix		
	A	B	C	A	B	C	A	B	C
1	11.2	11.2	11.5	6.7	6.8	7.0	4.8	4.8	5.0
	11.1	11.5	11.4	6.8	6.7	7.0	4.8	4.9	4.9
2	11.8	12.3	12.3	7.3	7.5	7.5	5.3	5.4	5.3
	11.8	12.3	11.9	7.2	7.7	7.3	5.3	5.2	5.3
3	11.5	12.3	12.7	6.6	7.1	7.8	5.0	5.3	5.2
	11.6	12.0	12.5	6.9	7.2	7.3	5.0	5.0	5.0
4	11.5	11.8	12.7	7.2	6.7	7.1	4.5	4.7	4.5
	11.3	11.7	12.7	6.9	7.0	7.0	4.6	4.5	4.5

5.12 What is there in the results of Problem 5.11 that would lead you to question some assumption about the experiment?

5.13 Tomato plants were grown in a greenhouse under treatments consisting of combinations of soil type (factor A) and fertilizer type (factor B). A completely randomized two-factor design was used with two replications per cell. The following data on the yield Y (in kilograms) of tomatoes were obtained for the 30 plants under study.

| Soil Type, A | Fertilizer Type, B | | | |
	1	2	3	$T_{i.}$
I	5, 7	5, 5	3, 5	30
II	5, 9	1, 3	2, 2	22
III	6, 8	4, 8	2, 4	32
IV	7, 11	6, 9	3, 7	44
V	6, 9	4, 6	3, 5	33
$T_{.j}$	73	52	36	161

Note: $\sum_k \sum_j \sum_i Y_{ijk}^2 = 1043.$

Complete the following ANOVA table.

Source	df	Sum of Squares	Mean Square
Soil type (A)			
Fertilizer type (B)			
Interaction (AB)			
Error			

5.14 For the ANOVA in Problem 5.13, carry out tests on main effects and interactions *in an appropriate order,* using levels of significance for each test that seem appropriate to you. For each test state H_0 and H_1, show the rejection region, and give your conclusion. After doing all tests, summarize your results in words.

5.15 On the basis of your results in Problem 5.14, state how you would carry out Newman–Keuls comparisons to find out the best combination (or combinations) of soil type and fertilizer type to achieve maximum mean yield of tomatoes. Then carry out these comparisons (at an α of 0.05) and state your conclusions.

5.16 Each of the following tables represent the cell means ($\bar{Y}_{ij.}$) for a two-factor completely randomized balanced experiment. For each table tell which (if any) of the following sums of squares would be zero for that table: SS_A, SS_B, SS_{AB}. (There may be none; there may be more than one.)

a.

b.

c.

A

B	1	3
	5	5

= 0

d.

A

B	5	3
	5	3

= 0

5.17 In each two-factor design table below, the numbers in the cells of the table are the population means of the observations for those cells. For each table, indicate whether there is interaction between the factors and justify your assertion.

a.

		Factor B			
		1	2	3	4
Factor A	1	7	6	5	2
	2	9	8	7	4

Is there interaction? *Yes/No* (circle one). Explain.

b.

		Factor B		
		1	2	3
Factor A	1	7	6	5
	2	5	6	7

Is there interaction? *Yes/No* (circle one). Explain.

c.

		Factor B	
		1	2
	1	8	10
Factor A	2	7	9
	3	4	7

Is there interaction? *Yes/No* (circle one). Explain.

5.18 An industrial engineer presented two types of stimuli (two-dimensional and three-dimensional films) of two different jobs (1 and 2) to each of five analysts. Each analyst was presented each job stimulus film twice, and the order of the whole experiment was considered to be completely randomized. The engineer was interested in the consistency of analyst ratings of four sequences within each job stimulus presentation. Since log variance is more likely to be normally distributed than variance, the log variance was recorded for the four sequences. The following tabulation resulted.

	Stimulus			
	Two-Dimensional		Three-Dimensional	
Analyst	Job 1	Job 2	Job 1	Job 2
1	1.42	1.40	1.00	0.92
	1.25	1.44	0.90	0.93
2	1.59	1.83	1.46	1.43
	2.09	2.02	1.68	1.02

3	1.26	1.80	0.85	1.33
	1.48	1.75	1.43	1.48
4	1.76	0.98	1.21	0.73
	1.47	1.23	1.25	1.22
5	1.43	1.17	1.60	1.31
	1.72	1.18	2.03	1.12

Do a complete analysis and summarize your results.

5.19 a. Make further tests on Problem 5.18 as suggested by the ANOVA results.
   b. Write the model and assumptions. Then, assess that model.

5.20 The following results were reported on a study of "factors that affect the salary of high school teachers of commercial subjects":

Source	df	SS
Sex	1	239,763
Size of school	2	423,056
Years in position	5	564,689
Sex × size interaction	2	18,459
Sex × years interaction	5	85,901
Size × years interaction	10	240,115
Sex × size × years interaction	10	151,394
Within classes	153	1,501,642

Indicate the mathematical model for this study. Show a possible data layout. Complete the ANOVA table and comment on any significant results and interpret.

5.21 Four factors are studied for their effect on the luster of plastic film. These factors are film thickness (1 or 2 mils), drying conditions (regular or special), length of wash (20, 30, 40, or 60 minutes), and temperature of wash (92 or 100°C). Two observations of film luster are taken under each set of conditions.
   a. Assuming complete randomization, analyze the data in the table.
   b. Conduct a residual analysis.

Minutes	Regular Dry				Special Dry			
	92°C		100°C		92°C		100°C	
1-mil Thickness								
20	3.4	3.4	19.6	14.5	2.1	3.8	17.2	13.4
30	4.1	4.1	17.5	17.0	4.0	4.6	13.5	14.3
40	4.9	4.2	17.6	15.2	5.1	3.3	16.0	17.8
60	5.0	4.9	20.9	17.1	8.1	4.3	17.5	13.9

2-mil Thickness

20	5.5	3.7	26.6	29.5	4.5	4.5	25.6	22.5
30	5.7	6.1	31.6	30.2	5.9	5.9	29.2	29.8
40	5.6	5.6	30.5	30.2	5.5	5.8	32.6	27.4
60	7.2	6.0	31.4	29.6	8.0	9.9	33.5	29.5

5.22 Plot and discuss any significant interactions found in Problem 5.21. What conditions would you recommend for maximum luster and why?

5.23 To study the effects of four types of wax applied to floors for polishing times of three different lengths, it was decided to try each combination twice and to completely randomize the order in which floors received which treatments. The criterion was a gloss index.
a. Outline an ANOVA table for this problem with proper degrees of freedom.
b. Explain, in words, what a significant interaction would mean in this situation.

5.24 A manufacturer of combustion engines is concerned about meeting EPA standards that limit the percentage of smoke emitted by such engines. An experiment was conducted involving engines with three timing levels (T), three throat diameters (D), two volume ratios in the combustion chamber (V), and two injection systems (I). Thirty-six engines, representing all possible combinations of these four factors, were then operated in a completely random order, and percentage of smoke was recorded to the nearest tenth of a percent. Results were as follows.

				T							
				1 **D**			**2** **D**			**3** **D**	
			1	**2**	**3**	**1**	**2**	**3**	**1**	**2**	**3**
V	1	*I*	1	0.4	3.8	7.0	1.8	4.8	8.1	1.7	4.9	8.2
			2	0.5	1.6	4.0	0.7	3.0	5.6	1.2	3.2	10.0
	2	*I*	1	0.7	1.8	6.0	1.4	1.0	1.5	7.0	3.1	3.4
			2	0.1	0.1	0.2	0.6	0.4	1.4	4.0	1.8	2.0

Do a complete analysis of these data and recommend the combination or combinations of factor levels that should give a minimum percentage of smoke emitted.

5.25 Students' scores on a psychomotor test were recorded for various sized targets at which the student must aim, different machines used, and the level of background illumination. Results are shown in the following ANOVA table.

Source	df	SS	MS
Target size (A)	3	235.200	78.400
Machine (B)	2	86.467	43.233

Level of illumination (C)	1	76.800	76.800
$A \times B$	6	104.200	17.367
$A \times C$	3	93.867	31.289
$B \times C$	2	12.600	6.300
$A \times B \times C$	6	174.333	29.056
Within sets	96	1198.000	12.478
Totals	119	1981.467	

a. Using a 5% significance level, indicate which effects are statistically significant.
b. Assuming that the original data were at your disposal, explain briefly the next steps you would take in analyzing the data from this experiment.
c. Devise a data layout sheet that could be used to collect each item of data for this experiment.
d. Assuming that the mean score for target size A_3, machine B_2, and level of illumination C_1 is 9.6, set 90% confidence limits on the true mean for this treatment combination.

5.26 When Problem 3.19 was run on bonding machines A, B, C, and D the bonding took place in three different positions (1, 2, 3) on the piece being bonded. Data including this position factor follows.

		Bonder		
Position	A	B	C	D
1	204	197	264	248
	181	223	226	138
	201	206	228	273
	203	232	249	220
	214	213	246	186
2	262	207	255	304
	246	259	186	330
	230	223	237	268
	256	195	236	295
	288	197	240	276
3	220	214	215	208
	232	248	176	248
	235	191	171	247
	231	197	208	220
	220	222	180	241

Analyze these data and discuss the results.

5.27 To determine the effect of cool-zone oxygen level and preheat oxygen level on the coefficient of variation in resistance on a 20 kΩ resistor, five cool-zone oxygen levels and two preheat oxygen levels were used. Three observations were taken per experiment.

Preheat Oxygen (ppm)	Cool-Zone Oxygen Level (ppm)				
	5	10	15	20	25
5	6.45	2.51	4.71	11.47	10.69
	2.71	4.42	4.91	9.31	9.25
	3.08	4.20	8.19	12.04	10.00
25	2.86	5.43	5.35	8.92	13.13
	5.51	8.37	4.20	7.57	12.01
	5.66	3.54	6.49	7.14	11.71

Analyze the data tabulated above and report your recommendations, assuming that a low coefficient of variation is desirable.

5.28 Using plots of data, residual plots, and other procedures as appropriate, assess the model used in Examples 1.6 and 5.1.

5.29 A laboratory uses a reaction injection molding (RIM) machine to mold very large poly-urethane part and measure the heat sag as a general measure of the quality of the material. For the RIM process, two liquid streams of chemicals flow into a mold cavity and react to form the polyurethane part. To complete the chemical reaction and drive off entrapped gases, the part is removed from the mold and postcured. It is important that the heat sag properties of parts molded in the lab be very similar to those of parts molded in the plant. The project engineer in charge of development claimed to have designed a completely randomized factorial experiment with postcure times of 60, 90, and 120 minutes and time delay (the amount of time between completion of molding and the taking of measurements) levels of 0, 12, and 24 hours. The intent was to obtain heat sag data for 12 parts at each treatment combination. However, equipment malfunctions resulted in unequal observations, as shown in the following data table.

Postcure Time (min)	Time Delay (h)					
	0		**12**		**24**	
60	0.55	0.93	0.71	0.90	0.90	0.67
	0.64	0.91	0.78	0.81	0.65	0.92
	0.77	1.08	1.22	0.92	0.81	0.72
	0.73	0.92	1.04	0.79	0.68	0.75
	0.82	1.07	1.02	1.02	0.66	0.67
	0.63	0.73	0.76	1.04	0.82	0.81
90	0.67	0.67	0.86	0.74	0.95	0.77
	0.76	0.90	0.83	0.82	1.35	1.44
	0.87	0.84	0.72	0.61	0.75	0.78
	0.54	0.75	0.79	0.92	0.56	1.03
	0.73	0.80	0.82	0.85	0.99	0.47
	0.56	0.66	0.63	0.67	0.68	0.98

120	0.47	0.68	0.86	0.67	0.53
	0.71	0.54	0.76	0.53	0.47
	0.67	0.55	0.62	0.88	0.19
	0.69	0.59	0.65	0.72	0.63
	0.65	0.72	0.88	0.77	0.46
	0.62	0.53	0.47	0.70	

a. Prepare an interaction plot. Interpret, assuming that the average heat sag reading is to be minimized.
b. Using the general linear models procedure in some statistical analysis program, prepare an ANOVA summary. The appropriate SS's are the adjusted or type III SS's.
c. Using data plots, residual plots, and other appropriate procedures, analyze further.
d. Based on data from a prior experiment, the project engineer decided that a postcure time of 90 minutes should be used with heat sag tests performed within 24 hours of molding. Do your findings support that decision? Comment.

5.30 Two manufacturers produce three types of bonding machine used in electronic component manufacturing. A completely randomized study designed to evaluate the effects of manufacturer and type of machine on bond strength produced the following coded response values.

	Machine Type		
Manufacturer	**1**	**2**	**3**
1	13.5	17.0	18.5
	14.0	16.0	18.0
	14.5	15.0	17.5
2	13.0	15.5	10.0
	12.0	14.5	12.0
	11.0	15.0	13.0

a. Write the model for this experiment.
b. Prepare an interaction plot of the six cell means. Comment.
c. Conduct a two-way analysis of variance. Summarize the results in an ANOVA table.
d. Prepare a Student–Newman–Keuls analysis of the cell means. Interpret.
e. Using plots of the residuals and other procedures as appropriate, assess the model.

Chapter **6**

# FIXED, RANDOM, AND MIXED MODELS

6.1 INTRODUCTION

We know from Chapter 1 that in the planning stages of an experiment, the experimenter must decide whether the levels of factors are to be set at fixed values or chosen at random from many possible levels. In the intervening chapters, except for Section 3.6, it has always been assumed that the factor levels were fixed. In practice it may be necessary to choose the levels of the factors at random, depending on the objectives of the experiment. Are the results to be judged for these levels alone, or are they to be extended to more levels, of which those in the experiment are but a random sample? In the case of some factors such as temperature, time, or pressure, it is usually desirable to pick fixed levels, often near the extremes and at some intermediate points, because a random choice might not cover the range in which the experimenter is interested. In such cases of fixed quantitative levels, we often feel safe in interpolating between the fixed levels chosen. Other factors such as operators, days, or batches may often be only a small sample of all possible operators, days, or batches. In such cases the particular operator, day, or batch may be less important than whether operators, days, or batches in general increase the variability of the experiment.

It is not reasonable to decide after the data have been collected whether the levels are to be considered fixed or random. This decision must be made prior to the running of the experiment, and if random levels are to be used, they must be chosen from all possible levels by a random process. In the case of random levels, it will be assumed that the levels are chosen from an infinite population of possible levels.

Types of Model

When all levels are fixed, the mathematical model of the experiment is called a *fixed model*. When all levels are chosen at random, the model is called a *random model*. When several factors are involved, some at fixed levels and others at random levels, the model is called a *mixed model*.

6.2 SINGLE-FACTOR MODELS

In the case of a single-factor experiment the factor may be referred to as a *treatment effect*, as in Chapter 3; and if the design is completely randomized, the model is

$$Y_{ij} = \mu + \tau_j + \varepsilon_{ij} \tag{6.1}$$

Whether the treatment levels are fixed or random, it is assumed in this model that μ is a fixed constant and the errors are normally and independently distributed with a zero mean and the same variance. That is, $\varepsilon_{ij} \sim \text{NID}(0, \sigma_\varepsilon^2)$ for all i, j pairs. The decision to fix or randomize the levels of the treatment will affect the assumptions about the treatment term τ_j.

Figure 6.1a shows three fixed means whose average is μ as these are the only means of concern and $\sum_j \tau_j = \sum_j (\mu_j - \mu) = 0$. Figure 6.1b shows three random means whose average is obviously not μ, inasmuch as these are but three means chosen at random from many possible means. These means and their corresponding τ_j's are assumed to form a normal distribution with a standard deviation of σ_τ.

The expected mean square (EMS) column of an ANOVA summary turns out to be extremely important in more complex experiments as an aid in deciding how to set up an F test for significance. The EMS for any term in the model is the long-range average of the calculated mean square when the Y_{ij} from the model is substituted in algebraic form into the mean square computation. The derivation of these EMS values is often complicated, but those for the single-factor model were derived in Chapter 3 and are summarized in Table 6.1. Those for two-factor models are derived in a later section of this chapter.

For the fixed model, if the hypothesis is true that $\tau_j = 0$ for all j, that is, all the k fixed treatment means are equal, then $\sum_j \tau_j^2 = 0$ and the EMS is σ_ε^2 for both τ_j and ε_{ij}. Hence the observed mean squares for treatments and error mean square are both estimates of the error variance, and they can be compared by means of an F test. If this F test shows a significantly high value, it must mean that $n \sum_j \tau_j^2/(k-1) = n\phi_\tau$ is not zero and the hypothesis is to be rejected.

For the random model, if the hypothesis is true that $\sigma_\tau^2 = 0$, that is, the variance among all treatment means is zero, then again each mean square is an estimate of the error variance. Again, an F test between the two mean squares is appropriate.

The different assumptions and other differences between the two models are summarized in Table 6.1. From that table, it is seen that for a single-factor experiment there is no difference in the test to be made after the analysis; the only difference is in the generality of

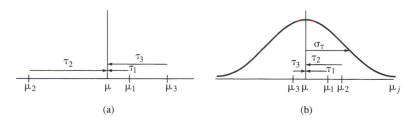

Figure 6.1   Assumed means in (a) fixed and (b) random models.

TABLE 6.1
Differences Among Fixed and Random Models for One-Factor Experiments

Fixed Model	Random Model
1. Assumptions: τ_j's are fixed constants.	1. Assumptions: τ_j's are random variables and are

$$\sum_{j=1}^{k} \tau_j = \sum_{j=1}^{k}(\mu_j - \mu) = 0 \qquad\qquad NID(0, \sigma_\tau^2)$$

Fixed Model	Random Model
(These add to zero because they are the only treatment means being considered.)	(Here σ_τ^2 represents the variance among all possible τ_j's or among the true treatment means μ_j. The τ_j's average to zero when averaged over all possible levels, but for the k levels of the experiment the τ_j's usually will not average to 0.)
2. Analysis: Procedures as given in Chapter 3 for computing sums of squares.	2. Analysis: Same as for the fixed model.
3. EMS: For n, the number of observations per treatment level and ϕ_τ a nonnegative constant:	3. EMS: For n the number of observations per treatment level:

Source	df	EMS		Source	df	EMS
τ_j	$k-1$	$\sigma_\varepsilon^2 + n\phi_\tau$		τ_j	$k-1$	$\sigma_\varepsilon^2 + n\sigma_\tau^2$
ε_{ij}	$k(n-1)$	σ_ε^2		ε_{ij}	$k(n-1)$	σ_ε^2

Fixed Model	Random Model
4. Hypothesis tested and test statistic: $H_0: \tau_j = 0;\ j = 1, \ldots, k$ $F = MS_{treatment}/MS_{error}$	4. Hypothesis tested and test statistic: $H_0: \sigma_\tau^2 = 0$ $F = MS_{treatment}/MS_{error}$

the conclusions. If H_0 is rejected, there is probably a difference among the k fixed treatment means for the fixed model; for the random model there is a difference among all treatments of which the k examined are but a random sample.

6.3 TWO-FACTOR MODELS

For two factors A and B the model in the general case is

$$Y_{ijk} = \mu + A_i + B_j + AB_{ij} + \varepsilon_{k(ij)}$$

with

$$i = 1, 2, \ldots, a \quad j = 1, 2, \ldots, b \quad k = 1, 2, \ldots, n$$

provided the design is completely randomized. In this model, it is again assumed that μ is a fixed constant and the $\varepsilon_{k(ij)}$'s are NID $(0, \sigma_\varepsilon^2)$. If both A and B are at fixed levels, the model is a fixed model. If both are at random levels, the model is a random model, and if one is at fixed levels and the other at random levels, the model is a mixed model. The assumptions, analysis, EMS data, and hypotheses of the models may be compared as follows.

Fixed	*Random*	*Mixed*
1. Assumptions: A_i's are fixed constants and	1. Assumptions: A_i's are NID $(0, \sigma_A^2)$	1. Assumptions: A_i's are fixed and

$$\sum_{i=1}^{a} A_i = 0$$

B_j's are fixed constants and

$$\sum_{j=1}^{b} B_j = 0$$

AB_{ij}'s are fixed constants and

$$\sum_i AB_{ij} = 0$$

$$\sum_j AB_{ij} = 0$$

2. Analysis: Procedures of Chapter 5 for sums of squares
3. EMS:

$$\sum_i^a A_i = 0$$

B_j's are NID $(0, \sigma_B^2)$

AB_{ij}'s are NID $(0, \sigma_{AB}^2)$

2. Analysis: Same
3. EMS:

$$\sum_i^a A_i = 0$$

B_j's are NID $(0, \sigma_B^2)$

AB_{ij}'s are NID $(0, \sigma_{AB}^2)$ but

$$\sum_i^a AB_{ij} = 0$$

$$\sum_j^b AB_{ij} \neq 0$$

(for A fixed, B random)

2. Analysis: Same
3. EMS:

Source	df	EMS (Fixed)	EMS (Random)	EMS (Mixed)
A_i	$a-1$	$\sigma_\varepsilon^2 + nb\phi_A$	$\sigma_\varepsilon^2 + n\sigma_{AB}^2 + nb\sigma_A^2$	$\sigma_\varepsilon^2 + n\sigma_{AB}^2 + nb\phi_A$
B_j	$b-1$	$\sigma_\varepsilon^2 + na\phi_B$	$\sigma_\varepsilon^2 + n\sigma_{AB}^2 + na\sigma_B^2$	$\sigma_\varepsilon^2 + na\sigma_B^2$
AB_{ij}	$(a-1)(b-1)$	$\sigma_\varepsilon^2 + n\phi_{AB}$	$\sigma_\varepsilon^2 + n\sigma_{AB}^2$	$\sigma_\varepsilon^2 + n\sigma_{AB}^2$
$\varepsilon_{k(ij)}$	$ab(n-1)$	σ_ε^2	σ_ε^2	σ_ε^2

---

4. Hypotheses tested:
   $H_1 : A_i = 0$ for all i
   $H_2 : B_j = 0$ for all j
   $H_3 : AB_{ij} = 0$ for all i and j

4. Hypothesis tested:
   $H_1 : \sigma_A^2 = 0$
   $H_2 : \sigma_B^2 = 0$
   $H_3 : \sigma_{AB}^2 = 0$

4. Hypotheses tested:
   $H_1 : A_i = 0$ for all i
   $H_2 : \sigma_B^2 = 0$
   $H_3 : \sigma_{AB}^2 = 0$

It is assumed for the mixed model that summing the interaction term over the fixed factor ($\sum_i$) is zero but summing it over the random factor ($\sum_j$) is not zero; this assumption affects the expected mean squares, as seen in the partial ANOVA summaries of item 3.

Determining the Test Statistic

For the fixed model the mean squares for A, B, and AB are each compared with the error mean square to test the respective hypotheses, as should be clear from an examination of the EMS column when the hypotheses are true. For the random model the third hypothesis of no interaction is tested by comparing the mean square for interaction with the mean square for error; but the first and second hypotheses are tested by comparing the mean square for the main effect (A_i or B_j) with the mean square for the interaction as seen by the expected mean square value in each case. For a mixed model the interaction hypothesis is tested by comparing the interaction mean square with the error mean square. The random effect B_j is also tested by comparing its mean square with the error mean square. The fixed effect (A_i), however, is tested by comparing its mean square with the interaction mean square.

These observations on a two-factor experiment indicate that the EMS column can be used to see how the tests of hypotheses should be run. It is also important to note that these EMS expressions should be determined prior to the running of the experiment. This will indicate whether a valid test of a hypothesis exists. In some cases the proper test indicated by the EMS column will not have enough degrees of freedom to provide adequate sensitivity, in which case the investigator may wish to change the experiment (e.g., by choosing more levels of some factors or more observations per treatment combination, or by changing from random to fixed levels of some factors).

6.4 EMS RULES

Sections 6.2 and 6.3 showed the importance of the EMS column in determining what tests of significance are to be run after the analysis is completed. Because of the importance of the EMS column in these and more complex models, it is often useful to have a simple method of determining these values from the model for the given experiment. A set of rules can be stated for the *balanced case* that will determine the EMS column very rapidly, without recourse to their derivation. The rules will be illustrated on the two-factor mixed model of Section 6.3. To determine the EMS column for any model:

1. Write the variable terms in the model as row headings in a two-way table.

A_i
B_j
AB_{ij}
$\varepsilon_{k(ij)}$

2. Write the subscripts in the model as column headings; over each superscript write F if the factor levels are fixed, R if random. Also write the number of observations each subscript is to cover.

	a F i	b R j	n R k
A_i			
B_j			
AB_{ij}			
$\varepsilon_{k(ij)}$			

3. For each row (each term in the model) copy the number of observations under each subscript, provided the subscript does not appear in the row heading.

	a F i	b R j	n R k
A_i		b	n
B_j	a		n
AB_{ij}			n
$\varepsilon_{k(ij)}$			

4. For any parenthetical subscripts in the model, place a 1 under the subscripts given in parentheses.

	a *F* *i*	*b* *R* *j*	*n* *R* *k*
A_i		b	n
B_j	a		n
AB_{ij}			n
$\varepsilon_{k(ij)}$	1	1	

5. Fill each remaining cell with a 0 or a 1, depending on whether the subscript represents a fixed F or a random R factor.

	a *F* *i*	*b* *R* *j*	*n* *R* *k*
A_i	0	b	n
B_j	a	1	n
AB_{ij}	0	1	n
$\varepsilon_{k(ij)}$	1	1	1

6. To find the expected mean square for any term in the model:

    a. Cover the entries in the column (or columns) that contain nonparenthetical subscript letters in this term in the model (e.g., for A_i, cover column i; for $\varepsilon_{k(ij)}$, cover column k).

    b. Multiply the remaining numbers in each row. Each of these products is the coefficient for its corresponding term in the model, provided the subscript on the term is also a subscript on the term whose expected mean square is being determined. The sum of these coefficients multiplied by the variance of their corresponding terms (ϕ_τ or σ_τ^2) is the EMS of the term being considered (e.g., for A_i, cover column i). For our example, the products of the remaining coefficients are bn, n, n, and 1, but the first n is not used, as there is no i in its term (B_j). The resulting EMS is then $bn\phi_A + n\sigma_{AB}^2 + 1 \cdot \sigma_\varepsilon^2$. For all terms, these rules give the following partial ANOVA table.

	a *F* *i*	*b* *R* *j*	*n* *R* *k*	EMS
A_i	0	b	n	$\sigma_\varepsilon^2 + n\sigma_{AB}^2 + nb\phi_A$
B_j	a	1	n	$\sigma_\varepsilon^2 + na\sigma_B^2$
AB_{ij}	0	1	n	$\sigma_\varepsilon^2 + n\sigma_{AB}^2$
$\varepsilon_{k(ij)}$	1	1	1	σ_ε^2

These results are seen to be in agreement with the EMS values for the mixed model in Section 6.3. Here ϕ_A is, of course, a fixed type of variance:

$$\phi_A = \frac{\sum_i A_i^2}{a-1}$$

Although the rules seem rather involved, they become very easy to use with a bit of practice. Two examples will illustrate the concept.

■ **Example 6.1**

The viscosity of a slurry is to be determined by four randomly selected laboratory technicians. Material from each of five mixing machines is bottled and divided in a way that provides two samples for each technician to test. These five mixers are the only machines of interest, and the samples can be presented to the technicians in a completely randomized order.

The model here assumes four randomly selected technicians and five fixed mixing machines. Each technician measures samples of each machine twice. The model is shown as the first column of Table 6.2, and the remainder of the table shows how the EMS column is determined.

TABLE 6.2
EMS for Example 6.1

Source	df	4 R i	5 F j	2 R k	EMS
T_i	3	1	5	2	$\sigma_\varepsilon^2 + 10\sigma_T^2$
M_j	4	4	0	2	$\sigma_\varepsilon^2 + 2\sigma_{TM}^2 + 8\phi_M$
TM_{ij}	12	1	0	2	$\sigma_\varepsilon^2 + 2\sigma_{TM}^2$
$\varepsilon_{k(ij)}$	20	1	1	1	σ_ε^2

The proper F tests are quite obvious from Table 6.2 and all tests have adequate degrees of freedom for a reasonable test. For example $F = \mathrm{MS}_M/\mathrm{MS}_{TM}$ with $\nu_1 = 4$, and $\nu_2 = 12$ is the test statistic for $H_0 : \tau_j = 0$ with $j = 1, 2, 3, 4, 5$.

■ **Example 6.2**

An industrial engineering student wished to determine the effect of five different clearances on the time required to position and assemble mating parts. Since all such experiments involve operators, it was natural to consider a random sample of operators to perform the experiment. The student also decided that the part should be assembled directly in front of the operator and at arm's length from the operator. He tried four different angles, from 0° directly in front of the operator through 30°, 60°, and 90° from this position. Thus four factors were involved, any one of which might affect the time required to position and assemble the part. The experimenter decided to replicate each set up six times and to randomize completely the order of experimentation. Here operators O_i were at random levels (six being chosen), angles A_j at four fixed levels ($0°$, $30°$, $60°$, $90°$), clearances C_k at five fixed levels, and locations L_m fixed either in front of or at arm's length from the operator. This is a $6 \times 4 \times 5 \times 2$ factorial experiment with six replications, run in a completely randomized design. The expected mean square values can be determined from the rules given in Section 6.4 as shown in Table 6.3.

TABLE 6.3
EMS for Clearance Problem

Source	df	6 R i	4 F j	5 F k	2 F m	6 R q	EMS
O_i	5	1	4	5	2	6	$\sigma_\varepsilon^2 + 240\sigma_O^2$
A_j	3	6	0	5	2	6	$\sigma_\varepsilon^2 + 60\sigma_{OA}^2 + 360\phi_A$
OA_{ij}	15	1	0	5	2	6	$\sigma_\varepsilon^2 + 60\sigma_{OA}^2$
C_k	4	6	4	0	2	6	$\sigma_\varepsilon^2 + 48\sigma_{OC}^2 + 288\phi_C$
OC_{ik}	20	1	4	0	2	6	$\sigma_\varepsilon^2 + 48\sigma_{OC}^2$
AC_{jk}	12	6	0	0	2	6	$\sigma_\varepsilon^2 + 12\sigma_{OAC}^2 + 72\phi_{AC}$
OAC_{ijk}	60	1	0	0	2	6	$\sigma_\varepsilon^2 + 12\sigma_{OAC}^2$
L_m	1	6	4	5	0	6	$\sigma_\varepsilon^2 + 120\sigma_{OL}^2 + 720\phi_L$
OL_{im}	5	1	4	5	0	6	$\sigma_\varepsilon^2 + 120\sigma_{OL}^2$
AL_{jm}	3	6	0	5	0	6	$\sigma_\varepsilon^2 + 30\sigma_{OAL}^2 + 180\phi_{AL}$
OAL_{ijm}	15	1	0	5	0	6	$\sigma_\varepsilon^2 + 30\sigma_{OAL}^2$
CL_{km}	4	6	4	0	0	6	$\sigma_\varepsilon^2 + 24\sigma_{OCL}^2 + 144\phi_{CL}$
OCL_{ikm}	20	1	4	0	0	6	$\sigma_\varepsilon^2 + 24\sigma_{OCL}^2$
ACL_{jkm}	12	6	0	0	0	6	$\sigma_\varepsilon^2 + 6\sigma_{OACL}^2 + 36\phi_{ACL}$
$OACL_{ijkm}$	60	1	0	0	0	6	$\sigma_\varepsilon^2 + 6\sigma_{OACL}^2$
$\varepsilon_{q(ijkm)}$	1200	1	1	1	1	1	σ_ε^2

It is easily seen from Table 6.3 all interactions involving operators and the operator main effect are tested against the error mean square at the bottom of the table. All interactions and main effects involving the fixed factors are tested by the mean square just below them in the table. For example, the test statistic for $H_{01} : \sigma_{OA}^2 = 0$ is $F_1 = \mathrm{MS}_{OA}/\mathrm{MS}_{\mathrm{error}}$ with $\nu_1 = 5$ and $\nu_2 = 1200$; and that for H_{02}: "The treatment effect of C is 0 at the 5 levels of C" is $F_2 = \mathrm{MS}_C/\mathrm{MC}_{OC}$ with $\nu_1 = 4$ and $\nu_2 = 20$.

The rules given in this section generally suffice for the most complex designs, as will be seen in later chapters.

6.5 EMS DERIVATIONS

Single-Factor Experiment

The EMS expressions for a single-factor experiment were derived in Section 3.6 (see Equations 3.19–3.26).

Two-Factor Experiment

Using the definitions of expected values given in Chapter 2 and the procedures of Section 3.6 the EMS expressions may be derived for the two-factor experiment.

For a two-factor experiment, the model is

$$Y_{ijk} = \mu + A_i + B_j + AB_{ij} + \varepsilon_{k(ij)} \tag{6.2}$$

with

$$i = 1, 2, \ldots, a \quad j = 1, 2, \ldots, b \quad k = 1, 2, \ldots, n$$

The sum of squares for factor A is

$$SS_A = \sum_{i=1}^{a} nb(\bar{Y}_{i..} - \bar{Y}_{...})^2$$

From the model in Equation (6.2), we can write

$$\bar{Y}_{i..} = \sum_{j}^{b} \sum_{k}^{n} Y_{ijk}/bn = \sum_{j}^{b} \sum_{k}^{n} (\mu + A_i + B_j + AB_{ij} + \varepsilon_{k(ij)})/bn$$

$$\bar{Y}_{i..} = \mu + A_i + \sum_{j}^{b} B_j/b + \sum_{j}^{b} AB_{ij}/b + \sum_{j}^{b} \sum_{k}^{n} \varepsilon_{k(ij)}/bn$$

$$\bar{Y}_{...} = \sum_{i}^{a} \sum_{j}^{b} \sum_{k}^{n} Y_{ijk}/abn$$

$$= \sum_{i}^{a} \sum_{j}^{b} \sum_{k}^{n} (\mu + A_i + B_j + AB_{ij} + \varepsilon_{k(ij)})/abn$$

$$\bar{Y}_{...} = \mu + \sum_{i}^{a} A_i/a + \sum_{j}^{b} B_j/b + \sum_{i}^{a} \sum_{j}^{b} AB_{ij}/ab + \sum_{i}^{a} \sum_{j}^{b} \sum_{k}^{n} \varepsilon_{k(ij)}/nab$$

Subtracting gives

$$\bar{Y}_{i..} - \bar{Y}_{...} = \left(A_i - \sum_{i}^{a} A_i/a \right) + \left(\sum_{j}^{b} AB_{ij}/b - \sum_{i}^{a} \sum_{j}^{b} AB_{ij}/ab \right)$$

$$+ \left(\sum_{j}^{b} \sum_{k}^{n} \varepsilon_{k(ij)}/bn - \sum_{i}^{a} \sum_{j}^{b} \sum_{k}^{n} \varepsilon_{k(ij)}/abn \right)$$

Squaring and adding gives

$$SS_A = nb \sum_{i=1}^{a} \left(A_i - \sum_{i}^{a} A_i/a \right)^2$$

$$+ nb \sum_{i=1}^{a} \left(\sum_{j}^{b} AB_{ij}/b - \sum_{i}^{a}\sum_{j}^{b} AB_{ij}/ab \right)^2$$

$$+ nb \sum_{i=1}^{a} \left(\sum_{j}^{b}\sum_{k}^{n} \varepsilon_{k(ij)}/bn - \sum_{i}^{a}\sum_{j}^{b}\sum_{k}^{n} \varepsilon_{k(ij)}/abn \right)^2$$

$$+ \text{cross-product terms}$$

Next we take the expected value for a fixed model, where

$$\sum_{i} A_i = 0 \qquad \sum_{i \text{ or } j} AB_{ij} = 0$$

The result is

$$E(\text{SS}_A) = nb \sum_{i}^{a} A_i^2 + 0 + \frac{nb}{n^2 b^2}(abn - bn)\sigma_\varepsilon^2$$

Thus,

$$E(\text{MS}_A) = E[\text{SS}_A/(a-1)] = nb \sum_{i=1}^{a} A_i^2/(a-1) + \sigma_\varepsilon^2 = nb\phi_A + \sigma_\varepsilon^2$$

which agrees with Section 6.3 for the fixed model.

If now the levels of A and B are random,

$$\sum_{i} A_i \neq 0 \qquad \sum_{i \text{ or } j} AB_{ij} \neq 0$$

and

$$E(\text{SS}_A) = nb(a-1)\sigma_A^2 + \frac{nb}{b^2}(ab - b)\sigma_{AB}^2 + \frac{nb}{n^2 b^2}(abn - bn)\sigma_\varepsilon^2$$

Thus,

$$E(\text{MS}_A) = nb\sigma_A^2 + n\sigma_{AB}^2 + \sigma_\varepsilon^2$$

as stated in Section 6.3 for a random model.

If the model is mixed with A fixed and B random,

$$\sum_{i}^{a} A_i = 0 \qquad \text{and} \qquad \sum_{i}^{a} AB_{ij} = 0$$

But

$$\sum_{j}^{b} AB_{ij} \neq 0$$

So,

$$E(\text{SS}_A) = nb \sum_i^a A_i^2 + \frac{nb}{b^2}(ab - b)\sigma_{AB}^2 + \frac{nb}{n^2 b^2}(nab - nb)\sigma_\varepsilon^2$$

and

$$E(\text{MS}_A) = nb \sum_i^a A_i^2/(a - 1) + n\sigma_{AB}^2 + \sigma_\varepsilon^2 = nb\phi_A + n\sigma_{AB}^2 + \sigma_\varepsilon^2$$

which agrees with the value stated in Section 6.3 for a mixed model.

One can use these expected value methods to derive all EMS values given in Section 6.3. Note that if A were random in the mixed model, we would write

$$\sum_j^a AB_{ij} = 0$$

and the interaction term would not appear in the factor A sum of squares. This is true of B in the mixed model of Section 6.3.

These few derivations should be sufficient to show the general method of derivation and to demonstrate the advantages of the simple rules in Section 6.4 in determining these EMS values.

6.6 THE PSEUDO–F TEST

Occasionally the EMS column for a given experiment indicates that there is no exact F test for one or more factors in the design model. Consider the following example.

■ **Example 6.3**

Two days in a given month were randomly selected for the running of an experiment. Three operators were also selected at random from a large pool of available operators. The experiment consisted of measuring the dry-film thickness of varnish for three different gate settings: 2, 4, and 6 mils. Two determinations were made by each operator each day and at each of the three gate settings. Results are shown in Table 6.4.

Assuming that days and operators are random effects, gate settings are fixed, and the design is completely randomized, the analysis yields Table 6.5.

All F tests are clear from the EMS column except for the test on gate setting, which is probably the most important factor in the experiment. Two interactions show significance. It is obvious from the results that gate setting is the most important factor, but how can it be tested? If the $D \times G$ interaction is assumed to be zero, then the gate setting can be tested against the $O \times G$ interaction term. On the other hand, if the $O \times G$ interaction is assumed to be zero, gate setting can be tested against the $D \times G$ interaction term. Although neither of these interactions is significant at the 5% level, both are numerically larger than the $D \times O \times G$ interaction against which they

TABLE 6.4
Dry-Film Thickness Experiment

Gate Setting	Day 1 Operator A	Operator B	C	Day 2 Operator A	Operator B	C
2	0.38	0.39	0.45	0.40	0.39	0.41
	0.40	0.41	0.40	0.40	0.43	0.40
4	0.63	0.72	0.78	0.68	0.77	0.85
	0.59	0.70	0.79	0.66	0.76	0.84
6	0.76	0.95	1.03	0.86	0.86	1.01
	0.78	0.96	1.06	0.82	0.85	0.98

TABLE 6.5
Analysis of Dry-Film Thickness Experiment

Source	df	SS	MS	EMS
Days D	1	0.0010	0.0010	$\sigma_\varepsilon^2 + 6\sigma_{DO}^2 + 18\sigma_D^2$
Operators O	2	0.1121	0.0560	$\sigma_\varepsilon^2 + 6\sigma_{DO}^2 + 12\sigma_O^2$
$D \times O$ interaction	2	0.0060	0.0030**	$\sigma_\varepsilon^2 + 6\sigma_{DO}^2$
Gate setting G	2	1.5732	0.7866**	$\sigma_\varepsilon^2 + 2\sigma_{DOG}^2 + 4\sigma_{OG}^2$ $+ 6\sigma_{DG}^2 + 12\phi_G$
$D \times G$ interaction	2	0.0113	0.0056	$\sigma_\varepsilon^2 + 2\sigma_{DOG}^2 + 6\sigma_{DG}^2$
$O \times G$ interaction	4	0.0428	0.0107	$\sigma_\varepsilon^2 + 2\sigma_{DOG}^2 + 4\sigma_{OG}^2$
$D \times O \times G$ interaction	4	0.0099	0.0025**	$\sigma_\varepsilon^2 + 2\sigma_{DOG}^2$
Error	18	0.0059	0.0003	σ_ε^2
Totals	35	1.7622		

**Two asterisks indicate significance at the 1% level.

are tested. In this case any test on G is contingent upon these tests on interaction. One method for testing hypotheses in such situations was developed by Satterthwaite.

The scheme consists of constructing a mean square as a linear combination of the mean squares in the experiment, where the EMS for this mean square includes the same terms as in the EMS of the term being tested, except for the variance of that term. For example, to test the gate-setting effect G in Table 6.5, a mean square is to be constructed whose expected value is

$$\sigma_\varepsilon^2 + 2\sigma_{DOG}^2 + 4\sigma_{OG}^2 + 6\sigma_{DG}^2$$

This can be found by the linear combination

$$MS = MS_{DG} + MS_{OG} - MS_{DOG}$$

because its expected value is

$$E(MS) = \sigma_\varepsilon^2 + 2\sigma_{DOG}^2 + 4\sigma_{OG}^2 + \sigma_\varepsilon^2 + 2\sigma_{DOG}^2 + 6\sigma_{DG}^2 - \sigma_\varepsilon^2 - 2\sigma_{DOG}^2$$
$$= \sigma_\varepsilon^2 + 2\sigma_{DOG}^2 + 4\sigma_{OG}^2 + 6\sigma_{DG}^2$$

An F test can now be constructed using the mean square for gate setting as the numerator and this mean square as the denominator. Such a test is called a pseudo–F test, or F' test. The real problem here is to determine the degrees of freedom for the denominator mean square. According to classical methods, if

$$MS = a_1(MS)_1 + a_2(MS)_2 + \cdots$$

and $(MS)_1$ is based on ν_1 df, $(MS)_2$ is based on ν_2 df, and so on, then the degrees of freedom for MS are

$$\nu = \frac{(MS)^2}{a_1^2[(MS)_1^2/\nu_1] + a_2^2[(MS)_2^2/\nu_2] + \cdots}$$

In the case of testing for the gate-setting effect above, $a_1 = 1$, $a_2 = 1$, and $a_3 = -1$ and the degrees of freedom are $\nu_1 = 4$, $\nu_2 = 2$, and $\nu_3 = 4$. Here MS = 0.0107 + 0.0056−0.0025 = 0.0138 and its df is

$$\nu = \frac{(0.0138)^2}{(1)^2[(0.0107)^2/4] + (1)^2[(0.0056)^2/2] + (-1)^2[(0.0025)^2/4]}$$

$$= \frac{1.9044 \times 10^{-4}}{0.4586 \times 10^{-4}} = 4.2$$

Hence the F' test is

$$F' = \frac{MS_G}{MS} = \frac{0.7866}{0.0138} = 57.0$$

with 2 and 4.2 df, which is significant at the 1% level of significance based on F with 2 and 4 or 2 and 5 df.

6.7 EXPECTED MEAN SQUARES VIA STATISTICAL COMPUTING PACKAGES

Two types of analysis of variance model exist for mixed models having some random and some fixed effects. The model used in this and most other texts (often called a *restricted model*) requires that an interaction term for which at least one factor is random and another is fixed sum to 0 over the subscripts associated with the fixed *factor*. The other, *unrestricted model* does not include that requirement.

Some statistical computing packages (e.g., JMP and SAS) use the unrestricted model. EMS reports and default tests produced by those packages may differ from those for the restricted model when the model is mixed.

The **Balanced ANOVA** procedure in Minitab uses the unrestricted model by default but allows the user to choose the restricted model. When the random factors are specified, the restricted model is chosen, and a listing of the expected mean squares is requested, a text report like that in Table 6.6 is displayed. The expected mean squares are the same as

TABLE 6.6
Minitab Text Report for Dry-Film Thickness Experiment

```
Analysis of Variance (Balanced Designs)
Factor          Type        Levels      Values
Day             random        2         1     2
Operator        random        3         1     2     3
Gate            fixed         3         1     2     3

Analysis of Variance for Thickness

Source              DF          SS          MS         F         P
Day                  1       0.00100     0.00100      0.34     0.621
Operator             2       0.11207     0.05604     18.77     0.051
Gate                 2       1.57317     0.78659       *
Day*Operator         2       0.00597     0.00299      9.19     0.002
Day*Gate             2       0.01134     0.00567      2.29     0.218
Operator*Gate        4       0.04284     0.01071      4.32     0.093
Day*Operator*Gate    4       0.00991     0.00248      7.62     0.001
Error               18       0.00585     0.00033
Total               35       1.76216

* No exact F-test can be calculated.

Source              Variance    Error    Expected Mean Square
                    component    term     (using restricted model)
1 Day               -0.00011       4      (8) + 6(4) + 18(1)
2 Operator           0.00442       4      (8) + 6(4) + 12(2)
3 Gate                             *      (8) + 2(7) + 4(6) + 6(5) + 12Q[3]
4 Day*Operator       0.00044       8      (8) + 6(4)
5 Day*Gate           0.00053       7      (8) + 2(7) + 6(5)
6 Operator*Gate      0.00206       7      (8) + 2(7) + 4(6)
7 Day*Operator*Gate  0.00108       8      (8) + 2(7)
8 Error              0.00033              (8)

* No exact F-test can be calculated.
```

those in Table 6.5 and the p values of the F tests that are clear from the EMS column are reported. As before, there is no clear test on gate setting.

In the Gate row and Expected Mean Square column of Table 6.6, the entry

$$(8) + 2(7) + 4(6) + 6(5) + 12Q[3]$$

tells us that the expected mean square for the gate effect is the variance of the 8th source of variation plus twice the variance of the 7th source of variation plus 4 times the variance of the 6th source of variation plus 6 times the 5th source of variation plus 12 times the component of the 3rd source of variation, which is a fixed effect. That is,

$$E[\mathrm{MS}_G] = \sigma_\varepsilon^2 + 2\sigma_{DOG}^2 + 4\sigma_{OG}^2 + 6\sigma_{DG}^2 + 12\phi_G$$

as noted in Table 6.5.

From the Day and Day*Operator rows of Table 6.6, we find that $E[\mathrm{MS}_D] = \sigma_\varepsilon^2 + 6\sigma_{DO}^2 + 18\sigma_D^2 = E[\mathrm{MS}_{DO}] + 18\sigma_D^2$. Substituting the observed mean squares and solving for σ_D^2 gives $\sigma_D^2 \approx -0.00011$, as reported in the Variance component column of the Day row. Since variances must be nonnegative, 0 should be used as the sample estimate.

6.8 REMARKS

The examples in this chapter show the importance of the EMS column in deciding just what mean squares should be compared in an F test of a given hypothesis. This EMS column is also useful (usually in random models) to solve for components of variance as illustrated in Section 3.6.

One special case is of interest. In a two-factor factorial when there is but one observation per cell ($n = 1$), the EMS columns of Section 6.3 reduce to those in Table 6.7.

A glance at these EMS values will show that there is no test for the main effects A and B in a fixed model, since interaction and error are hopelessly confounded. The only test possible is to assume that there is no interaction; then $\phi_{AB} = 0$, and the main effects are tested against the error. If a no-interaction assumption is not reasonable from information outside the experiment, the investigator should replicate the data in a fixed model instead of running one observation per cell.

For a random model, both main effects can be tested regardless of whether interaction is present. For a mixed model there is a test for the fixed effect A but no test for the random effect B. This may not be a serious drawback, since the fixed effect is often the most important; the B effect is included chiefly for reduction of the error term. Such a situation is seen in a randomized block design where treatments are fixed, but blocks may be chosen at random.

In the discussion of the single-factor experiment it was assumed that there were equal sample sizes n for each treatment. If this is not the case, the expected treatment mean square is

$$\sigma_\varepsilon^2 + n_0\sigma_\tau^2 \quad \text{or} \quad \sigma_\varepsilon^2 + n_0\phi_\tau$$

where

$$n_0 = \frac{N^2 - \sum_{j=1}^{k} n_j^2}{(k-1)N}$$

and

$$N = \sum_{j=1}^{k} n_j$$

The test for treatment effect is to compare the treatment mean square with the error mean square; the use of n_0 is primarily for computing components of variance.

TABLE 6.7
EMS for One Observation per Cell

Source	EMS (Fixed)	EMS (Random)	EMS (Mixed)
A_i	$\sigma_\varepsilon^2 + b\phi_A$	$\sigma_\varepsilon^2 + \sigma_{AB}^2 + b\sigma_A^2$	$\sigma_\varepsilon^2 + \sigma_{AB}^2 + b\phi_A$
B_j	$\sigma_\varepsilon^2 + a\phi_B$	$\sigma_\varepsilon^2 + \sigma_{AB}^2 + a\sigma_B^2$	$\sigma_\varepsilon^2 + a\sigma_B^2$
AB_{ij} or ε_{ij}	$\sigma_\varepsilon^2 + \phi_{AB}$	$\sigma_\varepsilon^2 + \sigma_{AB}^2$	$\sigma_\varepsilon^2 + \sigma_{AB}^2$

6.9 REPEATABILITY AND REPRODUCIBILITY FOR A MEASUREMENT SYSTEM

When measurements of manufactured devices are obtained by more than one operator, the measurement process is often assessed by conducting a completely randomized, two-factor experiment in which n representative parts are measured r times by each of k operators. Part and operator are random effects, since they are assumed to be randomly obtained from large universes of available parts and operators. The partial ANOVA summary of Table 6.8 is appropriate for this situation.

Using the expected mean squares in Table 6.8 and substituting the observed values of the mean squares, sample estimates of the variances associated with operator, part, interaction, and measurement error are, respectively:

$$s_P^2 = \frac{MS_P - MS_{OP}}{(kr)}$$

$$s_O^2 = \frac{MS_O - MS_{OP}}{(nr)}$$

$$s_{OP}^2 = \frac{MS_{OP} - MS_E}{r}$$

$$s_E^2 = MS_E$$

Since $\sigma_Y^2 = \sigma_P^2 + \sigma_O^2 + \sigma_{OP}^2 + \sigma_\varepsilon^2$, an estimate of the total variation σ_Y^2 is given by $s_{total}^2 = s_P^2 + s_O^2 + s_{OP}^2 + s_\varepsilon^2$.

Since we are dealing with sample data, sample estimates of the variances associated with operator, part, and/or interaction may be obtained in negative form. In such cases, zero should be used as the sample estimate.

Studies of this type are called *repeatability and reproducibility (R&R) studies.* The basic measure of spread in an R&R study is 5.15σ, since 99% of the values of a normally distributed random variable are within $5.15/2 = 2.575$ standard deviations of the mean. The estimated 99% spreads associated with this model are summarized in Table 6.9.

Let LSL and USL denote the upper and lower specification limits, respectively, of the measurement. The value of %R&R$_{tolerance}$ = $100(R\&R)/(USL - LSL)$ is used to estimate the percent of the specification spread consumed by the 99% spread associated with the reproducibility and repeatability of the measurement system. An estimate of the percentage of the total variation consumed by the measurement system is %R&R$_{process}$

TABLE 6.8
Partial ANOVA Summary for Repeatability and Reproducibility Study: Completely Randomized Design

Source	df	SS	MS	EMS
Part, P	$n - 1$	SS_P	MS_P	$\sigma_\varepsilon^2 + r\sigma_{OP}^2 + kr\sigma_P^2$
Operator, O	$k - 1$	SS_O	MS_O	$\sigma_\varepsilon^2 + r\sigma_{OP}^2 + nr\sigma_O^2$
Interaction, OP	$(n-1)(k-1)$	SS_{OP}	MS_{OP}	$\sigma_\varepsilon^2 + r\sigma_{OP}^2$
Gage error, E	$nk(r-1)$	SS_E	MS_E	σ_ε^2

TABLE 6.9
99% Spreads Associated with a Completely Randomized Model for R&R Studies

Repeatability (equipment variation)	$EV = 5.15\sqrt{s_\varepsilon^2}$
Reproducibility (appraiser variation)	$AV = 5.15\sqrt{s_O^2}$
Part	$PV = 5.15\sqrt{s_P^2}$
Interaction	$I = 5.15\sqrt{s_{OP}^2}$
Repeatability and reproducibility	$R\&R = \sqrt{(EV)^2 + (AV)^2 + I^2}$
Total variation	$TV = 5.15\sqrt{s_{\text{total}}^2}$

$= 100(R\&R)/(TV)$. Values of $\%R\&R_{\text{tolerance}}$ and $\%R\&R_{\text{process}}$ that do not exceed 10% are considered acceptable. Values exceeding 30% are usually unacceptable.

6.10 SAS PROGRAMS FOR RANDOM AND MIXED MODELS

The default F tests and p values in ANOVA tables for completely randomized factorial experiments analyzed with SAS are those for a fixed model. Test statistics and p values for tests having a denominator other than the error mean square are obtained by including instructions in the command file as to which tests are appropriate. This is done by adding a **TEST** statement of the form

TEST H= *term(s) to be tested*

E= *factor associated with the proper divisor MS*

where H denotes *hypothesis* and E denotes *error*. This is illustrated in the last two lines of Table 6.10—a SAS command file for the dry-film thickness study. The line

TABLE 6.10
SAS Command File for Film Thickness Study

```
OPTIONS LINESIZE=80;                        restricts width of output
DATA FILM;                                  names the data set
INPUT D O $ G THCKNSS @@;                    names the variables in order
CARDS;                                      signals that data follow
  1 A 2 0.38 1 A 2 0.40 1 A 4 0.63 1 A 4 0.59
  1 A 6 0.76 1 A 6 0.78 1 B 2 0.39 1 B 2 0.41
  1 B 4 0.72 1 B 4 0.70 1 B 6 0.95 1 B 6 0.96
  1 C 2 0.45 1 C 2 0.40 1 C 4 0.78 1 C 4 0.70
  1 C 6 1.03 1 C 6 1.06 2 A 2 0.40 2 A 2 0.40
  2 A 4 0.68 2 A 4 0.66 2 A 6 0.86 2 A 6 0.82
  2 B 2 0.39 2 B 2 0.43 2 B 4 0.77 2 B 4 0.76
  2 B 6 0.86 2 B 6 0.85 2 C 2 0.41 2 C 2 0.40
  2 C 4 0.85 2 C 4 0.84 2 C 6 1.01 2 C 6 0.98
;                                           signals the end of data
PROC GLM;                                   identifies the procedure
CLASS D O G;                                identifies the treatment variable
MODEL THCKNSS=D|O|G;                        identifies the model
TEST H=D O E=D*O;                           request for 2 special tests
TEST H=D*G O*G E=D*O*G;                     request for 2 more special tests
```

$$\text{TEST H=D 0 E=D*0;}$$

instructs SAS to test $H_{01} : D_i = 0; i = 1, 2$ by using $F = \text{MS}_D/\text{MS}_{DO}$ as the test statistic and to test $H_{02} : O_j = 0; j = 1, 2, 3$ using $F = \text{MS}_O/\text{MS}_{DO}$. The line

$$\text{TEST H=D*G 0*G E=D*0*G;}$$

instructs SAS to test $H_{03} : DG_{ik} = 0; i = 1, 2; k = 1, 2, 3$ by using $F = \text{MS}_{DG}/\text{MS}_{DOG}$ as the test statistic and to test $H_{04} : OG_{jk} = 0; j = 1, 2, 3$ using $F = \text{MS}_{OG}/\text{MS}_{DOG}$. The outputs are given in Figure 6.2.

Since sample sizes are equal and the experiment is a completely randomized factorial, corresponding type I and type III sums of squares in Figure 6.2 are equal. The seven F tests following the whole model ANOVA are those for a fixed model. From the EMS column of Table 6.5, we know that, of those, only the tests on the $D \times O$ and $D \times O \times G$ interactions are correct. There is no direct test on G, and the other four tests are those summarized at the bottom of Figure 6.2. A pseudo–F test on G is obtained manually, as illustrated in Section 6.6.

```
                    GENERAL LINEAR MODELS PROCEDURE
DEPENDENT VARIABLE: THCKNSS
                              SUM OF           MEAN
SOURCE                DF      SQUARES          SQUARE     F VALUE    PR > F
MODEL                 17    1.75631389      0.10331253    317.38     0.0001
ERROR                 18    0.00585000      0.00032500
CORRECTED TOTAL       35    1.76216389
                   R-SQUARE         C.V.      ROOT MSE       THCKNSS MEAN
                   0.996680       2.6436     0.01802776        0.6819444
SOURCE                DF     TYPE I SS    MEAN SQUARE    F VALUE    PR > F
D                      1    0.00100278    0.00100278       3.09    0.0960
O                      2    0.11207222    0.05603611     172.42    0.0001
D*O                    2    0.00597222    0.00298611       9.19    0.0018
G                      2    1.57317222    0.78658611    2420.26    0.0001
D*G                    2    0.01133889    0.00566944      17.44    0.0001
O*G                    4    0.04284444    0.01071111      32.96    0.0001
D*O*G                  4    0.00991111    0.00247778       7.62    0.0009
SOURCE                DF    TYPE III SS   MEAN SQUARE    F VALUE    PR > F
D                      1    0.00100278    0.00100278       3.09    0.0960
O                      2    0.11207222    0.05603611     172.42    0.0001
D*O                    2    0.00597222    0.00298611       9.19    0.0018
G                      2    1.57317222    0.78658611    2420.26    0.0001
D*G                    2    0.01133889    0.00566944      17.44    0.0001
O*G                    4    0.04284444    0.01071111      32.96    0.0001
D*O*G                  4    0.00991111    0.00247778       7.62    0.0009
TESTS OF HYPOTHESES USING THE TYPE III MS FOR D*O AS AN ERROR TERM
SOURCE                DF    TYPE III SS    F VALUE      PR > F
D                      1    0.00100278       0.34      0.6208
O                      2    0.11207222      18.77      0.0506
TESTS OF HYPOTHESES USING THE TYPE III MS FOR D*O AS AN ERROR TERM
SOURCE                DF    TYPE III SS    F VALUE      PR > F
D*G                    2    0.01133889       2.29      0.2175
O*G                    4    0.04284444       4.32      0.0926
```

Figure 6.2   SAS outputs for the dry-film thickness experiment.

6.11 FURTHER READING

Anderson, V. L., and R. A. McLean, *Design of Experiments: A Realistic Approach.* New York: Marcel Dekker, 1974. See Chapters 2 and 3 for discussions of expected mean squares and issues in experimentation.

Lorenzen, T. J., and V. L. Anderson, *Design of Experiments: A No Name Approach.* New York: Marcel Dekker, 1993. See Chapter 3 for a discussion of expected mean squares.

Mason, R. L., R. F. Gunst, and J. L. Hess, *Statistical Design and Analysis of Experiments with Applications to Engineering and Science.* New York: John Wiley & Sons, 1989. See pages 353–364 for a discussion of expected mean squares.

Pendleton, O., M. Von Tress, and R. Bremer, "Interpretation of the Four Types of Analysis of Variance Tables in SAS." Report SAR-3 for the Texas Transportation Institute and the Texas A&M University System, June 1985.

Searle, S. R., *Linear Models.* New York: John Wiley & Sons, 1971, Chapter 9.

PROBLEMS

6.1 In an experiment on the effects of three randomly selected operators and five fixed aluminizers on the aluminum thickness of a TV tube, two readings are made for each operator–aluminizer combination. The following ANOVA table is compiled.

Source	df	SS	MS
Operators	2	107,540	53,770
Aluminizers	4	139,805	34,951
$O \times A$ interaction	8	84,785	10,598
Error	15	230,900	15,393
Totals	29	563,030	

Assuming complete randomization, determine the EMS column for this problem and make the indicated significance tests.

6.2 Consider a three-factor experiment where factor A is at a levels, factor B at b levels, and factor C at c levels. The experiment is to be run in a completely randomized manner with n observations for each treatment combination. Assuming that factor A is run at a random levels and both B and C at fixed levels, determine the EMS column and indicate what tests would be made after the analysis.

6.3 Repeat Problem 6.2 with A and B at random levels, but C at fixed levels.

6.4 Repeat Problem 6.2 with all three factors at random levels.

6.5 Consider the completely randomized design of a four-factor experiment similar to Example 6.2. Assuming factors A and B are at fixed levels and C and D are at random levels, set up the EMS column and indicate the tests to be made.

6.6 Assuming three factors are at random levels and one is at fixed levels in Problem 6.5, work out the EMS column and the tests to be run.

6.7 A physical education experiment will be conducted to investigate the effects on heart rate of four types of exercise. Five subjects are chosen at random from a physical education class. A subject does each exercise twice. The 40 measurements are done in a completely randomized order, and each subject is allowed at least 10 minutes' rest between exercises.
   a. Give the model. State the parameters of its random variables.
   b. Work out the expected mean squares of all the effects. Also give the degrees of freedom for the effects.
   c. Tell how to form the F's for the tests that can be made.

6.8 Four elementary schools are chosen at random in a large city system and two methods of instruction are tried in the third, fourth, and fifth grades.
   a. Outline an ANOVA layout to test the effect of schools, methods, and grades on gain in reading scores assuming ten students per class.
   b. Work out the EMS column and indicate what tests are possible in this situation.

6.9 As part of an experiment on testing viscosity, four operators were chosen at random and four instruments were randomly selected from many instruments available. Each operator was to test a sample of product twice on each instrument. From the following results, determine the EMS column and find the percentage of the total variance in these readings that is attributable to each term in the model.

Source	df	MS
O_i	3	1498
I_j	3	1816
OI_{ij}	9	218
$\varepsilon_{k(ij)}$	16	67

6.10 Six different formulas or mixes of concrete are to be purchased from five competing suppliers. Tests of crushing strength are to be made on two blocks constructed according to each formula–supplier combination. One block will be poured, hardened, and tested for each combination before the second block is formed, thus making two replications of the whole experiment. Assuming mixes and suppliers fixed and replications random, set up a mathematical model for this situation, outline its ANOVA table, and show what F tests are appropriate.

6.11 Determine the EMS column for Problem 3.6 and solve for components of variance.

6.12 Derive the expression for n_0 in the EMS column of a single-factor completely randomized experiment where the n's are unequal.

6.13 Verify the numerical results of Table 6.5, using the methods of Chapter 5.

6.14 Verify the EMS column of Table 6.5 by the method of Section 6.4.

6.15 Set up F' tests for Problem 6.4 and explain how the df would be determined.

6.16 Set up F' tests for Problem 6.5 and explain how the df would be determined.

6.17 A manufacturer of direct ignition systems (DISs) uses one of two probe cards and one of four hot testers to test the circuits of a DIS. The probe cards provide the interface between the test and the DIS circuits. The tester measures the current, in milliamperes, under specified conditions. A completely randomized experiment using 16 DISs produced the following data.

Card	Tester 1	2	3	4
1	40.498	40.235	40.252	40.458
	40.513	40.253	40.277	40.503
2	40.382	40.158	40.185	40.460
	40.375	40.131	40.215	40.507

Assuming that the cards and testers used in the experiment were randomly obtained from large universes of cards and testers, estimate the percentage of the total variation in current due to cards and that due to testers.

6.18 For a particular type of resistor, specifications are LSL = 0.000 and USL = 0.700 Ω. A representative sample of 10 resistors is selected for use in a measurement system evaluation. Three operators are then asked to use an *LCR* (inductance, capacitance, resistance) meter to obtain two measurements for each resistor. The measurements, obtained in a completely random order, follow. Determine the values of %R&R$_{tolerance}$ and %R&R$_{process}$ for this study (see Section 6.9). Is the measurement system acceptable?

Part	Operator 1	2	3
1	0.417	0.394	0.404
	0.419	0.398	0.401
2	0.417	0.387	0.398
	0.417	0.399	0.402
3	0.423	0.389	0.407
	0.418	0.407	0.402
4	0.412	0.389	0.407
	0.410	0.405	0.411
5	0.407	0.386	0.400
	0.409	0.405	0.410
6	0.408	0.382	0.405
	0.413	0.400	0.410
7	0.409	0.385	0.407
	0.408	0.400	0.400
8	0.408	0.384	0.402
	0.411	0.401	0.405
9	0.412	0.387	0.412
	0.408	0.401	0.405
10	0.410	0.386	0.418
	0.404	0.407	0.404

6.19 An engineer wants to study the effects of region of the country (R), type of lens coating (L), type of housing coating (H), and type of sealer (S) on the reflectivity of an automobile's rear tail-lamp lens. Two levels of each of L, H, and S will be used. Fifteen lenses will be prepared for each of the $2 \times 2 \times 2 = 8$ treatment combinations. By random assignment, five lenses of each type will be installed on vehicles (a total of 40) destined for the Southeast region of the United States, five of each type will be installed on vehicles destined for the Southwest region, and five of each type will be installed on vehicles destined for the Midwest. The reflectivity of each lens will be measured before shipment and again after a period of approximately 6 months. The difference between the two measurements will be used as a measure of the change in reflectivity (Y). Determine an appropriate statistical model and the associated EMS table.

Chapter 7

NESTED AND NESTED-FACTORIAL EXPERIMENTS

7.1 INTRODUCTION

In a recent in-plant training course, the members of the class were assigned a final problem. Each person was to go into the plant and set up an experiment using the techniques that had been discussed in class. One engineer wanted to study the strain readings of glass cathode supports from five different machines. Each machine had four "heads" on which the glass was formed, and the engineer decided to take four samples from each head. She treated this experiment as a 5×4 factorial with four replications per cell. Complete randomization of the testing for strain readings presented no problem. Her model was

$$Y_{ijk} = \mu + M_i + H_j + MH_{ij} + \varepsilon_{k(ij)}$$

with

$$i = 1, 2, \ldots, 5 \quad j = 1, \ldots, 4 \quad k = 1, \ldots, 4$$

Her data and analysis appear in Table 7.1. In this model she assumed that both machines and heads were fixed, and she used the 10% significance level. The results indicated no significant machine or head effect on strain readings, but there was a significant interaction at the 10% level of significance.

Upon presenting her findings, the engineer was asked whether the four heads were actually removed from machine A and mounted on machine B, then on C, and so on. Of course, the answer was "no," since each machine had its own four heads. Thus, machines and heads did not form a factorial experiment; rather, the heads on each machine were unique for that particular machine. In such a case the experiment is called a *nested experiment:* levels of one factor are nested within, or are subsamples of, levels of another factor. Such experiments are also sometimes called *hierarchical* experiments. When factors are arranged in a factorial experiment as in Chapter 5, they are often referred to as *crossed factors*.

7.2 NESTED EXPERIMENTS

The data in Table 7.1 should be analyzed by treating the experiment as a nested experiment, since heads are nested within machines. Such a factor may be represented in the model as

TABLE 7.1
Data and ANOVA for Strain-Reading Problem

Head		Machine				
		A	B	C	D	E
1		6	10	0	11	1
		2	9	0	0	4
		0	7	5	6	7
		8	12	5	4	9
2		13	2	10	5	6
		3	1	11	10	7
		9	1	6	8	0
		8	10	7	3	3
3		1	4	8	1	3
		10	1	5	8	0
		0	7	0	9	2
		6	9	7	4	2
4		7	0	7	0	3
		4	3	2	8	7
		7	4	5	6	4
		9	1	4	5	0

Source	df	SS	MS	EMS	F	$F_{0.90}$
M_i	4	45.08	11.27	$\sigma_\varepsilon^2 + 16\phi_M$	1.05	2.04
H_j	3	46.45	15.48	$\sigma_\varepsilon^2 + 20\phi_H$	1.45	2.18
MH_{ij}	12	236.42	19.70	$\sigma_\varepsilon^2 + 4\phi_{MH}$	1.84	1.66
$\varepsilon_{k(ij)}$	60	642.00	10.70			
Totals	79	969.95				

$H_{j(i)}$, where j covers all levels, 1, 2, . . . within the ith level of M_i. The number of levels of the nested factor need not be the same for all levels of the other factors. They are all equal in this problem; that is, $j = 1, 2, 3, 4$ for all i. The errors, in turn, are nested within the levels of i and j; $\varepsilon_{k(ij)}$ and $k = 1, 2, 3, 4$ for all i and j.

To emphasize that the heads on each machine are different heads, the data layout in Table 7.2 shows heads 1, 2, 3, and 4 on machine A; heads 5, 6, 7, and 8 on machine B; and so on.

Because the heads that are mounted on the machine can be chosen from many possible heads, we might consider the four heads that could be used on a given machine as a random sample of heads. If such heads were selected at random for the machines, the model would be

$$Y_{ijk} = \mu + M_i + H_{j(i)} + \varepsilon_{k(ij)} \tag{7.1}$$

with

$$i = 1, \ldots, 5 \qquad j = 1, \ldots, 4 \qquad k = 1, \ldots, 4$$

This nested model has no interaction present, since the heads are not crossed with the five machines. If heads are considered random and machines fixed, the proper EMS values can be determined by the rules given in Chapter 6, as shown in Table 7.3.

This breakdown shows that the head effect is to be tested against the error, and the machine effect is to be tested against the heads-within-machines effect.

TABLE 7.2
Data for Strain-Reading Problem in a Nested Arrangement

| | | | | | | | | | | Machine | | | | | | | | | | |
| | A | | | | B | | | | | C | | | | D | | | | E | | |
Head	1	2	3	4	5	6	7	8	9	10	11	12	13	14	15	16	17	18	19	20
	6	13	1	7	10	2	4	0	0	10	8	7	11	5	1	0	1	6	3	3
	2	3	10	4	9	1	1	3	0	11	5	2	0	10	8	8	4	7	0	7
	0	9	0	7	7	1	7	4	5	6	0	5	6	8	9	6	7	0	2	4
	8	8	6	9	12	10	9	1	5	7	7	4	4	3	4	5	9	3	2	0
Head totals	16	33	17	27	38	14	21	8	10	34	20	18	21	26	22	19	21	16	7	14
Machine totals		93				81					82				88				58	

TABLE 7.3
EMS for Nested Experiment

| | 5 | 4 | 4 | |
| | F | R | R | |
Source	i	j	k	EMS
M_i	0	4	4	$\sigma_\varepsilon^2 + 4\sigma_H^2 + 16\phi_M$
$H_{j(i)}$	1	1	4	$\sigma_\varepsilon^2 + 4\sigma_H^2$
$\varepsilon_{k(ij)}$	1	1	1	σ_ε^2

Minitab Results for the Strain-Reading Problem

Many statistical computing packages will directly analyze nested experiments like this one. For example, when the data in Table 7.2 are analyzed using Minitab's **Balanced ANOVA . . .** module, the results in Table 7.4 are obtained. Notice that the heads are labeled 1, 2, 3, and 4 for each machine with head input as a nested, random factor. Also notice that the included EMS table is identical to that of Table 7.3 (the symbol Q[1] denotes ϕ_M).

From Table 7.4, machines appear to have no significant effect on strain readings (p value = 0.670), but the effect of heads within machines is significant for any $\alpha \geq 0.0625$. This example shows the importance of recognizing the difference between a nested experiment and a factorial experiment, since what the experimenter took as head effect (3 df) and interaction effect (12 df) is really heads-within-machines effect (15 df). These results might suggest a more careful adjustment between heads on the same machine.

Further Analysis of the Heads Within Machines

Since there are significant differences between heads within machines, the heads-within-machines sum of squares of 282.88 might well be analyzed further. To do so, one can look at its partitioning into the five sums of squares associated with the five machines. This can

TABLE 7.4
Minitab ANOVA for Nested Strain-Reading Problem

```
Analysis of Variance (Balanced Designs)
```

Factor	Type	Levels	Values				
Machine	fixed	5	1	2	3	4	5
Head(Machine)	random	4	1	2	3	4	

Analysis of Variance for Reading

Source	DF	SS	MS	F	P
Machine	4	45.08	11.27	0.60	0.670
Head(Machine)	15	282.88	18.86	1.76	0.063
Error	60	642.00	10.70		
Total	79	969.95			

Source	Variance component	Error term	Expected Mean Square (using restricted model)
1 Machine		2	(3) + 4(2) + 16Q[1]
2 Head(Machine)	2.040	3	(3) + 4(2)
3 Error	10.700		(3)

be accomplished manually or by means of a computer program that will give a one-factor ANOVA summary for each machine. For review purposes, a manual analysis is presented.

The following calculations for the sums of squares of the heads within a machine were derived from the totals in Table 7.2. Since each machine has four heads, each sum of squares has 3 degrees of freedom.

$$\text{Machine A: } SS_H = \frac{(16)^2 + (33)^2 + (17)^2 + (27)^2}{4} - \frac{(93)^2}{16}$$

$$= 590.750 - 540.5625 = 50.1875$$

$$\text{Machine B: } SS_H = \frac{(38)^2 + (14)^2 + (21)^2 + (8)^2}{4} - \frac{(81)^2}{16}$$

$$= 536.250 - 410.0625 = 126.1875$$

$$\text{Machine C: } SS_H = \frac{(10)^2 + (34)^2 + (20)^2 + 18^2}{4} - \frac{(82)^2}{16}$$

$$= 495.000 - 420.2500 = 74.7500$$

$$\text{Machine D: } SS_H = \frac{(21)^2 + (26)^2 + (22)^2 + (19)^2}{4} - \frac{(88)^2}{16}$$

$$= 490.500 - 484.0000 = 6.5000$$

$$\text{Machine E: } SS_H = \frac{(21)^2 + (16)^2 + (7)^2 + (14)^2}{4} - \frac{(58)^2}{16}$$

$$= 235.500 - 210.2500 = 25.2500$$

Adding these totals to find the sum of squares between heads within machines, we have

$$SS_{H(M)} = 50.1875 + 126.1875 + 74.7500 + 74.7500 + 6.5000 + 25.2500$$

$$= 282.8750 \approx 282.88$$

with degrees of freedom $5 \times 3 = 15$ as given in Table 7.4. Further breakdown and tests of Table 7.4 can be made, as in Table 7.5. From that table, using a 10% significance level, we see a significant difference between heads on machine B (p value $= 0.0125$) and machine C (p value $= 0.0833$).

The total, machine, and error sums of squares in Table 7.5 are those obtained by computer and summarized in Table 7.4. For computer programs that do not allow the user to input a nested model, a one-factor analysis of variance on machines can be used to obtain the total sum of squares and the machine sum of squares. Manual calculations using the individual data and machine totals in Table 7.2 give

$$SS_{total} = 6^2 + 2^2 + \cdots + 4^2 + 0^2 - \frac{(402)^2}{80} = 969.950$$

and

$$SS_{machine} = \frac{(93)^2 + (81)^2 + (82)^2 + (88)^2 + (58)^2}{16} - \frac{(402)^2}{80} = 45.075$$

Since $SS_{H(M)} = 282.875$ and $SS_{total} = SS_{machine} + SS_{H(M)} + SS_{error}$, we find that

$$SS_{error} = SS_{total} - SS_{machine} - SS_{H(M)} = 969.950 - 45.075 - 282.875 = 642.000$$

Further Analysis of the Heads on Machines B and C

Since there are significant differences ($\alpha = 0.10$) between heads on machines B and C, one might examine the head means within each of these machines in a Newman–Keuls sense. Statistical Tables E.1 and E.2 include only upper percentiles of the Studentized range for $\alpha = 0.01$ and $\alpha = 0.05$, so we must estimate or find a more complete table. Using Appendix 6 in Anderson and McLean [1, p. 296] with 60 df, we find values of 2.363, 2.959, and 3.312 for $p = 2$, 3, and 4, respectively.

Assuming that the error mean square (10.70) is our best estimate of the error variance, the standard error of a mean associated with one of the machine heads is $s_{\bar{y}} = \sqrt{10.70/4} \approx 1.636$. Multiplying each range in the preceding paragraph by this standard error, we find that the least significant ranges are

TABLE 7.5
Detailed ANOVA for Strain-Reading Data

Source	df	SS	MS	F	p
M_i	4	45.0750	11.26875	0.60	0.6700
$H_{j(i)}$	15	282.8750	18.85833	1.76	0.0630
$H_{j(A)}$	3	50.1875	16.72917	1.56	0.2084
$H_{j(B)}$	3	126.1875	42.06250	3.93	0.0125
$H_{j(C)}$	3	74.7500	24.91667	2.33	0.0833
$H_{j(D)}$	3	6.5000	2.16667	0.20	0.8960
$H_{j(E)}$	3	25.2500	8.41667	0.79	0.5042
$\varepsilon_{k(ij)}$	60	642.0000	10.70000		
Totals	79	969.9500			

p:	2	3	4
LSR:	3.866	4.841	5.418

The sample means for machine B are 9.50, 3.50, 5.25, and 2.00 for heads 5, 6, 7, and 8, respectively. Comparing the ranges between means with the least significant ranges produces the groups in Figure 7.1. This indicates that one should examine head 5, which has the true mean that is the largest of the head means.

The sample means for machine C are 2.50, 8.50, 5.00, and 4.50 for heads 9, 10, 11, and 12, respectively. Comparing the ranges between means with the least significant ranges produces the groups in Figure 7.2, indicating that the true mean for head 9 is less than that for head 10.

Assessing the Model

In our nested model the five sums of squares between heads within each machine were "pooled," or added, to give 282.875. This assumes that these sums of squares (which are proportional to variances) within each machine are of about the same magnitude. Such an assumption might be questioned, however, inasmuch as the observed sum of squares for heads within machine B exceeds that of machine D by a factor in excess of 19. This discrepancy may indicate the need for work on each machine with an aim toward more homogeneous strain readings between heads on each machine.

To further investigate the assumption of equal variances, the range chart of Figure 7.3 was prepared. The chart provides no strong evidence of unequal variances. [**Note:** The residual for the kth reading on the jth head of the ith machine is the difference between that reading and the mean of the four readings on that head.]

Most tests for homogeneity of variances are highly susceptible to nonnormality. However, Levene's test, which can be used with any continuous distribution, is robust to deviations from normality and is based on an assumption of independent random samples. Using the residuals for our nested model, Minitab reports a p value of 0.607 for a test of equal variances among machines and one of 0.833 for a test of equal variances among heads. When these results are combined with data from the range chart of Figure 7.3, it appears that the equal variances assumption is adequately satisfied.

Figure 7.1   Newman–Keuls test for machine B.

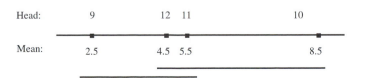

Figure 7.2   Newman–Keuls test for machine C.

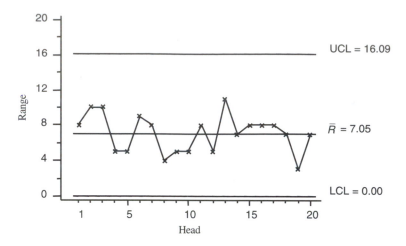

Figure 7.3   Range chart of the residuals: strain reading study.

If Levene's test is not provided in the particular software being used, the data can be subjected to a simple coding to make the test possible. Suppose k independent random samples are to be used to test for equality of variances. Let y_{ij} denote the ith observation in the jth sample, m_j the median of that sample, and $w_{ij} = |y_{ij} - m_j|$ the absolute deviation of y_{ij} from m_j. The test statistic is the F ratio for a one-factor analysis of variance on the k samples of absolute deviations. The test is an upper-tailed test, and a large p value supports an assumption of equal variances.

To investigate the normality assumption, consider the normal probability plot of Figure 7.4. The plotted points form a reasonably linear pattern and the p value of a Shapiro–Wilk test for normality exceeds 0.10. These results indicate that a normality assumption is reasonable.

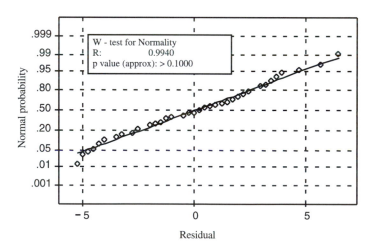

Figure 7.4   Minitab normal probability plot of strain-reading residuals.

7.3 ANOVA RATIONALE

To see that the sums of squares computed in Section 7.2 were correct, consider the nested model

$$Y_{ijk} = \mu + A_i + B_{j(i)} + \varepsilon_{k(ij)}$$

or

$$Y_{ijk} = \mu + (\mu_i - \mu) + (\mu_{ij} - \mu_i) + (Y_{ijk} - \mu_{ij})$$

which is an identity.

Using the best estimates of these population means from the sample data, we write the following sample model

$$Y_{ijk} \equiv \bar{Y}_{...} + (\bar{Y}_{i..} - \bar{Y}_{...}) + (\bar{Y}_{ij.} - \bar{Y}_{i..}) + (Y_{ijk} - \bar{Y}_{ij.})$$

with

$$i = 1, 2, \ldots, a \quad j = 1, 2, \ldots, b \quad k = 1, 2, \ldots, n$$

Transposing $\bar{Y}_{...}$ to the left of this expression, squaring both sides, and adding over i, j, and k gives

$$\sum_i^a \sum_j^b \sum_k^n (Y_{ijk} - \bar{Y}_{...})^2 = \sum_{i=1}^a nb(\bar{Y}_{i..} - \bar{Y}_{...})^2 + \sum_i^a \sum_j^b n(\bar{Y}_{ij.} - \bar{Y}_{i..})^2$$
$$+ \sum_i^a \sum_j^b \sum_k^n (Y_{ijk} - \bar{Y}_{ij.})^2$$

(since the sums of cross products equal zero). This equation expresses the idea that the total sum of squares is equal to the sum of squares between levels of A, plus the sums of squares between levels of B within each level of A, plus the sum of the squares of the errors. The degrees of freedom are

$$(abn - 1) \equiv (a - 1) + a(b - 1) + ab(n - 1)$$

Dividing each independent sum of squares by its corresponding degrees of freedom gives estimates of population variance as usual. For computing purposes, the sum of squares as given above should be expanded in terms of totals, with the general results shown in Table 7.6.

This is essentially the form followed in the strain-reading problem of the last section. Note that

$$SS_B = \sum_i^a \sum_j^b \frac{T_{ij.}^2}{n} - \sum_i^a \frac{T_{i..}^2}{nb} = \sum_i^a \left(\sum_j^b \frac{T_{ij.}^2}{n} - \frac{T_{i..}^2}{nb} \right)$$

which shows how the sum of squares was calculated in the last section: by getting the sum of squares between levels of B for each level of A and then pooling over all levels of A.

TABLE 7.6
General ANOVA for a Nested Experiment

Source	df	SS	MS
A_i	$a - 1$	$\sum\limits_{i}^{a} \dfrac{T_{i..}^2}{nb} - \dfrac{T_{...}^2}{nab}$	$\dfrac{SS_A}{a-1}$
$B_{j(i)}$	$a(b-1)$	$\sum\limits_{i}^{a}\sum\limits_{j}^{b} \dfrac{T_{ij.}^2}{n} - \sum\limits_{i}^{a} \dfrac{T_{i..}^2}{nb}$	$\dfrac{SS_B}{a(b-1)}$
$\varepsilon_{k(ij)}$	$ab(n-1)$	$\sum\limits_{i}^{a}\sum\limits_{j}^{b}\sum\limits_{k}^{n} Y_{ijk}^2 - \sum\limits_{i}^{a}\sum\limits_{j}^{b} \dfrac{T_{ij.}^2}{n}$	$\dfrac{SS_\varepsilon}{ab(n-1)}$
Totals	$abn - 1$	$\sum\limits_{i}^{a}\sum\limits_{j}^{b}\sum\limits_{k}^{n} Y_{ijk}^2 - \dfrac{T_{...}^2}{nab}$	

7.4 NESTED-FACTORIAL EXPERIMENTS

Many multiple-factor experiments involve both factors that are crossed with others and factors that are nested within levels of the others. When this happens—that is, when both factorial and nested factors appear in the same experiment—the term *nested-factorial experiment* is applied. The analysis of such an experiment is simply an extension of the methods of Chapter 5 and this chapter. Care must be exercised, however, in computing some of the interactions. Levels of both factorial and nested factors may be either fixed or random. The methods of Chapter 6 can be used to determine the EMS values and the proper tests to be run.

■ **Example 7.1**

An investigator who wished to increase the number of rounds per minute that could be fired from a naval gun devised a new loading method (method I) with the intent of improving performance in this task over that obtained with the existing method of loading (method II). Realizing that the general physique of a person might affect the speed with which a gun could be loaded, the investigator selected artillery teams of persons in three general groupings (slight, average, and heavy). The classification of such persons was on the basis of an Armed Services classification table. Three teams were randomly chosen to represent each of the three physique groupings.

   Once these three teams were selected to represent each of the physique groups, the nine teams set up a schedule for coming to a gym to run the test. In each case, a coin was tossed when a team arrived at the gym to decide which of the two methods would be used that day. This was repeated for three more random days with the restriction that in the four days each team should be assigned to each method twice.

To collect the data, a team would be timed each day for 20 minutes. Due to start-up concerns and fatigue concerns, only the middle 10 minutes would be used to give the number of rounds per minute for that run. The model for this experiment was:

$$Y_{ijkm} = \mu + M_i + G_j + MG_{ij} + T_{k(j)} + MT_{ik(j)} + \varepsilon_{m(ijk)}$$

where

$$M_i = \text{methods}, \ i = 1, 2$$

$$G_j = \text{groups}, \ j = 1, 2, 3$$

$$T_{k(j)} = \text{teams within groups}, \ k = 1, 2, 3, \text{for all } j$$

$$\varepsilon_{m(ijk)} = \text{random error}, \ m = 1, 2, \text{for all } i, j, k$$

The EMS values are shown in Table 7.7, which indicates the proper F tests to run, and the data appear in Table 7.8.

An ANOVA summary for this experiment is displayed in Table 7.9. Notice that the expected mean squares summarized are identical to those in Table 7.7. The results show a very significant method effect (p value = 0.000): as indicated in Table 7.8, the new method (I) averaged 23.58 rounds per minute and the old method (II) averaged only 15.08 rounds per minute. The results also show a significant difference among teams within groups (p value = 0.040), which points out individual differences in personnel. No other effects or interactions are significant.

TABLE 7.7
EMS for Gun-Loading Problem

Source	2 F i	3 F j	3 R k	2 R m	EMS
M_i	0	3	3	2	$\sigma_\varepsilon^2 + 2\sigma_{MT}^2 + 18\phi_M$
G_j	2	0	3	2	$\sigma_\varepsilon^2 + 4\sigma_T^2 + 12\phi_G$
MG_{ij}	0	0	3	2	$\sigma_\varepsilon^2 + 2\sigma_{MT}^2 + 6\phi_{MG}$
$T_{k(j)}$	2	1	1	2	$\sigma_\varepsilon^2 + 4\sigma_T^2$
$MT_{ik(j)}$	0	1	1	2	$\sigma_\varepsilon^2 + 2\sigma_{MT}^2$
$\varepsilon_{m(ijk)}$	1	1	1	1	σ_ε^2

TABLE 7.8
Data and ANOVA for Gun-Loading Problem

Team	Group I 1	2	3	Group II 4	5	6	Group III 7	8	9
Method I	20.2	26.2	23.8	22.0	22.6	22.9	23.1	22.9	21.8
	24.1	26.9	24.9	23.5	24.6	25.0	22.9	23.7	23.5
Method II	14.2	18.0	12.5	14.1	14.0	13.7	14.1	12.2	12.7
	16.2	19.1	15.4	16.1	18.1	16.0	16.1	13.8	15.1

TABLE 7.9
Minitab ANOVA for Gun-Loading Problem

```
Analysis of Variance (Balanced Designs)
```

Factor	Type	Levels	Values		
Method	fixed	2	1	2	
Group	fixed	3	1	2	3
Team(Group)	random	3	1	2	3

```
Analysis of Variance for Speed
```

Source	DF	SS	MS	F	P
Method	1	651.951	651.951	364.84	0.000
Group	2	16.052	8.026	1.23	0.358
Method*Group	2	1.187	0.594	0.33	0.730
Team(Group)	6	39.258	6.543	2.83	0.040
Method*Team(Group)	6	10.722	1.787	0.77	0.601
Error	18	41.590	2.311		
Total	35	760.760			

Source	Variance component	Error term	Expected Mean Square (using restricted model)
1 Method		5	(6) + 2(5) + 18Q[1]
2 Group		4	(6) + 4(4) + 12Q[2]
3 Method*Group		5	(6) + 2(5) + 6Q[3]
4 Team(Group)	1.0581	6	(6) + 4(4)
5 Method*Team(Group)	-0.2618	6	(6) + 2(5)
6 Error	2.3106		(6)

A further analysis of the significant differences among teams within groups I, II, and III shows the sums of squares to be 35.74, 1.62, and 1.90, respectively. Since the error mean square in Table 7.9 is 2.31, we see that the significant difference between teams within the three groups is concentrated in group I.

Since the effect of teams within groups is tested by using the error mean square in the denominator of the F ratio, the sample standard error of a mean associated with a team in group I is $s_{\bar{Y}} = \sqrt{2.311/4} \approx 0.76$. Using statistical Table E.2 with $\alpha = 0.01$ and $n_2 = 18$, we find the upper percentiles of the Studentized range to be 4.07 and 4.70 for $p = 2$ and 3, respectively. Multiplying these percentiles by 0.76 yields least significant ranges of 3.09 and 3.57 for $p = 2$ and 3, respectively. Comparing the ranges between means with the least significant ranges produces the groups in Figure 7.5, indicating that team 2 is exceptionally faster than the other two teams.

Team	1	3		2
Mean	18.675	19.15		22.55

Figure 7.5   Newman–Keuls test for I.

Manual Calculation of Sums of Squares

It would be well for the reader to verify that the sums of squares of Table 7.9 are correct. The only term that differs somewhat in this model from those previously handled is $MT_{ik(j)}$, that is, the interaction between methods and teams within groups. The safest way to compute this term is to compute the $M \times T$ interaction sums of squares within each of the three groups separately and then pool these sums of squares. (See Tables 7.10–7.12.)

To compute the $M \times T$ interaction for group I, we have

$$SS_{cell} = \frac{(44.3)^2 + (53.1)^2 + (48.7)^2 + (30.4)^2 + (37.1)^2 + (27.9)^2}{2}$$

$$- \frac{(241.5)^2}{12} = 5116.385 - 4860.1875 = 256.1975$$

$$SS_{method} = \frac{(146.1)^2 + (95.4)^2}{6} - 4860.1875 = 214.2075$$

$$SS_{team} = \frac{(74.7)^2 + (90.2)^2 + (76.6)^2}{4} - 4860.1875 = 35.7350$$

TABLE 7.10
Data on Gun-Loading Problem for Group I

	Team						Method Totals
	1		**2**		**3**		
Method I	20.2		26.2		23.8		
	24.1		26.9		24.9		
		44.3		53.1		48.7	146.1
Method II	14.2		18.0		12.5		
	16.2		19.1		15.4		
		30.4		37.1		27.9	95.4
Team totals		74.7		90.2		76.6	241.5

TABLE 7.11
Data on Gun-Loading Problem for Group II

	Team						Method Totals
	4		**5**		**6**		
Method I	22.0		22.6		22.9		
	23.5		24.6		25.0		
		45.5		47.2		47.9	140.6
Method II	14.1		14.0		13.7		
	16.1		18.1		16.0		
		30.2		32.1		29.7	92.0
Team totals		75.7		79.3		77.6	232.6

TABLE 7.12
Data on Gun-Loading Problem for Group III

	Team						Method Totals
	7		8		9		
Method I	23.1		22.9		21.8		
	22.9		23.7		23.5		
		46.0		46.6		45.3	137.9
Method II	14.1		12.2		12.7		
	16.1		13.8		15.1		
		30.2		26.0		27.8	84.0
Team totals		76.2		72.6		73.1	221.9

$$SS_{M \times T \text{ interaction}} = 256.1975 - 214.2075 - 35.7350 = 6.2550$$

For group II:

$$SS_{\text{cell}} = \frac{(45.5)^2 + (47.2)^2 + (47.9)^2 + (30.2)^2 + (32.1)^2 + (29.7)^2}{2}$$

$$- \frac{(232.6)^2}{12} = 4708.5200 - 4508.5633 = 199.9567$$

$$SS_{\text{method}} = \frac{(140.6)^2 + (92.0)^2}{6} - 4508.5633 = 196.8300$$

$$SS_{\text{team}} = \frac{(75.7)^2 + (79.3)^2 + (77.6)^2}{4} - 4508.5633 = 1.6217$$

$$SS_{M \times T \text{ interaction}} = 199.9567 - 196.8300 - 1.6217 = 1.5050$$

For group III:

$$SS_{\text{cell}} = \frac{(46.0)^2 + (46.6)^2 + (45.3)^2 + (30.2)^2 + (26.0)^2 + (27.8)^2}{2}$$

$$- \frac{(221.9)^2}{12} = 4350.2650 - 4103.3008 = 246.9642$$

$$SS_{\text{method}} = \frac{(137.9)^2 + (84.0)^2}{6} - 4103.3008 = 242.1009$$

$$SS_{\text{team}} = \frac{(76.2)^2 + (72.6)^2 + (73.1)^2}{4} - 4103.3008 = 1.9017$$

$$SS_{M \times T \text{ interaction}} = 246.9642 - 242.1009 - 1.9017 = 2.9616$$

Pooling for all three groups gives

$$SS_{M \times T} = 6.2550 + 1.5050 + 2.9616 \approx 10.722$$

which is recorded in Table 7.9.

Graphical Representation of the Gun-Loading Experiment

The gun-loading experiment discussed in Example 7.1 is represented graphically in Figure 7.6, which depicts the difference between crossed and nested factors. The lines extending from the levels of method to those of group cross, indicating that each level of method can be used in combination with any level of group. On the other hand, the lines originating at a particular group extend directly to the teams within that group and do not cross over to teams within another group. This is because the teams within a group are unique to that group, whereas the grouping levels can occur in both methods.

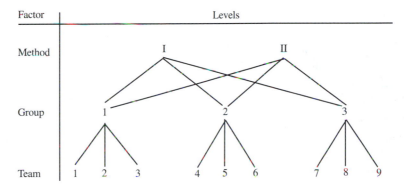

Figure 7.6   The crossed and nested factors of Example 7.1.

7.5 REPEATED-MEASURES DESIGN AND NESTED-FACTORIAL EXPERIMENTS

Many statisticians who work with psychologists and educators on the design of their experiments treat repeated-measures designs as a unique topic in these fields of applied statistics. This section shows that these designs are but a special case of factorial and nested-factorial experiments. Numerical examples and mathematical models are used to illustrate the correspondence between these designs.

In one of the simplest cases, a pretest and a posttest are given to the same group of subjects after a certain time lapse, during which some special instruction may have been administered.

■ **Example 7.2**

In a recent study at Purdue University, seven subjects were evaluated for physical strength before and after a specified training period. Results (in pounds) are shown in Table 7.13.

TABLE 7.13
Pretest and Posttest Measures for Example 7.2

Subjects	Pretest	Post-test
1	100	115
2	110	125
3	90	105
4	110	130
5	125	140
6	130	140
7	105	125

Since the same seven subjects were given each test, we have two repeated measures on each subject. This is an experiment that can be handled as a between-subjects, within-subjects, repeated-measures design (e.g., according to Winer [35]), with results reported as in Table 7.14.

TABLE 7.14
ANOVA for Table 7.13

Source of Variation	df	SS	MS	F
Between subjects, S_i	6	2084.71	347.45	
Within subjects	7	901.00	—	
Tests, T_j	1	864.29	864.29	145
Residual	6	35.71	5.96	
Totals	13	2985.71		

This analysis implies that the subject model is

$$Y_{ij} = \mu + S_i + \varepsilon_{j(i)}$$

$$\underbrace{6 \qquad 7}_{df}$$

And then the within-subjects' data are further broken down into tests and residual. This residual is actually the subject by test interaction so the model becomes

$$Y_{ij} = \mu + S_i + T_j + ST_{ij}$$

$$\underbrace{6 \quad 1 \quad 6}_{df}$$

This model contains no error term, since only one pretest score and one posttest score are available on each subject.

Now if this problem were considered in more general terms instead of the so-called repeated-measures format, one could treat the data layout as a two-factor factorial experiment with one observation per treatment. Subjects would then be chosen at random and tests would be fixed. Using the algorithm given in Chapter 6, we would produce the ANOVA layout shown in Table 7.15, where $k = 1$ and a separate error $\varepsilon_{k(ij)}$ is not retrievable.

TABLE 7.15
EMS Determination for Table 7.14

Source	df	7 R i	2 F j	1 R k	EMS
S_i	6	1	2	1	$\sigma_\varepsilon^2 + 2\sigma_S^2$
T_j	1	7	0	1	$\sigma_\varepsilon^2 + \sigma_{ST}^2 + 7\phi_T$
ST_{ij}	6	1	0	1	$\sigma_\varepsilon^2 + \sigma_{ST}^2$
$\varepsilon_{k(ij)}$	0	1	1	1	σ_ε^2 (not retrievable)

A glance at the EMS column of Table 7.15 indicates that the only proper F test is to compare the test mean square with the $S \times T$ interaction mean square, which was the test as shown in Table 7.14.

Of course, another approach to this example is to take differences between post- and pretest scores on each individual and test the hypothesis that the true mean of the difference is zero. Using this scheme, a t of 12.05 is found and it is well known that this t when squared equals the F of our ANOVA table:

$$t^2 = (12.05)^2 = 145 = F$$

This is, of course, equivalent to using a randomized complete block design with subjects treated as blocks and testing only for a treatment effect (see Example 4.1).

■ **Example 7.3**

A slightly more complex problem involves three factors in which the subjects are nested within groups. These groups could be classes or experimental conditions, with each subject again subjected to repeated measures. The study cited on physical strength was extended to incorporate three groups of subjects: two being subjected to special experimental training and the third acting as a control with no special training. Again, each subject was given a pre- and a posttest, but this time the measurement was velocity of a baseball throw, in meters per second. The resulting data are given in Table 7.16. If these data are treated as a repeated-measures design, the results are usually presented as shown in Table 7.17.

TABLE 7.16
Pretest and Posttest Throwing Velocities (m/s)
of Three Groups of Subjects

Group	Subject	Pretest	Posttest
I	1	26.25	29.50
	2	24.33	27.62
	3	22.52	25.71
	4	29.33	31.55
	5	28.90	31.35
	6	25.13	29.07
	7	29.33	31.15
II	8	27.47	28.74
	9	25.19	26.11
	10	23.53	25.45
	11	24.57	25.58
	12	26.88	27.70
	13	27.86	28.82
	14	28.09	28.99
III	15	22.27	22.52
	16	21.55	21.79
	17	23.31	23.53
	18	30.03	30.21
	19	28.17	28.65
	20	28.09	28.33
	21	27.55	27.86

If the layout of Table 7.16 is treated as a nested-factorial experiment in which subjects are nested within groups and then tests are factorial on both groups and subjects, the model, which includes a nonretrievable error term, can be written as follows:

$$Y_{ijk} = \mu + G_i + S_{j(i)} + T_k + GT_{ik} + TS_{kj(i)} + \varepsilon_{m(ijk)}$$

$$\underbrace{2 \quad\quad 18 \quad\quad 1 \quad 2 \quad\quad 18 \quad\quad 0}_{df}$$

TABLE 7.17
ANOVA of Table 7.16

Source	df		SS	MS	EMS	F
Between subjects	20		271.05	—		
Groups, G_i		2	28.14	14.07	$\sigma_e^2 + 2\sigma_S^2 + 14\phi_G$	1.04
Subjects within groups, $S_{j(i)}$		18	242.91	13.50	$\sigma_e^2 + 2\sigma_S^2$	
Within subjects	21		35.73	—		
Tests, T_k		1	21.26	21.26	$\sigma_e^2 + \sigma_{TS}^2 + 21\phi_T$	183
$G \times T$		2	12.38	6.19	$\sigma_e^2 + \sigma_{TS}^2 + 7\phi_{GT}$	53
$T \times S_{kj(i)}$		18	2.09	0.116	$\sigma_e^2 + \sigma_{TS}^2$	

TABLE 7.18
EMS Determination for Table 7.16

Source	df	F 3 i	R 7 j	F 2 k	R 1 m	EMS
G_i	2	0	7	2	1	$\sigma_e^2 + 2\sigma_S^2 + 14\phi_G$
$S_{j(i)}$	18	1	1	2	1	$\sigma_e^2 + 2\sigma_S^2$
T_k	1	3	7	0	1	$\sigma_e^2 + \sigma_{TS}^2 + 21\phi_T$
GT_{ik}	2	0	7	0	1	$\sigma_e^2 + \sigma_{TS}^2 + 7\phi_{GT}$
$TS_{kj(i)}$	18	1	1	0	1	$\sigma_e^2 + \sigma_{TS}^2$
$\varepsilon_{m(ijk)}$	0	1	1	1	1	σ_e^2 (not retrievable)

Use of this model and the algorithm described in Chapter 6 gives Table 7.18.

This scheme generates the EMS column that was simply reported in Table 7.17, and that column indicates the proper tests to be made. The numerical results are, of course, the same.

The nested-factorial experiment covers many situations in addition to the repeated-measures design. The factor in the nest may be farms within townships, classes within schools, heads within machines, samples within batches, and so on. It treats all three-factor situations in which at least one is nested and at least one is factorial. Usually in a repeated-measures design the subjects are chosen at random and the other two factors are considered to be fixed. By setting up the model and using the EMS algorithm, one may have factors that are random as well as fixed. For example, classes could very well be chosen at random from a given grade level and then the students within the classes chosen at random for the experiment.

As more factors are added in an experiment, one need only expand the mathematical model and determine the proper tests to make based on the EMS column. It is not necessary to think in terms of repeat measures. To illustrate further, we consider two cases of three-factor experiments with repeated measures. (This material was originally presented by Winer [40].) Case I has two of the factors (fixed) crossing (or repeated on) all subjects, and in case II subjects are nested within two factors (fixed) and the third factor crosses (or is repeated on) all subjects. Both these cases, and more complex ones, can easily be handled as nested-factorial experiments by writing the proper model. Handling them in this way does not require that all three factors be fixed. It simply requires that the subjects be treated as another factor nested within one or more of the other factors.

Models for cases I and II show the relationship between the two approaches.

CASE I

Y_{ijkm}

$$= \mu + \overbrace{A_i + S_{j(i)}}^{\text{between subjects}}$$

$$+ \overbrace{B_k + AB_{ik} + BS_{kj(i)} + C_m + AC_{im} + CS_{mkj(i)} + BC_{km} + ABC_{ikm} + BCS_{mkj(i)}}^{\text{within subjects}}$$

CASE II

$$Y_{ijkm} = \mu + \overbrace{A_i + B_j + AB_{ij} + S_{k(ij)}}^{\text{between subjects}}$$

$$+ \overbrace{C_m + AC_{im} + BC_{jm} + ABC_{ijm} + CS_{mk(ij)}}^{\text{within subjects}}$$

No error terms have been included because they are not retrievable.

These examples, it is hoped, are sufficient to remove any mystery surrounding the repeated-measures designs by showing that they fit into more general models familiar to statisticians in all fields.

7.6 SAS PROGRAMS FOR NESTED AND NESTED-FACTORIAL EXPERIMENTS

For nested and nested-factorial experiments, two new concepts are added to the SAS program:

1. Nesting notation. If A is one factor and the levels of a factor B are to be nested within levels of A, we write $B(A)$, and the model is RDG = $A \, B(A)$. In more complex problems we may wish to use the upright pipe to break out the interactions. If the pipe is inserted between A and $B(A)$, the program, recognizing the nested notation, will not try to compute an $A * B$ interaction. For example, with a class statement including A, B, and C, a model might be RDG = $A|B(A)|C$, and the output will include A, $B(A)$, C, $A*C$, $C*B(A)$.

2. Because nested factors are almost always run at random levels, it is necessary to determine the EMS appropriate for a given problem before the error term to be used for each main effect and interaction can be chosen. The SAS program will test all effects against the error term if one exists in the problem, so the program must be told which tests are appropriate. This is done by adding one or more **TEST** commands where H (for hypothesis) = term to be tested, and E (for error) = proper divisor for

testing H. This again shows the importance of performing the EMS determination *before* proceeding to an analysis.

SAS Analysis of a Nested Experiment

A SAS command file for the data in Table 7.2 is given in Table 7.19. We used the ANOVA procedure, but the GLM procedure can be used if preferred.

The outputs for the Table 7.19 command file are displayed in Figure 7.7. Results of the test on machine (p value = 0.6700) and head nested within machine (p value = 0.0625) agree with those of Table 7.4.

TABLE 7.19
SAS Command File for Strain-Reading Study

```
OPTIONS LINESIZE=80;
DATA STRAIN;
INPUT MACHINE $ HEAD RDG @@;
CARDS;
A  1  6 A  1  2 A  1 0 A  1  8 A  2 13 A  2  3 A  2 9 A  2  8
A  3  1 A  3 10 A  3  0 A  3  6 A  4  7 A  4  4 A  4 7 A  4  9
B  5 10 B  5  9 B  5  7 B  5 12 B  6  2 B  6  1 B  6 1 B  6 10
B  7  4 B  7  1 B  7  7 B  7  9 B  8  0 B  8  3 B  8 4 B  8  1
C  9  0 C  9  0 C  9 5 C  9  5 C 10 10 C 10 11 C 10 6 C 10  7
C 11  8 C 11  5 C 11 0 C 11  7 C 12  7 C 12  2 C 12 5 C 12  4
D 13 11 D 13  0 D 13 6 D 13  4 D 14  5 D 14 10 D 14 8 D 14  3
D 15  1 D 15  8 D 15 9 D 15  4 D 16  0 D 16  8 D 16 6 D 16  5
E 17  1 E 17  4 E 17 7 E 17  9 E 18  6 E 18  7 E 18 0 E 18  3
E 19  3 E 19  0 E 19 2 E 20  2 E 20  3 E 20  7 E 20 4 E 20  0
;
PROC ANOVA;
CLASS MACHINE HEAD;
MODEL RDG=MACHINE HEAD(MACHINE);
TEST H=MACHINE E=HEAD(MACHINE);
```

ANALYSIS OF VARIANCE PROCEDURE

DEPENDENT VARIABLE: RDG

SOURCE	DF	SUM OF SQUARES	MEAN SQUARE	F VALUE
MODEL	19	327.95000000	17.26052632	1.61
ERROR	60	642.00000000	10.70000000	PR > F
CORRECTED TOTAL	79	969.95000000		0.0823

R-SQUARE	C.V.	ROOT MSE	RDG MEAN
0.338110	65.0962	3.27108545	5.02500000

SOURCE	DF	ANOVA SS	F VALUE	PR > F
MACHINE	4	45.07500000	1.05	0.3876
HEAD(MACHINE)	15	282.87500000	1.76	0.0625

TESTS OF HYPOTHESES USING THE ANOVA MS FOR HEAD (MACHINE) AS AN ERROR TERM

SOURCE	DF	ANOVA SS	F VALUE	PR > F
MACHINE	4	45.07500000	0.60	0.6700

Figure 7.7   SAS outputs for the strain-reading experiment.

In Section 7.2, the five sums of squares associated with the heads within the five machines were obtained manually. Those sums of squares are easily obtained in SAS if the following lines are added to the command file of Table 7.19:

```
PROC SORT;
  BY MACHINE;
PROC ANOVA;
  CLASS MACHINE HEAD;
  MODEL RDG=HEAD;
  BY MACHINE;
```

Five ANOVA summaries are produced, one for each machine. The reported sums of squares for heads in a machine are 50.18750000, 126.18750000, 74.75000000, 6.50000000, and 25.25000000 for machines A, B, C, D, and E, respectively. These results agree with those in Table 7.5.

SAS Analysis of a Nested-Factorial Experiment

A SAS command file for the gun-loading experiment of Example 7.1 (data in Table 7.9) is given in Table 7.20. The ANOVA procedure was used because, for balanced designs, it is faster and uses less storage than GLM.

The outputs for the Table 7.20 command file are displayed in Figure 7.8. Results show a very significant method effect (p value = 0.0001) and a significant difference between teams within groups for any $\alpha \geq 0.0403$. As in Example 7.1, no other effects or interactions are significant.

TABLE 7.20
SAS Command File for Gun-Loading Study

```
OPTIONS LINESIZE=80;
DATA ROUNDS;
INPUT GROUP TEAM METHOD RDG @@;
CARDS;
1 1 1 20.2 1 1 1 24.1 1 1 2 14.2 1 1 2 16.2 1 2 1 26.2 1 2 1 26.9
1 2 2 18.0 1 2 2 19.1 1 3 1 23.8 1 3 1 24.9 1 3 2 12.5 1 3 2 15.4
2 4 1 22.0 2 4 1 23.5 2 4 2 14.1 2 4 2 16.1 2 5 1 22.6 2 5 1 24.6
2 5 2 14.0 2 5 2 18.1 2 6 1 22.9 2 6 1 25.0 2 6 2 13.7 2 6 2 16.0
3 7 1 23.1 3 7 1 22.9 3 7 2 14.1 3 7 2 16.1 3 8 1 22.9 3 8 1 23.7
3 8 2 12.2 3 8 2 13.8 3 9 1 21.8 3 9 1 23.5 3 9 2 12.7 3 9 2 15.1
;
PROC ANOVA;
CLASS GROUP TEAM METHOD;
MODEL RDG=METHOD|GROUP|TEAM(GROUP);
TEST H=METHOD METHOD*GROUP E=METHOD*TEAM(GROUP);
TEST H=GROUP E=TEAM(GROUP);
```

```
                      ANALYSIS OF VARIANCE PROCEDURE
DEPENDENT VARIABLE: RDG

SOURCE                 DF     SUM OF SQUARES    MEAN SQUARE        F VALUE
MODEL                  17      719.17000000    42.30411765          18.31
ERROR                  18       41.59000000     2.31055556         PR > F
CORRECTED TOTAL        35      760.76000000                        0.0001

R-SQUARE             C.V.          ROOT MSE       RDG MEAN
0.945331           7.8623        1.52005117    19.33333333

SOURCE                 DF           ANOVA SS      F VALUE         PR > F
METHOD                  1       651.95111111       282.16         0.0001
GROUP                   2        16.05166667         3.47         0.0530
TEAM(GROUP)             6        39.25833333         2.83         0.0403
GROUP*METHOD            2         1.18722222         0.26         0.7762
TEAM*METHOD(GROUP)      6        10.72166667         0.77         0.6009

TESTS OF HYPOTHESES USING THE ANOVA MS FOR TEAM*METHOD(GROUP)
   AS AN ERROR TERM

SOURCE                 DF           ANOVA SS      F VALUE         PR > F
METHOD                  1       651.95111111       364.84         0.0001
GROUP*METHOD            2         1.18722222         0.33         0.7296

TESTS OF HYPOTHESES USING THE ANOVA MS FOR TEAM(GROUP) AS AN ERROR TERM

SOURCE                 DF           ANOVA SS      F VALUE         PR > F
GROUP                   2        16.05166667         1.23         0.3576
```

Figure 7.8   SAS outputs for the gun-loading experiment.

7.7 SUMMARY

The summary at the end of Chapter 5 may now be extended for Part II.

Experiment	Design	Analysis
II. Two or more factors A. Factorial (crossed)	1. Completely randomized $Y_{ijk} = \mu + A_i + B_j + AB_{ij} + \varepsilon_{k(ij)}, \dots$ for more factors General case	1. ANOVA with interactions
B. Nested (hierarchical)	1. Completely randomized $Y_{ijk} = \mu + A_i + B_{j(i)} + \varepsilon_{k(ij)}$	1. Nested ANOVA
C. Nested factorial	1. Completely randomized $Y_{ijkm} = \mu + A_i + B_{j(i)} + C_k + AC_{ik} + BC_{jk(i)} + \varepsilon_{m(ijk)}$	1. Nested-factorial ANOVA

7.8 FURTHER READING

Anderson, V. L., and R. A. McLean, *Design of Experiments: A Realistic Approach.* New York: Marcel Dekker, 1974. See Chapter 6: "Nested (Hierarchical) and Nested Factorial Designs."

Gunter, B., "Process Capability Studies. Part 4: Applications of Experimental Design," *Quality Progress,* August 1991, pp. 123–126, 131–132. The use of nested designs to study process performance is considered.

Hogg, R. V., and J. Ledolter, *Applied Statistics for Engineers and Physical Scientists,* 2nd ed. New York: Macmillan Publishing Company, 1992. See pages 311–316 for a discussion of nested designs.

Mason, R. L., R. F. Gunst, and J. L. Hess, *Statistical Design and Analysis of Experiments with Applications to Engineering and Science.* New York: John Wiley & Sons, 1989. Nested designs are considered on pages 190–195, 364–368.

Nelson, L. S., "Using Nested Designs: I. Estimation of Standard Deviations," *Journal of Quality Technology,* Vol. 27, No. 2, April 1995, pp. 169–171.

Rasmussen, S., *An Introduction to Statistics with Data Analysis.* Pacific Grove, CA: Brooks/Cole Publishing Company, 1992. See pages 487–497 for a discussion of inferences about two or more variances.

PROBLEMS

7.1   Porosity readings on condenser paper were recorded for paper from four rolls taken at random from each of three lots. The results were as follows. Analyze these data, assuming that lots are fixed and rolls are random.

| | | Lot | | | | | | | | | | |
| | | I | | | | II | | | | III | | |
Roll	1	2	3	4	5	6	7	8	9	10	11	12
	1.5	1.5	2.7	3.0	1.9	2.3	1.8	1.9	2.5	3.2	1.4	7.8
	1.7	1.6	1.9	2.4	1.5	2.4	2.9	3.5	2.9	5.5	1.5	5.2
	1.6	1.7	2.0	2.6	2.1	2.4	4.7	2.8	3.3	7.1	3.4	5.0

7.2   In Problem 7.1 how would the results change (if they do) if the lots were chosen at random?

7.3   Set up the EMS column and indicate the proper tests to make if A is a fixed factor at five levels, if B is nested within A at four random levels for each level of A, if C is nested within B at three random levels, and if two observations are made in each "cell."

7.4   Repeat Problem 7.3 for A and B crossed or factorial and C nested within the A, B cells.

7.5   Two types of machine are used to wind coils. One type is hand operated; two machines of this type are available. The other type is power operated; two machines of this type are available. Three coils are wound on each machine from two different wiring stocks. Each coil is then measured for the outside diameter of the wire at a middle position on the coil. The results were as follows. (Units are 10^{-5} inch.)

	Machine Type			
	Hand		Power	
	Machine Number			
Stock	2	3	5	8
1	3279	3527	1904	2464
	3262	3136	2166	2595
	3246	3253	2058	2303
2	3294	3440	2188	2429
	2974	3356	2105	2410
	3157	3240	2379	2685

Set up the model for this problem and determine what tests can be run.

7.6 Do a complete analysis of Problem 7.5.

7.7 If the outside diameter readings in Problem 7.5 were taken at three fixed positions on the coil (tip, middle, and end), what model would now be appropriate and what tests could be run?

7.8 An experiment is to be designed to compare faculty morale in junior high school organizations for grades 6–8, grades 7–9, and grades 7–8. Two schools are randomly selected to represent each of these three types of organization, and data are to be collected from several teachers in each school. Someone suggests that male and female teachers might differ with respect to morale, so it is agreed to choose random samples of five male and five female teachers from each school.

a. Set up a mathematical model for this experiment, outline into ANOVA, and indicate what tests can be made.

b. Make any recommendations that seem reasonable based on your answers to part a, assuming that the data have not yet been collected.

7.9 In an attempt to study the effectiveness of two corn varieties (V_1 and V_2), two corn-producing counties (C) in Iowa were chosen at random. Four farms (F) were then randomly selected within each county. Seeds from both varieties were sent to the four farms and planted in random plots. At harvest, the number of bushels of corn per acre was recorded for each variety on each farm.

a. Write a mathematical model for this situation.

b. Determine the EMS for varieties in this experiment.

c. Assuming that only counties (C) showed significantly different average yields, and $MS_C = 130$ and $MS_F = 10$ (farms or error), find the percentage of the variance in this experiment that can be attributed to county differences.

7.10 An educator proposes a new teaching method and wishes to compare the achievement of students using his method with that of students using a traditional method. Twenty students are randomly placed into two groups of 10 students per group. Tests are given to all 20 students at the beginning of a semester, at the end of the semester, and 10 weeks after the end of the semester. The educator wishes to determine whether students exposed to the two methods show differences in the average achievement at each of the three time periods.

a. Write a mathematical model for this situation.

b. Set up an ANOVA table and show the F tests that can be made.

c. If the educator's method is really better than the traditional method, what would you expect to find if you graphed the results at each time period? Show by sketch.

d. If method A—the new method—gave a mean achievement score across all time periods of 60.0, explain in some detail how you would set confidence limits on this method mean.

7.11 In Example 7.3, the ANOVA of Table 7.17 suggests that further analysis is needed. Using the data of Table 7.16, make further analyses and state your conclusions. (A graph might also be helpful.)

7.12 In studying the intellectual attitude toward science as measured on the Bratt attitude scale, a science educator gave a pretest and a posttest to an experimental group exposed to a new curriculum and to a control group that was not so exposed. There were 15 students in each group. Set up a mathematical model for this situation and determine what tests are appropriate.

7.13 In Problem 7.12 a strong interaction was found between the groups and the tests with means the following: pretest control = 57.00, pretest experimental = 54.96, posttest control = 54.66, posttest experimental = 65.66. Assuming that the sum of squares for the error term used to test for this interaction was 492.61, make further tests and state some conclusions based on this information.

7.14 A researcher wished to test the figural fluency of three groups of Egyptian students: those taught by the Purdue Creative Thinking Program (PCTP) in a restricted atmosphere, those taught by PCTP in a permissive atmosphere, and a control group taught in a traditional manner with no creativity training. Four classes were randomly assigned to each of the three methods. Before training began, all pupils were administered a Group Embedded Figures Test and classified as either field-dependent or field-independent students. After 6 weeks of training, the following results on the averages of students within these classes on this one subtest (figural fluency) were found.

Source	df	SS	MS	F	Prob.
Methods (M_i)		124.60			
Classes within method ($C_{j(i)}$)		35.10			
Cognitive style (F_k)		2.76			
$M \times F$ interaction (MF_{ik})		25.76			
$F \times C$ interaction ($FC_{kj(i)}$)		20.97			

Complete the table and state your conclusions.

7.15 Fifty-four randomly selected students are assigned to one of three curricula (C) of science lessons. Over a period of a semester, six lessons (L) were presented to all students in these three groups according to their prescribed curricula. The first two curricula were experimental (A taught question-asking skills with evaluation feedback, and B taught with question-asking skills without the evaluation feedback). The third curriculum was used as a control with no instruction in question-asking skill or feedback. Results on the proportion of high-level questions asked on a test were as follow.

Source	df	SS	MS	F	Prob.
C_i	2	0.556	0.278	4.14	0.0211
$S_{k(i)}$	51	3.417	0.067		
L_j	5	0.250	0.050	2.25	0.0493
CL_{ij}	10	0.430	0.043	1.92	0.0431
$LS_{jk(i)}$	255	5.610	0.022		

Discuss these results and explain how the F tests were made.

7.16 For Problem 7.15 it was known in advance that the experimenter wished to compare the two experimental curricula results with the control, and also to compare the two experimental curricula results with each other. Set up proper a priori tests for this purpose and test the results based on the following means and the table given in Problem 7.15.

Mean proportion for curriculum A: 0.2224

Mean proportion for curriculum B: 0.1929

Mean proportion for curriculum C: 0.1236

7.17 In a study of the characteristics associated with guidance competence versus counseling competence, 144 students were divided into nine groups of 16 each. These nine groups represented all combinations of three levels of guidance ranking (high, medium, low) and three levels of counseling ranking (high, medium, low). All subjects were then given nine subtests. Assuming the rankings as two fixed factors, the subtests as fixed, and the subjects within the nine groups as random, set up a data layout and mathematical model for this experiment.

7.18 Determine the ANOVA layout and the EMS column for Problem 7.17 and indicate the proper tests that can be made.

7.19 Three days of sampling in which each sample was subjected to two types of size graders gave the following results, coded by subtracting 4% moisture and multiplying by 10.

	Day					
	1		2		3	
	Grader					
Sample	A	B	A	B	A	B
1	4	11	5	11	0	6
2	6	7	17	13	−1	−2
3	6	10	8	15	2	5
4	13	11	3	14	8	2

5	7	10	14	20	8	6
6	7	11	11	19	4	10
7	14	16	6	11	5	18
8	12	10	11	17	10	13
9	9	12	16	4	16	17
10	6	9	−1	9	8	15
11	8	13	3	14	7	11

Assuming graders fixed, days random, and samples within days random, set up a mathematical model for this experiment and determine the EMS column.

7.20 Complete the ANOVA for Problem 7.19 and comment on the results.

7.21 In a filling process two random runs are made. For each run six hoppers are used, and a different set of six hoppers is used in each run. Assume that hoppers are a random sample of possible hoppers. Samples are taken from the bottom, middle, and top of each hopper in order to check on a possible position effect on filling. Two observations are made within positions and runs on all hoppers. Set up a mathematical model for this problem and the associated EMS column.

7.22 The data below are from the experiment described in Problem 7.21.

	Run 1						Run 2					
Hopper	**Bottom**		**Middle**		**Top**		**Bottom**		**Middle**		**Top**	
1	19	0	31	25	25	13	6	6	0	6	0	0
2	13	0	19	38	31	25	0	13	0	31	6	0
3	13	0	31	19	13	0	28	6	25	19	13	28
4	13	0	19	31	25	13	28	0	6	0	28	25
5	19	19	19	38	19	25	16	13	13	19	0	0
6	6	0	0	0	0	0	13	6	6	19	0	19

Do a complete analysis of these data.

7.23 Some research on the abrasive resistance of filled epoxy plastics was carried out by molding blocks of plastic using five fillers: iron oxide, iron filings, copper, alumina, and graphite. Two concentration ratios were used for the ratio of filler to epoxy resin: 0.5:1 and 1:1. Three sample blocks were made up of each of the 10 combinations above. These blocks were then subjected to a reciprocating motion in which gritcloth was used as the abrasive material in all tests. After 10,000 cycles, each sample block was measured in three places at each of three fixed positions on the plates: I, II, and III. These blocks had been measured before cycling, which allowed the difference in thickness to be used as the measured variable. Measurements were made to the nearest 0.0001 inch. Assuming complete randomization of the order of testing of the 30 blocks, set up a mathematical model for this situation and set up the ANOVA table with an EMS column. We are interested in the effects on the block of concentration ratio, fillers, and positions. The three observations at each position will be considered as error, and there may well be differences between the average of the three sample blocks within each treatment combination.

7.24 The tables on pages 218–19 show data for Problem 7.23: for each filler material, the associated concentrations (0.5:1 and 1:1) and positions (I, II, and III) are given. Complete an ANOVA for these data and state your conclusions.

7.25 From the results of Problem 7.24, what filler–concentration combination would you recommend if you were interested in a minimum amount of wear? The 1:1 ratio is cheaper than the 0.5:1 combination.

7.26 Glass lenses were produced on a production line that required 10 handling stations. The company was concerned about scratched-lens damage, which might occur at any one of these 10 stations. Two trays of lenses were randomly selected at each station. The number of defective lenses found in samples of 64 out of 576 lenses from each of two positions (inner and outer sections) of each tray were reported. Write a mathematical model for this experiment to check whether the stations differ, whether the two positions in the trays differ, or whether the trays within the stations differ in the number of defective lenses found in the samples of 64.

7.27 For the model in Problem 7.26, determine the appropriate tests that can be run by examining the EMS column for this model.

7.28 Data from Problem 7.26 are given as follows:

	Lens Damage						Lens Damage		
OBS	POS	Station	Tray	RDG	OBS	POS	Station	Tray	RDG
1	1	1	1	0	21	2	1	1	0
2	1	1	2	1	22	2	1	2	0
3	1	2	3	3	23	2	2	3	1
4	1	2	4	1	24	2	2	4	0
5	1	3	5	0	25	2	3	5	0
6	1	3	6	3	26	2	3	6	1
7	1	4	7	0	27	2	4	7	0
8	1	4	8	0	28	2	4	8	1
9	1	5	9	0	29	2	5	9	1
10	1	5	10	0	30	2	5	10	0
11	1	6	11	0	31	2	6	11	0
12	1	6	12	2	32	2	6	12	0
13	1	7	13	1	33	2	7	13	0
14	1	7	14	1	34	2	7	14	0
15	1	8	15	0	35	2	8	15	0
16	1	8	16	4	36	2	8	16	2
17	1	9	17	1	37	2	9	17	4
18	1	9	18	0	38	2	9	18	1
19	1	10	19	0	39	2	10	19	0
20	1	10	20	0	40	2	10	20	2

Do a complete analysis of these data and state your conclusions.

Sample	Alumina						Graphite					
	0.5:1			1:1			0.5:1			1:1		
	I	II	III	I	II	III	I	II	III	I	II	III
1	7.0	5.4	6.4	7.1	5.0	6.4	8.8	5.5	9.0	18.0	14.4	18.5
	7.3	5.1	5.7	6.9	5.7	6.6	8.3	5.9	11.0	17.6	12.4	19.2
	6.6	5.0	7.7	6.5	4.0	6.2	6.9	4.7	11.3	18.4	12.6	19.4
	20.9	15.5	19.8	20.5	14.7	19.2	24.0	16.1	31.3	54.0	39.4	57.1
2	6.8	5.1	4.6	6.8	4.0	7.5	7.7	6.8	10.1	15.6	11.6	17.6
	6.9	5.5	5.8	7.2	5.2	7.8	6.5	7.0	11.2	14.6	12.8	18.7
	6.2	4.8	5.8	4.8	4.4	6.0	5.9	7.2	10.8	15.1	13.4	19.3
	19.9	15.4	16.2	18.8	13.6	21.3	20.1	21.0	32.1	45.3	37.8	55.6
3	7.3	6.3	5.1	6.2	4.4	5.7	7.2	7.0	7.7	13.3	11.7	15.5
	7.6	5.3	6.5	6.6	4.8	6.6	6.8	6.9	9.3	13.8	13.4	18.0
	5.6	4.8	7.1	4.4	3.8	5.6	6.2	5.2	9.8	13.6	12.1	19.6
	20.5	16.4	18.7	17.2	13.0	17.9	20.2	19.1	26.8	40.7	37.2	53.1
Totals	163.3			156.2			210.7			420.2		

Sample	Iron Filings						Iron Oxide						Copper					
	0.5:1			1:1			0.5:1			1:1			0.5:1			1:1		
	I	II	III	I	II	III	I	II	III	I	II	III	I	II	III	I	II	III
1	2.1	1.1	1.3	3.6	1.6	2.3	3.0	1.1	3.2	1.8	1.3	2.4	2.7	1.4	2.8	2.8	1.5	2.4
	2.1	1.1	1.7	3.8	1.0	2.5	3.8	1.8	4.1	2.1	1.0	1.9	2.9	2.2	3.8	2.6	1.1	2.1
	1.0	0.9	1.7	2.9	1.6	2.4	3.0	1.1	3.3	1.9	1.6	1.8	3.0	1.8	3.4	1.9	1.4	2.0
	5.2	3.1	4.7	10.3	4.2	7.2	9.8	4.0	10.6	5.8	3.9	6.1	8.6	5.4	10.0	7.3	4.0	6.5
2	1.7	1.1	1.7	2.1	1.1	1.7	3.1	1.6	2.2	2.1	1.0	1.5	2.5	1.8	2.4	2.1	1.0	1.5
	1.8	1.0	2.0	2.6	1.1	1.0	3.0	1.7	3.0	2.0	0.8	2.0	3.0	2.6	3.3	2.5	1.0	1.6
	1.3	0.8	2.0	1.6	0.9	1.4	2.7	1.2	2.7	2.3	1.0	2.0	2.0	2.1	2.4	1.5	1.2	1.7
	4.8	2.9	5.7	6.3	3.1	4.1	8.8	4.5	7.9	6.4	2.8	5.5	7.5	6.5	8.1	6.1	3.2	4.8
3	3.2	0.8	1.4	2.3	1.1	1.8	2.1	1.5	2.0	1.6	1.2	1.6	3.3	2.1	3.4	2.6	1.3	2.6
	2.8	0.8	2.0	2.6	1.2	2.0	2.9	1.9	2.5	1.6	1.1	1.4	3.2	2.0	4.1	2.9	1.5	3.3
	2.0	0.5	1.9	2.1	0.7	2.4	1.6	2.5	2.4	1.9	1.2	2.1	3.2	2.0	3.8	2.8	1.2	2.7
	8.0	2.1	5.3	7.0	3.0	6.2	6.6	5.9	6.9	5.1	3.5	5.1	9.7	6.1	11.3	8.3	4.0	8.6
Totals		41.8			51.4			65.0			44.2			73.2			52.8	

7.29 In electronics manufacture, devices called *dice* are affixed to metallic wafers. On individual wafers, gold wires are bonded to each die at locations called *bond positions*. To study the effect of the method of processing wafers on the strength of the bonds, three processing methods will be used, with two wafers processed for each method. The pull strength of wire bonds at each of four (fixed) bond positions and five (fixed) die positions will then be measured. The wafers and bond positions are arranged as shown.

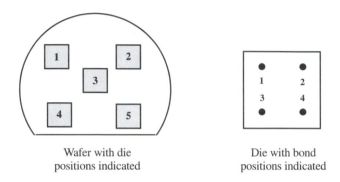

Wafer with die
positions indicated

Die with bond
positions indicated

a. Prepare a data layout.
b. Write a statistical model for this experiment.
c. Prepare a partial ANOVA summary. Include the expected mean squares.
d. Determine the appropriate tests that can be run and comment.

7.30 A study is to be conducted to determine customer preference for short-travel push buttons on an automobile radio. There are two types: tact and rubber. The tact push buttons have a hard, firm feel, whereas the rubber push buttons have a soft, smooth feel. Two displays will be prepared—one with tact push buttons and one with rubber push buttons. Random samples of 50 female and 50 male subjects will be selected. Each subject will be asked to rate both displays on a scale from 1 to 10, with 10 denoting the highest level of satisfaction with performance of the push buttons.
a. Prepare a data layout.
b. Write a mathematical model for this experiment.
c. Prepare a partial ANOVA summary. Include the expected mean squares.
d. Determine the appropriate tests that can be run and comment.

7.31 A consumer of chrome-plated parts wants to compare parts produced by each of four suppliers. Twenty parts will be randomly selected from the production of each supplier. One reading will be taken at each of three positions on each of three critical surfaces of each part. Thus, nine readings will be taken on each of the 80 parts. Determine an appropriate statistical model, the associated EMS table, and the appropriate tests. Comment.

7.32 In integrated circuit manufacture, the surface of a metallic wafer is etched by wetting that surface with an acid mix. The etch depth across the wafer should be uniform. That uniformity is checked by measuring the etch depth of a wafer at each of five fixed wafer positions (see left-hand figure of Problem 7.29). The natural logarithm of the sample standard deviation of those five measurements is then taken as the measure (Y) of uniformity. Thus, smaller values of Y indicate better uniformity in the etch depth.

A project team hopes to determine which combination of acid temperature (T) and acid mix (M) produces the best uniformity. The temperature controller only cools the acid mix, so the temperature will be set at 19°C, then lowered to 16°C, and then lowered to 13°C. Preparation of an acid mix is expensive and time-consuming, so all three temperatures will be used with a particular acid mix before the next acid mix is tested. Three different acid mixes will be considered. Once a particular combination of acid mix and temperature has been achieved, a cassette of wafers will be processed. One wafer will then be selected from each of three fixed locations (L) in the cassette. The value of Y will then be calculated for each wafer.

a. Prepare a layout for this experiment.
b. Write an appropriate statistical model.
c. Prepare a partial ANOVA table. Include the expected mean squares.
d. Analyze completely and comment on the following data, obtained by the project team.

Rep	T	A	L	y	Rep	T	A	L	y	Rep	T	A	L	y
1	1	1	1	1.18	2	1	1	1	0.67	3	1	1	1	0.81
1	1	1	2	1.48	2	1	1	2	0.81	3	1	1	2	1.83
1	1	1	3	2.20	2	1	1	3	1.58	3	1	1	3	2.13
1	1	2	1	−0.45	2	1	2	1	0.81	3	1	2	1	0.81
1	1	2	2	0.94	2	1	2	2	1.28	3	1	2	2	1.28
1	1	2	3	1.58	2	1	2	3	1.39	3	1	2	3	1.67
1	1	3	1	−1.02	2	1	3	1	−0.21	3	1	3	1	0.00
1	1	3	2	0.36	2	1	3	2	0.94	3	1	3	2	0.36
1	1	3	3	1.06	2	1	3	3	1.28	3	1	3	3	0.81
1	2	1	1	1.67	2	2	1	1	1.28	3	2	1	1	1.75
1	2	1	2	1.91	2	2	1	2	1.83	3	2	1	2	1.83
1	2	1	3	1.99	2	2	1	3	2.06	3	2	1	3	2.13
1	2	2	1	−0.21	2	2	2	1	0.52	3	2	2	1	0.36
1	2	2	2	0.36	2	2	2	2	0.52	3	2	2	2	0.52
1	2	2	3	0.81	2	2	2	3	0.52	3	2	2	3	0.81
1	2	3	1	0.36	2	2	3	1	0.19	3	2	3	1	0.52
1	2	3	2	0.67	2	2	3	2	0.52	3	2	3	2	0.52
1	2	3	3	0.94	2	2	3	3	1.06	3	2	3	3	0.81
1	3	1	1	1.06	2	3	1	1	1.48	3	3	1	1	1.18
1	3	1	2	1.58	2	3	1	2	1.75	3	3	1	2	1.39
1	3	1	3	1.83	2	3	1	3	1.99	3	3	1	3	1.83
1	3	2	1	0.52	2	3	2	1	0.19	3	3	2	1	−0.21
1	3	2	2	0.52	2	3	2	2	0.19	3	3	2	2	0.52
1	3	2	3	1.48	2	3	2	3	0.36	3	3	2	3	0.67
1	3	3	1	0.81	2	3	3	1	1.06	3	3	3	1	1.06
1	3	3	2	1.28	2	3	3	2	1.28	3	3	3	2	1.18
1	3	3	3	1.48	2	3	3	3	1.75	3	3	3	3	1.28

Chapter  8

EXPERIMENTS OF TWO OR MORE FACTORS: RESTRICTIONS ON RANDOMIZATION

8.1 INTRODUCTION

In the discussion of factorial and nested experiments it was assumed that the whole experiment was performed in a completely randomized manner. In practice, however, it may not be feasible to have the entire experiment run several times in one day, or run by one experimenter, and so forth. It then becomes necessary to abandon the idea of complete randomization and block the experiment in the same manner used for the single-factor experiment in Chapter 4. Instead of running several replications of the experiment all at one time, it may be possible to run one complete replication on one day, a second complete replication on another day, a third replication on a third day, and so on. In this case each replication is a block, and the design is a randomized block design with a complete factorial or nested experiment randomized within each block.

Occasionally a second restriction on randomization is necessary, in which case a Latin square design might be used, with the treatments in the square representing a complete factorial or nested experiment.

8.2 FACTORIAL EXPERIMENT IN A RANDOMIZED BLOCK DESIGN

Just as the factorial arrangements were extracted from the treatment effect in Equations (5.1) and (5.2), so can these arrangements be extracted from the treatment effect in the randomized complete block design of Chapter 4 (Section 4.2). For the completely randomized design, you will recall that

$$Y_{ij} = \mu + \tau_{ij} + \varepsilon_{k(ij)}$$

can be subdivided into

$$Y_{ijk} = \mu + A_i + B_j + AB_{ij} + \varepsilon_{k(ij)}$$

222

where $A_i + B_j + AB_{ij} = \tau_{ij}$ of the first model.

When a complete experiment is replicated several times and R_k represents the blocks or replications, the randomized block model is

$$Y_{ijkm} = \mu + R_k + \tau_{ij} + \varepsilon_{m(ijk)}$$

where R_k represents the blocks or replications and τ_{ij} represents the treatments. If the treatments are formed from a two-factor factorial experiment, then

$$\tau_{ij} = A_i + B_j + AB_{ij}$$

and the complete model is

$$Y_{ijk} = \mu + R_k + A_i + B_j + AB_{ij} + \varepsilon_{ijk} \tag{8.1}$$

Here it is assumed that there is no interaction between replications (blocks) and treatments, which is the usual assumption underlying a randomized block design. Any such interaction is confounded in the error term ε_{ijk}.

To get a more concrete picture of this type of design, consider three levels of factor A, two levels of B, and four replications. The treatment combinations may be written as $A_1 B_1$, $A_1 B_2$, $A_2 B_1$, $A_2 B_2$, $A_3 B_1$, and $A_3 B_2$. Such a design assumes that all six of these treatment combinations can be run on a given day, if days are the replications or blocks. A layout in which these six treatment combinations are randomized within each replication might be

Replication I:	$A_1 B_2$	$A_3 B_1$	$A_3 B_2$	$A_2 B_1$	$A_1 B_1$	$A_2 B_2$
Replication II:	$A_2 B_2$	$A_1 B_1$	$A_3 B_2$	$A_2 B_1$	$A_1 B_2$	$A_3 B_1$
Replication III:	$A_2 B_1$	$A_3 B_2$	$A_1 B_2$	$A_3 B_1$	$A_2 B_2$	$A_1 B_1$
Replication IV:	$A_1 B_1$	$A_3 B_1$	$A_3 B_2$	$A_2 B_1$	$A_1 B_2$	$A_2 B_2$

An analysis breakdown is shown in Table 8.1.

To see how the interaction of replications and treatments is used as error, consider the expanded model and the corresponding EMS terms exhibited in Table 8.2. Here the factors are fixed and replications are considered as random.

Since the degrees of freedom are quite low for the tests indicated in Table 8.2, and since there is no separate estimate of error variance, it is customary to assume that $\sigma_{RA}^2 = \sigma_{RB}^2 = \sigma_{RAB}^2 = 0$ and to pool these three terms to serve as the error variance.

TABLE 8.1
Two-Factor Experiment in a Randomized Block Design

Source	df	
Replications (R_k)	3	
Treatments (τ_{ij})	5	
$\quad A_i$		2
$\quad B_j$		1
$\quad AB_{ij}$		2
Error (ε_{ijk})	15	
Total	23	

TABLE 8.2
EMS Terms for Two-Factor Experiment in a Randomized Block

Source	df	3 F i	2 F j	4 R k	1 R m	EMS
R_k	3	3	2	1	1	$\sigma_\varepsilon^2 + 6\sigma_R^2$
A_i	2	0	2	4	1	$\sigma_\varepsilon^2 + 2\sigma_{RA}^2 + 8\phi_A$
RA_{ik}	6	0	2	1	1	$\sigma_\varepsilon^2 + 2\sigma_{RA}^2$
B_j	1	3	0	4	1	$\sigma_\varepsilon^2 + 3\sigma_{RB}^2 + 12\phi_B$
RB_{jk}	3	3	0	1	1	$\sigma_\varepsilon^2 + 3\sigma_{RB}^2$
AB_{ij}	2	0	0	4	1	$\sigma_\varepsilon^2 + \sigma_{RAB}^2 + 4\phi_{AB}$
RAB_{ijk}	6	0	0	1	1	$\sigma_\varepsilon^2 + \sigma_{RAB}^2$
$\varepsilon_{m(ijk)}$	0	1	1	1	1	σ_ε^2 (not retrievable)
Total	23					

The reduced model is then as shown in Table 8.3. Here all effects are tested against the 15-df error term. Although we usually pool the interactions of replications with treatment effects for the error term, this is not always necessary. It will depend on the degrees of freedom available for testing various hypotheses in Table 8.2, and also on whether any repeat measurements can be taken within a replication. If the latter is possible, a separate estimate of σ_ε^2 is available and all tests in Table 8.2 may be made.

TABLE 8.3
EMS Terms for Two-Factor Experiment,
Randomized Block Design, Reduced Model

Source	df	EMS
R_k	3	$\sigma_\varepsilon^2 + 6\sigma_R^2$
A_i	2	$\sigma_\varepsilon^2 + 8\phi_A$
B_j	1	$\sigma_\varepsilon^2 + 12\phi_B$
AB_{ij}	2	$\sigma_\varepsilon^2 + 4\phi_{AB}$
ε_{ijk}	15	σ_ε^2
Total	23	

■ **Example 8.1**

A researcher was interested in determining how nozzle types and operators affected the rate of fluid flow, in cubic centimeters, through these nozzles. The researcher considered three fixed nozzle types and five randomly chosen operators. Each of these 15 combinations was to be run in random order on each of three days. A 3-day set would count as three replications of the complete 3×5 factorial experiment. After subtracting 96.0 cm³ from each reading and multiplying all readings by 10, the researcher obtained the data layout and observed readings (in cubic centimeters) recorded in Table 8.4.

TABLE 8.4
Nozzle Example Data

	Operator														
	1			2			3			4			5		
	Nozzle														
Replication	A	B	C	A	B	C	A	B	C	A	B	C	A	B	C
I	6	13	10	26	4	−35	11	17	11	21	−5	12	25	15	−4
II	6	6	10	12	4	0	4	10	−10	14	2	−2	18	8	10
III	−15	13	−11	5	11	−14	4	17	−17	7	−5	−16	25	1	24

The model for this example is

$$Y_{ijk} = \mu + R_k + N_i + O_j + NO_{ij} + \varepsilon_{ijk}$$

with

$$k = 1, 2, 3 \qquad i = 1, 2, 3 \qquad j = 1, 2, \ldots, 5$$

The R_k levels are random, O_j levels are random, and N_i levels are fixed. If m repeat measurements are considered in the model, but $m = 1$, the EMS values are those appearing in Table 8.5. These values indicate that all main effects and the interaction can be tested with reasonable precision (df).

TABLE 8.5
EMS for Nozzle Example

Source	df	3 F i	5 R j	3 R k	1 R m	EMS
R_k	2	3	5	1	1	$\sigma_\varepsilon^2 + 15\sigma_R^2$
N_i	2	0	5	3	1	$\sigma_\varepsilon^2 + 3\sigma_{NO}^2 + 15\phi_N$
O_j	4	3	1	3	1	$\sigma_\varepsilon^2 + 9\sigma_O^2$
NO_{ij}	8	0	1	3	1	$\sigma_\varepsilon^2 + 3\sigma_{NO}^2$
$\varepsilon_{m(ijk)}$	28	1	1	1	1	σ_ε^2
Total	44					

For $\alpha = 0.05$, the results of tests of hypotheses in Table 8.6 show only a significant nozzle × operator effect. A plot of the sample means (Figure 8.1) shows this interaction graphically. From that plot, it appears that the average flow rate is least when operator 2 uses nozzle C. Further analysis in the Newman–Keuls sense may shed more light on this claim.

TABLE 8.6
Minitab ANOVA for Nozzle Example

```
Analysis of Variance (Balanced Designs)
Factor        Type       Levels      Values
R             random          3           1        2        3
N             fixed           3           1        2        3
O             random          5           1        2        3        4        5
Analysis of Variance for Flow
Source        DF               SS             MS             F              P
R              2           328.84         164.42           1.70          0.201
N              2          1426.98         713.49           3.13          0.099
O              4           798.80         199.70           2.06          0.112
N*O            8          1821.47         227.68           2.35          0.045
Error         28          2709.16          96.76
Total         44          7085.24
Source              Variance          Error         Expected Mean Square
                    component          term          (using restricted model)
  1 R                  4.511            5            (5)  + 15(1)
  2 N                                   4            (5)  + 3(4)  + 15Q[2]
  3 O                 11.438            5            (5)  + 9(3)
  4 N*O               43.643            5            (5)  + 3(4)
  5 Error             96.756                         (5)
```

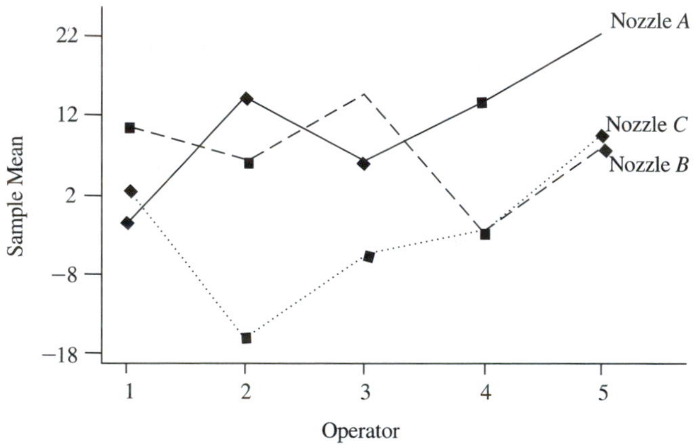

Figure 8.1   Interaction plot for Example 8.1: nozzle × operator.

Further Analysis of the Nozzle Data

If desirable, the model of Example 8.1 can be expanded to include an estimate of replication by operator and replication by nozzle interactions using the three-way interaction ($N \times O \times R$) as the error. The results of such a breakdown are recorded in Table 8.7.

The F test values are shown in Table 8.7 except for the test on nozzle that is labeled with an asterisk. The $N \times O$ interaction does not show significance at the 5% level as it did in Table 8.6 because the test is less sensitive as a result of the reduction in degrees of freedom from 28 to 16 in the error mean square.

TABLE 8.7
Minitab ANOVA for Nozzle Example with Replication Interactions

```
Analysis of Variance (Balanced Designs)
```

Factor	Type	Levels	Values				
R	random	3	1	2	3		
N	fixed	3	1	2	3		
O	random	5	1	2	3	4	5

```
Analysis of Variance for Flow
```

Source	DF	SS	MS	F	P
R	2	328.8	164.4	1.86	0.216
N	2	1427.0	713.5	*	
R*N	4	272.2	68.1	0.63	0.649
O	4	798.8	199.7	2.26	0.151
R*O	8	705.6	88.2	0.82	0.600
N*O	8	1821.5	227.7	2.10	0.098
Error	16	1731.3	108.2		
Total	44	7085.2			

```
* No exact F-test can be calculated.
```

Source	Variance component	Error term	Expected Mean Square (using restricted model)
1 R	5.081	5	(7) + 3(5) + 15(1)
2 N		*	(7) + 3(6) + 5(3) + 15Q[2]
3 R*N	-8.031	7	(7) + 5(3)
4 O	12.389	5	(7) + 3(5) + 9(4)
5 R*O	-6.669	7	(7) + 3(5)
6 N*O	39.825	7	(7) + 3(6)
7 Error	108.208		(7)

```
* No exact F-test can be calculated.
```

Because there is no direct test on nozzle, a pseudo–F test is used as described in Chapter 6. From Table 8.7, we find that

$$E[\mathrm{MS}_{RN}] = \sigma_\varepsilon^2 + 5\sigma_{RN}^2$$

and

$$E[\mathrm{MS}_{NO}] = \sigma_\varepsilon^2 + 3\sigma_{NO}^2$$

So,

$$
\begin{aligned}
E[\mathrm{MS}_{RN} + \mathrm{MS}_{NO} - \mathrm{MS}_{\text{error}}] &= (\sigma_\varepsilon^2 + 5\sigma_{RN}^2) + (\sigma_\varepsilon^2 + 3\sigma_{NO}^2) - \sigma_\varepsilon^2 \\
&= \sigma_\varepsilon^2 + 3\sigma_{NO}^2 + 5\sigma_{RN}^2
\end{aligned}
$$

To obtain the F' statistic, we note that

$$E[\mathrm{MS}_N] = \sigma_\varepsilon^2 + 3\sigma_{NO}^2 + 5\sigma_{RN}^2 + 15\phi_N$$

and let

$$\mathrm{MS} = \mathrm{MS}_{RN} + \mathrm{MS}_{NO} - \mathrm{MS}_{\text{error}}$$

Here the coefficients are 1, 1, and -1, and the corresponding observed mean squares and degrees of freedom are

$$\text{MS}_{RN} = 68.1 \quad \text{with } \nu_1 = 4 \text{ df}$$

$$\text{MS}_{NO} = 227.7 \quad \text{with } \nu_2 = 8 \text{ df}$$

$$\text{MS}_{\text{error}} = 108.2 \quad \text{with } \nu_3 = 16 \text{ df}$$

Since the observed value of MS_N is 713.5, we can write

$$
\begin{aligned}
f' &= \frac{\text{MS}_N}{\text{MS}_{NR} + \text{MS}_{NO} - \text{MS}_{\text{error}}} \\
&= \frac{713.5}{68.1 + 227.7 - 108.2} \\
&= \frac{713.5}{187.6} = 3.8
\end{aligned}
$$

and

$$\nu = \frac{(187.6)^2}{(1)^2[(68.1)^2/4] + (1)^2[(227.7)^2/8] + (-1)^2[(108.2)^2/16]} = 4.2$$

With 2 and 4 df the 5% F values is 6.94, and with 2 and 5 df it is 5.79. Thus, the observed value of F' is not significant at the 5% level.

Because both the $R \times N$ and $R \times O$ interaction F tests have p values in excess of 0.5, we are probably justified in assuming that σ_{RN}^2 and σ_{RO}^2 are negligible and in pooling these with σ_{RNO}^2 for a 28-df error mean square as in Table 8.6.

8.3 FACTORIAL EXPERIMENT IN A LATIN SQUARE DESIGN

We know from Chapter 4 that if there are two restrictions on the randomization, and the number of restriction levels equals the number of treatment levels, a Latin square design may be used. The model for this design is

$$Y_{ijk} = \mu + R_i + \tau_j + \gamma_k + \varepsilon_{ijk}$$

where τ_j represents the treatments, and both R_i and γ_k represent restrictions on the randomization. If a 4×4 Latin square is used for a given experiment, such as the one in Section 4.5, the treatments A, B, C, and D could represent the four treatment combinations of a 2×2 factorial (Table 8.8). Here

$$\tau_j = A_m + B_q + AB_{mq}$$

with its 3 df. Thus a factorial experiment ($A \times B$) is run in a Latin square design. For the problem of Section 4.5, instead of four tire brands, the four treatments could consist of two brands (A_1 and A_2) and two types (B_1 and B_2). The types might be black- and white-walled tires. The complete model would then be

$$Y_{ikmq} = \mu + R_i + \gamma_k + A_m + B_q + AB_{mq} + \varepsilon_{ikmq}$$

and the degree-of-freedom breakdown would be that shown in Table 8.9.

TABLE 8.8
4 × 4 Latin Square (Same as Table 4.10)

	Car			
Position	I	II	III	IV
1	C	D	A	B
2	B	C	D	A
3	A	B	C	D
4	D	A	B	C

TABLE 8.9
Degree-of-Freedom Breakdown for Table 8.8

Source	df
R_i	3
γ_k	3
A_m	1
B_q	1
AB_{mq}	1
ε_{ikmq}	6
Total	15

(with braces grouping A_m, B_q, AB_{mq}: $\}$ 3 df for τ_j)

This idea could be extended to running a factorial in a Graeco–Latin square, and so on.

8.4 REMARKS

The preceding sections considered factorial experiments when the design of the experiment was a randomized block or a Latin square. It is also possible to run a nested experiment or a nested-factorial experiment in a randomized block or Latin square design. In the case of a nested experiment, the treatment effect might be broken down as follows:

$$\tau_m = A_i + B_{j(i)}$$

and when this experiment has one restriction on the randomization, the model is

$$Y_{ijk} = \mu + R_k + \underbrace{A_i + B_{j(i)}}_{\tau_m} + \varepsilon_{ijk}$$

If a nested factorial is repeated on several different days, the model would be

$$Y_{ijkm} = \mu + R_k + \underbrace{A_i + B_{j(i)} + C_m + AC_{im} + BC_{mj(i)}}_{\tau_m} + \varepsilon_{ijkm}$$

where the whole nested factorial is run in a randomized block design. The analysis of such designs follows from the methods of Chapter 7.

8.5 SAS PROGRAMS

Methods discussed previously can be used to obtain SAS outputs for the nozzle experiment considered in Section 8.2. To simplify data input, DO loops are introduced.

The Reduced Model Analysis of Example 8.1

In Example 8.1, the three interactions involving replicates were pooled with the error term. A SAS command file for that situation is presented in Table 8.10 and the resulting outputs are included in Figure 8.2.

The DO loops in Table 8.10 allow us to input the data exactly as according to the Replications section of Table 8.4. As SAS begins to process the data, REP is set to 1, operator to 1, nozzle to 3, and a flow rate of 6 is read. This produces the same result as using class statements and having SAS read 1 1 3 6 as the first data point. The **OUTPUT** statement in the DO loop for nozzle has SAS store the data points in the data set XMPL as they are read.

Notice the importance of knowing the expected mean squares before writing the command file. Knowing that the nozzle effect is tested by comparing the nozzle mean square with that for the interaction between nozzle and operator makes it possible to include that test in the command file.

The Full Model Analysis of Example 8.1

SAS will output sums of squares for a full model that includes the three-factor interaction if the model statement in Table 8.10 is modified appropriately. Since no degrees of freedom

---

TABLE 8.10
SAS Command File for Nozzle Example: Reduced Model

```
OPTIONS LINESIZE=80;
DATA XMPL;
DO REP = 1 TO 3;
  DO OPERATOR = 1 TO 5;
    DO NOZZLE = 1 TO 3;
      INPUT FLOWRATE @@;
      OUTPUT;
      END;
  END;
END;
CARDS;
   6 13   10 26    4 -35 11 17   11 21  -5   12 25 15  -4
   6  6   10 12    4   0  4 10  -10 14   2   -2 18  8  10
 -15 13  -11  5   11 -14  4 17  -17  7  -5  -16 25  1  24
;
PROC GLM;
CLASS REP OPERATOR NOZZLE;
MODEL FLOWRATE = REP NOZZLE|OPERATOR;
  TEST H = NOZZLE E = NOZZLE*OPERATOR;
```

```
                   GENERAL LINEAR MODELS PROCEDURE
DEPENDENT VARIABLE: FLOWRATE
                               SUM OF           MEAN
SOURCE                 DF      SQUARES         SQUARE    F VALUE    PR > F
MODEL                  16   4376.08888889   273.50555556    2.83    0.0078
ERROR                  28   2709.15555556    96.75555556
CORRECTED TOTAL        44   7085.24444444
                  R-SQUARE          C.V.      ROOT MSE      FLOWRATE MEAN
                  0.617634       178.4838    9.83644019       5.51111111
SOURCE                 DF    TYPE I SS     MEAN SQUARE    F VALUE    PR > F
REP                     2    328.84444444  164.42222222    1.70    0.2011
NOZZLE                  2   1426.97777778  713.48888889    7.37    0.0027
OPERATOR                4    798.80000000  199.70000000    2.06    0.1124
OPERATOR*NOZZLE         8   1821.46666667  227.68333333    2.35    0.0448
SOURCE                 DF    TYPE III SS   MEAN SQUARE    F VALUE    PR > F
REP                     2    328.84444444  164.42222222    1.70    0.2011
NOZZLE                  2   1426.97777778  713.48888889    7.37    0.0027
OPERATOR                4    798.80000000  199.70000000    2.06    0.1124
OPERATOR*NOZZLE         8   1821.46666667  227.68333333    2.35    0.0448

TESTS OF HYPOTHESES USING THE TYPE III MS FOR OPERATOR*NOZZLE
   AS AN ERROR TERM
SOURCE                 DF    TYPE III SS    F VALUE    PR > F
NOZZLE                  2   1426.97777778    3.13     0.0989
```

Figure 8.2   SAS output for the nozzle example: reduced model.

are left for the error term, no F tests are possible. However, knowledge of the expected mean squares permits us to specify tests as appropriate. To obtain the outputs in Figure 8.3, one can place the statements in Table 8.11 after the last statement in Table 8.10; or, the procedure in Table 8.10 can be replaced with that in Table 8.11.

Tests of replicates and two-way interactions involving replicates were omitted from the analysis in Figure 8.3. This was done because our primary concern is with nozzle, operator, and the interaction between nozzle and operator. Test results for replicate, the interaction between replicate and nozzle, and the interaction between replicate and operator can be obtained by adding more **TEST** statements to the procedure in Table 8.11. As in Section 8.2, a pseudo–F test must be used to test for the nozzle effect.

TABLE 8.11
Additions for Table 8.10 That Produce a Full Model Analysis

```
PROC GLM;
  CLASS REP OPERATOR NOZZLE;
  MODEL FLOWRATE = REP|NOZZLE|OPERATOR;
    TEST H = OPERATOR E = REP*OPERATOR;
    TEST H = NOZZLE*OPERATOR E = REP*NOZZLE*OPERATOR;
```

```
                          GENERAL LINEAR MODELS PROCEDURE
          DEPENDENT VARIABLE: FLOWRATE
                                      SUM OF           MEAN
          SOURCE               DF     SQUARES          SQUARE   F VALUE   PR > F
          MODEL                44   7085.24444444   161.02828283    .         .
          ERROR                 0      0.00000000     0.00000000
          CORRECTED TOTAL      44   7085.24444444
                          R-SQUARE             C.V.      ROOT MSE    FLOWRATE MEAN
                          1.000000           0.0000    0.00000000       5.51111111
          SOURCE               DF    TYPE I SS     MEAN SQUARE   F VALUE   PR > F
          REP                   2    328.84444444  164.42222222    .         .
          NOZZLE                2   1426.97777778  713.48888889    .         .
          REP*NOZZLE            4    272.22222222   68.05555556    .         .
          OPERATOR              4    798.80000000  199.70000000    .         .
          REP*OPERATOR          8    705.60000000   88.20000000    .         .
          OPERATOR*NOZZLE       8   1821.46666667  227.68333333    .         .
          REP*OPERATOR*NOZZLE  16   1731.33333333  108.20833331    .         .

          SOURCE               DF   TYPE III SS    MEAN SQUARE   F VALUE   PR > F
          REP                   2    328.84444444  164.42222222    .         .
          NOZZLE                2   1426.97777778  713.48888889    .         .
          REP*NOZZLE            4    272.22222222   68.05555556    .         .
          OPERATOR              4    798.80000000  199.70000000    .         .
          REP*OPERATOR          8    705.60000000   88.20000000    .         .
          OPERATOR*NOZZLE       8   1821.46666667  227.68333333    .         .
          REP*OPERATOR*NOZZLE  16   1731.33333333  108.20833333    .         .
          TESTS OF HYPOTHESES USING THE TYPE III MS FOR REP*NOZZLE AS AN ERROR TERM
          SOURCE               DF   TYPE III SS      F VALUE   PR > F
          OPERATOR              4    798.80000000       2.26   0.1512
          TESTS OF HYPOTHESES USING THE TYPE III MS FOR REP*OPERATOR*NOZZLE
              AS AN ERROR TERM
          SOURCE               DF   TYPE III SS      F VALUE   PR > F
          OPERATOR*NOZZLE       8   1821.46666667      2.10   0.0977
```

Figure 8.3   SAS output for the nozzle example: full model.

8.6 SUMMARY

The summary at the end of Chapter 7 may now be extended.

Experiment	Design	Analysis
II. Two or more factors A. Factorial (crossed)		
	1. Completely randomized $Y_{ijk} = \mu + A_i + B_j + AB_{ij} + \varepsilon_{k(ij)}, \ldots$ for more factors a. General case	1. a. ANOVA with interactions

2. Randomized block
  a. Complete
$$Y_{ijk} = \mu + R_k + A_i + B_j + AB_{ij} + \varepsilon_{ijk}$$
3. Latin square
  a. Complete
$$Y_{ijkm} = \mu + R_k + \gamma_m + A_i + B_j + AB_{ij} + \varepsilon_{ijkm}$$

2.
  a. Factorial ANOVA with replications R_k

3.
  a. Factorial ANOVA with replications and positions

B. Nested (hierarchical)

1. Completely randomized
$$Y_{ijk} = \mu + A_i + B_{j(i)} + \varepsilon_{k(ij)}$$
2. Randomized block

  a. Complete
$$Y_{ijk} = \mu + R_k + A_i + B_{j(i)} + \varepsilon_{ijk}$$
3. Latin square
  a. Complete
$$Y_{ijkm} = \mu + R_k + \gamma_m + A_i + B_{j(i)} + \varepsilon_{ijkm}$$

1. Nested ANOVA

2. Nested ANOVA with blocks R_k
  a. Nested ANOVA with blocks R_k

3.
  a. Nested ANOVA with blocks and positions

C. Nested factorial

1. Completely randomized
$$Y_{ijkm} = \mu + A_i + B_{j(i)} + C_k + AC_{ik} + BC_{kj(i)} + \varepsilon_{m(ijk)}$$
2. Randomized block
  a. Complete
$$Y_{ijkm} = \mu + R_k + A_i + B_{j(i)} + C_m + AC_{im} + BC_{mj(i)} + \varepsilon_{ijkm}$$
3. Latin square
  a. Complete
$$Y_{ijkmq} = \mu + R_k + \gamma_m + A_i + B_{j(i)} + C_q + AC_{iq} + BC_{qj(i)} + \varepsilon_{ijkmq}$$

1. Nested-factorial ANOVA

2.
  a. ANOVA with blocks R_k

3.
  a. ANOVA with blocks and positions

PROBLEMS

8.1 Data on the glass rating of tubes taken from two fixed stations and three shifts are recorded each week for three weeks. With each week considered to be a block, the six treatment combinations (two stations by three shifts) were tested in a random order each week, with results as follow:

Week	Station 1 Shift			Station 2 Shift		
	1	2	3	1	2	3
1	3	3	3	6	3	6
	6	4	6	8	9	8
	6	7	7	11	11	13
2	14	8	11	4	15	4
	16	8	12	6	15	7
	19	9	17	7	17	10

3	2	2	2	2	2	10
	3	3	4	5	4	12
	6	4	6	7	6	13

Analyze these data as a factorial run in a randomized block design.

8.2 In the example of Section 4.5 (Table 4.12) consider tire brands A, B, C, D as four combinations of two factors, ply of tires and type of tread, where

$$A = P_1 T_1 \qquad B = P_1 T_2 \qquad C = P_2 T_1 \qquad D = P_2 T_2$$

represent the four combinations of two ply values and two tread types. Analyze the data for this revised design—a 2^2 factorial run in a Latin square design.

8.3 If in Problem 7.5 the stock factor is replaced by "days" (considered to be random) and the rest of the experiment is performed in a random manner on each of the two days, use the same numerical results and reanalyze with this restriction.

8.4 Again, in Problem 7.5 assume that the whole experiment (24 readings) were repeated on five randomly selected days. Set up an outline of this experiment, including the EMS column, df, and so on.

8.5 Three complete replications were run in a study of aluminum thickness. Samples were taken from five machines and two types of slug—rivet and staple—in each replicate. Considering replicates to be random and machines and type of slug fixed, set up a model for this problem and determine what tests can be made. Comment on the adequacy of the various tests. Data are as follows:

Replication	Slug	Machine 1	2	3	4	5
I	Rivet	175	95	180	170	155
	Staple	165	165	175	185	130
II	Rivet	190	185	180	200	190
	Staple	170	160	175	165	190
III	Rivet	185	165	175	195	200
	Staple	190	160	200	185	200

8.6 Use a computer program to analyze the data of Problem 8.5 and state your conclusions.

8.7 Data were collected for three consecutive years on the number of days after planting that the first flowering of a plant occurred. Each year five varieties of plants were planted at six different stations. Considering the repeat years to be random and varieties and stations fixed, outline the ANOVA for this problem.

8.8 The data for Problem 8.7 follow.

	Variety				
Station/Year	**Hazel**	**Coltsfoot**	**Anemone**	**Blackthorn**	**Mustard**
Broadchalke					
1932	57	67	95	102	123
1933	46	72	90	88	101
1934	28	66	89	109	113
Total	131	205	274	299	337
Bratton					
1932	26	44	92	96	93
1933	38	68	89	89	110
1934	20	64	106	106	115
Total	84	176	287	291	318
Lenham					
1932	48	61	78	99	113
1933	35	60	89	87	109
1934	48	75	95	113	111
Total	131	196	262	299	333
Dorstone					
1932	50	68	85	117	124
1933	37	65	74	93	102
1934	19	61	80	107	118
Total	106	194	239	317	344
Coaley					
1932	23	74	105	103	120
1933	36	47	85	90	101
1934	18	69	85	105	111
Total	77	190	275	298	332
Ipswich					
1932	39	57	91	102	112
1933	39	61	82	93	104
1934	43	61	98	98	112
Total	121	179	271	293	328

Do a complete ANOVA on these data and discuss the results.

8.9 Puffed cereal was packaged from a special machine with six heads, and the filling could come from any one of three positions—top, middle, or bottom of a hopper. Data were collected on a measure of filling capacity from each of these 18 possible combinations of head and position. Measurements were taken on each of two cycles within each of the combinations above. These cycles within treatments can be considered random. The whole experiment was repeated in a second run at a later date. Assuming runs are random, set up an analysis of variance table for this problem and also make up a data layout table for the collection of the data.

8.10 Data for Problem 8.9 are given below for two runs, three positions (B, M, T), six heads (H), and two cycles (C) with each combination:

	Run 1						Run 2					
	B		M		T		B		M		T	
H	C_1	C_2	C_1	C_2	C_1	C_2	C_1	C_2	C_1	C_2	C_1	C_2
1	19	0	31	25	25	13	6	6	0	6	0	0
2	13	0	19	38	31	25	0	13	0	31	6	0
3	13	0	31	19	13	0	28	6	25	19	13	28
4	13	0	19	31	25	13	28	0	6	0	28	25
5	19	19	19	38	19	25	16	13	13	19	0	0
6	6	0	0	0	0	0	13	6	6	19	0	19

Do a complete analysis of the data. Could terms in the table be combined as was done in the text example? Explain why or why not.

8.11 Discuss the similarities and differences between a nested experiment with two factors and a randomized block design with one factor.

8.12 Assuming that the nested experiment involving machines and heads in Section 7.2 was completely repeated on three subsequent randomly selected days, outline its mathematical model and ANOVA table complete with the EMS column.

8.13 For Problem 8.12, show a complete model with *all* interactions and the reduced model. How might this experiment be made less costly and still give about the same information?

8.14 For the nested-factorial example in Section 7.4 on gun loading, set up the model assuming that the whole experiment is repeated on four more randomly selected days.

8.15 Examine both the full and reduced models for Problem 8.14 and comment.

8.16 Suppose the same hoppers are used in both runs of the experiment described in Problem 7.21.
a. Prepare a table that includes the model, expected mean squares, and tests.
b. Using the data in Problem 7.22, conduct a complete analysis.

8.17 Coal supplied to a large steel plant is tested twice for ash content, first at the supplier's laboratory and again at the plant's laboratory. To compare the results of ash content measurements (Y) obtained by the two laboratories, 10 coal samples are prepared at the supplier's laboratory. The prepared samples are split to give two equivalent sets of 10 samples. One set of samples is measured at the supplier's laboratory; the other set is measured at the plant's laboratory. Data for two replicates of this experiment follow.

	Replicate				
	1			**2**	
Sample	**Supplier**	**Plant**	**Sample**	**Supplier**	**Plant**
1	5.43	5.21	11	5.47	5.13
2	5.35	5.48	12	5.31	5.46
3	5.53	5.48	13	5.46	5.54
4	5.55	5.61	14	5.55	5.54
5	5.93	5.96	15	5.93	6.00
6	5.95	6.09	16	5.97	5.99
7	6.34	6.49	17	6.32	6.43
8	6.05	6.12	18	6.02	6.13
9	5.87	6.04	19	5.87	6.00
10	5.60	5.70	20	5.58	5.67

a. Prepare a table that includes the model, expected mean squares, and tests.
b. Conduct a complete analysis. Do the results indicate a difference between the average ash content readings at the two laboratories? If so, find a 99% confidence interval for $\mu_2 - \mu_1$.

8.18 One step in the manufacture of a solid state sensor device involves the suspension of a wafer over the column of a fountain etcher. Etchant fluid flows up the column and etches the back side of the wafer while the front remains dry. A project team intends to conduct an experiment to study the effects on the uniformity of the etch rate across the wafer of two variables: A, the distance of the wafer from the top of the column, and B, the amount of fluid flowing out the top of the column. Three levels of each variable will be used. One wafer will be etched for each of the $3 \times 3 = 9$ treatment combinations, and the etch depth will be measured at each of seven fixed locations on that wafer. The dependent variable (Y) is to be the coefficient of variation ($S/\bar{X}$) for that sample of measurements. Two replicates of the experiment will be run. Formulate a statistical model with EMS table and statistical tests.

8.19 Data for Problem 8.18 follow.

Rep	A	B	y	Rep	A	B	y
1	1	1	0.07463	2	1	1	0.06309
1	1	2	0.03439	2	1	2	0.03186
1	1	3	0.01690	2	1	3	0.03583
1	2	1	0.05039	2	2	1	0.07674
1	2	2	0.03632	2	2	2	0.05507
1	2	3	0.02696	2	2	3	0.05115
1	3	1	0.05637	2	3	1	0.76818**
1	3	2	0.06165	2	3	2	0.11930
1	3	3	0.04683	2	3	3	0.07999

** The wafer that produced this value was not etched properly.

a. Analyze these data. A smaller coefficient of variation indicates better uniformity of the etch rate across the wafer.

b. Remove the aberrant value in the data table and use the General Linear Model module in a statistical computing program to analyze the remaining data. Comment.

8.20 After investigating the data in Problem 8.19 the project team found that the wafer for which $y = 0.76818$ was obtained had not been etched completely across the wafer. However, the etch was good at four of the seven locations on that wafer. Calculating the coefficient of variation for the sample of readings at the four good locations, the team substituted 0.05969 for the value used in Problem 8.19. Conduct a complete analysis of the data with this new value. [*Note:* Different columns were used for each replicate. After studying the statistical results, further investigation revealed contaminates in one of the columns.]

Chapter 9

2^f FACTORIAL EXPERIMENTS

9.1 INTRODUCTION

In Chapter 5 we considered factorial experiments and gave a general method for their analysis. Among the special situations that are of considerable interest in future designs is the case of f factors in which each factor is at just two levels. These levels might be two extremes of temperature, two extremes of pressure, two time values, two machines, and so on. Although the case may seem rather trivial, since only two levels are involved, it is, nevertheless, very useful for at least two reasons: to introduce notation and concepts that will apply to more involved designs, and to illustrate the main effects and interactions really present in this simple case. It is also true that in practice many experiments are run at just two levels of each factor. For example, the ceramic tool-cutting study of Chapters 1 and 5 is a $2 \times 2 \times 2$, or 2^3 factorial, with four observations per cell. Throughout this chapter the two levels will be considered to be fixed levels. This is quite reasonable because the two levels are chosen at points near the extremes, rather than at random.

9.2 2^2 FACTORIAL

In the simplest case, two factors are of interest and each factor is set at just two levels. This is a $2 \times 2 = 2^2$ factorial, and the design will be considered to be completely randomized. Our investigation of current flow in an integrated circuit (Section 5.1) is of this type. The factors affecting current flow are temperature and altitude, and each is set at two levels: temperature at 25 and 55°C, and altitude at 0 and 3 K. Responses for the four treatment combinations (reproduced from Figure 5.3) are displayed in Figure 9.1. The response variable is the current flow.

To generalize a bit, consider temperature as factor A and altitude as factor B. The model for this completely randomized design is

$$Y_{ij} = \mu + A_i + B_j + AB_{ij} + \varepsilon_{k(ij)} \tag{9.1}$$

where $i = 1, 2$, $j = 1, 2$, and $k = 1, \ldots, n$, with n the number of observations at each treatment combination. Since $n = 1$, the experiment was a completely randomized factorial, and no other data were obtained, no assessment of interaction can be made independent of error.

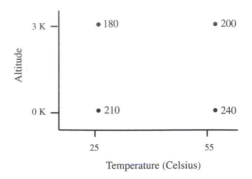

Figure 9.1   Responses (milliamperes) for the power flow study ($K = 10,000$ feet).

Notations for the Treatment Combinations

Subscripts may be placed on AB to indicate the levels of factors at each treatment combination. The notation $A_0 B_0$ is often used to indicate that both temperature and altitude are at their low levels, whereas $A_1 B_1$ indicates that both factors are at their high levels. Following this notation, $A_1 B_0$ means that A is at its high level and B is at its low level; similarly, $A_0 B_1$ indicates that A is at its low level and B is at its high level. This notation can be applied to any number of factors having two levels each.

Since 2^f experiments are encountered so often in the literature, many authors use the subscripts on AB as exponents on the small letters ab. In this notation, if both factors are at their low levels, $a^0 b^0 = (1)$, and (1) represents the treatment combination for which each factor is at its low level. Likewise, $a^0 b^1 = b$ represents B at its high level and A at its low level, $a^1 b^0 = a$ signifies the treatment combination with A at its high level and B at its low level, and $a^1 b^1 = ab$ signifies the treatment combination for which each factor is at its high level.

The "lowercase" notation generalizes to any 2^f factorial. If, for example, $f = 5$ with A, B, C, D, and E the factors, acd denotes the treatment combination for which A, C and D are set at their high levels with B and E set at their low levels. Notice that the absence of a letter indicates that the corresponding factor is at its low level.

For the data in Figure 9.1, the treatment combinations can be represented by the vertices of a square as in Figure 9.2. In Figure 9.2 the low and high levels of factors A and B are represented by -1 and $+1$, respectively, on the A and B axes. The intersection of these levels in the plane of the figure shows the four treatment combinations. Thus, $(-1, -1) = (1)$ represents both factors at their low levels, $(+1, -1) = a$ represents A high with B low, $(-1, +1) = b$ represents A low with B high, and $(+1, +1) = ab$ represents both factors at their high levels. These symbols do not represent quantities; they are merely mnemonic devices to simplify the design and its analysis.

Normal Order

The *normal order* for writing the treatment combinations for a 2^2 factorial experiments is (1), *a, b, ab.* Note that (1) is written first, then the high level of each factor with the low

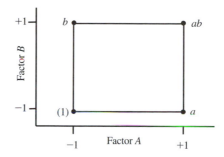

Figure 9.2 Treatment combinations for a 2^2 factorial experiment.

level of the other; the fourth symbol in the list is the algebraic product of the second and third symbols in the list. When a third factor is introduced, it is placed at the end of this sequence and then multiplied by all its predecessors. For example, if factor C is also present at two levels, the treatment combinations (in normal order) are

$$(1), a, b, ab, c, ac, bc, abc$$

which can be represented as the vertices of a cube. Adding D as a factor at two levels gives

$$(1), a, b, ab, c, ac, bc, abc, d, ad, bd, abd, cd, acd, bcd, abcd$$

Effect of a Factor

Returning to Figures 9.1 and 9.2, let T_1, T_2, T_3, and T_4 denote the totals of the n observations at the vertices (1), a, b, and ab, respectively (in this example, $n = 1$). The *effect of a factor* is defined as the change of response produced by a change in the level of that factor. At the low level of B, the observed effect of A is then $240 - 210$ or $T_2 - T_1$, whereas the observed effect of A at the high level of B is $200 - 180$ or $T_4 - T_3$. The average effect of moving from the low level to the high level of A is

$$A_{\text{eff}} = \tfrac{1}{2}[(T_2 - T_1) + (T_4 - T_3)] = \tfrac{1}{2}[-T_1 + T_2 - T_3 + T_4] \tag{9.2}$$

Notice that all the coefficients on this estimator are $+1$ when A is at its high level in the treatment combinations (i.e., in a and ab) and all are -1 when A is at its low level [i.e., in (1) and b]. Note also that the numerator

$$C_A = -T_1 + T_2 - T_3 + T_4 \tag{9.3}$$

is a contrast among the treatment totals as defined in Section 3.4 (the sum of its coefficients is 0). This concept will be useful, since the sum of squares due to this contrast can easily be determined.

Since the observed effect of B at the low level of A is $180 - 210 = T_3 - T_1$ and that at the high level of A is $200 - 240 = T_4 - T_2$, the average effect of moving from the low level of B to the high level is

$$B_{\text{eff}} = \tfrac{1}{2}[(T_3 - T_1) + (T_4 - T_2)] = \tfrac{1}{2}[-T_1 - T_2 + T_3 + T_4] \tag{9.4}$$

and again it is seen that the same four responses are used, but $+1$ coefficients are on the treatment totals for the high level of B and -1 coefficients for the low level of B. Also,

$$C_B = -T_1 - T_2 + T_3 + T_4 \tag{9.5}$$

is a linear contrast among the treatment totals.

To determine the effect of the interaction between A and B, note that at the high level of B the A effect is $T_4 - T_3$, and at the low level of B the A effect is $T_2 - T_1$. If these two effects differ, there is interaction between A and B. Thus, the average interaction effect is the average *difference* between these two differences. That is,

$$AB_{\text{eff}} = \tfrac{1}{2}[(T_4 - T_3) - (T_2 - T_1)] = \tfrac{1}{2}[T_1 - T_2 - T_3 + T_4] \tag{9.6}$$

Here again the same four responses are used with a different combination of coefficients, and

$$C_{AB} = T_1 - T_2 - T_3 + T_4 \tag{9.7}$$

is a linear contrast among the treatment totals. The interaction effect takes the responses on one diagonal of the square with $+1$ coefficients and the responses on the other diagonal with -1 coefficients.

Summarizing the contrasts of Equations (9.3), (9.5), and (9.7) in normal order gives

$$C_A = -T_1 + T_2 - T_3 + T_4$$
$$C_B = -T_1 - T_2 + T_3 + T_4$$
$$C_{AB} = T_1 - T_2 - T_3 + T_4$$

Notice that C_A, C_B, and C_{AB} are mutually orthogonal. Note also that the coefficients for the interaction contrast can be found by multiplying the corresponding coefficients of the contrasts for the two main effects.

Orthogonal Tables

The design for a completely randomized 2^2 factorial experiment and the coefficients associated with the orthogonal contrasts among the treatment totals are summarized in Table 9.1. Since the contrasts defined by any two columns are orthogonal, such tables are called *orthogonal tables*. The A and B columns define the treatment combinations, and the coefficients in the AB column are the products of the corresponding coefficients in those columns.

Each linear contrast defined by Table 9.1 is of the form

TABLE 9.1
Orthogonal Table for a 2^2 Factorial Experiment

Treatment Combination	Linear Contrast Coefficients			Total
	A	B	AB	
(1)	-1	-1	$+1$	T_1
a	$+1$	-1	-1	T_2
b	-1	$+1$	-1	T_3
ab	$+1$	$+1$	$+1$	T_4

$$C = c_1 T_1 + c_2 T_2 + c_3 T_3 + c_4 T_4 \qquad (9.8)$$

with

$$\sum_{i=1}^{4} c_i^2 = \sum_{i=1}^{4} (\pm 1)^2 = 4$$

Under the usual assumptions of normality, independence, and common variance σ^2, C/n is a normally distributed random variable with mean

$$E[C/n] = c_1 \mu_1 + c_2 \mu_2 + c_3 \mu_3 + c_4 \mu_4 = \Psi \qquad (9.9)$$

and variance

$$\text{Var}(C/n) = \frac{4\sigma^2}{n} \qquad (9.10)$$

where μ_1, μ_2, μ_3, and μ_4 are the means of the populations associated with the treatment combinations (1), a, b, and ab, respectively. The constant Ψ is the *true effect* of the factor or interaction associated with C, and the random variable C/n is an estimator of that effect, where n is the number of observations at each treatment combination.

Recalling Equation (3.11) of Section 3.4.1, we now write

$$SS_C = \frac{C^2}{n \sum_{i=1}^{4} c_i^2} = \frac{C^2}{4n} \qquad (9.11)$$

The usefulness of this formula is illustrated in Example 9.1. Note also that this sum of squares is the same as the sum of squares for the treatment effect being estimated. For example, the sum of squares for A and the sum of squares for the contrast associated with A are equal.

■ **Example 9.1 (Analysis of the Current Flow Experiment)**

Table 9.2 is an orthogonal table for the current flow experiment. For that experiment, only one observation was obtained at each treatment combination. Thus, the totals are the observations. The observed value of the contrast associated with A is

$$c_A = (-1)(210) + (+1)(240) + (-1)(180) + (+1)(200) = 50$$

Notice that this is the dot product of the column vector associated with A and that associated with the treatment totals. Using this value with Equation (9.11) and $n = 1$, we find that the observed value of SS_A is

$$SS_A = SS_{C_A} = \frac{(c_A)^2}{[4(1)]} = \frac{(50)^2}{4} = 625$$

Proceeding in like manner with B and AB, the observed contrasts are

$$c_B = (-1)(210) + (-1)(240) + (+1)(180) + (+1)(200) = -70$$

$$c_{AB} = (+1)(210) + (-1)(240) + (-1)(180) + (+1)(200) = -10$$

and the sums of squares are

$$SS_B = SS_{C_B} = \frac{(c_B)^2}{[4(1)]} = \frac{(-70)^2}{4} = 1225$$

$$SS_{AB} = SS_{C_{AB}} = \frac{(c_{AB})^2}{[4(1)]} = \frac{(-10)^2}{4} = 25$$

Dividing each observed contrast by 2 [see Equations (9.2), (9.4), and (9.6)] gives the observed estimated effects in Table 9.2.

Since each effect and the interaction has but 1 degree of freedom, and only one observation was obtained per treatment combination, the interaction effect is confounded with error and

$$SS_{total} = SS_A + SS_B + SS_{AB} = 1875$$

The sums of squares calculated here could be placed in an ANOVA summary, but the lack of a measure of error and a total of 3 degrees of freedom make such a summary trivial. We will be content to note that even though no separate measure of error is available, the sums of squares indicate that both main effects are large compared with the interaction effect. [**Note:** When more treatment combinations are involved, a normal quantile plot of the estimated effects can be used to analyze experiments such as this one, as is illustrated in Section 9.6.]

TABLE 9.2
Orthogonal Table for the Current Flow Experiment

Treatment Combination	Linear Contrast Coefficients			Total
	A	B	AB	
(1)	−1	−1	+1	210
a	+1	−1	−1	240
b	−1	+1	−1	180
ab	+1	+1	+1	200
Estimated effect	25	−35	−5	830

■ Example 9.2

A study of the effects of two factors (A and B) on the force required to shear a glued component from a circuit board produced the coded data in Table 9.3. The observed contrasts among the treatment totals are

$$c_A = (-1)(265) + (+1)(650) + (-1)(1165) + (+1)(1475) = 695$$

$$c_B = (-1)(265) + (-1)(650) + (+1)(1165) + (+1)(1475) = 1725$$

$$c_{AB} = (+1)(265) + (-1)(650) + (-1)(1165) + (+1)(1475) = -75$$

Since $n = 5$, the associated sums of squares are

$$SS_A = SS_{C_A} = \frac{(c_A)^2}{[4(5)]} = \frac{(695)^2}{20} = 24{,}141.25$$

$$SS_B = SS_{C_B} = \frac{(c_B)^2}{[4(5)]} = \frac{(1725)^2}{20} = 148{,}781.25$$

$$SS_{AB} = SS_{C_{AB}} = \frac{(c_{AB})^2}{[4(5)]} = \frac{(-75)^2}{20} = 281.25$$

The variance of the sample of 20 observations is $s_Y^2 \approx 12{,}390.72$, so the total sum of squares is

$$SS_{total} = 19 s_Y^2 \approx 19(12{,}390.72) = 235{,}423.68$$

It follows that

$$SS_{error} = SS_{total} - (SS_A + SS_B + SS_{AB}) = 235{,}423.68 - 173{,}203.75 = 62{,}219.93$$

Placing these sums of squares in an ANOVA summary gives Table 9.4.

The interaction is not significant (p value $= 0.7947$), so we investigate the main effects. The sample average at the low level of A is $(265 + 1165)/10 = 143$ and that at the high level is $(650 + 1475)/10 = 212.5$. The sample averages at the low and high levels of B are $(265 + 650)/10 = 91.5$ and $(1165 + 1475)/10 = 264$, respectively. Assuming that bigger is better, the process should be run with A and B at their high levels.

TABLE 9.3
Orthogonal Table for the Shear Force Experiment: Example 9.2

Treatment Combination	Linear Contrast Coefficients			Sample Observations					Total
	A	B	AB	1	2	3	4	5	
(1)	−1	−1	+1	125	50	40	0	50	265
a	+1	−1	−1	175	150	50	100	175	650
b	−1	+1	−1	150	250	200	240	325	1165
ab	+1	+1	+1	250	275	200	350	400	1475
Sample contrast	695	1725	−75						3555

TABLE 9.4
ANOVA for the Shear Force Data of Table 9.3

Source	df	SS	MS	F	p
A	1	24,141.25	24,141.24	6.21	0.0241
B	1	148,781.25	148,781.25	38.26	0.0000
AB	1	281.25	281.25	0.07	0.7947
Error	16	62,219.93	3,888.75		
Totals	19	235,423.68			

9.3 2³ FACTORIAL

If we add a third factor C, also at two levels, the experiment will be a $2 \times 2 \times 2$ or 2^3 factorial, again run in a completely randomized manner. The treatment combinations are now (1), a, b, ab, c, ac, bc, abc, and they may be represented as vertices of a cube as in Figure 9.3. When JMP (or Minitab) is used to generate such cubes, the sample means are recorded at each vertex. (The plot shown here is a modified JMP graphic.) In the discussions that follow, we will denote the totals at (1), a, b, ab, c, ab, bc, and abc by T_1, T_2, T_3, T_4, T_5, T_6, T_7, and T_8, respectively.

With these $8 = 2^3$ observations, or $8n$ if there are replications, we now show that the main effects and each interaction may be estimated by adding the appropriate multiples (-1 or $+1$) of the totals at the eight vertices of the cube plot. Notice that A is at its high level for the treatment combinations that determine the right-hand plane (a, ab, ac, abc) of the cube plot in Figure 9.3 and is at its low level for those in the left-hand plane [(1), b, bc, c]. The effect of moving from the low level of A to the high level of A is $T_2 - T_1$ when B and C are at their low levels, $T_4 - T_3$ when B is high and C is low, $T_6 - T_5$ when B is low and C is high, and $T_8 - T_7$ when B and C are at their high levels. Thus, the average effect of moving from the low level to the high level of A is

$$A_{\text{eff}} = \frac{(T_2 - T_1) + (T_4 - T_3) + (T_6 - T_5) + (T_8 - T_7)}{4}$$

$$= \frac{-T_1 + T_2 - T_3 + T_4 - T_5 + T_6 - T_7 + T_8}{4}$$

and the numerator

$$C_A = -T_1 + T_2 - T_3 + T_4 - T_5 + T_6 - T_7 + T_8$$

is a contrast among the treatment totals.

A similar argument for factor B (using responses in the lower and higher planes of the cube) reveals that the average effect of moving from the low level of B to the high level is

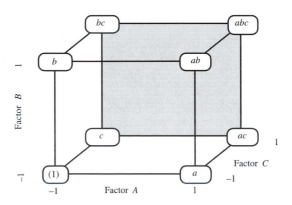

Figure 9.3 Cube plot for a 2^3 factorial arrangement.

$$B_{\text{eff}} = \frac{-T_1 - T_2 + T_3 + T_4 - T_5 - T_6 + T_7 + T_8}{4}$$

with

$$C_B = -T_1 - T_2 + T_3 + T_4 - T_5 - T_6 + T_7 + T_8$$

a contrast in the treatment totals.

Likewise (using responses in the back plane of the cube and those in the front plane), the average effect of moving from the low to the high level of C is

$$C_{\text{eff}} = \frac{-T_1 - T_2 - T_3 - T_4 + T_5 + T_6 + T_7 + T_8}{4}$$

with

$$C_C = -T_1 - T_2 - T_3 - T_4 + T_5 + T_6 + T_7 + T_8$$

a contrast in the treatment totals.

The AB interaction is determined by the difference in the A effect from the low level of B to the high level of B, regardless of C. At the low level of B, the average effect of moving from the low level of A to the high level of A is $[(T_2 - T_1) + (T_6 - T_5)]/2$; and, at the high level of B that average effect is $[(T_4 - T_3) + (T_8 - T_7)]/2$. The average difference in these is

$$\begin{aligned}
AB_{\text{eff}} &= \frac{[(T_4 - T_3) + (T_8 - T_7)]/2 - [(T_2 - T_1) + (T_6 - T_5)]/2}{2} \\
&= \frac{T_1 - T_2 - T_3 + T_4 + T_5 - T_6 - T_7 + T_8}{4}
\end{aligned}$$

with

$$C_{AB} = T_1 - T_2 - T_3 + T_4 + T_5 - T_6 - T_7 + T_8$$

a contrast in the treatment totals.

Proceeding with the AC and BC interactions as we did with the AB interaction, we find

$$AC_{\text{eff}} = \frac{+T_1 - T_2 + T_3 - T_4 - T_5 + T_6 - T_7 + T_8}{4}$$

with

$$C_{AC} = T_1 - T_2 + T_3 - T_4 - T_5 + T_6 - T_7 + T_8$$

and

$$BC_{\text{eff}} = \frac{+T_1 + T_2 - T_3 - T_4 - T_5 - T_6 + T_7 + T_8}{4}$$

with

$$C_{BC} = T_1 + T_2 - T_3 - T_4 - T_5 - T_6 + T_7 + T_8$$

The effect of the BC interaction at the low level of A is $[(T_7 - T_3) - (T_5 - T_1)]/2$ and that at the high level of A is $[(T_8 - T_4) - (T_6 - T_2)]/2$. Any difference in these is an ABC interaction. Thus,

$$ABC_{\text{eff}} = \frac{[(T_8 - T_4) - (T_6 - T_2)]/2 - [(T_7 - T_3) - (T_5 - T_1)]/2}{2}$$

$$= \frac{-T_1 + T_2 + T_3 - T_4 + T_5 - T_6 - T_7 + T_8}{4}$$

with

$$C_{ABC} = -T_1 + T_2 + T_3 - T_4 + T_5 - T_6 - T_7 + T_8$$

a linear contrast in the treatment totals.

A summary of the preceding contrasts is presented in Table 9.5. Notice that the A, B, and C columns define the treatment combinations. For example, ac is the treatment combination for which A is at the high level (+1), B is at the low level (−1), and C is at the high level (+1). Thus, the A, B, and C columns can be generated by first listing the treatment combinations in normal order and then recording the corresponding coefficients. For column C, if the letter c is present in the description of the treatment combination record +1 in the corresponding row; otherwise, record −1. Also notice that coefficients for contrasts associated with interaction effects are the products of corresponding coefficients for the main effects. For example, entries in the AB column are the products of corresponding entries in the A column and the B column. This algorithm generalizes to any 2^f factorial experiment.

Each linear contrast defined in Table 9.5 is of the form

$$C = c_1 T_1 + c_2 T_2 + c_3 T_3 + c_4 T_4 + c_5 T_5 + c_6 T_6 + c_7 T_7 + c_8 T_8$$

with

$$\sum_{i=1}^{8} c_i^2 = \sum_{i=1}^{8} (\pm 1)^2 = 8$$

Letting n denote the number of observations at a treatment combination, the sum of squares associated with C (which is the sum of squares for the factor associated with C) is

$$SS_C = \frac{C^2}{n \sum_{i=1}^{8} c_i^2} = \frac{C^2}{8n} \tag{9.12}$$

TABLE 9.5
Orthogonal Table for a 2^3 Factorial Experiment

Treatment Combination	A	B	C	AB	AC	BC	ABC	Total
(1)	−1	−1	−1	+1	+1	+1	−1	T_1
a	+1	−1	−1	−1	−1	+1	+1	T_2
b	−1	+1	−1	−1	+1	−1	+1	T_3
ab	+1	+1	−1	+1	−1	−1	−1	T_4
c	−1	−1	+1	+1	−1	−1	+1	T_5
ac	+1	−1	+1	−1	+1	−1	−1	T_6
bc	−1	+1	+1	−1	−1	+1	−1	T_7
abc	+1	+1	+1	+1	+1	+1	+1	T_8

and the estimated effect of the main effect or interaction associated with C is

$$\text{effect} = \frac{C}{4n} \tag{9.13}$$

Example 9.3 (The Ceramic Tool Experiment Revisited)

In Chapters 1 and 5 we analyzed in detail a problem on power requirements for cutting with ceramic tools. From Table 1.2 we recognize this as a 2^3 factorial with four replications per treatment combination. We will use the coded data of Table 5.7 to illustrate the special methods of this chapter.

Letting A denote the tool type, B denote the bevel angle, and C denote the type of cut, we use the totals of the coded data in Table 5.7 to obtain Table 9.6. Using column A and the corresponding totals, we find

$$\begin{aligned} c_A &= (-1)(2) + (+1)(-5) + (-1)(15) + (+1)(13) + (-1)(-12) + (+1)(-17) \\ &\quad + (-1)(-2) + (+1)(-7) = -19 \end{aligned}$$

Since $n = 4$, the estimated effect of A is

$$A_{\text{eff}} = \frac{c_A}{4(4)} = \frac{-19}{16} = -1.1875$$

from Equation (9.13). Using Equation (9.12), the associated sum of squares is

$$SS_A = \frac{(c_A)^2}{8n} = \frac{(-19)^2}{8(4)} = 11.2815 \approx 11.28$$

which agrees with Table 5.8. The other contrasts, estimated effects, and sums of squares summarized in Table 9.6 are calculated in like manner.

TABLE 9.6
Orthogonal Table for the Ceramic Tool Experiment

Treatment Combination	Linear Contrast Coefficients							Total
	A	B	C	AB	AC	BC	ABC	
(1)	−1	−1	−1	+1	+1	+1	−1	2
a	+1	−1	−1	−1	−1	+1	+1	−5
b	−1	+1	−1	−1	+1	−1	+1	15
ab	+1	+1	−1	+1	−1	−1	−1	13
c	−1	−1	+1	+1	−1	−1	+1	−12
ac	+1	−1	+1	−1	+1	−1	−1	−17
bc	−1	+1	+1	−1	−1	+1	−1	−2
abc	+1	+1	+1	+1	+1	+1	+1	−7
Contrast	−19	51	−63	5	−1	−11	−5	
Estimated effect	−1.1875	3.1875	−3.9375	0.3125	−0.0625	−0.6875	−0.3125	
SS	11.281	81.281	124.031	0.781	0.031	3.781	0.781	

We assume that computer software will usually be used to obtain sums of squares and ANOVA summaries. For completeness, however, we note that the variance for the sample of 32 coded values in Table 5.7 is $s_Y^2 \approx 14.055444$, so $\text{SS}_{\text{total}} = 31 \, s_Y^2 \approx 435.719$. Thus, $\text{SS}_{\text{error}} \approx 435.719 - (11.281 + 81.281 + \cdots + 3.781 + 0.781) = 213.752$. The difference between this latter value and that of Table 5.8 is due to rounding.

9.4 2^f REMARKS

The methods shown for 2^2 and 2^3 factorials may be extended to 2^f factorials, where f factors are each considered at two levels. The orthogonal contrasts for main effects are easily determined by constructing an orthogonal table containing $f + 2$ columns for which:

1. The first column is a listing of the treatment combinations $[(1), a, b, ab, c, \ldots]$ in normal order.
2. The second column is a listing of the coefficients of the corresponding coefficients of the contrast associated with A, where $+1$ is entered if the treatment combination contains a and -1 is entered otherwise.
3. Columns for the coefficients of contrasts associated with the remaining main effects $(B, C, \ldots)$ are formed in the same manner as that for A. That is, $+1$ is entered if the lowercase letter associated with that effect is in the description of the treatment combination and -1 is entered otherwise.
4. The right-most column is a listing of the totals to be obtained at each treatment combination. Those entries are $T_1, T_2, T_3, \ldots$.

Orthogonal contrasts in the treatment totals are obtained by finding the dot product of the column vector associated with the main effect and that of the treatment totals. Each such contrast is of the form

$$C = c_1 T_1 + c_2 T_2 + \cdots + c_k T_k = \sum_{i=1}^{k} c_i T_i \tag{9.14}$$

where

$$\sum_{i=1}^{k} c_i^2 = \sum_{i=1}^{k} (\pm 1)^2 = k = 2^f \tag{9.15}$$

Contrasts in the treatment totals for an interaction effect are formed by expanding the orthogonal table to include a column for that effect and obtaining entries in that column by multiplying corresponding coefficients in the columns associated with letters in the interaction effect. These contrasts also satisfy the conditions expressed by Equations (9.14) and (9.15).

Letting Eff denote the effect associated with a contrast C, the general relationships for 2^f factorials with n observations per treatment are

$$C = n(2^{f-1})(\text{Eff}) \tag{9.16}$$

or

$$\text{Eff} = \frac{C}{2^{f-1} \times n} \tag{9.17}$$

and

$$\text{SS}_C = \frac{C^2}{2^f \times n} \tag{9.18}$$

A general ANOVA would be as in Table 9.7, but not in "normal" order.

TABLE 9.7
ANOVA for a 2^f Factorial with n Replications

Source		df	
Main effects	A	1	
	B	1	
	C	1	$\Big\} f$
	$\vdots$	$\vdots$	
Two-factor interactions	AB	1	
	AC	1	
	BC	1	$\Big\}\ _fC_2 = \frac{f(f-1)}{2}$
	$\vdots$	$\vdots$	
Three-factor interactions	ABC	1	
	ABD	1	
	BCD	1	$\Big\}\ _fC_3 = \frac{f(f-1)(f-2)}{6}$
	$\vdots$	$\vdots$	
Four-factor interactions, and so on			
Sum of all treatment combinations		$2^f - 1$	
Residual or error		$2^f \times (n-1)$	
Total		$(n \times 2^f) - 1$	

9.5 THE YATES METHOD

The sums of squares and tests for significance considered in Examples 9.1 and 9.2 can be obtained with software designed for statistical computing. The methods considered here are included to introduce the reader to some new notations, orthogonal tables, and a procedure for calculating estimated effects in 2^f factorial experiments. These are most useful when fractions of such experiments are used to screen for the most active factors (see Chapter 12). We now introduce a rather simple scheme, developed by Yates [37], as an alternative procedure for computing the contrasts of a 2^f factorial experiment.

Begin by preparing a table, the first column of which contains the 2^f treatment combinations (1), a, b, ab, . . . listed in normal order. Place the total responses to each of these treatment combinations (T_1, T_2, T_3, T_4, . . .) in the second column. Then form a third column, labeled *cycle 1*, for which the first 2^{f-1} entries are $T_1 + T_2$, $T_3 + T_4$, . . . and the second 2^{f-1} entries are $T_2 - T_1$, $T_4 - T_3$, ... Next, a fourth column labeled *cycle 2* is prepared by means of the same procedure used on the *cycle 1* entries. Proceed in this manner until the column labeled *cycle f* has been determined. The entry in row (1) of *cycle f* is the grand total of all readings, that in row a is c_A, that in row b is c_B, and so forth. Estimated effects and sums of squares can be obtained using Equations (9.17) and (9.18) of Section 9.4.

■ Example 9.4 (Using the Yates Method with the Shear Force Data)

In Example 9.2 we studied the effects of two factors on the force required to shear a glued component from a circuit board. We now illustrate the use of the Yates method to calculate the contrasts for that example.

In Table 9.8, list all treatment combinations in the first column. Place the total response to each of these treatment combinations in corresponding rows of the second column. For the third column, labeled *cycle 1*, add the adjacent responses in nonoverlapping pairs to obtain the first two entries (i.e., $265 + 650 = 915$ and $1165 + 1475 = 2640$); then, subtract the responses in nonoverlapping pairs, always subtracting the first from the second, to obtain the last two entries (i.e., $650 - 265 = 385$ and $1475 - 1165 = 310$). Entries for *cycle 2* are obtained in the same manner as those for *cycle 1*, using the *cycle 1* results: $915 + 2640 = 3555$, $385 + 310 = 695$, $2640 - 915 = 1725$, and $310 - 385 = -75$. In this case, $f = 2$, so only two cycles are calculated in this manner.

The first entry in the *cycle 2* column is the sum of the 20 sample values in Table 9.3. The entries 695, 1725, and -75 are the observed sample contrasts associated with A, B, and AB, respectively. These results are identical to those in Table 9.3.

TABLE 9.8
The Yates Method on the Shear Force Data

Treatment Combination	Sample Total	cycle 1	cycle 2
(1)	265	915	3555 = grand total
a	650	2640	695 = c_A
b	1165	385	1725 = c_B
ab	1475	310	$-75 = c_{AB}$

As proof for the Yates method on a 2^2 factorial, we need only go through the steps used in Example 9.4 using T_1, T_2, T_3, and T_4 to denote the treatment totals. This gives the results in Table 9.9.

TABLE 9.9
The Yates Method on a 2^2 Factorial in General

Treatment Combination	Sample Total	cycle 1	cycle 2
(1)	T_1	$T_1 + T_2$	$T_1 + T_2 + T_3 + T_4 =$ grand total
a	T_2	$T_3 + T_4$	$-T_1 + T_2 - T_3 + T_4 = C_A$
b	T_3	$T_2 - T_1$	$-T_1 - T_2 + T_3 + T_4 = C_B$
ab	T_4	$T_4 - T_3$	$T_1 - T_2 - T_3 + T_4 = C_{AB}$

■ **Example 9.5 (Using the Yates Method with the Ceramic Tool Data)**

In Example 9.3 we used orthogonal tables to calculate contrasts for the ceramic tool data first considered in Chapter 1. Now consider the first and second columns of Table 9.10, which contain the treatment combinations and corresponding response totals for those data. The reader can verify that the Yates method produces the entries in the *cycle 1, cycle 2,* and *cycle 3* columns of that table. Since $f = 3$, only three cycles are required. The contrasts are identical to those obtained in Example 9.3.

TABLE 9.10
The Yates Method on the Coded Ceramic Data

Treatment Combination	Sample Total	cycle 1	cycle 2	cycle 3
(1)	2	-3	25	$-13 =$ grand total
a	-5	28	-38	$-19 = c_A$
b	15	-29	-9	$51 = c_B$
ab	13	-9	-10	$5 = c_{AB}$
c	-12	-7	31	$-63 = c_C$
ac	-17	-2	20	$-1 = c_{AC}$
bc	-2	-5	5	$-11 = c_{BC}$
abc	-7	-5	0	$-5 = c_{ABC}$

9.6 ANALYSIS OF 2^f FACTORIALS WHEN $n = 1$

Cost, time considerations, and physical limitations may make it necessary to conduct a 2^f factorial experiment with the number of observations at each treatment combination limited to 1. When a complete model is to be used, an estimate of the experimental error will not be available. Thus, F or t ratios cannot be employed to search for significant effects. One procedure that has proven extremely successful is based on a normal quantile plot of the estimated effects.

Under the usual assumptions, the contrasts associated with each treatment effect are normally distributed with common variance. If all true effects are 0, the observed values can be treated as a random sample from a normal population with mean 0. In this case, a normal quantile plot of the estimated effects should approximate a straight line through the origin. A factor associated with an estimated effect that deviates extremely from that line is not likely to be due to chance and is treated as being significantly different from 0.

■ **Example 9.6**

A chip maker uses a certain process to etch a trench in the silicon layer of an integrated circuit component. A study of the effects of three factors (A = flow rate of gas 1, B = flow rate of gas 2, and C = power setting) produced the coded data in Table 9.11. Time constraints forced the experimenters to use only one observation per treatment combination, so the values in the total column of that table are individual observations. A target of 0 has been set for the true average of the coded trench heights.

The contrasts in Table 9.11 can be obtained by means of the method illustrated in Section 9.3 or the Yates method of Section 9.5. Dividing each contrast by $1 \times 2^2 = 4$ produces the effects in Table 9.11 [see Equation (9.17)].

The effects of the three-factor interaction and factor A fall well off the solid line in the normal quantile plot in Figure 9.4a. Combining this information with the interaction plot of Figure 9.4b leads us to conclude that the average trench height will be nearest the target value of 0 near the high level of gas 2, and the low power level. Setting both gas 2 and power at their low levels or their high levels should be avoided.

TABLE 9.11
Orthogonal Table for Trench Height Study

| Treatment Combination | \multicolumn{7}{c}{Linear Contrast Coefficients} | Total |
	A	B	C	AB	AC	BC	ABC	
(1)	−1	−1	−1	+1	+1	+1	−1	112
a	+1	−1	−1	−1	−1	+1	+1	−136
b	−1	+1	−1	−1	+1	−1	+1	−21
ab	+1	+1	−1	+1	−1	−1	−1	17
c	−1	−1	+1	+1	−1	−1	+1	−49
ac	+1	−1	+1	−1	+1	−1	−1	−63
bc	−1	+1	+1	−1	−1	+1	−1	64
abc	+1	+1	+1	+1	+1	+1	+1	−103
Contrast	−391	93	−123	133	29	53	−439	
Effect	−97.75	23.25	−30.75	33.25	7.25	13.25	−109.75	

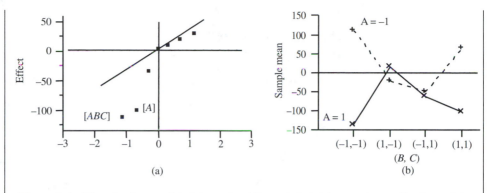

Figure 9.4 Plots for the trench height study of Example 9.6: (a) normal quantile plot and (b) interaction plot.

9.7 SOME COMMENTS ABOUT COMPUTER USE

As computer software improves, more techniques become available. To use them effectively, one must be aware of their possibilities. The following comments are made to make the reader aware of a few such possibilities.

Data from a study of the height of trenches etched into the silicon layer of an integrated circuit were considered in Example 9.6. Entering those data into a Minitab worksheet and selecting **Stat ▶ DOE ▶ Fit Factorial Model . . .** produces the outputs in Figure 9.5.

The estimated effects in Figure 9.5 are identical to those in Table 9.11. Multiplying each effect by 4 will produce the contrasts summarized in that table [see Equation (9.16)].

Selecting the **Display effects plot** option in Minitab produced the normal quantile plot in Figure 9.5. Such a plot was used in Example 9.5 to identify any active factors. The Minitab plot has a different shape from that in Figure 9.4 because the axes are reversed.

Using the **Design Experiment . . .** platform in the Tables menu of JMP with these same data produces the outputs in Figure 9.6. The estimated effects are listed in the Original column of the Transformed Parameter Estimates text box. The entry in the Intercept row is the mean of the sample of eight observations. Doubling the entries associated with $A, B, \ldots$, and $A*B*C$ will give the estimated effects in Table 9.11. The effects recorded by JMP are based on sample means rather than totals and are used as estimates of the parameters for a linear regression model (to be discussed in a later chapter).

The normal quantile plot in Figure 9.6 provides the same information as the plot in Figure 9.4. The plotted points and the line through the origin are those for a procedure due to Lenth [19], which is used to eliminate the subjectivity of drawing a line of unknown slope that passes through (0, 0) and "fits" the plotted effects. JMP labels only estimates that deviate substantially from the plotted line.

Fractional Factorial Fit

Estimated Effects and Coefficients for Rdg

Term	Effect	Coef
Constant		-22.37
A	-97.75	-48.87
B	23.25	11.63
C	-30.75	-15.38
A*B	33.25	16.63
A*C	7.25	3.63
B*C	13.25	6.63
A*B*C	-109.75	-54.87

Analysis of Variance for Rdg

Source	DF	Seq SS	Adj SS	Adj MS	F	P
Main Effects	3	22082	22082	7360.8	**	
2-Way Interactions	3	2667	2667	889.1	**	
3-Way Interactions	1	24090	24090	24090.1	**	
Residual Error	0	0	0	0.0		
Total	7	48840				

```
 Nscores-
        -
        -                                              *   AB
        -
   1.0+
        -
        -                                         *   B
        -
        -                                     *   BC
        -
  -0.0+
        -                                  *   AC
        -
        -                      *   C
        -
        -        *   A
  -1.0+
        -
        -    *   ABC
        -
        ----+---------+---------+---------+---------+--Effects
          -105      -70       -35        0        35
```

Figure 9.5   Minitab outputs for the trench height study of Example 9.6.

Response: Y

The estimates are not correlated.

The estimates have the same variance.

Transformed Parameter Estimates

Term	Original	Scaled	Normalized
Intercept	−22.375	0.000	−63.286
A	−48.875	−52.250	−138.239
B	11.625	12.428	32.880
C	−15.375	−16.437	−43.487
B*A	16.625	17.773	47.023
C*A	3.625	3.875	10.253
C*B	6.625	7.082	18.738
A*B*C	−54.875	−58.664	−155.210

Lenth PSE

65.230601

(a)

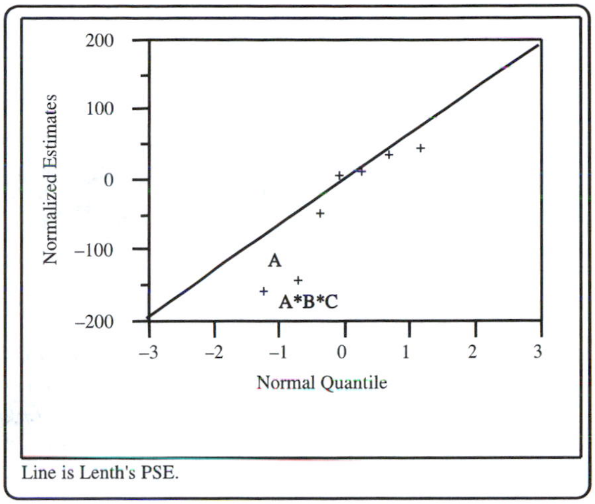

(b)

Figure 9.6 JMP outputs for trench height study of Example 9.6: (a) effect screening and (b) normal plot.

9.8 SUMMARY

The summary at the end of Chapter 8 may now be extended for part II.A.1.

Experiment	Design	Analysis
II. Two or more factors A. Factorial (crossed)		
	1. Completely randomized $Y_{ijk} = \mu + A_i + B_j$ $\quad + AB_{ij} + \varepsilon_{k(ij)}, \ldots$ for more factors	1.
	a. General case	a. ANOVA with interactions
	b. 2^f	b. General ANOVA or Orthogonal tables or Yates method; use (1), a, b, ab, ...

9.9 FURTHER READING

Anderson, V. L., and R. A. McLean, *Design of Experiments: A Realistic Approach.* New York: Marcel Dekker, 1974. See pages 225–238 for a discussion of two-level factorials and two-level factorials in randomized complete block designs.

Benski, H. C., "Use of a Normality Test to Identify Significant Effects in Factorial Designs," *Journal of Quality Technology,* Vol. 21, 1989, pp. 174–178.

Box, G. E. P., W. G. Hunter, and J. S. Hunter, *Statistics for Experimenters: An Introduction to Design, Data Analysis, and Model Building.* New York: John Wiley & Sons, 1978. See pages 329–334 for discussions of normal probability plots of estimated effects for unreplicated factorials and diagnostic checks.

DeVor, R. E., T. Chang, and J. W. Sutherland, *Statistical Quality Design and Control: Contemporary Concepts and Methods.* New York: Macmillan Publishing Company, 1992. See pages 542–572 and 585–606 for a detailed discussion of two-level factorial designs and their analysis.

Kasperski, W. J., and H. Schneider, "Using Normal Probability Plots to Determine Significant Factors in Unreplicated Factorial Designs," *Journal of Quality Technology,* Vol. 9, No. 3, 1997, pp. 449–456.

Mason, R. L., R. F. Gunst, and J. L. Hess, *Statistical Design and Analysis of Experiments with Applications to Engineering and Science.* New York: John Wiley & Sons, 1989. See pages 126–133 for a discussion of effects for two-level factors in completely randomized factorial experiments.

Vardeman, S. B., *Statistics for Engineering Problem Solving.* Boston: PWS Publishing Company, 1994. See pages 477–497 for a discussion of two-level factorial experiments and their analysis.

PROBLEMS

9.1 For the 2^2 factorial with three observations per cell given below (hypothetical data), do an analysis by the method of Chapter 5.

	Factor A	
Factor B	A_1	A_2
B_1	0	4
	2	6
	1	2
B_2	−1	−1
	−3	−3
	1	−7

9.2 a. Redo Problem 9.1 by the method of orthogonal tables.
   b. Redo Problem 9.1 by the Yates method.

9.3 In an experiment on chemical yield, three factors were studied, each at two levels. The experiment was completely randomized and the factors were known only as *A, B,* and *C*. The results are listed below. Analyze by the methods of Chapter 5.

A_1				A_2			
B_1		B_2		B_1		B_2	
C_1	C_2	C_1	C_2	C_1	C_2	C_1	C_2
1595	1745	1835	1838	1573	2184	1700	1717
1578	1689	1823	1614	1592	1538	1815	1806

9.4 a. Redo Problem 9.3 by the method of orthogonal tables.
b. Redo Problem 9.3 by the Yates method.

9.5 Plot any results from Problem 9.3 that might be meaningful from a management point of view.

9.6 The results of Problem 9.3 lead to another experiment with four factors each at two levels, with the following data.

A_1							
B_1				B_2			
C_1		C_2		C_1		C_2	
D_1	D_2	D_1	D_2	D_1	D_2	D_1	D_2
1985	2156	1694	2184	1765	1923	1806	1957
1592	2032	1712	1921	1700	2007	1758	1717

A_2							
B_1				B_2			
C_1		C_2		C_1		C_2	
D_1	D_2	D_1	D_2	D_1	D_2	D_1	D_2
1595	1578	2243	1745	1835	1863	1614	1917
2067	1733	1745	1818	1823	1910	1838	1922

Analyze these data by the general methods of Chapter 5.

9.7 Analyze Problem 9.6 using the Yates method.

9.8 Plot from Problem 9.6 any results you think are meaningful.

9.9 Consider a four-factor experiment with each factor at two levels.
a. Write out the treatment combinations in this experiment in normal order.
b. Assuming that there is only one observation per treatment, outline an ANOVA for this example and describe and justify what you would use for an error term to test for significance of various factors and interactions.
c. Explain how you would determine the AD interaction from the responses to the treatment combinations given in part a.

9.10 An experimenter is interested in the effects of five factors—A, B, C, D, and E—each at two levels on some response variable Y. Assuming that data are available for each of the possible treatment combinations in this experiment, answer the following.

a. How would you determine the effect of factor C and its sum of squares based on data taken at each of the treatment combinations?

b. Set up an ANOVA table for this problem assuming n observations for each treatment combination.

c. If $n = 2$ for part b, at what numerical value of F would you reject the hypothesis of no ACE interaction?

d. How would you propose testing for ACE interaction and what F would be necessary for rejection in this problem if $n = 1$?

9.11 Consider an experiment having five factors A, B, C, D, and E, each at two levels.

a. How many treatment combinations are there?

b. Outline an ANOVA table to test all main effects and all two-way and all three-way interactions, if there is only one observation per treatment combination.

c. Explain how you would find the proper signs on the treatment combinations for the ACE interaction.

d. An experimenter comments that "since all factors are at only two levels, it is unnecessary to run a Newman–Keuls test—one can simply look and see which level gives the greater mean response." Explain situations in which the comment would be true for your results and situations in which it would not be true.

9.12 Three factors are studied for their effect on the horsepower necessary to remove a cubic inch of metal per minute. The factors are feed rate at two levels, tool condition at two levels, and tool type at two levels. Three observations are taken for each treatment combination, and all 24 observations of horsepower are made in a completely randomized order. The data appear below.

	Feed Rate			
	0.011		**0.015**	
	Tool Condition		**Tool Condition**	
Tool Type	*Same*	*New*	*Same*	*New*
Utility	0.576	0.548	0.514	0.498
	0.576	0.555	0.515	0.519
	0.565	0.540	0.518	0.504
Precision	0.526	0.547	0.494	0.521
	0.542	0.524	0.504	0.480
	0.548	0.525	0.530	0.494

Do a complete ANOVA on these data as a 2^f experiment and state your conclusions.

9.13 Minimum horsepower is desired for the experiment given in Problem 9.12. Recommend the best set of factor conditions on the basis of your results in Problem 9.12.

9.14 Any factor whose levels are powers of 2 such as 4, 8, 16, . . . , may be considered as pseudofactors in the form 2^2, 2^3, 2^4, ... The Yates method can then be used if all factor levels are 2 or powers of 2 and the resulting sums of squares for the pseudofactors can be added to give the sum of squares for the actual factors and appropriate interactions. Try out this scheme on the data of Problem 5.21.

9.15 A systematic test was made to determine the effects on the coil breakdown voltage of the following six variables, each at two levels as indicated.
   a. Firing furnace: number 1 or 3
   b. Firing temperature: 1650°C or 1700°C
   c. Gas humidification: yes or no
   d. Coil outside diameter: large or small
   e. Artificial chipping: yes or no
   f. Sleeve: number 1 or 2

Assuming complete randomization, devise a data sheet for this experiment, outline its ANOVA, and explain your error term if only one coil is to be used for each of the treatment combinations.

9.16 If seven factors are to be studied at two levels each, outline the ANOVA for this experiment in a table similar to Table 9.7 and, assuming $n = 1$, suggest a reasonable error term.

9.17 In a steel mill an experiment was designed to study the effect of dropout temperature, back-zone temperature, and type of atmosphere on heat slivers in samples of steel. This response variable is a mean surface factor obtained for each type of sliver. Results are as follows:

	Dropout Temperature (°C)			
	2145		2165	
	Back-Zone Temperature (°C)			
Atmosphere	2040–2140	2060–2160	2040–2140	2060–2160
Oxidizing	7.8	4.3	4.5	8.2
Reducing	6.8	5.9	11.0	6.0

   a. Assuming complete randomization, analyze these results. To get an error term for testing, three of the data points were repeated to give an SS_{error} of 1.11 based on 3 degrees of freedom.
   b. Prepare an orthogonal table, estimate the seven effects, and prepare a normal quantile plot of those effects. Interpret.

9.18 Further analyze the results of Problem 9.17 and make any recommendations you can for minimizing heat slivers.

9.19 At the same time that data were collected on heat slivers in Problem 9.17, data were also collected on grinder slivers. The results were found to be as follows:

	Dropout Temperature (°C)			
	2145		**2165**	
	Back-Zone Temperature (°C)			
Atmosphere	**2040–2140**	**2060–2160**	**2040–2140**	**2060–2160**
Oxidizing	0.0	4.3	9.8	4.0
Reducing	0.0	4.4	10.1	3.9

a. For these data the $SS_{error} = 13.22$ based on three repeated measurements. Completely analyze these data.
b. Assuming that no estimate of the error variance is available, prepare a normal quantile plot of the estimated effects. Interpret and compare with the results in part a.

9.20 Sketch rough graphs for the following situations in which each factor is at two levels only:
   a. Main effects A and B both significant, AB interaction not significant.
   b. Main effect A significant, B not significant, and AB significant.
   c. Both main effects not significant but interaction significant.

9.21 For a 2^3 factorial, can you sketch a graph (or graphs) to show no significant two-factor interactions, but a significant three-factor interaction? Explain or illustrate.

9.22 The effect of chip lot (A or B) and temperature (cold, room) on the failure rate of integrated circuits in parts per million (ppm) was studied at one plant. Two lots were taken at each of the four treatment conditions, and tested with results as follows:

	Chip Lot	
Temperature	**A**	**B**
Cold	3,548	6,704
	3,932	6,656
Room	10,107	34,653
	9,562	31,097

Use any method you like to analyze these data and state your conclusions.

9.23 The company wished to determine whether it could eliminate the cold test. How would you answer this query in light of the results in Problem 9.22?

9.24 Are you bothered by the results in Problem 9.22? If so, what do you think may have happened in this experiment?

9.25 A steel company wished to determine which of three factors might be affecting the rating of carbon edge on some coils. The factors that it manipulated were type of anneal (shelf or regular), stack position (*A* or *B*), and coil position (north or south edge). Carbon edge ratings were determined on four coils in each of the eight treatments. Analyze the data below and state your conclusions.

Edge	Type of Anneal			
	Regular Position		Shelf Position	
	A	B	A	B
North	0	0	0	0
	0	0	0	0
	81	0	8	0
	36	28	36	0
South	81	0	81	0
	108	0	36	0
	72	0	36	0
	108	0	36	0

9.26 Consider the following orthogonal table for a hypothetical 2^2 factorial experiment.

Treatment Combination	Linear Contrast Coefficients			Observations	
	A	B	AB	1	2
(1)	−1	−1	+1	4	6
a	+1	−1	−1	2	−2
b	−1	+1	−1	3	7
ab	+1	+1	+1	−4	−6

    a. Calculate the contrast, estimated effect, and sum of squares for each factor.
    b. Prepare an interaction plot of the treatment means. Interpret

9.27 A 2^5 factorial experiment is to be conducted with one observation per treatment combination. The factors suggest that the effects of the three-factor and higher order interactions are zero.
    a. How many degrees of freedom are associated with the error sum of squares?
    b. If all factors are at fixed levels, what is an appropriate test statistic for a formal test on factor *D*?

9.28 In one stage of circuit board manufacture, capacitors are glued to the board prior to soldering. Each end of the capacitor is positioned over a small, elevated region (a *solder pad*) to which the solder adheres. Using the current technique, capacitors are glued to small, elevated regions between solder pads (called *checkers*). It has been suggested that the checkers could be omitted without affecting the strength of the glue bond prior to soldering. The process engineer is also concerned about the effect of the capacitor size (small or large) on the glue bond. Shear forces (in inch-ounces) follow for a completely randomized 2^2 factorial experiment.

	Checkers			
Capacitor Size	No		Yes	
Small	15.0	10.5	24.5	10.5
	28.0	21.0	21.0	14.0
	23.0	24.5	17.5	14.0
	25.0	21.0	31.5	21.0
	20.0	10.5	21.0	21.0
Large	39.0	31.5	38.5	38.5
	42.0	31.5	31.5	42.0
	53.0	42.0	42.0	42.0
	30.0	35.0	38.5	42.0
	44.0	21.0	42.0	42.0

a. Conduct a complete analysis of these data.
b. Assess the model assumptions. Use residual plots and other techniques as appropriate.

9.29 Using the following "cube" plot, determine the effects of the factors used in the associated 2^2 factorial experiment. Recorded values are totals for 2 observations.

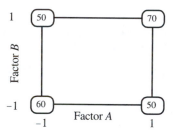

9.30 Data for a study of the effects of country in which a bonding machine was manufactured (factor A, where $-1 =$ domestic; $1 =$ foreign) and machine design (factor B, where $-1 =$ standard; $1 =$ new) on the bond strength (in grams) of an electronic component are as follows.

Treatment Combination	Linear Contrast Coefficients			Observations				
	A	B	AB	1	2	3	4	5
(1)	−1	−1	+1	204	181	201	203	214
a	+1	−1	−1	244	270	208	235	216
b	−1	+1	−1	264	226	228	249	246
ab	+1	+1	+1	197	223	206	232	213

a. Calculate the contrast, estimated effect, and sum of squares for each effect.
b. Prepare an ANOVA summary and interaction plot of the means. Interpret.
c. Further analyze the results in an effort to determine the treatment combination for which the true, average bond strength is greatest. Assume that increased bond strength does not introduce problems such as brittleness.
d. Assess the model assumptions. Use residual plots and other techniques as appropriate.

9.31 Using the following interaction plot, determine the effects of the factors used in the associated 2^2 factorial experiment. Recorded values are the means of samples of size 5.

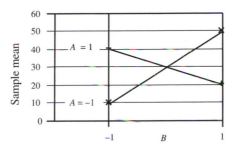

9.32 A study of the effects of the power setting, SF6 rate, and oxygen rate on the height of a trench etched in the silicon layer of an integrated circuit produced the following cube plot. Recorded values (in micron meters) are the totals for samples of size 4 for a completely randomized 2^3 factorial experiments. A target of 3 μm has been set for the true average trench height.

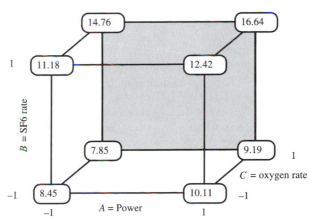

a. Calculate the values of the contrasts associated with the design of this experiment.
b. Using the contrast values in part a, determine the associated sums of squares.
c. The total sum of squares for the 32 observations is 17.43195. Using this fact and the results in part b, prepare an ANOVA summary for this experiment. Include F ratios and p values.
d. Using formal and graphical procedures as appropriate, further analyze these data. Based on your results, what factor levels should be used in future processing?

9.33 When desiging a 2^3 factorial experiment without replication, a project team proposes using the treatment combinations defined in the following orthogonal table. To accommodate time constraints, the data will be obtained according to the schedule in the two right-hand columns of that table.

Run	A	B	C	AB	AC	BC	ABC	y	Shift	Day

Run	A	B	C	AB	AC	BC	ABC	y	Shift	Day
1	−1	−1	−1	+1	+1	+1	−1	y_1	1	1
2	+1	−1	−1	−1	−1	+1	+1	y_2	1	1
3	−1	+1	−1	−1	+1	−1	+1	y_3	2	1
4	+1	+1	−1	+1	−1	−1	−1	y_4	2	1
5	−1	−1	+1	+1	−1	−1	+1	y_5	1	2
6	+1	−1	+1	−1	+1	−1	−1	y_6	1	2
7	−1	+1	+1	−1	−1	+1	−1	y_7	2	2
8	+1	+1	+1	+1	+1	+1	+1	y_8	2	2

(The top three column groups are labeled "Linear Contrast Coefficients" and "When to Run".)

a. What difficulties with interpretation of the results might occur?
b. What might the team do to alleviate those difficulties?

9.34 Suppose the data in Problem 9.6 were not obtained in completely random order. Rather, suppose the experiment was conducted as a completely randomized 2^4 factorial without replication during each of two weeks. Using the data in Problem 9.6, suppose that the first data entry in each cell was obtained during the first week and the other data entry was obtained during the second week.
a. Compare the mathematical model for the experimental design of Problem 9.6 with that for the experimental design described here.
b. Analyze the data for the situation described here. How do the results compare with those obtained in Problem 9.6 (or Problem 9.7)?

9.35 The surface temperature of a compact disc played in an automobile radio–CD unit is of concern to design engineers. Temperatures above 50°C can burn the skin on contact. Temperatures above 70°C accelerate degradation of the CD mechanism and may make the compact disc unusable. The engineers hope to determine the best way to reduce that surface temperature (Y), by conducting a completely randomized factorial experiment involving four factors at two levels each: A, convector (unpainted or painted black); B, cooling holes (as designed or additional holes included); C, voltage regulator placement (under the CD mechanism or on the side of the radio unit); and D, cooling fan (not used or mounted on the side of the chassis). Analyze completely the following data.

Run	tc	A	B	C	D	y
11	(1)	0	0	0	0	43.84
13	a	1	0	0	0	28.47
2	b	0	1	0	0	41.74
1	ab	1	1	0	0	38.92
12	c	0	0	1	0	33.57
7	ac	1	0	1	0	28.79
3	bc	0	1	1	0	35.74

10	abc	1	1	1	0	32.35
14	d	0	0	0	1	28.29
15	ad	1	0	0	1	21.05
4	bd	0	1	0	1	22.67
16	abd	1	1	0	1	20.85
8	cd	0	0	1	1	20.31
5	acd	1	0	1	1	20.10
9	bcd	0	1	1	1	25.06
6	abcd	1	1	1	1	18.09

Chapter 10

3^f FACTORIAL EXPERIMENTS

10.1 INTRODUCTION

Just as 2^f factorial experiments represent an interesting special case of factorial experimentation, so also do 3^f factorial experiments. The 3^f factorials consider f factors at three levels; thus there are 2 degrees of freedom between the levels of each of these factors. The 3^f factorials play an important role in more complicated design problems, which are discussed in subsequent chapters. For this chapter, we assume that the design is a completely randomized design and that the levels of the factors considered are fixed levels. Such levels may be either qualitative or quantitative.

For a three-level factor, one can partition the 2 degrees of freedom into two orthogonal constrasts, each with 1 degree of freedom, as shown in Chapter 3.

If an experiment can be designed such that the levels of an effect are quantitative and equispaced, two orthogonal contrasts can be found that will have physical meaning in terms of the equispaced variable, say X.

Linear and Quadratic Effects of X

Figure 10.1 shows three equispaced levels of a factor X (namely, X_1, X_2, and X_3) and the corresponding responses of a Y variable at these three points. The Y totals are shown as $T_{.1}$, $T_{.2}$, and $T_{.3}$. In Figure 10.1, if X increases from X_1 to X_3, the increase in response totals goes from $T_{.1}$ to $T_{.3}$, or $T_{.3} - T_{.1}$ becomes a linear contrast if there really is a straight-line response between X_1 and X_3.

The linear effect from X_1 to X_2 is $T_{.2} - T_{.1}$ and from X_2 to X_3 is $T_{.3} - T_{.2}$. The cumulative *linear effect* is then the sum of these, or $T_{.2} - T_{.1} + T_{.3} - T_{.2} = T_{.3} - T_{.1}$. If, however, the response is not linear, the linear effect from X_1 to X_2 is different from the linear effect from X_2 to X_3, and this difference produces a curvature in the response or a quadratic effect. This is then given by $(T_{.3} - T_{.2}) - (T_{.2} - T_{.1}) = T_{.3} - 2T_{.2} + T_{.1}$, which is another contrast called the *quadratic effect*. Note then that the linear and quadratic contrasts are

$$C_L = T_{.3} \qquad - T_{.1}$$
$$C_Q = T_{.3} - 2T_{.2} + T_{.1}$$

and these are orthogonal, allowing one to "pull out" a linear and quadratic effect from every quantitative, equispaced three-level factor.

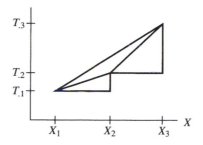

Figure 10.1   Linear and quadratic contrasts.

10.2 3^2 FACTORIAL

If just two factors are crossed in an experiment and each of the two is set at three levels, there are $3 \times 3 = 9$ treatment combinations. Because each factor is at three levels, the notation of Chapter 9 will no longer suffice. Now each factor has a low, an intermediate, and a high level, which may be designated as 0, 1, and 2. A model for this arrangement would be

$$Y_{ij} = \mu + A_i + B_j + AB_{ij} + \varepsilon_{ij}$$

where $i = 1, 2, 3$, $j = 1, 2, 3$, and the error term is confounded with the AB interaction unless there are some replications in the nine cells, in which case,

$$Y_{ijk} = \mu + A_i + B_j + AB_{ij} + \varepsilon_{k(ij)}$$

and $k = 1, 2, \ldots, n$ for n replications.

Some Notation

To introduce some notation for treatment combinations when three levels are involved, we present in Figure 10.2 a data layout in which two digits are used to describe each of the nine treatment combinations. The first digit indicates the level of factor A, and the second digit the level of factor B. Thus, 12 means A at its intermediate level and B at its highest level. This notation can easily be extended to more factors and as many levels as are necessary. It could have been used for 2^f factorials as 00, 10, 01, and 11, corresponding respectively to (1), a, b, and ab. The only reason for not using this digital notation on 2^f factorials is that so much of the literature includes this (1), a, b, $\ldots$ notation. When coefficients on these treatment combinations are properly chosen, the linear and quadratic effects of both A and B can be determined, as well as their interactions, such as $A_L \times B_L$, $A_L \times B_Q$, $A_Q \times B_L$, and $A_Q \times B_Q$.

A Hypothetical Example

Suppose the responses in Table 10.1 were recorded for the treatment combinations indicated in the upper left-hand corner of each cell. These data will be used to illustrate the methods of analysis of a 3^2 factorial.

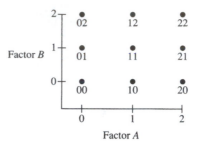

Figure 10.2   A 3^2 data layout.

TABLE 10.1
3^2 Factorial with Responses and Totals

Factor B	Factor A 0		1		2		$T_{.j}$
0	00	1	10	−2	20	3	2
1	01	0	11	4	21	1	5
2	02	2	12	−1	22	2	3
$T_{i.}$	3		1		6		10

A General Two-Factor Analysis

A two-factor ANOVA summary for Table 10.1 is given in Table 10.2. No p values are included because only one observation was obtained per treatment combination, leaving no degrees of freedom for error (residual).

If needed, calculational procedures considered in Chapter 5 can be used to obtain the sums of squares for a 3^2 factorial. Using the totals in Table 10.1 with the formulas in Chapter 5 gives

$$SS_{total} = 1^2 + 0^2 + 2^2 + \cdots + 2^2 - \frac{(10)^2}{9} = 28.889$$

$$SS_A = \frac{3^2 + 1^2 + 6^2}{3} - \frac{(10)^2}{9} = 4.222$$

$$SS_B = \frac{2^2 + 5^2 + 3^2}{3} - \frac{(10)^2}{9} = 1.556$$

$$SS_{error} \text{ (or } SS_{AB}) = 28.889 - 4.222 - 1.556 = 23.111$$

when rounded to three decimal places. These results agree with those of Table 10.2.

TABLE 10.2
SPSS ANOVA for Data in Table 10.1

Source of Variation	Sum of Squares	DF	Mean Square	F	Sig of F
Main Effects	5.778	4	1.444		
A	4.222	2	2.111		
B	1.556	2	.778		
2-Way Interactions	23.111	4	5.778		
A B	23.111	4	5.778		
Explained	28.889	8	3.611		
Residual	.000	0	.000		
Total	28.889	8	3.611		

Orthogonal Contrasts

A further breakdown of the analysis in Table 10.2 is now possible. One simply recalls that coefficients of $-1, 0, +1$ applied to the responses at low, intermediate, and high levels of a factor will measure its linear effect, whereas coefficients of $+1, -2, +1$ applied to these same responses will measure the quadratic effect of this factor. As in the case of 2^f factorials, products of coefficients will give the proper coefficients for various interactions. This can best be shown by Table 10.3, which indicates the coefficients for each effect to be used with the nine treatment combinations. Notice that coefficients in column $A_L B_L$ are products of corresponding coefficients in columns A_L and B_L.

From Table 10.3 it can be seen that A_L compares all highest levels of $A(+1)$ with all lowest levels of $A(-1)$, and A_Q compares the extreme levels with twice the intermediate levels. Both these effects are taken across *all* levels of B. Now B_L compares

TABLE 10.3
Orthogonal Table for the 3² Factorial Experiment: Data in Table 10.1

Treatment Combination	A_L	A_Q	B_L	B_Q	$A_L B_L$	$A_L B_Q$	$A_Q B_L$	$A_Q B_Q$	Total = y_{ij}
00	−1	+1	−1	+1	+1	−1	−1	+1	1
01	−1	+1	0	−2	0	+2	0	−2	0
02	−1	+1	+1	+1	−1	−1	+1	+1	2
10	0	−2	−1	+1	0	0	+2	−2	−2
11	0	−2	0	−2	0	0	0	+4	4
12	0	−2	+1	+1	0	0	−2	−2	−1
20	+1	+1	−1	+1	−1	+1	−1	+1	3
21	+1	+1	0	−2	0	−2	0	−2	1
22	+1	+1	+1	+1	+1	+1	+1	+1	2
$\sum c_i^2$	6	18	6	18	4	12	12	36	
Contrast	3	7	1	−5	−2	0	−2	28	
SS	1.50	2.72	0.17	1.39	1.00	0.00	0.33	21.78	

the highest versus the lowest level of B at the 0 level of A, then at level 1 of A, then at level 2 of A, reading down the B_L column. Similarly, B_Q compares the extreme levels of B with twice the intermediate level at all three levels of A. The coefficients for interaction are found by multiplying corresponding main-effect coefficients. To make these coefficients seem plausible, examine them in the light of the interactions that *should* be there.

In an orthogonal table such as Table 10.3, the value of a contrast associated with a particular effect is the dot product of the column of coefficients for that effect and the total column. For example, the observed value of the contrast associated with A_L is

$$c_{A_L} = (-1)(1) + (-1)(0) + (-1)(2) + (0)(-2)$$
$$+ (0)(4) + (0)(-1) + (+1)(3) + (+1)(1) + (+1)(2) = 3$$

The other contrast values in Table 10.3 are obtained in the same manner.

For $C = c_1 T_1 + \cdots + c_k T_k$ a linear contrast in treatment totals for samples of n values,

$$SS_C = \frac{C^2}{n \sum c_i^2} \tag{10.1}$$

Since $n = 1$, $\sum c_i^2 = 6$ for A_L, and $c_{A_L} = 3$, this gives

$$SS_{A_L} = \frac{(3)^2}{1(6)} = 1.50$$

The remaining sums of squares, obtained in like manner, are summarized in Table 10.3. When these sums of squares are added to a two-factor ANOVA summary like that of Table 10.2, we obtain Table 10.4.

The results of this analysis will not be tested, since there is no separate measure of error, and the interaction effect is obviously large compared with other effects. The numbers used here are purely hypothetical, presented only to show how such data can be analyzed and how the notation can be used.

TABLE 10.4
ANOVA Breakdown or 3^2 Factorial

Source	df		SS	
A	2		4.22	
$\quad A_L$		1		1.50
$\quad A_Q$		1		2.72
B	2		1.56	
$\quad B_L$		1		0.17
$\quad B_Q$		1		1.39
AB	4		23.11	
$\quad A_L B_L$		1		1.00
$\quad A_L B_Q$		1		0.00
$\quad A_Q B_L$		1		0.33
$\quad A_Q B_Q$		1		21.78
Totals	8		28.89	

Interaction Plots

It may be instructive to examine graphically the meaning of the high $A_Q B_Q$ interaction in Table 10.4, because its dwarfs all main effects and other interactions in the example.

If the $A_L B_L$ interaction is graphed using only the extreme levels of each factor, the results are as shown in Figure 10.3a. Although the lines do cross, there is a very small interaction. Note also that the average change in response for the lowest level of A to the highest level of A (shown by $\times$) is very small, indicating a very small A_L effect as the data show. Plotting $A_L B_Q$ means using all three levels of B, giving Figure 10.3b, which again shows little interaction.

For $A_Q B_L$ the results are as shown in Figure 10.4, which once again shows little interaction. This graph does indicate that the quadratic effect of A is more pronounced than its linear effect, an emphasis that is borne out by the data.

Replotting Figure 10.3b with B as abscissa gives Figure 10.5, which shows the same lack of serious interaction as in Figure 10.3b, but does show a slight quadratic bend in factor B over the linear effect.

So far no startling results have been seen. Now plot the middle level of B on Figure 10.4 to obtain Figure 10.6. Note the way this "curve" reverses its trend compared with the other two. This shows that not until both factors are considered at all three of their levels does this $A_Q B_Q$ interaction show up. It can also be seen by adding the middle level of A on Figure 10.5 (see Figure 10.7).

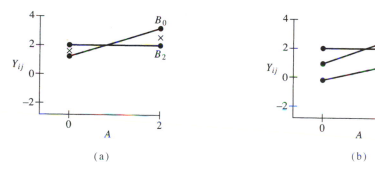

(a)   (b)

Figure 10.3   Two 3^f factorial experiments: (a) $A_L B_L$ interaction and (b) $A_L B_Q$ interaction.

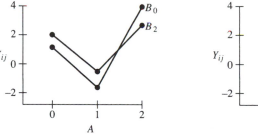
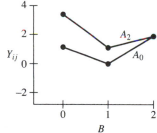

Figure 10.4 $A_Q B_L$ interaction.   **Figure 10.5** $A_L B_Q$ interaction again.

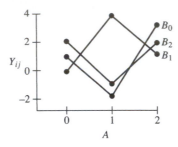

Figure 10.6 $A_Q B_Q$ interaction.

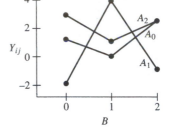

Figure 10.7 $A_Q B_Q$ interaction again.

■ **Example 10.1 (Initial Analysis of the Gas Emission Data in Table 10.5)**

The effect of two factors—prechamber volume ratio (V) and injection timing (T)—on the amount (in parts per million) of noxious gas emitted from an engine was to be studied. The volume ratio was set at three equispaced levels and the time was also set at three equispaced levels. Two engines were built for each of the nine treatment combinations, and the amount of gas emitted was recorded for the 18 engines as shown in Table 10.5.

TABLE 10.5
Gas Emission Data of Example 10.1

Time (T)	Low	Volume (V) Medium	High
Short	6.27	8.08	7.34
	5.43	8.04	7.87
Medium	6.94	7.48	8.61
	6.51	7.52	8.32
Long	7.22	8.65	9.02
	7.05	8.97	9.07

An ANOVA summary for Table 10.5, assuming a completely randomized design, is given in Table 10.6. That summary shows strong main effects and an interaction significant for any $\alpha \geq 0.029$. Since both factors are quantitative and equispaced, we can treat these data as we did the hypothetical data of Table 10.3. The resulting orthogonal contrasts are summarized in Table 10.7.

Using Equation (10.1) with $n = 2$, we find

$$SS_{V_L} = \frac{(c_{V_L})^2}{n \sum c_i^2} = \frac{(10.81)^2}{2(6)} = 9.738$$

Sums of squares for the remaining contrasts are obtained in like manner. Rounding to thousandths and adding these sums of squares to Table 10.6 gives Table 10.8, where we see that volume ratio produced both a strong linear effect and quadratic effect on the amount of gas emitted. The timing shows a strong linear effect, and a significant interaction between the two quadratic effects is shown.

TABLE 10.6
SPSS ANOVA for Gas Emission Data

Source of Variation	Sum of Squares	DF	Mean Square	F	Sig of F
Main Effects	15.607	4	3.902	50.435	.000
T	4.166	2	2.083	26.925	.000
V	11.441	2	5.721	73.946	.000
2-Way Interactions	1.385	4	.346	4.477	.029
T V	1.385	4	.346	4.477	.029
Explained	16.992	8	2.124	27.456	.000
Residual	.696	9	.077		
Total	17.688	17	1.040		

TABLE 10.7
Orthogonal Table for Example 10.1: Gas Emission Study

Treatment Combination	V_L	V_Q	T_L	T_Q	V_LT_L	V_LT_Q	V_QT_L	V_QT_Q	Total
00	-1	$+1$	-1	$+1$	$+1$	-1	-1	$+1$	11.70
01	-1	$+1$	0	-2	0	$+2$	0	-2	13.45
02	-1	$+1$	$+1$	$+1$	-1	-1	$+1$	$+1$	14.27
10	0	-2	-1	$+1$	0	0	$+2$	-2	16.12
11	0	-2	0	-2	0	0	0	$+4$	15.00
12	0	-2	$+1$	$+1$	0	0	-2	-2	17.62
20	$+1$	$+1$	-1	$+1$	-1	$+1$	-1	$+1$	15.21
21	$+1$	$+1$	0	-2	0	-2	0	-2	16.93
22	$+1$	$+1$	$+1$	$+1$	$+1$	$+1$	$+1$	$+1$	18.09
$\sum c_i^2$	6	18	6	18	4	12	12	36	
Contrast	10.81	-7.83	6.95	2.25	0.31	0.37	2.45	-8.97	
SS	9.738	1.703	4.025	0.141	0.012	0.006	0.250	1.118	

An interaction plot of the nine treatment means is presented in Figure 10.8. Careful study of that plot confirms the reasonableness of the results given in Table 10.8. Note that without T_2 there would be little obvious interaction.

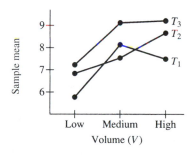

Figure 10.8   Treatment means for Example 10.1.

TABLE 10.8
ANOVA Breakdown for Example 10.1

Source		df		SS		MS	f	p value
V		2		11.441		5.721		
	V_L		1		9.738	9.738	126.468	0.000
	V_Q		1		1.703	1.703	22.117	0.001
T		2		4.166		2.083		
	T_L		1		4.025	4.025	52.273	0.000
	T_Q		1		0.141	0.141	1.831	0.209
VT		4		1.385		0.346		
	$V_L T_L$		1		0.012	0.012	0.156	0.702
	$V_L T_Q$		1		0.006	0.006	0.078	0.786
	$V_Q T_L$		1		0.250	0.250	3.247	0.105
	$V_Q T_Q$		1		1.118	1.118	14.519	0.004
Error		9		0.696		0.077		
Totals		17		17.688				

Values of contrasts and their significance tests were obtained manually for our hypothetical study and Example 10.1. Such results can also be obtained using computer software such as JMP, SAS, and SPSS. If the contrasts for Example 10.1 are entered into SPSS, the results in Table 10.9 are obtained. SPSS uses contrasts in the treatment means, so the contrast values in Table 10.7 are twice those in Table 10.9. Such differences do not affect the sums of squares, however.

When $T \sim t_{(v)}$ and $F \sim F_{(1,v)}$, $T^2 = F$. Thus, the two-tailed t tests used by SPSS are equivalent to the upper-tailed F tests in the ANOVA summary. Differences are due to rounding.

TABLE 10.9
SPSS Tests of Effects in Example 10.1

		Pooled Variance Estimate			
	Value	S. Error	T Value	D.F.	T Prob.
Contrast 1	5.4050	.4818	11.220	9.0	.000
Contrast 2	−3.9150	.8344	−4.692	9.0	.001
Contrast 3	3.4750	.4818	7.213	9.0	.000
Contrast 4	1.1250	.8344	1.348	9.0	.211
Contrast 5	.1550	.3933	.394	9.0	.703
Contrast 6	.1850	.6813	.272	9.0	.792
Contrast 7	1.2250	.6813	1.798	9.0	.106
Contrast 8	−4.4850	1.1800	−3.801	9.0	.004

■ **Example 10.2 (Further Analysis of the Gas Emission Data in Table 10.5)**

Since the chief concern in the gas emission example is to reduce the amount of noxious gas to a minimum and a significant interaction has been detected, one could examine all nine treatment means in a Newman–Keuls sense and look for conditions that might result in a minimum mean.

For each of the nine sample means, $s_{\bar{Y}} = \sqrt{\text{MSE}/2} = \sqrt{0.077/2} \approx 0.20$. Consulting Statistical Table E.1 for $n_2 = 9$ degrees of freedom, we find:

p:	2	3	4	5	6	7	8	9
q:	3.20	3.95	4.42	4.76	5.02	5.24	5.43	5.60
LSR:	0.64	0.79	0.88	0.95	1.00	1.05	1.09	1.12

Using these least significant ranges with the sample means placed in ascending order produces Figure 10.9. The sample mean of 5.85 shares no line with any other sample mean, leading us to conclude that the true average emission is least when the prechamber volume ratio and the injection time are set at their low levels.

Treatment combination:	00	01	02	11	20	10	21	12	22
Mean:	5.85	6.72	7.13	7.50	7.60	8.06	8.47	8.81	9.04

Figure 10.9 Newman–Keuls graphic for gas emission data.

The I and J Components of Interaction

We now consider a partitioning of the interaction sum of squares for a 3^2 factorial into two sums of squares, each with two degrees of freedom. The procedure is illustrated using the hypothetical data in Table 10.1. We begin by placing a copy of the Treatment columns of Table 10.1 to the right of that table, thereby producing Table 10.10. The procedure involves adding data by diagonals rather than by rows or columns.

First consider the diagonals downward from left to right (as indicated by the arrows). The sum along the main diagonal is $1+4+2 = 7$, the next one to the right is $-2+1+2 = 1$, and the last sum taken in this manner is $3 + 0 + (-1) = 2$. Notice that the additional copy of Table 10.1 makes it easy to obtain these sums. Since the sum of the nine observations is 10, the sum of squares between these three diagonals is

TABLE 10.10
Diagonal Computations

Factor B	Factor A			Factor A		
	$\searrow 0$	$\searrow 1$	$\searrow 2$	0	1	2
0	1	-2	3	1	-2	3
1	0	4	1	0	4	1
2	2	-1	2	2	-1	2

$$\frac{(7)^2 + (1)^2 + (2)^2}{3} - \frac{(10)^2}{9} = 6.89$$

If the diagonals are now considered downward and to the left, their totals are

$$3 + 4 + 2 = 9$$

$$1 + 1 - 1 = 1$$

$$-2 + 0 + 2 = 0$$

and their sum of squares is

$$\frac{(9)^2 + (1)^2 + (0)^2}{3} - \frac{(10)^2}{9} = 16.22$$

These two somewhat artificial sums of squares (i.e., 6.89 and 16.22) are seen to add up to the interaction sum of squares

$$6.89 + 16.22 = 23.11$$

These two components of interaction have no physical significance, but simply illustrate another way to extract two orthogonal components of interaction. Testing each of these separately for significance has no meaning, but this arbitrary breakdown is very useful in more complex designs. Some authors refer to these two components as the I and J *components of interaction*

$$I(AB) = 6.89 \quad 2 \text{ df}$$

$$J(AB) = 16.22 \quad 2 \text{ df}$$

$$\text{total } A \times B = 23.11 \quad 4 \text{ df}$$

Each such component carries 2 degrees of freedom, and sometimes the two are called, respectively, the AB^2 and AB components of $A \times B$ interaction. In this notation, effects can be multiplied together using a modulus of 3, since this is a 3^f factorial. A *modulus* of 3 means that the resultant number is equal to the remainder when the number in the usual base of 10 is divided by 3. Thus $4 = 1$ in modulus 3, because 1 is the remainder when 4 is divided by 3. The following associations also hold:

$$\text{numbers:} \quad 0 \quad 3 = 0 \quad 6 = 0 \quad 9 = 0$$

$$1 \quad 4 = 1 \quad 7 = 1 \quad 10 = 1$$

$$2 \quad 5 = 2 \quad 8 = 2 \quad 11 = 2$$

$$\vdots$$

When the form $A^p B^q$ is used, it is postulated that the only exponent allowed on the first letter in the expression is a 1. To make it a 1, the expression can be squared and reduced, modulus 3. For example,

$$A^2B = (A^2B)^2 = A^4B^2 = AB^2$$

Hence AB and AB^2 are the only components of the $A \times B$ interaction with 2 df each. Here the two types of notation are related as follows:

$$I(AB) = AB^2$$

$$J(AB) = AB$$

To summarize this simple experiment, all effects can be expressed with 2 df each, as in Table 10.11.

It will be found very useful to break such an experiment down into 2-df effects when more complex designs are considered. Example 10.3 presents this breakdown merely to show another way to partition the interaction effect.

TABLE 10.11
3^2 Factorial by 2-df Analysis

Source	df	SS
A_i	2	4.22
B_j	2	1.56
$I(AB) = AB^2$	2	6.89
$J(AB) = AB$	2	16.22
Totals	8	28.89

■ **Example 10.3 (2-df Interaction Components for the Gas Emission Study)**

As an exercise, let us pull out the VT and VT^2 components of interaction for the gas emission study considered in Example 10.1. Following the procedure illustrated for the hypothetical data of Table 10.1, we add a copy of the treatment columns in Table 10.5 to obtain Table 10.12. Adding downward from left to right along the main diagonal, we obtain the sum $6.27 + 5.43 + 7.48 + 7.52 + 9.02 + 9.07 = 44.79$. The reader can verify that the totals for the other two such diagonals are 47.32 and 46.28. The sum of these three totals gives a grand total of $44.79 + 47.32 + 46.28 = 138.39$. Thus,

$$SS_{VT^2} = \frac{(44.79)^2 + (47.32)^2 + (46.28)^2}{6} - \frac{(138.39)^2}{18} \approx 0.539$$

Adding downward from right to left, the sums on the diagonals are found to be 44.48, 47.66, and 46.25. Thus,

$$SS_{VT} = \frac{(44.48)^2 + (47.66)^2 + (46.25)^2}{6} - \frac{(138.39)^2}{18} \approx 0.846$$

Notice that $SS_{VT^2} + SS_{VT} = 0.539 + 0.846 = 1.385$, which is the TV interaction sum of squares reported in Table 10.6.

TABLE 10.12
Data for Diagonal Computations in Example 10.3

Time, *T*	Low	Volume, *V* Medium	High	Low	Volume, *V* Medium	High
Short	6.27	8.08	7.34	6.27	8.08	7.34
	5.43	8.04	7.87	5.43	8.04	7.87
Medium	6.94	7.48	8.61	6.94	7.48	8.61
	6.51	7.52	8.32	6.51	7.52	8.32
Long	7.22	8.65	9.02	7.22	8.65	9.02
	7.05	8.97	9.07	7.05	8.97	9.07

10.3 3^3 FACTORIAL

Suppose an experimenter has three factors, each at three levels, or a $3 \times 3 \times 3 = 3^3$ factorial. There are several ways to break down the effects of factors *A, B,* and *C* and their associated interactions. If the order of experimentation is completely randomized, the model for such an experiment is

$$Y_{ijk} = \mu + A_i + B_j + AB_{ij} + C_k + AC_{ik} + BC_{jk} + ABC_{ijk} + \varepsilon_{ijk}$$

with the last two terms confounded unless there is replication within the cells. In this model $i = 1, 2, 3, j = 1, 2, 3,$ and $k = 1, 2, 3,$ making 27 treatment combinations. These 27 treatment combinations may be as shown in Table 10.13.

Association of the proper coefficients on these 27 treatment combinations would allow the Table 10.14 breakdown of an ANOVA if all effects were set at quantitative levels.

In an actual problem, these three-way interactions would be hard to explain. Thus quite often, the *ABC* interaction is left with its 8 df for use as an error term to test the main effects *A, B, C,* and the two-way interactions.

These effects also can be partitioned in terms of 2-df effects, by means of *I* and *J* components on *AB, AC,* and *BC* interactions. These could be designated as $AB, AB^2, AC, AC^2,$

TABLE 10.13
3^3 Factorial Treatment Combinations

Factor *B*	Factor *C*	Factor *A* 0	1	2
0	0	000	100	200
	1	001	101	201
	2	002	102	202
1	0	010	110	210
	1	011	111	211
	2	012	112	212
2	0	020	120	220
	1	021	121	221
	2	022	122	222

and BC, BC^2, each with 2 df. However, the three-way interaction with its 8 df may need a further breakdown. Sometimes ABC is broken into four 2-df components called $X(ABC)$, $Y(ABC)$, $Z(ABC)$, and $W(ABC)$, or, using the notation of the last section, AB^2C, ABC^2, ABC, and AB^2C^2. Here again no first letter is squared, and $A^2BC = (A^2BC)^2 = A^4B^2C^2 = AB^2C^2$ modulus 3. Such a partitioning would yield Table 10.15.

TABLE 10.14
3³ Factorial Analysis for Linear and Quadratic Effects

Source	df		Source	df
A_i	2		A_QC_L	1
A_L		1	A_QC_Q	1
A_Q		1	BC_{jk}	4
B_j	2		B_LC_L	1
B_L		1	B_LC_Q	1
B_Q		1	B_QC_L	1
AB_{ij}	4		B_QC_Q	1
A_LB_L		1	ABC_{ijk}	8
A_LB_Q		1	$A_LB_LC_L$	1
A_QB_L		1	$A_LB_LC_Q$	1
A_QB_Q		1	$A_LB_QC_L$	1
C_k	2		$A_LB_QC_Q$	1
C_L		1	$A_QB_LC_L$	1
C_Q		1	$A_QB_LC_Q$	1
AC_{ik}	4		$A_QB_QC_L$	1
A_LC_L		1	$A_QB_QC_Q$	1
A_LC_Q		1		
			Total	26

TABLE 10.15
3³ Factorial in 2-df Analyses

Source	df	
A	2	
B	2	
AB	2	} 4
AB^2	2	
C	2	
AC	2	} 4
AC^2	2	
BC	2	} 4
BC^2	2	
ABC	2	
ABC^2	2	} 8
AB^2C	2	
AB^2C^2	2	
Total	26	

■ **Example 10.4**

Now we turn to a problem involving the effect of three factors, each at three levels: the measured variable is yield and the factors that might affect this response are days, operators, and concentrations of solvent. We choose three days, three operators, and three concentrations: days and operators are qualitative effects; concentrations are quantitative and are set at 0.5, 1.0, 2.0. Although these are not equispaced, the logarithms of these three levels are equispaced, and the logarithms can be used if a curve fitting is warranted. For the purposes of this chapter, all levels of all factors are considered to be fixed and the design completely randomized. We decide to take three replications of each of the $3^3 = 27$ treatment combinations. The data, after coding by subtracting 20.0, are as presented in Table 10.16.

If these data are analyzed on a purely qualitative basis, the methods of Chapter 5 can be used. The resulting ANOVA is shown in Table 10.17.

The model for this example is merely:

$$Y_{ijkm} = \mu + D_i + O_j + DO_{ij} + C_k + DC_{ik} + OC_{jk} + DOC_{ijk} + \varepsilon_{m(ijk)}$$

From this analysis the concentration effect is tremendous, and the days, operators, and day × operator interaction are all significant at the 1% level.

TABLE 10.16
Example Data on 3^3 Factorial with Three Replications

Concentration, C_k	Day, D_i 5/14			5/15			5/16		
	Operator, O_j								
	A	B	C	A	B	C	A	B	C
0.5	1.0	0.2	0.2	1.0	1.0	1.2	1.7	0.2	0.5
	1.2	0.5	0.0	0.0	0.0	0.0	1.2	0.7	1.0
	1.7	0.7	−0.3	0.5	0.0	0.5	1.2	1.0	1.7
1.0	5.0	3.2	3.5	4.0	3.2	3.7	4.5	3.7	3.7
	4.7	3.7	3.5	3.5	3.0	4.0	5.0	4.0	4.5
	4.2	3.5	3.2	3.5	4.0	4.2	4.7	4.2	3.7
2.0	7.5	6.0	7.2	6.5	5.2	7.0	6.7	7.5	6.2
	6.5	6.2	6.5	6.0	5.7	6.7	7.5	6.0	6.5
	7.7	6.2	6.7	6.2	6.5	6.8	7.0	6.0	7.0

TABLE 10.17
Minitab ANOVA for Example 10.3

```
Source    DF         SS          MS             F            P
D          2      3.483       1.742          9.43        0.000
O          2      6.142       3.071         16.63        0.000
D*O        4      4.072       1.018          5.51        0.001
C          2    468.985     234.493       1269.65        0.000
D*C        4      0.586       0.147          0.79        0.534
O*C        4      0.894       0.223          1.21        0.317
D*O*C      8      1.094       0.137          0.74        0.655
Error     54      9.973       0.185
Total     80    495.231
```

Because concentrations are at quantitative levels, the linear and quadratic effects of concentrations may be computed, as well as the interactions between linear effect of concentration and days, quadratic effect of concentration and days, linear effect of concentration and operators, and quadratic effect of concentration and operators. It is seldom worthwhile to extract three-way interaction in this way. Instead, it is usually helpful to calculate these qualitative effects by constructing some two-way tables for the interactions that are being computed. Two such tables are shown (Table 10.18a and Table 10.18b).

Applying the linear and quadratic coefficients to the concentration totals of Table 10.18, we have

Sums of Squares

$$C_L = -1(18.6) + 0(105.6) + 1(177.5) = 158.9$$

$$SS_{C_L} = \frac{(158.9)^2}{27(2)} = 467.58$$

$$C_Q = +1(18.6) - 2(105.6) + 1(177.5) = -15.1$$

$$SS_{C_Q} = \frac{(-15.1)^2}{27(6)} = 1.41$$

$$\overline{SS_C = 468.99}$$

TABLE 10.18
Cell Totals for $D \times C$ and $O \times C$ Interactions

	10.18a Concentration				10.18b Concentration		
Day	0.5	1.0	2.0	Operator	0.5	1.0	2.0
5/14	5.2	34.5	60.5	A	9.5	39.1	61.6
5/15	4.2	33.1	56.6	B	4.3	32.5	55.3
5/16	9.2	38.0	60.4	C	4.8	34.0	60.6
Totals	18.6	105.6	177.5 301.7	Totals	18.6	105.6	177.5 301.7

For the $D \times C$ interactions use Table 10.18a, and consider each level of days separately. At

$$5/14: \quad C_L = -1(5.2) + 0(34.5) + 1(60.5) = 55.3$$
$$5/15: \quad C_L = -1(4.2) + 0(33.1) + 1(56.6) = 52.4$$
$$5/16: \quad C_L = -1(9.2) + 0(38.5) + 1(60.4) = 51.2$$

The $D \times C_L$ $SS_{\text{interaction}}$ is then

$$\frac{(55.3)^2 + (52.4)^2 + (51.2)^2}{9(2)} - \frac{(158.9)^2}{27(2)} = 0.49$$

For quadratic effects, at

$$5/14: \quad C_Q = +1(5.2) - 2(34.5) + 1(60.5) = -3.3$$
$$5/15: \quad C_Q = +1(4.2) - 2(33.1) + 1(56.6) = -5.4$$
$$5/16: \quad C_Q = +1(9.2) - 2(38.5) + 1(60.4) = -6.4$$

The $D \times C_Q$ $SS_{\text{interaction}}$ is then

$$\frac{(-3.3)^2 + (-5.4)^2 + (-6.4)^2}{9(6)} - \frac{(-15.1)^2}{27(6)} = 0.09$$

and

$$SS_{D \times C} = SS_{D \times C_L} + SS_{D \times C_Q} = 0.49 + 0.09 = 0.58$$

When the same procedure is applied to the data of Table 10.18b, we have

$$SS_{O \times C_L} = \frac{(52.1)^2 + (51.0)^2 + (55.8)^2}{9(2)} - \frac{(158.9)^2}{27(2)} = 0.70$$

$$SS_{O \times C_Q} = \frac{(-7.1)^2 + (-5.4)^2 + (-2.6)^2}{9(6)} - \frac{(-15.1)^2}{27(6)} = 0.19$$

and

$$SS_{O \times C} = SS_{O \times C_L} + SS_{O \times C_Q} = 0.70 + 0.19 = 0.89$$

The resulting ANOVA can now be shown in Table 10.19.

This second analysis shows that the linear effect and the quadratic effect of concentration are extremely significant. Figure 10.10a shows the effect of operators, days, and $D \times O$ interaction. Figure 10.10b indicates that the linear effect of concentration far outweighs the quadratic effect and there is no significant interaction. If a straight line or three straight lines were fit to these data, the logs of the concentrations would be used, since the logs are equispaced.

TABLE 10.19
Second ANOVA for Example 10.4

Source	df	SS	MS
D_i	2	3.48	1.74**
O_j	2	6.14	3.07**
DO_{ij}	4	4.07	1.02**
C_L	1	467.58	467.58***
C_Q	1	1.41	1.41**
$D \times C_L$	2	0.49	0.24
$D \times C_Q$	2	0.09	0.04
$O \times C_L$	2	0.70	0.35
$O \times C_Q$	2	0.19	0.09
DOC_{ijk}	8	1.09	0.14
$\varepsilon_{m(ijk)}$	54	9.98	0.18
Totals	80	495.22	

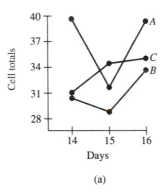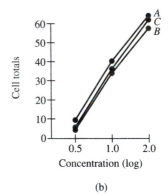

Figure 10.10   Plots of the 3^3 example of Table 10.16: (a) effect of operators, days, and $D \times O$ interaction and (b) linear effect of concentration versus quadratic effects.

The Diagonal Components of the Interaction Terms

Although analysis of the yield data in Table 10.16 would usually conclude with that of Example 10.3, we break down each interaction into its diagonal, or I and J, components for the sake of illustrating the technique. To compute the two diagonal components of the two-factor interactions, the two parts of Table 10.18 can be used, along with a similar table for the $D \times O$ cells (see Table 10.20). From Table 10.20 the diagonal components of the $D \times O$ interaction are

$$I(D \times O) = \frac{(39.5 + 28.6 + 34.8)^2 + (30.2 + 34.1 + 39.5)^2 + (30.5 + 33.3 + 31.2)^2}{27}$$

$$- \frac{(301.7)^2}{81} = 1.74 \qquad (\text{call it } DO^2)$$

and

$$J(D \times O) = \frac{(30.5 + 28.6 + 39.5)^2 + (96.2)^2 + (106.9)^2}{27} - \frac{(301.7)^2}{81} = 2.33$$

(call it DO)

These total $1.74 + 2.33 = 4.07$, the $D \times O$ interaction sum of squares.

Applying the same technique to the two parts of Table 10.18 gives

$$I(DC) = DC^2 = \frac{(98.7)^2 + (102.7)^2 + (100.3)^2}{27} - \frac{(301.7)^2}{81} = 0.30$$

$$J(DC) = DC \ = \frac{(102.8)^2 + (99.8)^2 + (99.1)^2}{27} - \frac{(301.7)^2}{81} = 0.29$$

$$D \times C = 0.59$$

$$I(OC) = OC^2 = \frac{(102.6)^2 + (99.9)^2 + (99.2)^2}{27} - \frac{(301.7)^2}{81} = 0.24$$

$$J(OC) = OC \ = \frac{(98.9)^2 + (104.0)^2 + (98.8)^2}{27} - \frac{(301.7)^2}{81} = 0.65$$

$$O \times C = 0.89$$

To break down the 8 degrees of freedom of the $D \times O \times C$ interaction, form an $O \times D$ table showing each of the three levels of concentration C as in Table 10.21.

For each of these concentration levels, find the I and J effect totals, for example, at C_1,

I components are 8.1, 3.3, 7.2

J components are 5.0, 6.1, 7.5

TABLE 10.20
Cell Totals for $D \times O$ Interactions

Day	A	Operator B	C
5/14	39.5	30.2	30.5
5/15	31.2	28.6	34.1
5/16	39.5	33.3	34.8

TABLE 10.21
Cells Totals for $D \times O$ Interaction at Each Level of Concentration

O_i	D_i at C_1			D_i at C_2			D_i at C_3		
A	3.9	1.5	4.1	13.9	11.0	14.2	21.7	18.7	21.2
B	1.4	1.0	1.9	10.4	10.2	11.9	18.4	17.4	19.5
C	−0.1	1.7	3.2	10.2	11.9	11.9	20.4	20.5	19.7

Now form a table with these I and J components at each level of C (see Table 10.22).

TABLE 10.22
Diagonal Totals for Each Level of Concentration

C_k	I(DO)			J(DO)		
	i_0	i_1	i_2	j_0	j_1	j_2
0.5	8.1	3.3	7.2	5.0	6.1	7.5
1.0	36.0	33.1	36.5	34.6	33.3	37.7
2.0	58.8	58.6	60.1	59.0	56.8	61.7

Treat each half of Table 10.22 as a simple interaction and compute the I and J components. Thus,

$$DO^2C^2 = I[C \times I(DO)] = \frac{(101.3)^2 + (98.6)^2 + (101.8)^2}{27} - \frac{(301.7)^2}{81} = 0.22$$

$$DO^2C = J[C \times I(DO)] = \frac{(99.1)^2 + (99.4)^2 + (103.2)^2}{27} - \frac{(301.7)^2}{81} = 0.39$$

$$DOC^2 = I[C \times J(DO)] = \frac{(100.0)^2 + (102.8)^2 + (98.9)^2}{27} - \frac{(301.7)^2}{81} = 0.30$$

$$DOC = J[C \times J(DO)] = \frac{(99.8)^2 + (102.4)^2 + (99.5)^2}{27} - \frac{(301.7)^2}{81} = 0.19$$

$$\text{total } D \times O \times C = 1.10$$

compared with 1.09 in Table 10.19. This last breakdown into four parts could also have been accomplished by considering the $C \times O$ interaction at three levels of D_i, or the $C \times D$ interaction at three levels of O_j. The resulting analysis is summarized in Table 10.23. This analysis is in substantial agreement with Tables 10.17 and 10.19. No new tests would be performed on the data in Table 10.23 because they represent only an arbitrary breakdown of the interactions into 2-df components. The purpose of such a breakdown is discussed in subsequent chapters. For testing hypotheses on interaction, these components are added together again.

TABLE 10.23
Third ANOVA for 3^3 Experiment

Source	df	SS
D_i	2	3.48
O_j	2	6.14
DO	2	2.33
DO^2	2	1.74
C_k	2	468.99
DC	2	0.29
DC^2	2	0.30
OC	2	0.65
OC^2	2	0.24
DOC	2	0.19
DOC^2	2	0.30
DO^2C	2	0.39
DO^2C^2	2	0.22
$\varepsilon_{m(ijk)}$	54	9.98
Totals	80	495.24

10.4 COMPUTER PROGRAMS

As illustrated for Example 10.1, computer programs such as those in JMP, SPSS, and SAS may be used to test for linear and quadratic effects, provided one can set up the proper contrast coefficients. Since such contrasts are used when the factor is quantitative and equispaced, a regression program may offer the best means of obtaining the equation of "best fit" in such cases. Regression techniques are considered in Chapter 15.

10.5 SUMMARY

The summary at the end of Chapter 9 may now be extended for Part II.A.1.

Experiment	Design	Analysis
II. Two or more factors A. Factorial (crossed)		
	1. Completely randomized $Y_{ijk} = \mu + A_i + B_J + AB_{ij}$ $+\varepsilon_{k(ij)}, \ldots$ for more factors	1.
	a. General case	a. ANOVA with interactions
	b. 2^f case	b. Yates method or general ANOVA; use (1), a, b, ab, ...

c. 3^f case

c. General ANOVA; use
00, 10, 20, 01, 11, . . . and
$A \times B = AB + AB^2, \ldots$
for interaction

PROBLEMS

10.1 The following table presents data describing pull-off force in pounds on glued boxes at three temperatures and three humidities with two observations per treatment combination in a completely randomized experiment. Do a complete analysis of this problem by the general methods of Chapter 5.

		Temperature A	
Humidity B	Cold	Ambient	Hot
50%	0.8	1.5	2.5
	2.8	3.2	4.2
70%	1.0	1.6	1.8
	1.6	1.8	1.0
90%	2.0	1.5	2.5
	2.2	0.8	4.0

10.2 Assuming that the temperatures in Problem 10.1 are equispaced, extract linear and quadratic effects of both temperature and humidity as well as all components of interaction.

10.3 From Problem 10.1 extract the AB and AB^2 components of interaction.

10.4 Develop a "Yates method" for this 3^2 experiment and check the results with those above.

10.5 When the data on noxious gas were collected in Example 10.1, data were also recorded on percent smoke emitted by the 18 engines. Results were as given below. Do an ANOVA by the methods of Chapter 5.

		Volume (V)	
Time (T)	Low	Medium	High
Short	0.3	0.1	0.4
	0.4	0.4	0.4
Medium	0.1	0.1	0.4
	0.4	0.2	1.2
Long	0.8	0.7	2.5
	2.0	1.6	3.6

10.6   From Problem 10.5 extract linear and quadratic components of both main effects and the interaction.

10.7   From Problem 10.5 extracts the VT and the VT^2 components of interaction.

10.8   Plot any results that stand out in Problem 10.6.

10.9   Run further tests to recommend the best volume and time combinations for minimizing smoke in Problem 10.5.

10.10 Compare the results obtained in Problem 10.8 with the results of Example 10.1 and make some overall recommendations about these engine designs.

10.11 A behavior variable on concrete pavements was measured for three surface thicknesses: 3, 4, and 5 inches; three base thicknesses: 0, 3, and 6 inches; and three subbase thicknesses: 4, 8, and 12 inches. Two observations were made under each of the 27 pavement conditions and complete randomization performed.

Do a complete ANOVA of this experiment by the methods of Chapter 5.

	Surface Thickness (in.)								
	3			4			5		
Subbase Thickness (in.)	Base Thickness (in.)			Base Thickness (in.)			Base Thickness (in.)		
	0	**3**	**6**	**0**	**3**	**6**	**0**	**3**	**6**
4	2.8	4.3	5.7	4.1	5.4	6.7	6.0	6.3	7.1
	2.6	4.5	5.3	4.4	5.5	6.9	6.2	6.5	6.9
8	4.1	5.7	6.9	5.3	6.5	7.7	6.1	7.2	8.1
	4.4	5.8	7.1	5.1	6.7	7.4	5.8	7.1	8.4
12	5.5	7.0	8.1	6.5	7.7	8.8	7.0	8.0	9.1
	5.3	6.8	8.3	6.7	7.5	9.1	7.2	8.3	9.0

10.12 Since all three factors in Problem 10.11 are quantitative and equispaced, determine linear and quadratic effects for each factor and all interaction breakdowns. Test for significance.

10.13 Break down the interactions of Problem 10.11 into 2-df components (AB, AB^2, ABC, AB^2C, etc.).

10.14 Use a Yates method to solve Problem 10.11 and check the results.

10.15 Plot any significant results of Problem 10.11.

10.16 The following data are on the wet-film thickness (in mils) of lacquer. The factors studied were type of resin (two types), gate-blade setting in mils (three settings), and weight fraction of nonvolatile material in the lacquer (three fractions).

Gate Setting	Resin Type 1			Resin Type 2		
	Weight Fraction			Weight Fraction		
	0.20	0.25	0.30	0.20	0.25	0.30
2	1.6	1.5	1.5	1.5	1.4	1.6
	1.5	1.3	1.3	1.4	1.3	1.4
4	2.7	2.5	2.4	2.4	2.6	2.2
	2.7	2.5	2.3	2.3	2.4	2.1
6	4.0	3.6	3.5	4.0	3.7	3.4
	3.9	3.8	3.4	4.0	3.6	3.3

Do an ANOVA of the data above and pull out any quantitative effects in terms of their components.

10.17 A study of the effects of temperature and current density on the coating thickness of an integrated circuit produced the following data. The experiment was conducted as a completely randomized factorial with temperatures of 90, 105, and 120°F and current densities of 90, 130, and 170 amperes per square foot. Thicknesses were measured to the nearest microinch.

Density	Temperature (°F)		
	90	105	120
90	220.52	231.59	268.93
	228.59	243.22	251.46
	233.41	245.52	252.58
130	325.83	329.87	343.68
	332.67	359.96	348.39
	342.34	341.53	343.65
170	400.83	434.13	438.51
	415.07	438.93	442.77
	395.92	426.77	453.98

a. Assuming fixed factors, do a complete analysis of this problem by the general methods of Chapter 5. Include an interaction plot and Newman–Keuls analysis.
b. Extract linear and quadratic effects of both factors. Test for significance and comment.

10.18 A study of the effects of three fixed factors, each at three equispaced levels, on the density of ceramic tiles produced the following coded data.

A	B	C 1	C 2	C 3
1	1	90	66	124
	2	138	88	134
	3	128	58	130
2	1	10	60	16
	2	72	110	34
	3	80	10	0
3	1	−16	96	102
	2	36	162	144
	3	−10	156	130

a. Assuming the true three-factor interaction effect to be 0, conduct a complete analysis by the general methods of Chapter 5.
b. Using the results in part a, extract linear and quadratic factors for the main effects. Test for significance and comment.

Chapter 11

FACTORIAL EXPERIMENT: SPLIT-PLOT DESIGN

11.1 INTRODUCTION

In many experiments in which a factorial arrangement is desired, it may not be possible to randomize completely the order of experimentation. In Chapter 8 restrictions on randomization were considered, and both randomized block and Latin square designs were discussed. There are still many practical situations in which it is not at all feasible to even randomize within a block. Under certain conditions these restrictions will lead to a split-plot design. The data in Table 11.1, which we shall use to show why such designs are quite common, summarize the effect of oven temperature T and baking time B on the life Y of an electrical component [37].

Looking only at Table 11.1, we might think of it to be a 4×3 factorial with three replications per cell and proceed to the analysis of Table 11.2. Here only the temperature has a significant effect on the life of the component at any reasonable significance level. Thus, we might explore the nature of the relationship between temperature and Y. Before embarking on such an analysis, however, we should question the order of experimentation.

TABLE 11.1
Electrical Component Life-Test Data

Baking Time, B (min)	Oven Temperature, T (°F)			
	580	600	620	640
5	217	158	229	223
	188	126	160	201
	162	122	167	182
10	233	138	186	227
	201	130	170	181
	170	185	181	201
15	175	152	155	156
	195	147	161	172
	213	180	182	199

TABLE 11.2
Minitab ANOVA for Electrical Component Life-Test Data as a Factorial

```
Analysis of Variance (Balanced Designs)

Factor      Type      Levels      Values
T           fixed        4          1        2        3        4
B           fixed        3          1        2        3

Analysis of Variance for Y

Source      DF           SS         MS           F            P
T            3        12494.3     4164.8        7.31        0.001
B            2          566.2      283.1        0.50        0.614
T*B          6         2600.4      433.4        0.76        0.607
Error       24        13670.0      569.6
Total       35        29331.0

Source            Variance          Error       Expected Mean Square
                  component         term        (using restricted model)
 1 T                                 4          (4) + 9Q[1]
 2 B                                 4          (4) + 12Q[2]
 3 T*B                               4          (4) + 3Q[3]
 4 Error            569.6                        (4)
```

11.2 A SPLIT-PLOT DESIGN

The data analysis of Table 11.2 assumes a completely randomized design. This means that one of the four temperatures was chosen at random and the oven was heated to this temperature; then a baking time was chosen at random, and an electrical component was inserted in the oven and baked for the time selected. After this run, the whole procedure was repeated until the data had been compiled. Now, was the experiment really conducted in this manner? Of course the answer is "no." Once an oven has been heated to temperature, all nine components are inserted; three are baked for 5 minutes, three for 10 minutes, and three for 15 minutes. We would argue that this is the only practical way to run the experiment. Complete randomization is too impractical as well as too expensive. Fortunately, we can handle this restriction on the complete randomization by means of what is called a *split-plot* design.

The four temperature levels are called *plots*. They could be called blocks, but in an earlier chapter blocks and replications were used for a complete rerun of the whole experiment. (The word "plots" was inherited from agricultural applications.) In such a setup, temperature—a main effect—is confounded with these plots. If conditions change from one oven temperature to another, the changes will show up as temperature differences. Thus, in such a design, a main effect is confounded with plots. This is necessary because it is often the most practical way to order the experiment. Now once a temperature has been set up by choosing one of the four temperatures at random, three components can be placed in the oven and one component can be baked for 5 minutes, another for 10 minutes, and the third one for 15 minutes. The specific component that is to be baked for 5, 10, and 15 minutes is again decided by a random selection. These three baking-time levels may be thought of as a splitting of the plot into three parts, one part for each baking time. This defines the three parts of a main plot that are called split plots. Note that only three components are placed in the oven, not nine. The temperature is then changed to another level and three more components are placed in the oven for 5, 10, and 15 minutes baking time. The same procedure is followed for all four temperatures; after this the whole experiment may be replicated. These replications may be run several days after the initial experiment; in fact, it is often desirable to collect data from two or three replications and then decide whether more replications are necessary.

The Data Layout

The life-test experiment was actually run as a split-plot experiment and laid out as shown in Table 11.3, where randomization restrictions are indicated by the differential use of lines. That is, the experiment starts by running a replication or block (double lines); then within a replication an oven temperature is chosen at random (single vertical line), and within that temperature is chosen at random (single vertical line), and within that temperature level baking time is randomized. Thus a set of observations with no lines separating them (e.g., 158, 138, 152 for temperature 600°F, replication I) indicates random order of these three observations. This notion of using straight lines to show randomization restrictions was suggested in Chapter 4 (Section 4.2). If data are presented as in Table 11.1, there is no way to tell how they were collected. Were they completely randomized—all 36 readings? Or were there randomization restrictions as shown in Table 11.3?

    In Table 11.3 temperatures are confounded with plots and the $R \times T$ cells are called whole plots. Inside a whole plot, the baking times are applied to one-third of the material. These plots associated with the three baking times are called split plots. Since one main effect is confounded with plots and the other main effect is not, it is usually desirable to place in the split the main effect we are most concerned about testing, because this factor is not confounded.

The Mathematical Model

We might think that factor B (baking time) is nested with the main plots. This is not the case, since the same levels of B are used in all plots. A model for this experiment would be

TABLE 11.3
Split-Plot Layout for Electrical Component Life-Test Data

Replication, R	Baking Time, B (min)	Oven Temperature, T (°F)			
		580	600	620	640
I	5	217	158	229	223
	10	233	138	186	227
	15	175	152	155	156
II	5	188	126	160	201
	10	201	130	170	181
	15	195	147	161	172
III	5	162	122	167	182
	10	170	185	181	201
	15	213	180	182	199

$$Y_{ijk} = \mu + \underbrace{R_i + T_j + RT_{ij}}_{\text{whole plot}} + \underbrace{B_k + RB_{ik} + TB_{jk} + RTB_{ijk}}_{\text{split plot}} \qquad (11.1)$$

The first three variable terms in this model represent the whole plot, and the RT interaction is often referred to as the *whole-plot error*. The usual assumption is that this interaction does not exist, that this term is really an estimate of the error within the main plot. The last four terms represent the split plot, and the RTB interaction is referred to as the *split-plot error*. Sometimes the RB term is also considered to be nonexistent and is combined with RTB as an error term. A separate error term might be obtained if it were feasible to repeat some observations within the split plot. The proper EMS values can be found by considering m repeat measurements where $m = 1$ (Table 11.4). Note that in Table 11.4 the degrees of freedom in the whole plot add to 11 because there are 12 whole plots (3 replications × 4 temperatures), and the degrees of freedom in the split plot add to 24 because there are 2 df within each whole plot, hence 2 × 12 within all whole plots.

Analysis of the Data

Because the error mean square cannot be isolated in this experiment, $\sigma_\varepsilon^2 + \sigma_{RTB}^2$ is taken as the split-plot error, and $\sigma_\varepsilon^2 + 3\sigma_{RT}^2$ is taken as the whole-plot error. The main effects and effect of TB can be tested, as seen from the EMS column of Table 11.4, although no exact tests exist for the replication effect nor for interactions involving the replications. This is not a serious disadvantage for this design, since tests on replication effects are not of interest but are isolated only to reduce the error variance.

A Minitab analysis of the data in Table 11.3 is shown in Table 11.5. This analysis follows the methods given in Chapter 5. When the full model of Equation (11.1) is used, Minitab outputs an error message and no sums of squares. To avoid this unwanted result and to obtain Table 11.5, the three-factor interaction was deleted from the model. The

TABLE 11.4
EMS for Split-Plot Electrical Component Life-Test Data

	Source	df	3 R i	4 F j	3 F k	1 R m	EMS
Whole plot	R_i	2	1	4	3	1	$\sigma_\varepsilon^2 + 12\sigma_R^2$
	T_j	3	3	0	3	1	$\sigma_\varepsilon^2 + 3\sigma_{RT}^2 + 9\phi_T$
	RT_{ij}	6	1	0	3	1	$\sigma_\varepsilon^2 + 3\sigma_{RT}^2$
Split plot	B_k	2	3	4	0	1	$\sigma_\varepsilon^2 + 4\sigma_{RB}^2 + 12\phi_B$
	RB_{ik}	4	1	4	0	1	$\sigma_\varepsilon^2 + 4\sigma_{RB}^2$
	TB_{jk}	6	3	0	0	1	$\sigma_\varepsilon^2 + \sigma_{RTB}^2 + 3\phi_{TB}$
	RTB_{ijk}	12	1	0	0	1	$\sigma_\varepsilon^2 + \sigma_{RTB}^2$
	$\varepsilon_{m(ijk)}$	—	1	1	1	1	σ_ε^2 (not retrievable)
	Total	35					

TABLE 11.5
Minitab ANOVA for Split-Plot Electrical Component Life-Test Data*

```
Analysis of Variance (Balanced Designs)
```

Factor	Type	Levels	Values			
R	random	3	1	2	3	
T	fixed	4	1	2	3	4
B	fixed	3	1	2	3	

```
Analysis of Variance for Y
```

Source	DF	SS	MS	F	P
R	2	1962.7	981.4	4.04	0.045
T	3	12494.3	4164.8	14.09	0.004
R*T	6	1773.9	295.7	1.22	0.362
. . . .	. . . .	. . . .	. . . .	. . . .	. . . .
B	2	566.2	283.1	0.16	0.856
R*B	4	7021.3	1755.3	7.23	0.003
T*B	6	2600.4	433.4	1.79	0.185
Error	12	2912.1	242.7		
Total	35	29331.0			

Source	Variance component	Error term	Expected Mean Square (using restricted model)
1 R	61.56	7	(7) + 12(1)
2 T		3	(7) + 3(3) + 9Q[2]
3 R*T	17.66	7	(7) + 3(3)
. . . .		. . . .	. . . .
4 B		5	(7) + 4(5) + 12Q[4]
5 R*B	378.16	7	(7) + 4(5)
6 T*B		7	(7) + 3Q[6]
7 Error	242.67		(7)

*Dotted lines have been added to delineate the whole and split plots.

differences between the EMSs given in Table 11.4 and those reported in Table 11.5 are due to this change in the model equation.

From the EMSs of Table 11.4, we see that no exact tests are available for testing the replication effect or replication interaction with other factors. Thus, test results reported in Table 11.5 for R, R*T, and R*B should be ignored. The temperature effect is significant for any $\alpha \geq 0.004$, but neither baking time (p value $= 0.856$) nor the T*B interaction ($p = 0.185$) is significant for any reasonable α.

The results of this split-plot analysis are not too different from the results using the incorrect method of Table 11.2, but one is not always so fortunate. This split-plot design does show, however, the need for careful consideration of the method of randomization before an experiment is begun. A split-plot design represents a restriction on the randomization over a complete randomization in a factorial experiment.

Restriction Errors

Some statisticians recommend introducing a restriction error into a model wherever such restrictions occur; see Anderson and McLean [2]. No such term has been introduced here, and so it is strongly recommended that (1) F tests be run only within the whole plot or within the split plot and (2) mean squares in the whole plot not be compared with mean squares in the split plot regardless of the EMS column. For example, the tests on R and R*T reported in Table 11.5 should be ignored even if the correct model equation is the one used to obtain that table.

Completely Randomized Factorials Versus Split-Plot Factorials

To emphasize a bit more the difference between a completely randomized 4 × 3 factorial design and a split-plot 4 × 3 factorial design, one might let the numbers 1 through 4 on one die (red) represent temperature and the numbers 1 through 3 on another die (green) represent baking times, and then toss the dice to effect the proper randomization.

In the case of complete randomization, both dice are tossed and the first six results are as indicated in Table 11.6, where the numbers 1, 2, . . . , 6 indicate the order of the experiment: first set temperature at 600°F, bake for 10 minutes; then set temperature at 620°F, bake for 15 minutes; and so on. The order is scattered throughout the 12 treatment combinations. In the case of the split-plot design, a temperature is randomly chosen (roll the red die) and then the baking times are chosen at random within that temperature (roll the green die). Table 11.7 shows one such randomization for the first six combinations. In Table 11.7 it should be more obvious that temperature is confounded with the order in which the experiment is run, since three treatments are run at 580°F first, then three are run at 620°F second, and so on. It is hoped that by replicating the whole experiment several times, we will cause any effect of order confounded with temperature to be averaged out. Nevertheless, in a split-plot design a main effect is confounded.

TABLE 11.6
Data for a Completely Randomized Factorial

Baking Time (min)	Oven Temperature (°F)			
	580	600	620	640
5	3		4	
10		1		5
15	6		2	

TABLE 11.7
Data for a Split-Plot Factorial

Baking Time (min)	Oven Temperature (°F)			
	580	600	620	640
5	3		4	
10	1		6	
15	2		5	

■ **Example 11.1**

A defense-related organization was to study the pull-off force necessary to separate boxes of chaff from a tape to which they are affixed. These boxes are made of a cardboard material, approximately 3 inches by 3 inches by 1 inch, and they are mounted on a strip of cloth tape that has an adhesive backing. The tape is 2 inches wide, and the boxes are placed 7 inches center to center on the strip. There are 75 boxes mounted on each strip.

The tape is pulled from the box at a 90° angle as it is wound onto a drum. During this separation process the portion of the tape still adhering to the box carries the box onto a platform. The box trips a microswitch, which energizes a plunger. The plunger then kicks the box out of the machine.

The problem was discussed with the plant engineers, and several factors that might affect this pull-off force were listed. The most important factors were temperature and humidity. It was agreed to use three fixed temperature levels, -55, 25, and 55°C, and three fixed humidity levels, 50, 70 and 90%, to give nine basic treatment combinations. Because there might be differences in pull-off force as a result of the strip selected, it was decided to choose five different strips at random. There might also be differences within a strip, so two boxes were chosen at random and cut from each strip for the test.

The test in the laboratory was accomplished by hand-holding the package, attaching a spring scale to the strip by means of a hole previously punched in the strip, and pulling the tape from the package in a direction perpendicular to the package.

It seemed best to set the climatic condition (a combination of temperature and humidity) at random from one of the $3 \times 3 = 9$ conditions and then test two boxes from each of the five strips while these conditions were maintained. Then another of the nine conditions would be set, and the results again determined on two boxes from each of the five strips. This is then a restriction on the randomization, and the resulting design is a split-plot design. It was agreed to replicate the complete experiment four times. A layout for replication I of this experiment is shown in Table 11.8.

TABLE 11.8
Split-Plot Design of Chaff Experiment: Template for Replications

			−55		Temperature, T (°C) 25			55		
					Humidity H (%)					
Strip, S	Box	50	70	90	50	70	90	50	70	90
1	1									
	2									
2	3									
	4									
3	5									
	6									
4	7									
	8									
5	9									
	10									

Replications II, III, and IV would each be a repeat of replication I, but with a new order of randomizing the nine atmospheric conditions in each replication. In this design atmospheric conditions and replications form the whole plot and strips are in the split-plot.

The model for this design and its associated EMS relation are set up in Table 11.9. From the EMS column, tests can be made on the effects of replications, replications by temperature interaction, replications by humidity interaction, replications by temperature by humidity interaction, strips, and strips by all other factor interactions. Unfortunately no exact tests are available for the factors of chief importance: temperature, humidity, and temperature–humidity interaction. Of course, we could first test the hypotheses that can be tested, and if some of these are not significant at a reasonably high level (say 25%), assume they are nonexistent and remove the corresponding terms from the EMS column. For example, if the TS interaction can be assumed to be zero ($\sigma_{TS}^2 = 0$), the temperature mean square can be tested against the RT interaction mean square with 2 and 6 df. If this is not a reasonable assumption, a pseudo–F (F') test can be used, as discussed in Chapter 6.

Unfortunately, data were available only for the first replication of this experiment, so its complete analysis cannot be given. The method of analysis is the same as given in Chapter 5, even though this experiment is rather complicated. It is presented here only to show another actual example of a split-plot design.

TABLE 11.9
EMS for Chaff Experiment

	Source	df	4 R i	3 F j	3 F k	5 R m	2 R q	EMS
Whole plot	R_i	3	1	3	3	5	2	$\sigma_\varepsilon^2 + 18\sigma_{RS}^2 + 90\sigma_R^2$
	T_j	2	4	0	3	5	2	$\sigma_\varepsilon^2 + 6\sigma_{RTS}^2 + 24\sigma_{TS}^2 + 30\sigma_{RT}^2 + 120\phi_T$
	RT_{ij}	6	1	0	3	5	2	$\sigma_\varepsilon^2 + 6\sigma_{RTS}^2 + 30\sigma_{RT}^2$
	H_k	2	4	3	0	5	2	$\sigma_\varepsilon^2 + 6\sigma_{RHS}^2 + 24\sigma_{HS}^2 + 30\sigma_{RH}^2 + 120\phi_H$
	RH_{ik}	6	1	3	0	5	2	$\sigma_\varepsilon^2 + 6\sigma_{RHS}^2 + 30\sigma_{RH}^2$
	TH_{jk}	4	4	0	0	5	2	$\sigma_\varepsilon^2 + 2\sigma_{RTHS}^2 + 8\sigma_{THS}^2 + 10\sigma_{RTH}^2 + 40\phi_{TH}$
	RTH_{ijk}	12	1	0	0	5	2	$\sigma_\varepsilon^2 + 2\sigma_{RTHS}^2 + 10\sigma_{RTH}^2$
Split plot	S_m	4	4	3	3	1	2	$\sigma_\varepsilon^2 + 18\sigma_{RS}^2 + 72\sigma_S^2$
	RS_{im}	12	1	3	3	1	2	$\sigma_\varepsilon^2 + 18\sigma_{RS}^2$
	TS_{jm}	8	4	0	3	1	2	$\sigma_\varepsilon^2 + 6\sigma_{RTS}^2 + 24\sigma_{TS}^2$
	RTS_{ijm}	24	1	0	3	1	2	$\sigma_\varepsilon^2 + 6\sigma_{RTS}^2$
	HS_{km}	8	4	3	0	1	2	$\sigma_\varepsilon^2 + 6\sigma_{RHS}^2 + 24\sigma_{HS}^2$
	RHS_{ikm}	24	1	3	0	1	2	$\sigma_\varepsilon^2 + 6\sigma_{RHS}^2$
	THS_{jkm}	16	4	0	0	1	2	$\sigma_\varepsilon^2 + 2\sigma_{RTHS}^2 + 8\sigma_{THS}^2$
	$RTHS_{ijkm}$	48	1	0	0	1	2	$\sigma_\varepsilon^2 + 2\sigma_{RTHS}^2$
	$\varepsilon_{q(ijkm)}$	180	1	1	1	1	1	σ_ε^2
	Total	359						

11.3 A SPLIT-SPLIT-PLOT DESIGN

A study of the cure rate index on some samples of rubber involved three laboratories, three temperatures, and three types of mix. Material for the three mixes was sent to one of the three laboratories, where the experiment was run on the three mixes at the three temperatures (145, 155, and 165°C). However, once a temperature was set, all three mixes were subjected to that temperature; then another temperature was set, and again all three mixes were tested, and finally the third temperature was set. Material was also sent to the second and third laboratories, where similar experimental procedures were performed. There are therefore two restrictions on randomization, since the laboratory is chosen, then the temperature is chosen, and then mixes can be randomized at that particular temperature and laboratory.

By complete replication of the whole experiment to achieve four replications, the three laboratories and four replications form the whole plots. Then the temperature levels may be randomized at each laboratory and in each replication to form a split plot. Then at each temperature–laboratory–replication combination, the three mixes are randomly applied, forming what is called a *split-split plot,* indicating that two main effects (laboratory and temperature) are confounded with blocks.

Table 11.10 shows one possible order of the first nine experiments in one replication of a completely randomized design, a split-plot design, and a split-split-plot design. The resultant data seldom reveal how the experiment was ordered, but the use of lines in the data given in Table 11.11 should help indicate the restrictive order of randomization.

The Mathematical Model and Expected Mean Squares

Letting R, L, T, and M denote replicate, laboratory, temperature, and mix, respectively, we can give the following mathematical model for this split-split-plot experiment:

TABLE 11.10
Order of First Nine Rubber Cure Rate Index Experiments for Three Designs

| | 145 | | | Temperature (°C) 155 | | | 165 | | | |
| | Mix | | | Mix | | | Mix | | | |
Laboratory	A	B	C	A	B	C	A	B	C	Design
1			3				4			Completely randomized
2	8	1			5		7		9	
3			2						6	
1										Split plot
2	4	6	1	7	9	3	2	8	5	
3										
1										Split-split plot
2	8	9	7	3	1	2	4	6	5	
3										

TABLE 11.11
Rubber Cure Rate Index Data

| | | Temperature (°C) | | | | | | | | |
| | | 145 | | | 155 | | | 165 | | |
Replication	Laboratory	Mix A	Mix B	C	Mix A	Mix B	C	Mix A	Mix B	C
I	1	18.6	14.5	21.1	9.5	7.8	11.2	5.4	5.2	6.3
	2	20.0	18.4	22.5	11.4	10.8	13.3	6.8	6.0	7.7
	3	19.7	16.3	22.7	9.3	9.1	11.3	6.7	5.7	6.6
II	1	17.0	15.8	20.8	9.4	8.3	10.0	5.3	4.9	6.4
	2	20.1	18.1	22.7	11.5	11.1	14.0	6.9	6.1	8.0
	3	18.3	16.7	21.9	10.2	9.2	11.0	6.0	5.5	6.5
III	1	18.7	16.5	21.8	9.5	8.9	11.5	5.7	4.3	5.8
	2	19.4	16.5	21.5	11.4	9.5	12.0	6.0	5.0	6.6
	3	16.8	14.4	19.3	9.8	8.0	10.9	5.0	4.6	5.9
IV	1	18.7	17.6	21.0	10.0	9.1	11.1	5.3	5.2	5.6
	2	20.0	16.7	21.3	11.5	9.7	11.5	5.7	5.2	6.3
	3	17.1	15.2	19.3	9.5	9.0	11.4	4.8	5.4	5.8

$$Y_{ijkmq} = \mu + \underbrace{R_i + L_j + RL_{ij}}_{\text{whole plot}} + \underbrace{T_k + RT_{ik} + LT_{jk} + RLT_{ijk}}_{\text{split plot}}$$

$$\underbrace{+ M_m + RM_{im} + LM_{jm} + RLM_{ijm} + TM_{km} + RTM_{ikm}}$$
$$\underbrace{+ LTM_{jkm} + RLTM_{ijkm} + \varepsilon_{q(ijkm)}}_{\text{split-split plot}} \qquad (11.2)$$

with $i = 1, 2, 3, 4$, $j = 1, 2, 3$, $k = 1, 2, 3$, $m = 1, 2, 3$, and $q = 1$. This model and its EMS values are given in Table 11.12.

The EMS column of Table 11.12 indicates that F tests can be made on all three fixed effects and their interactions without a separate error term. Because the effect of replication is often considered to have no great interest and the interaction of replication with the other factors is often assumed to be nonexistent, this appears to be a feasible experiment. If examination of the EMS column should indicate that some F tests have too few degrees of freedom in their denominator (some statisticians say < 6), the experimenter might consider increasing the number of replications to increase the precision of the test. In the experiment just described, if five replications were used, the whole plot error RL

TABLE 11.12
EMS for Rubber Cure Rate Index Experiment

	Source	df	4 R i	3 F j	3 F k	3 F m	1 R q	EMS
Whole plot	R_i	3	1	3	3	3	1	$\sigma_\varepsilon^2 + 27\sigma_R^2$
	L_j	2	4	0	3	3	1	$\sigma_\varepsilon^2 + 9\sigma_{RL}^2 + 36\phi_L$
	RL_{ij}	6	1	0	3	3	1	$\sigma_\varepsilon^2 + 9\sigma_{RL}^2$
Split plot	T_k	2	4	3	0	3	1	$\sigma_\varepsilon^2 + 9\sigma_{RT}^2 + 36\phi_T$
	RT_{ik}	6	1	3	0	3	1	$\sigma_\varepsilon^2 + 9\sigma_{RT}^2$
	LT_{jk}	4	4	0	0	3	1	$\sigma_\varepsilon^2 + 3\sigma_{RLT}^2 + 12\phi_{LT}$
	RLT_{ijk}	12	1	0	0	3	1	$\sigma_\varepsilon^2 + 3\sigma_{RLT}^2$
Split-split plot	M_m	2	4	3	3	0	1	$\sigma_\varepsilon^2 + 9\sigma_{RM}^2 + 36\phi_M$
	RM_{im}	6	1	3	3	0	1	$\sigma_\varepsilon^2 + 9\sigma_{RM}^2$
	LM_{jm}	4	4	0	3	0	1	$\sigma_\varepsilon^2 + 3\sigma_{RLM}^2 + 12\phi_{LM}$
	RLM_{ijm}	12	1	0	3	0	1	$\sigma_\varepsilon^2 + 3\sigma_{RLM}^2$
	TM_{km}	4	4	3	0	0	1	$\sigma_\varepsilon^2 + 3\sigma_{RTM}^2 + 12\phi_{TM}$
	RTM_{ikm}	12	1	3	0	0	1	$\sigma_\varepsilon^2 + 3\sigma_{RTM}^2$
	LTM_{jkm}	8	4	0	0	0	1	$\sigma_\varepsilon^2 + \sigma_{RLTM}^2 + 4\phi_{LTM}$
	$RLTM_{ijkm}$	24	1	0	0	0	1	$\sigma_\varepsilon^2 + \sigma_{RLTM}^2$
	$\varepsilon_{q(ijkm)}$	0	1	1	1	1	1	σ_ε^2 (not retrievable)
Total		107						

would have 8 df instead of 6, the split-plot error RLT would have 16 instead of 12, and so on.

Another technique that is sometimes used to increase the degrees of freedom is to pool all interactions with replications into the error term for that section of the table. Here one might pool RT and RLT and RM, RLM and RTM with $RLTM$ if additional degrees of freedom are believed to be necessary. One way to handle experiments with several replications is to stop after two or three replications, compute the results, and see whether significance has been achieved or, perhaps, that the F's are large even if not significant. Then add another replication, and so on, increasing the precision (and the cost) of the experiment in the hope of detecting significant effects if they are there.

The Analysis

A Minitab analysis of the data in Table 11.11 is shown in Table 11.13. Since the error term in model Equation (11.2) has no degrees of freedom, Minitab outputs an error message and no sums of squares when that model is used. To obtain Table 11.13, instead of an error message, and so on, we can delete the four-factor interaction from the model. The differences between the EMSs given in Table 11.12 and those reported in Table 11.13 are due to this change in the model equation.

Based on the EMSs of Table 11.12, tests on all effects involving R should be ignored. The laboratory effect is significant for any $\alpha \geq 0.023$, whereas temperature, mix, and the interaction between those are significant for any reasonable value of α. These results

TABLE 11.13
Minitab ANOVA for Rubber Cure Rate Index*

Source	DF	SS	MS	F	P
R	3	9.414	3.138	15.20	0.000
L	2	40.664	20.332	7.57	0.023
R*L	6	16.110	2.685	13.00	0.000
T	2	3119.509	1559.755	4528.73	0.000
R*T	6	2.066	0.344	1.67	0.172
L*T	4	4.936	1.234	1.89	0.176
R*L*T	12	7.817	0.651	3.15	0.008
M	2	145.718	72.859	133.38	0.000
R*M	6	3.278	0.546	2.65	0.041
L*M	4	0.339	0.085	0.35	0.842
R*L*M	12	2.941	0.245	1.19	0.346
T*M	4	43.687	10.922	58.75	0.000
R*T*M	12	2.231	0.186	0.90	0.560
L*T*M	8	1.077	0.135	0.65	0.727
Error	24	4.956	0.206		
Total	107	3404.743			

Source	Variance component	Error term	Expected Mean Square (using restricted model)
1 R	0.10858	15	(15) + 27(1)
2 L		3	(15) + 9(3) + 36Q[2]
3 R*L	0.27539	15	(15) + 9(3)
4 T		5	(15) + 9(5) + 36Q[4]
5 R*T	0.01532	15	(15) + 9(5)
6 L*T		7	(15) + 3(7) + 12Q[6]
7 R*L*T	0.14830	15	(15) + 3(7)
8 M		9	(15) + 9(9) + 36Q[8]
9 R*M	0.03775	15	(15) + 9(9)
10 L*M		11	(15) + 3(11) + 12Q[10]
11 R*L*M	0.01285	15	(15) + 3(11)
12 T*M		13	(15) + 3(13) + 12Q[12]
13 R*T*M	-0.00687	15	(15) + 3(13)
14 L*T*M		15	(15) + 4Q[14]
15 Error	0.20650		(15)

*Dotted lines have been added to delineate the whole, split, and split-split plots.

suggest that one should further analyze the effects of laboratory and interaction between temperature and mix.

To compare the three laboratories, we use MS_{RL} as the best estimate for error. For each of the three sample means, $s_{\bar{y}} = \sqrt{MS_{RL}/36} = \sqrt{2.685/36} \approx 0.27$. We use Statistical Table E.2 and $n_2 = 6$ df to find upper percentage points of 3.46 and 4.34 for $p = 2$ and $p = 3$, respectively. The corresponding least significant ranges are $(3.46)(0.27) = 0.93$ and $(4.34)(0.27) = 1.17$. When these least significant ranges are used with the sample means, placed in ascending order, we have Figure 11.1, which shows that the true average cure rate at laboratory 2 exceeds that of each of the other two laboratories.

Comparison of the nine temperature–mix–treatment combinations is left as an exercise. In that case, MS_{RTM} should be used as an error estimate.

Laboratory:	1	3	2
Mean:	11.217	11.247	12.533

Figure 11.1   Newman–Keuls graphic for cure rate data.

11.4 USING SAS TO ANALYZE A SPLIT-PLOT EXPERIMENT

Previously discussed methods can be used to obtain SAS outputs for split-plot and split-split-plot experiments. To illustrate, a command file and corresponding outputs for the life-test data Table 11.3 are given in Table 11.14 and Figures 11.2 and 11.3. From Table 11.3, the factors under consideration are replicate (R), baking time (B), and oven temperature (T).

The model statement in Table 11.14 is equivalent to Equation (11.1). Knowledge of the expected mean squares (Table 11.4) makes it possible for us to request the three tests that follow the model statement.

The last line of Table 11.14 requests a Newman–Keuls analysis on temperature. Such an analysis is a post hoc procedure, so any significant differences indicated by a Newman–Keuls analysis should be ignored when the initial ANOVA indicates that the factor is not significant.

Since temperature is a significant quantitative factor, a regression study of the relationship between component life and temperature is appropriate. The Newman–Keuls analysis, which is usually reserved for qualitative factors, has been included for illustration.

TABLE 11.14
SAS Command File for Life-Test Experiment

```
OPTIONS LINESIZE=80;
DATA SPLTPLOT;
DO R = 1 TO 3;
  DO B = 1 TO 3;
    DO T = 1 TO 4;
      INPUT LIFE @@;
      OUTPUT;
      END;
  END;
END
CARDS;
 217 158 229 223 233 138 186 227 175 152 155 156
 188 126 160 201 201 130 170 181 195 147 161 172
 162 122 167 182 170 185 181 201 213 180 182 199
;
PROC GLM;
  CLASS R B T;
    MODEL LIFE = R|T|B;
TEST H=T E=R*T;
TEST H=B E=R*B;
TEST H=T*B E=R*T*B;
MEANS T/SNK E=R*T;
```

```
                       GENERAL LINEAR MODELS PROCEDURE
                                 SUM OF              MEAN
SOURCE                    DF     SQUARES            SQUARE   F VALUE   PR > F
MODEL                     35  29330.97222222   838.02777778      •         •
ERROR                      0      0.00000000     0.00000000
CORRECTED TOTAL           35  29330.97222222
                    R-SQUARE              C.V.        ROOT MSE        LIFE MEAN
                    1.000000            0.0000      0.00000000      178.47222222
SOURCE                    DF       TYPE I SS    MEAN SQUARE   F VALUE   PR > F
R                          2   1962.72222222   981.36111111      •         •
T                          3  12494.30555556  4164.76851852      •         •
R*T                        6   1773.94444444   295.65740741      •         •
B                          2    566.22222222   283.11111111      •         •
R*B                        4   7021.27777778  1755.31944445      •         •
B*T                        6   2600.44444444   433.40740741      •         •
R*B*T                     12   2912.05555556   242.67129630      •         •
SOURCE                    DF     TYPE III SS    MEAN SQUARE   F VALUE   PR > F
R                          2   1962.72222222   981.36111111      •         •
T                          3  12494.30555556  4164.76851852      •         •
R*T                        6   1773.94444444   295.65740741      •         •
B                          2    566.22222222   283.11111111      •         •
R*B                        4   7021.27777778  1755.31944445      •         •
B*T                        6   2600.44444444   433.40740741      •         •
R*B*T                     12   2912.05555556   242.67129630      •         •
TESTS OF HYPOTHESES USING THE TYPE III MS FOR R*T AS AN ERROR TERM
SOURCE                    DF     TYPE III SS       F VALUE   PR > F
T                          3  12494.30555556         14.09   0.0040
TESTS OF HYPOTHESES USING THE TYPE III MS FOR R*B AS AN ERROR TERM
SOURCE                    DF     TYPE III SS       F VALUE   PR > F
B                          2    566.22222222          0.16   0.8563
TESTS OF HYPOTHESES USING THE TYPE III MS FOR R*B*T AS AN ERROR TERM
SOURCE                    DF     TYPE III SS       F VALUE   PR > F
B*T                        6   2600.44444444          1.79   0.1848
```

Figure 11.2   SAS outputs for the life-test study.

```
STUDENT-NEWMAN-KEULS TEST FOR VARIABLE:LIFE
NOTE: THIS TEST CONTROLS THE TYPE I EXPERIMENTWISE ERROR RATE UNDER THE
      COMPLETE NULL HYPOTHESIS BUT NOT UNDER PARTIAL NULL HYPOTHESES
                   ALPHA=0.05    DF=6    MSE=295.657
      NUMBER OF MEANS          2          3          4
      CRITICAL RANGE     19.8335    24.8714    28.0597
    MEANS WITH THE SAME LETTER ARE NOT SIGNIFICANTLY DIFFERENT
                   SNK GROUPING       MEAN   N   TEMP
                            A      194.889   9   1
                            A
                            A      193.556   9   4
                            A
                            A      176.778   9   3

                            B      148.667   9   2
```

Figure 11.3   SAS outputs for the life-test study (continued).

11.5 SUMMARY

The summary at the end of Chapter 8 may now be extended for Part II.A.2.

Experiment	Design	Analysis
II. Two or more factors A. Factorial (crossed)		
	1. Complete randomized . . .	1. . . .
	2. Randomized block a. Complete $Y_{ijk} = \mu + R_k + A_i$ $\quad + B_j + AB_{ij} + \varepsilon_{ijk}$ b. Incomplete, confounding: i. Main effect: split plot $Y_{ijk} = \mu + \underbrace{R_i + A_j + RA_{ij}}_{\text{whole plot}}$ $\quad + \underbrace{B_k + RB_{ik} + AB_{jk} + RAB_{ijk}}_{\text{split plot}}$	2. a. Factorial ANOVA with replications R_k b. i. Split-plot ANOVA

11.6 FURTHER READING

Anderson, V. L., and R. A. McLean, *Design of Experiments: A Realistic Approach.* New York: Marcel Dekker, 1974. See Chapter 7, "Split Plot Type Designs."

Mason, R. L., R. F. Gunst, and J. L. Hess, *Statistical Design and Analysis of Experiments with Applications to Engineering and Science.* New York: John Wiley & Sons, 1989. Split-plot designs are considered on pages 195–201.

Ostle, B., and L. C. Malone, *Statistics in Research: Basic Concepts and Techniques for Research Works,* 4th ed. Ames: Iowa State University Press, 1988. Split-plot designs are considered on pages 426–428.

Snedecor, G. W., and W. G. Cochran, *Statistical Methods,* 7th ed. Ames: Iowa State University Press, 1980. Split-plot and nested designs are considered on pages 325–329.

PROBLEMS

11.1 As part of the experiment discussed in Example 11.1, a chamber was set at 50% relative humidity and two boxes from each of three strips were inserted and the pull-off force determined. This was repeated for 70 and 90%, the order of humidities being randomized. Two replications of the whole experiment were made, and the results follow.

Replication	Strips	Humidity (%)					
		50		70		90	
I	1	1.12	1.75	3.50	0.50	1.00	0.75
	2	1.13	3.50	0.75	1.00	0.50	0.50
	3	2.25	3.25	1.75	1.88	1.50	0.00

II	1	1.75	1.88	1.75	0.75	1.50	0.75
	2	5.25	5.25	0.75	2.25	1.50	1.50
	3	1.50	3.50	1.62	2.50	0.75	0.50

Set up a split-plot model for this experiment and outline the ANOVA table, including EMS.

11.2 Analyze Problem 11.1.

11.3 The following data layout was proposed to handle a factorial experiment designed to determine the effects of orifice size and flow rate on environmental chamber pressure in millimeters of mercury.

Flow Rate	Orifice Size (in.)		
	8	10	12
3.5			
4.0			
4.5			

a. If this experiment is run as a completely randomized design, how might data be collected to fill in the above table?
b. If this experiment were run as a split-plot design, how might data be collected?
c. Assuming three complete replications of the run above as a split-plot design, outline its ANOVA table. (You need *not* include the EMS column.)
d. Suggest how this experiment might be improved.

11.4 Interest centered on the yields of alfalfa in tons per acre from three varieties following four dates of final cutting. A large section of land was divided into three strips, and each strip was planted with one of the three varieties of alfalfa. Once a strip had been selected, one-fourth of the plants were cut on the first cutting date, one-fourth on the next, and so on, for the four dates studied. This whole experiment could be replicated several times as necessary by choosing sections at random.
a. Identify the type of design implied here.
b. Set up an appropriate ANOVA table for this situation with the minimum number of replications you feel might be adequate to detect differences, if they exist.

11.5 A time study uses stimuli of two types—two- and three-dimensional film—and combinations of stimuli and job are repeated once to give eight strips of film. The order of these eight strips, completely randomized on one long film, is presented to five analysts for rating at one sitting. Since this condition represents a restriction on the randomization, the whole experiment is repeated at three sittings (four in all), and the experiment is considered to be a split-plot design. Set up a model for this experiment and indicate the tests to be made. (Consider analysts as random.)

11.6 Determine an F' test for any effects in Problem 11.5 that cannot be run directly.

11.7   Three replications are made of an experiment to determine the effect of days of the week and operators on screen-color difference of a TV tube in K. On a given day, each of four operators measured the screen-color difference on a given tube. The order of measuring by the four operators is randomized each day for five days of the week. The results are

Replication	Operator	Monday	Tuesday	Wednesday	Thursday	Friday
I	A	800	950	900	740	880
	B	760	900	820	740	960
	C	920	920	840	900	1020
	D	860	940	820	940	980
II	A	780	810	880	960	920
	B	820	940	900	780	820
	C	740	900	880	840	880
	D	800	900	800	820	900
III	A	800	1030	800	840	800
	B	900	920	880	920	1000
	C	800	1000	920	760	920
	D	900	980	900	780	880

Set up the model and outline the ANOVA table for this problem.

11.8   Analyze Problem 11.7, with days as fixed, operators as random, and replications as random.

11.9   Explain how a split-plot design differs from a nested design, since both have a factor within levels of another factor.

11.10 Using an available computer package, analyze the data of Table 11.11 on the rubber cure rate index experiment. Compare with the results of Table 11.13.

11.11 Conduct a multiple comparison analysis of the nine temperature–mix means of the cure rate index data in Table 11.11.

11.12 Redo Table 11.12, assuming that the three laboratories are chosen at random from a large number of possible laboratories.

11.13 Analyze the results of Table 11.11, based on the EMS values of Problem 11.12.

11.14 In a study of the effect of baking temperature and recipe (or mix) on the quality of cake, three mixes were used: A, B, and C. One of the three was chosen at random, whereupon the mix was made up and five cakes were poured, each placed in a different oven and baked at temperatures of 300, 325, 350, 375, and 400°F. After the five cakes had been baked, they were removed from the oven, and their quality (Y) was evaluated. Another mix was chosen, and again five cakes were poured and baked at the five temperatures. Then this was repeated on the third mix. This experiment was replicated r times.
a. Identify this design. Show its data layout.

b. Write the mathematical model for this situation and determine the ANOVA outline indicating tests to be made.

c. Recommend how many replications r you would run to make this an acceptable experiment and explain why.

11.15 If in Problem 11.14 only one oven was available but cakes of all three recipes could be inserted in the oven when set at a given temperature, how would the problem differ from the setup in Problem 11.14? Explain which of the two designs you would prefer and why.

11.16 Running two random replications of an experiment similar to that described in Problem 11.14, also aimed at determining the effect of temperature and mix on cake quality, resulted in the following data.

Replication	Mix	Temperatures (°F)					
		175	185	195	205	215	225
I	1	47	29	35	47	57	45
	2	35	46	47	39	52	61
	3	43	43	43	46	47	58
II	1	26	28	32	25	37	33
	2	21	21	28	26	27	20
	3	21	28	25	25	31	25

Outline the ANOVA table for this problem, show what tests can be run, and discuss these tests.

11.17 Analyze the data given in Problem 11.16 and state your conclusions.

11.18 What next steps would you recommend after your conclusions in Problem 11.17?

11.19 A large polyurethane part used in the automotive industry is produced using a reaction injection molding (RIM) machine. The heat sag (Y) of this part is a general measure of the quality of the molded material. A chemical engineer wants to investigate the effects on Y of two catalysts (A and B). Four concentrations of A (0.00, 0.02, 0.04, and 0.06) can be used in random order, whereas four concentrations of B (0.015, 0.030, 0.045, and 0.060) must be used in order from low to high. For each combination of A and B, a part will be molded and the heat sag will be measured at two positions (gate and ear).

a. Prepare a data layout for this situation.

b. The engineer plans to run three replicates of this experiment. Determine the mathematical model.

c. Based on your mathematical model, determine the expected mean squares and appropriate tests. Comment. If some tests are inadequate, suggest methods of improvement.

11.20 The effects of pressure, power, and temperature on the oxide removal rate in the manufacture of integrated circuits were studied by randomly setting the temperature at one of three levels and determining the removal rate for wafers randomly prepared at each of the $3 \times 3 = 9$ combinations of pressure and power. Replicating the experiment three times produced the following data.

Replicate	Temperature	Pressure 1 Power 1	Pressure 1 Power 2	Pressure 1 Power 3	Pressure 2 Power 1	Pressure 2 Power 2	Pressure 2 Power 3	Pressure 3 Power 1	Pressure 3 Power 2	Pressure 3 Power 3
1	1	146	202	264	189	229	298	181	225	267
	2	141	212	258	160	228	323	182	253	330
	3	133	180	247	159	184	250	173	199	270
2	1	167	212	258	188	231	311	180	222	264
	2	147	197	259	154	224	334	178	250	331
	3	120	177	242	160	189	242	174	201	276
3	1	161	223	260	187	231	309	179	204	261
	2	140	190	271	155	226	327	175	251	320
	3	136	182	248	159	188	244	179	203	278

a. Determine the mathematical model.
b. Determine the expected mean squares and appropriate tests.
c. Analyze completely.

11.21 During processing, electronic components are loaded into a furnace and heated to a specified temperature. A process engineering team is planning to investigate the effects of furnace, loading method, and position in the furnace on a quality characteristic of these components. Two furnaces, three loading methods, and three furnace positions (front, middle, back) are to be used. Only one furnace is available at a time, so each of the three loading methods will be tested in a furnace before a second furnace becomes available. The order in which the furnaces can be used and the order in which the loading methods are used with a furnace are to be completely randomized. Seven replicates of the experiment will be run.

a. Prepare a data layout for this situation.
b. Determine the mathematical model.
c. Determine the expected mean squares and appropriate tests. Comment.

Chapter 12

FACTORIAL EXPERIMENT: CONFOUNDING IN BLOCKS

12.1 INTRODUCTION

As we found in Chapter 11, there are many situations in which randomization is restricted. The split-plot design is an example of a restriction on randomization where a main effect is confounded with blocks. Sometimes these restrictions are necessary because the complete factorial cannot be run in one day, or in one environmental chamber. When such restrictions are imposed on the experiment, the researcher must decide what information may be sacrificed, and thus what is to be confounded. A simple example may help to illustrate this point.

A chemist was studying the disintegration of a chemical solution. He wished to determine whether the depth of the mixture in a beaker would affect the amount of disintegration and whether material from the center of the beaker would differ from material near the edge. He decided on a 2^2 factorial experiment—one factor being depth, the other radius. A cross section of the beaker (Figure 12.1) shows the four treatment combinations to be considered, where A and B are radius and depth, respectively.

In Figure 12.1, (1) represents radius zero, depth at lower level (near bottom of beaker); a represents the radius at near maximum (near the edge of the beaker) and a low depth level; b is radius zero, depth at a higher level; and ab is both radius and depth at higher levels. To get a reading on the amount of disintegration of material, samples of solution were to be drawn from these four positions. This was accomplished by dipping into the beaker

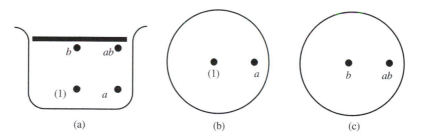

Figure 12.1   Beaker cross sections: (a) vertical, (b) at low level of B, and (c) at high level of B.

with an instrument designed to "capture" a small amount of solution. Since, however, the experimenter had only two hands, he could get only two samples at the same time. By the time he was ready to take the other two samples, conditions might have changed, and the amount of disintegration probably would be greater. Thus there is a physical restriction on this 2^2 factorial experiment—only two observations can be made at one time. The question is: Which two treatment combinations should be taken on the first dip? Considering the two dips as blocks, and recognizing that the whole factorial cannot be performed in one dip, we are forced to use an incomplete block design.

Three Possible Plans

Since $_4C_2 = 6$ blocks of 2 can be selected, $6/2 = 3$ block patterns are possible. These are depicted in Figure 12.2. If plan I is used, the blocks (dips) are confounded with the radius effect, since all zero-radius readings are in dip 1 and all maximum radius readings are in dip 2. This can also be seen by considering Table 12.1—an orthogonal table for this experiment. The contrast coefficients are -1 for the treatment combinations in dip 1 and $+1$ for those in dip 2. Since the contrast for A is

$$C_A = (-1)T_1 + (+1)T_2 + (-1)T_3 + (+1)T_4$$

a block effect could be substituted for the radius effect and the analysis would remain the same. That is, all information about one is contained in the information about the other.

In plan II the depth effect is confounded with the dips, as low-level depth readings are both in dip 1 and high-level depth readings are in dip 2. This confounding is also evident from Table 12.1, since the contrast for B is

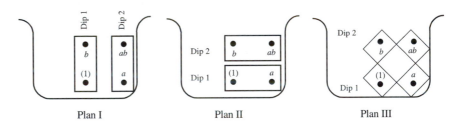

Plan I   Plan II   Plan III

Figure 12.2   Three possible blockings for a 2^2 factorial.

TABLE 12.1
Orthogonal Table for a 2^2 Factorial Experiment

Treatment Combination	Linear Contrast Coefficients			Total
	A	B	AB	
(1)	-1	-1	$+1$	T_1
a	$+1$	-1	-1	T_2
b	-1	$+1$	-1	T_3
ab	$+1$	$+1$	$+1$	T_4

$$C_B = (-1)T_1 + (-1)T_2 + (+1)T_3 + (+1)T_4$$

with -1 associated with the treatment combinations in dip 1 and $+1$ associated with those in dip 2.

In plan III neither main effect is confounded, but the interaction is confounded with dips. To see that the interaction is confounded, note that the contrast for AB is

$$C_{AB} = (+1)T_1 + (-1)T_2 + (-1)T_3 + (+1)T_4$$

In this case $+1$ is associated with the treatment combinations in dip 1 and -1 is associated with those in dip 2. Hence we cannot distinguish between a block effect (dips) and the interaction effect (AB).

In most cases it is better to confound an interaction than to confound a main effect. Thus, plan III is the best of the three plans. The hope is, of course, that there is no interaction and that information on the main effects can still be found from such an experiment.

12.2 CONFOUNDING SYSTEMS

To design an experiment in which the number of treatments that can be run in a block is less than the total number of treatment combinations, the experimenter must first decide on the effects to be confounded. If, as in the concentration example of Section 12.1, there is only one interaction in the experiment and that interaction can be confounded with blocks, the problem is simply one of deciding which treatment combinations to place in each block. There are three ways of arriving at these decisions.

The concentration example illustrates one nice feature of 2^f factorials. Once an orthogonal table containing the contrast coefficients for the factor to be confounded has been constructed (or the contrast for that factor has been determined), the treatment combinations associated with -1 can be used as one block. The second block consists of the remaining treatment combinations (those associated with $+1$). A more general method is necessary when the number of blocks or treatment levels increases.

Defining Relations

A *defining relation* (or identity relationship or defining contrast) is merely an expression stating which effects are to be confounded with blocks. In a 2^2 factorial like the concentration example, to confound AB with blocks, write

$$I = AB$$

as the defining relation. Once the defining relation has been set up, several methods are available for determining which treatment combinations will be placed in each block. One method has already been considered—one block consists of the treatment combinations associated with the -1 coefficients in the contrast for AB and the second block consists of the remaining treatment combinations.

A second method involves consideration of the treatment combinations (1), a, b, and ab. Those that have an even number of letters in common with the letters in the defining relation form one block (with lowercase and uppercase treated as the same) and all others form the second block. Here (1) has no letter in common with AB and ab has two letters in common with AB. Since 0 and 2 are even numbers, (1) and ab form one block. Both a and b have one letter (an odd number) in common with AB, so a and b form another block. The two blocks are depicted in Figure 12.3.

Defining Equations

One disadvantage of the preceding methods of determining blocks for confounding is that they can be used only to determine two blocks for 2^f factorials. We now turn to a method attributed to Kempthorne [12], which can be extended to other situations.

Suppose the experimental factors for a 2^f factorial are $A_1, \ldots, A_f$. Let 0 and 1 denote the low and high levels of A_i, respectively. For one of the independent defining relations (there may be more than one), let k_i denote the exponent on the ith factor appearing in that defining relation. If the factor does not appear, the exponent is 0. The linear expression

$$L = k_1 x_1 + k_2 x_2 + \cdots + k_f x_f, \text{ mod } 2$$

with x_i the level of the ith main effect appearing in a given treatment combination, is the *defining equation* associated with the defining relation. All treatment combinations with the same L value form a block.

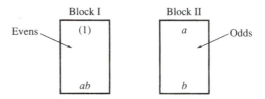

Figure 12.3 AB confounded in two blocks.

■ **Example 12.1 (A 2^2 Factorial in Two Blocks of Two)**

For a 2^2 factorial like the concentration example, the two-factor interaction AB is to be confounded with blocks. Using $I = AB$ as the defining relation, $k_1 = k_2 = 1$ and the defining equation is $L = 1 \cdot x_1 + 1 \cdot x_2 = x_1 + x_2$, mod 2.

When the treatment combination is (1), A and B are set at their low levels; so $x_1 = x_2 = 0$ and $L = 0 + 0 = 0$.

When the treatment combination is a, A is high and B is low, giving $x_1 = 1$ and $x_2 = 0$. In this case, $L = 1 + 0 = 1$.

At treatment combination b, A is low and B is high. Thus, $x_1 = 0$ and $x_2 = 1$. This gives $L = 0 + 1 = 1$ at b.

Finally, $x_1 = x_2 = 1$ when A and B are both at their high levels. Thus, $L = 1 + 1 = 0$ at ab. [**Note:** Calculations are in modulo 2 arithmetic.]

Since $L = 0$ at (1) and ab, those treatment combinations form one block. Treatment combinations a and b form the second block ($L = 1$). These blocks are the same as those in Figure 12.3.

■ **Example 12.2 (A 2^3 Factorial in Two Blocks of Four)**

A 2^3 factorial experiment is to be run in two blocks of four treatments each with the highest order interaction confounded with blocks. If $I = ABC$ is the defining relation, the exponents are $k_1 = k_2 = k_3 = 1$ and the defining equation is

$$L = 1 \cdot x_1 + 1 \cdot x_2 + 1 \cdot x_3 = x_1 + x_2 + x_3, \text{ mod } 2$$

Using modulo 2 arithmetic, values of L at each of the eight treatment combinations are

$$\begin{array}{llll}
(1): & L = 0 + 0 + 0 = 0 & c: & L = 0 + 0 + 1 = 1 \\
a: & L = 1 + 0 + 0 = 1 & ac: & L = 1 + 0 + 1 = 0 \\
b: & L = 0 + 1 + 0 = 1 & bc: & L = 0 + 1 + 1 = 0 \\
ab: & L = 1 + 1 + 0 = 0 & abc: & L = 1 + 1 + 1 = 1
\end{array}$$

The blocks are then

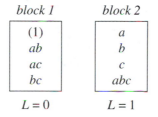

$$\begin{array}{cc}
\textit{block 1} & \textit{block 2} \\
\end{array}$$

block 1	block 2
(1)	a
ab	b
ac	c
bc	abc

$$\begin{array}{cc}
L = 0 & L = 1
\end{array}$$

The Principal Block

Consider the set $G = \{(1), a, b, ab, \ldots\}$ of treatment combinations for a 2^f factorial. If multiplication is defined on these elements so that (1) is the identity and the square of each element is (1), then G is a commutative group. When one of the methods of confounding described in this section is used, the block containing the treatment combination (1) is called the *principal block*. The treatment combinations in this block form a subgroup of G. Thus, another block can be generated from the principal block by selecting any element of G that is not in the principal block and determining the products of that element with each of the elements in the principal block. This procedure can be repeated until each element of G is in exactly one block and each block has the same number of elements. [**Note:** These blocks are known as *cosets* in group theory.] Thus, once the principal block has been determined, all other blocks for the confounding scheme may be obtained by multiplication.

Such multiplication is not needed when only two blocks are to be generated, since one block will be the principal block and the second will have the treatment combinations that are not in the principal block.

■ Example 12.3

To illustrate the preceding comments, consider a 2^3 factorial that is to be run in two blocks of four treatment combinations each. From Example 12.2, when ABC is confounded with blocks, the principal block is $\{(1), ab, ac, bc\}$. Since a is not in this block, we then multiply each element of the principal block by a to obtain $a \cdot (1) = a, a \cdot (ab) = a^2b = (1) \cdot b = b, a \cdot (ac) = a^2c = (1) \cdot c = c$, and $a(bc) = abc$. Thus, $\{a, b, c, abc\}$ forms the second block. These concepts extend to more complicated situations where more than two blocks are to be generated.

12.3 BLOCK CONFOUNDING, NO REPLICATION

In many cases an experimenter who cannot afford several replications of an experiment is further restricted in that the complete factorial cannot be run in one block or at one time. Again, the experiment may be blocked and information recovered on all but some high-order interactions, which may be confounded. The methods given in Section 12.2 can be used to determine the block composition for a specific confounding scheme. If only one replication is possible, one is run, and some of the higher order interaction terms must be used as experimental error unless some independent measure of error is available from earlier data. This type of design is used mostly with 2^f factorials with $f \geq 4$. In this section we consider two confounding schemes for 2^4 factorials with fixed levels.

A 2^4 Factorial in Two Blocks of Eight

Consider a 2^4 factorial where only eight treatment combinations can be run at one time. One possible confounding scheme is to confound the four-factor interaction with blocks. To do so, use

$$I = ABCD$$

as the defining relation and

$$L = x_1 + x_2 + x_3 + x_4, \ \text{mod } 2$$

as the defining equation. Since $L = 0$ for treatment combinations $(1), ab, bc, ac, abcd, cd, ad$, and bd, the two blocks are:

| block 1 | (1) | ab | bc | ac | abcd | cd | ad | bd | $L = 0$ |
| block 2 | a | b | abc | c | bcd | acd | d | abd | $L = 1$ |

A formal analysis would be as shown in Table 12.2. The three-way interactions may be pooled here to give 4 df for error, assuming that these interactions are actually nonexistent. All main effects and first-order interactions could then be tested with 1 and 4 df. An examination of the results might suggest a way to pool some of the two-way interactions that have small mean squares with the three-way interactions, to increase the degrees of freedom in the error term.

As an alternative to the analysis obtained by pooling, one could use an orthogonal table to obtain contrasts and estimated effects for the 15 factors and interactions in Table 12.2. A normal probability plot of the estimated effects could then be used to screen for any active effects.

A 2^4 Factorial in Four Blocks of Four

If only four treatment combinations can be run in a block, one must run the 2^4 factorial in four blocks of four treatment combinations each. Since 3 degrees of freedom are associated with the four blocks, three effects (each with 1 df) must be confounded with blocks. Slight modifications to the procedures considered in Section 12.2 allow us to accomplish such confounding.

If two interactions are confounded with blocks, the product (with modulo 2 arithmetic used on the exponents) of these two is also confounded. Modulo 2 arithmetic is used because the products of corresponding coefficients for the contrasts of two factors gives the coefficients for those of the product. Thus, if AB and CD are confounded, $(AB)(CD) =$

TABLE 12.2
2^4 Factorial in Two Blocks

Source	df	
A	1	
B	1	
C	1	
D	1	
AB	1	
AC	1	
AD	1	
BC	1	
BD	1	
CD	1	
ABC	1	
ABD	1	4 df for error
ACD	1	
BCD	1	
Blocks or ABCD	1	
Total	15 df	

$ABCD$ will also be confounded. It might be better, however, if only one two-factor interaction were confounded.

Suppose ABC and BCD are to be confounded with blocks. Then $(ABC)(BCD) = AB^2C^2D = AB^0C^0D = AD$, since modulo 2 arithmetic is used on the exponents. This gives

$$I = ABC = BCD = AD$$

as the defining relation.

Since the product of any two of ABC, BCD, and AD gives the third, only two defining equations are needed. We will use the defining equations associated with ABC and BCD. The reader can verify that the other two possible choices will yield the same blocking scheme.

The defining equations associated with ABC and BCD are

$$L_1 = 1 \cdot x_1 + 1 \cdot x_2 + 1 \cdot x_3 + 0 \cdot x_4 + = x_1 + x_2 + x_3, \text{mod } 2$$

and

$$L_2 = 0 \cdot x_1 + 1 \cdot x_2 + 1 \cdot x_3 + 1 \cdot x_4 + = x_2 + x_3 + x_4, \text{mod } 2$$

respectively. Since $L_1 = L_2 = 0$ for treatment combinations (1), bc, abd, and acd, these treatment combinations form the principal block. Treatment combination a is not in that block, so we multiply each treatment combination in the principal block by a and find that $a \cdot (1) = a, a \cdot bc = abc, a \cdot acd = a^2cd = a^0cd = cd$, and $a \cdot abd = a^2bd = a^0bd = bd$ form a second block. Now b is in neither of these two blocks, so we proceed in the same manner and find that

$$b \cdot (1) = b, b \cdot bc = c, b \cdot abd = ad \quad \text{and} \quad b \cdot acd = abcd$$

form a third block. The remaining treatment combinations (ab, ac, bcd, and d) form block 4. This gives the blocking plan depicted in Table 12.3, and the order of experimentation within each block is randomized, as is the order in which the blocks are run. The resulting design is a randomized incomplete block design in which blocks are confounded with the interactions ABC, BCD, and AD.

Instead of first determining the principal block and then generating the remaining blocks by multiplication, one could calculate the values of the defining equations for each treatment combination and place those having the same values for both equations in the same block. Thus, all treatment combinations for which $L_1 = 1$ and $L_2 = 0$ would be placed in the same block, as indicated in Table 12.3.

TABLE 12.3
Blocking Plan for a 2^4 Factorial That Confounds ABC, BCD, and AD with Blocks

block 1	(1)	bc	acd	abd	$L_1 = 0; L_2 = 0$	block 2	a	abc	cd	bd	$L_1 = 1; L_2 = 0$
block 3	b	c	abcd	ad	$L_1 = 1; L_2 = 1$	block 4	ab	ac	bcd	d	$L_1 = 0; L_2 = 1$

TABLE 12.4
Selected Block Designs for 2^f Factorials

f	Blocks	Defining Relation	Principal Block
2	2	$I = AB$	$\{(1), ab\}$
3	2	$I = ABC$	$\{(1), ab, ac, bc\}$
	4	$I = AB = AC = BC$	$\{(1), abc\}$
4	2	$I = ABCD$	$\{(1), ab, bc, ac, abcd, cd, ad, bd\}$
	4	$I = ABC = BCD = AD$	$\{(1), bc, acd, abd\}$
	8	$I = AB = BC = CD = AC = ABCD = BD = AD$	$\{(1), abcd\}$
5	2	$I = ABCDE$	$\{(1), ab, ac, ad, ae, bc, bd, be, cd, ce, de,$ $abcd, abce, abde, acde, bcde\}$
	4	$I = ABC = CDE = ABDE$	$\{(1), ab, de, acd, ace, bcd, bce, abde\}$
	8	$I = ABE = BCE = CDE = AC$ $= ABCD = BD = ADE$	$\{(1), ace, bde, abcd\}$
	16	$I = AB = AC = CD = DE = BC = ABCD = ABDE$ $= AD = ACDE = BD = BCDE = ABCE = AE$ $= BE = CE$	$\{(1), abcde\}$
6	2	$I = ABCDEF$	$\{(1), ab, ac, ad, ae, af, bc, bd, be, bf, cd,$ $ce, cf, de, df, ef, abcd, abce, abcf, abde,$ $abdf, abef, acde, acdf, acef, adef, bcde,$ $bcdf, bcef, bdef, cdef, abcdef\}$
	4	$I = ABCF = CDEF = ABDE$	$\{(1), ab, cf, de, acd, ace, adf, aef, bcd, bce,$ $bdf, bef, abcf, abde, cdef, abcdef\}$
	8	$I = ACE = ABEF = ABCD = BCF = BDE$ $= CDEF = ADF$	$\{(1), acf, ade, bce, bdf, abcd, abef, cdef\}$
	16	$I = ABF = ACF = CDF = DEF = BC = ABCD$ $= ABDE = AD = ACDE = BDF = BCDEF$ $= CE = ABCEF = AEF = BE$	$\{(1), adf, bcef, abcde\}$
	32	$I = AB = BC = CD = DE = EF = AC = ABCD$ $= ABDE = ABEF = BD = BCDE = BCEF = CE$ $= CDEF = DF = AD = ACDE = ACEF = ABCE$ $= ABCDEF = ABDF = BE = BDEF = BCDF = AE$ $= ADEF = ACDF = CF = ABCF = BF = AF$	$\{(1), abcdef\}$

For convenience, some defining relations and principal blocks for selected 2^f factorials are summarized in Table 12.4. Similar tables can be found in [9].

■ **Example 12.4 (Coil Breakdown Voltage Study: Blocks Confounded with Interaction)**

A systematic test was to be made on the effects of four variables, each at two levels, on coil breakdown voltage. The variables were A, firing furnace 1 or 3; B, firing temperature 1650 or 1700° C; C, gas humidification, no or yes; and D, coil outside diameter, small (< 0.0300 inch) or large (> 0.0305 inch). Since only four coils could be tested in a given short time interval, it was necessary to run four blocks of four coils for such testing. The experimenters decided to run a 2^4 factorial in four blocks of four, confounding ABC, BCD, and AD with the four blocks. The data of Table 12.5 were obtained from the blocking plan of Table 12.3.

TABLE 12.5
Coil Breakdown voltages for Example 12.4*

				Block			
	1		2		3		4
tc	*y*	*tc*	*y*	*tc*	*y*	*tc*	*y*
(1)	82	*a*	76	*b*	79	*ab*	85
bc	55	*abc*	74	*c*	71	*ac*	84
acd	81	*cd*	72	*abcd*	89	*bcd*	84
abd	88	*bd*	73	*ad*	79	*d*	80

*tc denotes "treatment combination."

When blocks (or ABD, BCD, and AD) serve as a factor while the remaining three- and four-way interactions are pooled for an error term, the following mathematical model is possible:

$$Y_{ijklm} = \mu + A_i + B_j + AB_{ij} + C_k + AC_{ik} + BC_{jk} + D_l + BD_{jl} + CD_{kl}$$
$$+ block_m + \varepsilon_{1(ijklm)}$$

When this model is used with the **General Linear Model** procedure in Minitab, the analysis of Table 12.6 is output.

Factor A (furnace) can be declared significant for any $\alpha \geq 0.061$. However, the p values associated with B (temperature) and the BC interaction are so large over the ranges used in this experiment (0.928 and 0.543, respectively) that we may be able to

TABLE 12.6
Minitab ANOVA for Example 12.4

```
General Linear Model
Factor      Levels     Values
A           2          1        2
B           2          1        2
C           2          1        2
D           2          1        2
Block       4          1        2        3        4
Analysis of Variance for Y
Source      DF      Seq SS      Adj SS      Adj MS        F        P
A           1       225.00      225.00      225.00      8.60    0.061
B           1         0.25        0.25        0.25      0.01    0.928
C           1        64.00       64.00       64.00      2.45    0.216
D           1       100.00      100.00      100.00      3.82    0.146
A*B         1        56.25       56.25       56.25      2.15    0.239
A*C         1        64.00       64.00       64.00      2.45    0.216
B*C         1        12.25       12.25       12.25      0.47    0.543
B*D         1       110.25      110.25      110.25      4.21    0.132
C*D         1       121.00      121.00      121.00      4.62    0.121
Block       3       199.50      199.50       66.50      2.54    0.232
Error       3        78.50       78.50       26.17
Total       15     1031.00
```

simplify the model by removing B and the interactions associated with B. This does not mean that the firing temperature is not important, but that its effect may be negligible over the range of temperatures under consideration. The resulting Minitab outputs for this reduced model are given in Table 12.7. In this case, A (furnace) is significant for any $\alpha \geq 0.043$ and no other factor is significant for $\alpha < 0.113$.

It appears that A may be the most important factor and B may have no effect on coil breakdown voltage over the experimental range. Since 2^f experiments are often run to get an overall picture of the important factors, each set at the extremes of its range, another experiment might now be planned without factor B. That experiment might be a 3^3 factorial experiment that requires blocking and includes factors A, C, and D.

The outputs of Table 12.7 include a warning that one residual is unusually large in absolute value, perhaps due to nonnormality or some other model inadequacy. A thorough residual analysis should be conducted before conclusions based on either Minitab ANOVA are finalized. Inadequacies in the model assumptions may invalidate any conclusions made from ANOVA summaries.

TABLE 12.7
Minitab ANOVA for Example 12.4: Reduced Model

```
General Linear Model
```

Factor	Levels	Values			
A	2	1	2		
C	2	1	2		
D	2	1	2		
Block	4	1	2	3	4

```
Analysis of Variance for Y
```

Source	DF	Seq SS	Adj SS	Adj MS	F	P
A	1	225.00	225.00	225.00	6.12	0.043
C	1	64.00	64.00	64.00	1.74	0.229
D	1	100.00	100.00	100.00	2.72	0.143
A*C	1	64.00	64.00	64.00	1.74	0.229
C*D	1	121.00	121.00	121.00	3.29	0.113
Block	3	199.50	199.50	66.50	1.81	0.233
Error	7	257.50	257.50	36.79		
Total	15	1031.00				

```
Unusual Observations for Y
```

Obs.	Y	Fit	Stdev.Fit	Residual	St.Resid
2	55.0000	63.5000	4.5488	-8.5000	-2.12R

```
R denotes an obs. with a large st. resid.
```

Analysis Using the Yates Method

If computer resources are not available, one might use the Yates method of Section 9.5 to analyze data like those in Table 12.5. The reader can verify the results in Table 12.8 for those data. The sums of squares associated with ABC, BCD, and AD are

$$SS_{ABC} = \frac{(c_{ABC})^2}{16 \times 1} = \frac{(-26)^2}{16} = 42.25$$

$$SS_{BCD} = \frac{(c_{BCD})^2}{16 \times 1} = \frac{(50)^2}{16} = 156.25$$

$$SS_{AD} = \frac{(c_{AD})^2}{16 \times 1} = \frac{(-4)^2}{16} = 1.00$$

and total $42.25 + 156.25 + 1.00 = 199.50$. But the effects of ABC, BCD, and AD are confounded with blocks, so $SS_{block} = 199.50$. This is easily confirmed by noting that the block totals are 306, 295, 318, and 333 with a grand total of 1252. Thus,

$$SS_{block} = \frac{(306)^2 + (295)^2 + (318)^2 + (333)^2}{4} - \frac{(1252)^2}{16}$$
$$= 98168.50 - 97969.00 = 199.50$$

The remaining sums of squares are calculated in like manner. If all three-way and four-way interactions that are not confounded are pooled for an error term, the sums of squares agree with those in Table 12.6.

TABLE 12.8
The Yates Method on the Coil Breakdown Voltage Data

Treatment Combination	Sample Total	cycle 1	cycle 2	cycle 3	cycle 4
(1)	82	158	322	606	$1252 =$ grand total
a	76	164	284	646	$60 = c_A$
b	79	155	320	32	$2 = c_B$
ab	85	129	326	28	$30 = c_{AB}$
c	71	159	0	-20	$-32 = c_C$
ac	84	161	32	22	$32 = c_{AC}$
bc	55	153	14	18	$-14 = c_{BC}$
abc	74	173	14	12	$-26 = c_{ABC}$
d	80	-6	6	-38	$40 = c_D$
ad	79	6	-26	6	$-4 = c_{AD}$
bd	73	13	2	32	$42 = c_{BD}$
abd	88	19	20	0	$-6 = c_{ABD}$
cd	72	-1	12	-32	$44 = c_{CD}$
acd	81	15	6	18	$-32 = c_{ACD}$
bcd	84	9	16	-6	$50 = c_{BCD}$
abcd	89	5	-4	-20	$-14 = c_{ABCD}$

12.4 BLOCK CONFOUNDING WITH REPLICATION

Whenever an experiment is restricted so that all treatments cannot appear in one block, some interaction is usually confounded with blocks. Interactions of *least interest* should be confounded with blocks. These need not be small—just of least interest. If several replications of the whole experiment (all blocks) are possible, as in the case of the split plot, the same interaction may be confounded in all replications. In this case, the design is said to be *completely confounded*. If, on the other hand, one interaction is confounded in the first replication, a different interaction is confounded in the second replication, and so on, the design is said to be *partially confounded*.

12.4.1 Complete Confounding

Considering a 2^3 factorial experiment in which only four treatment combinations can be finished in one day yields a 2^3 factorial run in two incomplete blocks of four treatment combinations each. To confound the highest order interaction ABC with blocks, the defining relation is

$$I = ABC$$

As seen in Section 12.2, the blocks would be

block 1	block 2
(1)	a
ab	b
ac	c
bc	abc

If this whole experiment (2^3 factorial in two blocks of four each) can be replicated, say three times, the order in which the blocks are run is randomized and the order within blocks is randomized, a data layout like that of Figure 12.4 will be obtained.

The confounding scheme (ABC confounded with blocks) is the same for all three replications in Figure 12.4, but the order of experimentation has been randomized within each replication. Also, the block that is to be run first in each replication is decided at random. An analysis layout for this experiment appears in Table 12.9. In such cases, the replication interaction with all three main effects and their interactions are usually taken as the error term for testing the important effects. The replication effect and the block (or ABC) effect could be tested against the replication $\times$ block interaction, but the degrees of freedom are low and the power of such a test is poor. The design is quite powerful, however, in testing the main effects A, B, and C and their first-order interactions. No clear information on the ABC interaction is possible.

TABLE 12.9
Analysis Layout for Completely Confounded 2^3 Factorial

Source	df	df
Replications	2	
Blocks (or ABC)	1	
Replications $\times$ block interactions	2	5 between plots
A	1	
B	1	
AB	1	
C	1	
AC	1	
BC	1	
Replications $\times$ all others	12	18 within plots
Totals	23	23

replication I		replication II		replication III	
run 1	*run 2*	*run 1*	*run 2*	*run 1*	*run 2*
ac	a	c	(1)	ab	c
(1)	c	abc	ac	(1)	b
ab	abc	b	bc	ac	abc
bc	b	a	ab	bc	a

Figure 12.4 Layout for three replications of a 2^3 factorial in two blocks of four.

■ **Example 12.5** (**2^3 Factorial in Two Blocks of Four with *ABC* Completely Confounded**)

Coils were wound from each of two winding machines using either of two wire stocks. Their outside diameters were measured at two positions on an optical comparator. Since only four readings could be made in a given time period, the $2^3 = 8$ treatment combinations were divided into two blocks of four. However, four replications were possible. At first the experimenter decided to confound *ABC* in all four replicates, where *A* represents machines, *B* stocks, and *C* positions. Each replicate is comprised of the two blocks determined at the beginning of this section.

A layout with data is given in Table 12.10. Analysis by means of Minitab gives Table 12.11. The Minitab summary shows only positions on the coil as having a highly significant effect on coil outside diameter. In that summary, two residuals are noted that have unusually large absolute values. A residual analysis should be conducted to ensure that the model assumptions are adequately satisfied before any final conclusions and recommendations are made.

TABLE 12.10
Layout and Data for Example 12.5: *ABC* Confounded in All Replicates*

replication I			replication II	
run 1	*run 2*		*run 1*	*run 2*
ab = 2173	b = 2300		a = 2228	ac = 3495
(1) = 2249	c = 3538		abc = 3592	(1) = 2094
ac = 3532	abc = 3524		c = 3116	bc = 2249
bc = 2948	a = 2319		b = 2386	ab = 2373

replication III			replication IV	
run 1	*run 2*		*run 1*	*run 2*
(1) = 2382	c = 3528		abc = 2996	ab = 2393
ab = 2240	b = 2118		c = 1934	bc = 2814
bc = 3495	abc = 3350		b = 2234	ac = 3400
ac = 2995	a = 2272		a = 2215	(1) = 2297

*All measurements multiplied by 10^5, so 2173 is an outside diameter of 0.02173 inch.

TABLE 12.11
Minitab ANOVA for Example 12.5

`General Linear Model`

Analysis of Variance for Y

Source	DF	Seq SS	Adj SS	Adj MS	F	P
R	3	409720	409720	136573	1.17	0.347
Block	1	8483	8483	8483	0.07	0.790
R*Block	3	515408	515408	171803	1.48	0.254
A	1	364445	364445	364445	3.13	0.094
B	1	5228	5228	5228	0.04	0.834
A*B	1	18964	18964	18964	0.16	0.691
C	1	6330572	6330572	6330572	54.44	0.000
A*C	1	302059	302059	302059	2.60	0.124
B*C	1	16699	16699	16699	0.14	0.709
Error	18	2093039	2093039	116280		
Total	31	10064614				

Unusual Observations for Y

Obs.	Y	Fit	Stdev.Fit	Residual	St.Resid
19	3495.00	2958.94	225.55	536.06	2.10R
26	1934.00	2645.62	225.55	−711.62	−2.78R

R denotes an obs. with a large st. resid.

12.4.2 Partial Confounding

In working Example 12.5, we might want a test on the ABC interaction. To do so, we could confound some interactions other than ABC in some of the replications. We might use four replications and confound ABC in the first one, AB in the second, AC in the third, and BC in the fourth. Thus, the four replications will yield full information on A, B, and C but three-fourths information on AB, AC, BC, and ABC, since we can compute an unconfounded interaction such as AB in three out of four of the replications. The proper block entries for each replication are given in Figure 12.5.

An analysis layout for the design of Figure 12.5 is shown in Table 12.12. The residual term with its 17 degrees of freedom can be explained as follows. In all four replications the main effects A, B, and C can be isolated. This gives $3 \times 1 = 3$ df for each replication by main-effect interaction, or a total of 9 df. The AB, AC, BC, and ABC interactions can be isolated in only three out of the four replications. Hence the replications by AC have

replication I confound ABC		replication II confound AB		replication III confound AC		replication IV confound BC	
(1)	a	(1)	a	(1)	a	(1)	b
ab	b	c	b	ac	c	bc	c
ac	c	ab	ac	abc	bc	a	ab
bc	abc	abc	bc	b	ab	abc	ac
$L_1 = x_1 + x_2 + x_3$		$L_2 = x_1 + x_2$		$L_3 = x_1 + x_3$		$L_4 = x_2 + x_3$	

Figure 12.5   Layout for four replications of a 2^3 factorial in two blocks of four: partial confounding

$2 \times 1 = 2$ df, or a total of 8 df for all such interactions. This gives $9 + 8 = 17$ df for effects by replications, which is usually taken as the error estimate. The blocks and replications are separated out in the top of Table 12.12 only to reduce the experimental error. The numerical analysis of problems such as this is carried out in the manner given in earlier chapters.

As with the completely confounded design of Section 12.4.1, the order of experimentation is randomized within each block. Also, the block to be run first in each replication is decided at random. Since the treatment combinations in a block differ among replications, blocks are nested in replications. Thus, we have a split-plot design with blocks nested within replications.

TABLE 12.12
ANOVA for Partially Confounded 2^3 Factorial

Source	df		df
Replications	3		
Blocks with replications	4		7 between plots
ABC (Replication I)		1	
AB (Replication II)		1	
AC (Replication III)		1	
BC (Replication IV)		1	
A		1	
B		1	
C		1	
AB		1	Only from
AC		1	replications
BC		1	where not
ABC		1	confounded
Replications × all effects	17		24 within plots
Totals	31		31

■ **Example 12.6 (2^3 Factorial in Two Blocks of Four with ABC Partially Confounded)**

In Example 12.5 a study involving winding machines was considered. When coils from two other winding machines were to be tested, it was decided to partially confound the interactions. The experimenter decided on a design like that in Figure 12.5 to obtain three-fourths information on each of ABC, AB, AC, and BC, where A represents machines, B stocks, and C positions.

A layout with data is given in Table 12.13. Analysis with Minitab gives Table 12.14. As in Example 12.5, the Minitab summary shows that position on the coil has a highly significant effect on coil outside diameter, but an AC interaction is also significant for $\alpha \leq 0.015$. In that summary, one unusually large residual is noted, and B has an unusually large p value (0.976). Before any final conclusions and recommendations are made, a residual analysis should be conducted to ensure that the model assumptions are adequately satisfied.

TABLE 12.13
Layout and Data for Example 12.6: Partial Confounding*

replication I (confounding ABC)		replication II (confounding AB)	
block 1	block 2	block 3	block 4

(1) = 2208	a = 2196	(1) = 2004	a = 2179
ab = 2133	b = 2086	ab = 2112	b = 2073
ac = 2459	c = 3356	c = 3073	ac = 3474
bc = 3096	abc = 2776	abc = 2631	bc = 3360

replication III (confounding AC)		replication IV (confounding BC)	
block 5	block 6	block 7	block 8

c = 2839	(1) = 1916	a = 2056	b = 1878
a = 2189	ac = 2979	(1) = 2010	ab = 2156
bc = 3522	b = 2151	bc = 3209	c = 3423
ab = 2095	abc = 2500	abc = 3066	ac = 2524

*All measurements multiplied by 10^5, so 2208 is an outside diameter of 0.02208 inch.

TABLE 12.14
Minitab ANOVA for Example 12.6

General Linear Model

Analysis of Variance for Y

Source	DF	Seq SS	Adj SS	Adj MS	F	P
R	3	38718	38718	12906	0.23	0.875
Block(R)	4	401060	264697	66174	1.17	0.359
A	1	224283	224283	224283	3.97	0.063
B	1	53	53	53	0.00	0.976
A*B	1	737	737	737	0.01	0.910
C	1	6886689	6886689	6886689	121.79	0.000
A*C	1	416067	416067	416067	7.36	0.015
B*C	1	2667	2667	2667	0.05	0.831
A*B*C	1	70742	70742	70742	1.25	0.279
Error	17	961283	961283	56546		
Total	31	9002297				

Unusual Observations for Y

Obs.	Y	Fit	Stdev.Fit	Residual	St.Resid
15	3474.00	3086.14	162.81	387.86	2.24R

R denotes an obs. with a large st. resid.

12.5 CONFOUNDING IN 3^f FACTORIALS

When a 3^f-factorial experiment cannot be completely randomized, it is usually blocked in blocks that are multiples of 3. The use of the I and J components of interaction introduced in Section 10.2 is helpful in confounding only part of an interaction with blocks. Kempthorne's rule also applies, using such interactions as AB, AB^2, ABC^2, ..., and treatment combinations in the form 00, 10, 01, 11 instead of (1), a, b, ab as in 2^f experiments. Recall that for a 3^f factorial, 0, 1, and 2 denote the low, medium, and high levels of a factor, respectively.

A 3^2 Factorial in Three Blocks of Three

If a 3^2 factorial is restricted so that only three of the nine treatment combinations can be run in one block, we usually confound AB or AB^2, each of which carries 2 degrees of freedom. When one is confounding AB^2, the defining relation is

$$I = AB^2$$

and the defining equation is

$$L = x_1 + 2x_2, \text{mod } 3$$

For the treatment combinations 00, 11, and 22, $0 + 2(0) = 0$, $1 + 2(1) = 3 = 0$ mod 3, and $2 + 2(2) = 6 = 0$ mod 3, respectively. Since $L = 0$ for each of these, the principal block is {00, 11, 22}.

Once the principal block has been obtained, another block can be generated by adding (modulo 3) to each element in the principal block an element that is not in that block. If we choose 10, which is not in the principal block, block 2 then contains $00 + 10 = 10$, $11 + 10 = 21$, and $22 + 10 = 32 = 02$ mod 3.

After two blocks have been determined, an element present in neither block can be added to each element of the principal block to obtain a third block. In this case, only three elements remain, so such addition is not necessary. The third block is {20, 01, 12}.

The blocks can also be determined by calculating the values of L for each treatment combination and then placing those treatment combinations having a common L value in the same block, as depicted in Figure 12.6.

When confounding AB for a 3^2 factorial, the defining relation is

$$I = AB$$

and the defining equation is

$$L = x_1 + x_2, \text{mod } 3$$

Calculating the value of L for each treatment combination and placing those treatment combinations having the same L value in a block, we have the blocking plan of Figure 12.7.

block 1 *L = 0*	*block 2* *L = 1*	*block 3* *L = 2*
00	10	20
11	21	01
22	02	12

Figure 12.6   Blocking plan for a 3^2 factorial that confounds AB^2 with blocks.

block 1 *L = 0*	*block 2* *L = 1*	*block 3* *L = 2*
00	10	20
12	22	02
21	01	11

Figure 12.7   Blocking plan for a 3^2 factorial that confounds AB with blocks.

■ **Example 12.7 (Analysis of a 3^2 Factorial with AB^2 Confounded with Blocks)**

Consider the hypothetical data in Table 12.15, where AB^2 is confounded with blocks. These are the same data given in Table 10.1. The totals for the three blocks are 7, 1, and 2, so the block sum of squares is

$$SS_{block} = \frac{7^2 + 1^2 + 2^2}{3} - \frac{(7 + 1 + 2)^2}{9} = 6.89$$

which is identical with the $I(AB)$ interaction component in Table 10.11, as it should be. The remainder of the analysis proceeds the same as in Section 10.2, and the resulting ANOVA table is that exhibited in Table 12.16.

If these were real data, the AB part of the interaction might be considered to be error, but then neither main effect is significant. The purpose here is only to illustrate the blocking of a 3^2 factorial.

TABLE 12.15
Layout and Data for Example 12.7

block 1	block 2	block 3
00 = 1	10 = −2	20 = 3
11 = 4	21 = 1	01 = 0
22 = 2	02 = 2	12 = −1

TABLE 12.16
ANOVA for Example 12.7

Source	df	SS	MS
Blocks or AB^2	2	6.89	3.44
A	2	4.22	2.11
B	2	1.56	0.78
AB	2	16.22	8.11
Totals	8	28.89	

A 3^3 Factorial in Three Blocks of Nine

Consider now a 3^3 experiment in which the 27 treatment combinations (such as given in Table 10.12) cannot all be completely randomized. If nine can be randomized and run on one day, nine on another day, and so on, we can use a 3^3 factorial in three blocks of nine treatment combinations each. This requires 2 df to be confounded with blocks. Since the ABC interaction with 8 df can be partitioned into ABC, ABC^2, AB^2C, and AB^2C^2, each with 2 df, one of these four could be confounded with blocks. If AB^2C is confounded, the identifying relation is

$$I = AB^2C$$

and the defining equation is

$$L = x_1 + 2x_2 + x_3, \text{mod } 3$$

Values of L are obtained using arithmetic modulo 3. Thus, the value of L associated with treatment combination 212 is $1(2) + 2(1) + 1(2) = 6 = 0 \text{ mod } 3$. Placing treatment combinations that share a common value of L in the same block, the three blocks are:

block 1	000	011	110	121	102	212	220	022	201	$L = 0$
block 2	100	111	210	221	202	012	020	122	011	$L = 1$
block 3	200	211	010	021	002	112	120	222	101	$L = 2$

A partial ANOVA summary for this design is given in Table 12.17. From that summary we see that this is a useful design, because if we are willing to pool the 6 df in ABC as error, we can retrieve all main effects (A, B, C) and all two-way interactions. The determination of sums of squares, and so forth, is the same as for the complete 3^3 factorial given in Example 10.4. Confounding other parts of the ABC interaction would, of course, yield different blocking arrangements, although the outline of the analysis would be the same. In practice, different blocking arrangements might yield different responses.

A 3^3 Factorial in Nine Blocks of Three

An experimenter wishes to run a 3^3 factorial in nine blocks of three treatment combinations each. First, however, a blocking plan must be determined and the implications of using that plan must be evaluated. It would be frustrating (and quite possibly embarrassing, if not career ending) to spend the time and money to run an experiment without any reasonable chance of gaining useful information about the factors of interest.

TABLE 12.17
Partial ANOVA Summary for a 3^3
Factorial in Three Blocks

Source	df
Blocks (or AB^2C)	2
A	2
B	2
AB	4
C	2
AC	4
BC	4
Error (or ABC, ABC^2, AB^2C^2)	6
Total	26

Here 8 df must be confounded with blocks. The defining relation

$$I = AB^2C^2 = AB = BC^2 = AC$$

can be used for this purpose. Using modulo 3 arithmetic on the exponents, we write $(AB^2C^2)(AB) = A^2C^2 = AC$ and $(AB)(BC^2) = AB^2C^2$. Thus, these are not all independent. In fact, any two of the expressions can be used to generate the other two. For this reason we need only two defining equations. Selecting AB^2C^2 and AB from the defining relation, the associated defining equations are

$$L_1 = x_1 + 2x_2 + 2x_3, \text{ mod } 3$$

$$L_2 = x_1 + x_2, \text{ mod } 3$$

These two defining equations will yield nine pairs of numbers—one pair for each block.

First we determine the principal block, where both L_1 and L_2 are zero. One treatment combination is obviously 000. Another is 211:

$$L_1 = 1(2) + 2(1) + 2(1) = 6 = 0 \text{ mod } 3 \qquad \text{and}$$

$$L_2 = 1(2) + 1(1) + 0(1) = 3 = 0 \text{ mod } 3$$

and a third is 122:

$$L_1 = 1(1) + 2(2) + 2(2) = 9 = 0 \text{ mod } 3 \qquad \text{and}$$

$$L_2 = 1(1) + 1(2) + 0(2) = 3 = 0 \text{ mod } 3$$

The other eight blocks can now be generated from this principal block, giving the blocking plan of Figure 12.8.

If the components of the three-way interaction that are not confounded with blocks can be assumed to equal 0 and pooled with error, an analysis layout would be like that shown in Table 12.18. Notice that this design might be reasonable if interest centered only in the main effects A, B, and C. Such a design might be necessary when three factors were involved, each at three levels, but only three treatment combinations could be run in one

block 1	block 2	block 3
000	001	022
211	212	210
122	120	121

block 4	block 5	block 6
010	020	100
221	201	011
102	112	222

block 7	block 8	block 9
200	110	101
111	021	012
022	202	220

Figure 12.8 Blocking plan for a 3^3 factorial that confounds AB^2C^2, AB, BC^2, and AC with blocks.

day. These three might involve an elaborate environmental test in which at best only three sets of environmental conditions—temperature, pressure, and humidity—can be simulated in one day.

12.6 SAS PROGRAMS

Previously discussed methods can be used to obtain SAS outputs for data analysis when interactions are confounded with blocks. In this section, command files and outputs are presented for the confounding situations considered in Examples 12.5 and 12.6.

The **ANOVA** procedure is used with Example 12.5, although **GLM** can also be used. We warn the reader that although **ANOVA** is more efficient, results may be erroneous if unbalanced data are involved.

Complete Confounding (Example 12.5)

We shall use the data for a 2^3 factorial in two blocks of four with ABC completely confounded with blocks given in Table 12.10. In this case A represents machines (1 or 2), B represents stocks (1 or 2), and C represents positions (1 or 2). The mathematical model

$$Y_{ijklm} = \mu + R_i + \beta_j + R\beta_j + A_k + B_l + AB_{kl} + C_m + AC_{km} + BC_{lm} + \varepsilon_{1(ijklm)}$$

where R denotes replicates and β denotes blocks can be expressed in a SAS command file as

```
MODEL Y = REP|BLK A B A*B C A*C B*C;
```

We include this statement in the command file of Table 12.19 to produce the outputs in Figure 12.9. Except for rounding, the results are identical to the Minitab results of Table 12.11.

TABLE 12.18
Partial ANOVA Summary for a 3^3 Factorial in Nine Blocks

Source	df
Blocks (or AB, BC^2, AC, AB^2C^2)	8
A	2
B	2
C	2
AB^2	2
AC^2	2
BC	2
Error (or ABC, ABC^2, AB^2C)	6
Total	26

TABLE 12.19

SAS Command File for Example 12.5: *ABC* Confounded in All Replicates

```
OPTIONS LINESIZE=80;
DATA COILS;
INPUT REP BLK A B C Y @@;
CARDS;
1 1 1 1 1 2249 1 1 2 2 1 2173 1 1 2 1 2 3532 1 1 1 2 2 2948
1 2 2 1 1 2319 1 2 1 2 1 2300 1 2 1 1 2 3538 1 2 2 2 2 3524
2 1 1 1 1 2094 2 1 2 2 1 2373 2 1 2 1 2 3495 2 1 1 2 2 2249
2 2 2 1 1 2228 2 2 1 2 1 2386 2 2 1 1 2 3116 2 2 2 2 2 3592
3 1 1 1 1 2382 3 1 2 2 1 2240 3 1 2 1 2 2995 3 1 1 2 2 3495
3 2 2 1 1 2272 3 2 1 2 1 2118 3 2 1 1 2 3528 3 2 2 2 2 3350
4 1 1 1 1 2297 4 1 2 2 1 2393 4 1 2 1 2 3400 4 1 1 2 2 2814
4 2 2 1 1 2215 4 2 1 2 1 2234 4 2 1 1 2 1934 4 2 2 2 2 2996
;
PROC ANOVA;
CLASS REP BLK A B C;
MODEL Y = REP|BLK A B A*B C A*C B*C;
```

ANALYSIS OF VARIANCE PROCEDURE

DEPENDENT VARIABLE: Y

SOURCE	DF	SUM OF SQUARES	MEAN SQUARE	F VALUE	PR > F
MODEL	13	7971574.90625000	613198.06971154	5.27	0.0008
ERROR	18	2093039.31250000	116279.96180556		
CORRECTED TOTAL	31	10064614.21875000			

	R-SQUARE	C.V.	ROOT MSE	Y MEAN
	0.792040	12.5744	340.99847772	2711.84375000

SOURCE	DF	ANOVA SS	F VALUE	PR > F
REP	3	409719.59375000	1.17	0.3471
BLK	1	8482.53125000	0.07	0.7902
REP*BLK	3	515407.84375000	1.48	0.2542
A	1	364444.53125000	3.13	0.0936
B	1	5227.53125000	0.04	0.8345
A*B	1	18963.78125000	0.16	0.6911
C	1	6330571.53125000	54.44	0.0001
A*C	1	302058.78125000	2.60	0.1244
B*C	1	16698.78125000	0.14	0.7092

Figure 12.9 SAS output for Example 12.5: *ABC* confounded in all replicates.

Partial Confounding (Example 12.6)

Here we shall use the data for a 2^3 factorial in two blocks of four with ABC partially confounded with blocks given in Table 12.14. As with Example 12.5, A represents machines (1 or 2), B represents stocks (1 or 2), and C represents positions (1 or 2). In this case, blocks are nested in replicates. The mathematical model

$$Y_{ijklm} = \mu + R_i + \beta_{j(i)} + A_k + B_l + AB_{kl} + C_m + AC_{km} + BC_{lm} + ABC_{klm} + \varepsilon_{1(ijklm)}$$

where R denotes replicates and β denotes blocks can be expressed in a SAS command file as

MODEL Y = REP BLK(REP) A|B|C;

We include this statement in the command file of Table 12.20 to produce the outputs in Figure 12.10. Except for rounding, the results are identical to the Minitab results of Table 12.14.

By default, the **GLM** procedure includes type I and type III sums of squares in the outputs. To conserve space, the type III sums of squares are not included in Figure 12.10 because the two sums of squares are identical for these data.

TABLE 12.20
SAS Command File for Example 12.6: Partial Confounding of ABC

```
OPTIONS LINESIZE=80;
DATA COILS;
INPUT REP BLK A B C Y @@;
CARDS;
1 1 1 1 1 2208 1 1 2 2 1 2133 1 1 2 1 2 2459 1 1 1 2 2 3096
1 2 2 1 1 2196 1 2 1 2 1 2086 1 2 1 1 2 3356 1 2 2 2 2 2776
2 3 1 1 1 2004 2 3 2 2 1 2112 2 3 1 1 2 3073 2 3 2 2 2 2631
2 4 2 1 1 2179 2 4 1 2 1 2073 2 4 2 1 2 3474 2 4 1 2 2 3360
3 5 2 1 1 2189 3 5 2 2 1 2095 3 5 1 1 2 2839 3 5 1 2 2 3522
3 6 1 1 1 1916 3 6 1 2 1 2151 3 6 2 1 2 2979 3 6 2 2 2 2500
4 7 1 1 1 2010 4 7 2 1 1 2056 4 7 1 2 2 3209 4 7 2 2 2 3066
4 8 2 2 1 2156 4 8 2 1 2 2524 4 8 1 2 1 1878 4 8 1 1 2 3423
;
PROC GLM;
CLASS REP BLK A B C;
MODEL Y = REP BLK(REP) A|B|C;
```

GENERAL LINEAR MODELS PROCEDURE

SOURCE	SUM OF DF	MEAN SQUARES	SQUARE	F VALUE
MODEL	14	8041013.85416667	574358.13244048	10.16
ERROR	17	961283.11458333	56546.06556373	PR > F
CORRECTED TOTAL	31	9002296.96875000		0.0001

R-SQUARE	C.V.	ROOT MSE	Y MEAN
0.893218	9.3105	237.79416638	2554.03125000

SOURCE	DF	TYPE I SS	F VALUE	PR > F
REP	3	38717.59375000	0.23	0.8754
BLK (REP)	4	401060.12500000	1.77	0.1808
A	1	224282.53125000	3.97	0.627
B	1	52.53125000	0.00	0.9760
A*B	1	737.04166667	0.01	0.9104
C	1	6886688.28125000	121.79	0.0001
A*C	1	416066.66666667	7.36	0.0148
B*C	1	2667.04166667	0.05	0.8307
A*B*C	1	70742.04166667	1.25	0.2789

Figure 12.10   SAS output for Example 12.6: Partial confounding of ABC.

12.7 SUMMARY

The summary at the end of Chapter 11 may now be extended.

Experiment	Design	Analysis
II. Two or more factors A. Factorial (crossed)		
	1. Completely randomized $\cdots$	1. $\cdots$
	2. Randomized block a. Complete $\quad Y_{ijk} = \mu + R_k + A_i$ $\quad\quad + B_j + AB_{ij} + \varepsilon_{ijk}$ b. Incomplete, confounding: $\quad$ i. Main effect—split plot $\quad\quad Y_{ijk} = \mu + R_i + A_j + RA_{ij}$ $\quad\quad\quad$ whole plot $\quad\quad + B_k + RB_{ik} + AB_{jk} + RAB_{ijk}$ $\quad\quad\quad$ split plot	2. a. Factorial ANOVA with replications R_k b. $\quad$ i. Split-plot ANOVA
	$\quad$ ii. Interactions in 2^f and 3^f $\quad\quad$ (1) Several replications $\quad\quad\quad Y_{ijkq} = \mu + R_i + \beta_j + R\beta_{ij}$ $\quad\quad\quad\quad + A_k + B_q$ $\quad\quad\quad\quad + AB_{kq} + \varepsilon_{ijkq}, \ldots$	$\quad$ ii. $\quad\quad$ (1) Factorial ANOVA with replications R_i and blocks β_j or confounded interaction
	$\quad\quad$ (2) One replication only $\quad\quad\quad Y_{ijk} = \mu + \beta_i + A_j$ $\quad\quad\quad\quad + B_k + AB_{jk}, \ldots$	$\quad\quad$ (2) Factorial ANOVA with blocks β_j or confounded interaction

12.8 FURTHER READING

Anderson, V. L., and R. A. McLean, *Design of Experiments: A Realistic Approach*. New York: Marcel Dekker, 1974. See pages 241–251, 281–292.

Montgomery, D. C., *Design and Analysis of Experiments*, 4th ed. New York: John Wiley & Sons, 1997. See pages 356–370, 449–456.

PROBLEMS

12.1 Assuming that a 2^3 factorial such as the problem of Chapters 1, 5, and 9 involving two tool types, two bevel angles, and two types of cut cannot all be run on one lathe in one time period but must be divided into four treatment combinations per lathe, set up a confounding scheme for confounding the AC interaction in these two blocks of four. Assuming that the results of one replication give $(1) = 5, a = 0, b = 4, ab = 2, c = -3, ac = 0, bc = -1, abc = -2$, show that the scheme you have set up does indeed confound AC with blocks.

12.2   Assuming three replications of the 2^3 experiment in Problem 12.1 can be run but each one must be run in two blocks of four, set up a scheme for the complete confounding of ABC in all three replicates and show its ANOVA layout.

12.3   Repeat Problem 12.2 using partial confounding of AB, AC, and BC.

12.4   Data involving four replications of the 2^3 factorial described in the foregoing problems and in Section 12.3 gave the following numerical results.

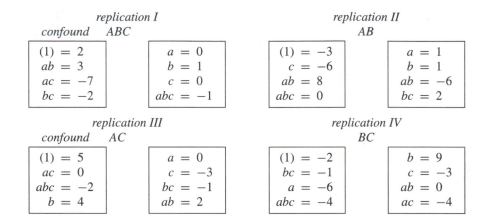

replication I
confound   ABC

(1) = 2	a = 0
ab = 3	b = 1
ac = −7	c = 0
bc = −2	abc = −1

replication II
AB

(1) = −3	a = 1
c = −6	b = 1
ab = 8	ab = −6
abc = 0	bc = 2

replication III
confound   AC

(1) = 5	a = 0
ac = 0	c = −3
abc = −2	bc = −1
b = 4	ab = 2

replication IV
BC

(1) = −2	b = 9
bc = −1	c = −3
a = −6	ab = 0
abc = −4	ac = −4

Analyze these data and state your conclusions.

12.5   A 2^3 factorial is to be used to study three factors and only four treatments can be run on each of two days. On the first day the following data are collected:

$$(1) = 2 \qquad a = 1 \qquad bc = 3 \qquad abc = 4$$

and on the next day:

$$c = 1 \qquad ac = 4 \qquad b = 6 \qquad ab = 0$$

By examining the variation between the two days and your knowledge of a 2^f experiment, find out which term in the model (main effect or interaction) is confounded with days.

12.6   For the data of Problem 9.6 consider running the 16 treatment combinations in four blocks of four (two observations per treatment). Confound ACD, BCD, and AB and then determine the block compositions.

12.7   From the results of Problem 9.6, show that the blocks are confounded with the interactions of Problem 12.6.

12.8   For Problem 10.11 consider this experiment as run in three blocks of nine with two replications for each treatment combination. Confound ABC (2 df) where A is surface thickness, B is base thickness, and C is subbase thickness, and determine the design. Also use the numerical results of Problem 10.13 to show that ABC is indeed confounded in your design.

12.9 Set up a scheme for confounding a 2^4 factorial in two blocks of eight each, confounding ABD with blocks.

12.10 Repeat Problem 12.9 and confound BCD.

12.11 Work out a confounding scheme for a 2^5 factorial, confounding in four blocks of eight each. Show the outline of an ANOVA table.

12.12 Repeat Problem 12.11 in four replications confounding different interactions in each replication. Show the ANOVA table outline.

12.13 Work out the confounding of a 3^4 factorial in three blocks of 27 each.

12.14 Repeat Problem 12.13 in nine blocks of nine each.

12.15 In Example 12.5, since ABC is completely confounded, the analysis could be easily run on a computer or by using the Yates method on sections A, B, C and general methods for replications and blocks. Verify the results given in Table 12.11.

12.16 Yates method can be used to obtain sums of squares for Example 12.6 if it is applied with care. Try to verify Table 12.14 results using a modified Yates procedure.

12.17 A systematic test was made to determine the effects on coil breakdown voltage of the following six factors, each at two levels as indicated:

a. Firing furnace	Number 1 or 3
b. Firing temperature	1650 or 1700°C
c. Gas humidification	No or yes
d. Coil outer diameter	< 0.0300 inch or > 0.0305 inch
e. Artificial chipping	No or yes
f. Sleeve	Number 1 or 2

Since all 64 experiments could not be performed under the same conditions, the whole experiment was run in eight subgroups of eight experiments. Set up a reasonable confounding scheme for this problem and outline its ANOVA.

12.18 An experiment was run on the effects of several annealing variables on the magnetic characteristic of a metal. The following factors were to be considered.

a. Temperature	1375, 1450, 1525, 1600°C
b. Time in hours	2, 4
c. Exogas ratio	6.5:1, 8.5:1
d. Dew point	10, 30°C
e. Cooling rate (hours)	4, 8
f. Core material	A, B

Since all factors are at two levels or multiples of two levels, outline a scheme for this experiment in two blocks of 64 observations.

12.19 Outline a Yates method for handling the data of Problem 12.18.

12.20 If X is the surface rating of a steel sample and material is tested from eight heats, two chemical treatments, and three positions on the ingot, the resulting factorial is an $8 \times 2 \times 3$ of $2^3 \times 2 \times 3$ or $2^4 \times 3$, which is called a mixed factorial in the form $2^m \cdot 3^n$. Set up an ANOVA table for such a problem treating the eight heats as 2^3 pseudofactors.

12.21 The materials in Problem 12.20 must be fired in furnaces that can handle only 24 steel samples at a time. Devise a confounding scheme for the experiment under these conditions.

12.22 Complete the analysis of Problem 12.21 for the following data.

| | Position | | | | | |
| | 1 | | 2 | | 3 | |
Heat	Chemical 1	Treatment 2	Chemical 1	Treatment 2	Chemical 1	Treatment 2
1	1	0	0	0	0	2
2	−3	2	−1	−2	−2	−2
3	1	2	0	3	0	4
4	0	0	−2	1	0	1
5	−3	1	0	2	−2	0
6	−2	1	−3	2	0	0
7	3	3	3	3	1	−1
8	1	1	0	5	1	2

12.23 Consider a 2^5 factorial experiment with all fixed levels to be run in four blocks of eight treatment cominations each. The whole experiment can be replicated in a total of three replications. The following interactions are to be confounded in each replication.

Replication I—$ABC, CDE, ABDE$

Replication II—$BCDE, ACD, ABE$

Replication III—$ABCD, BCE, ADE$

Determine the block compositions for each block in each replication and outline the ANOVA for this problem.

12.24 Conduct residual analyses on the two models considered in Example 12.4. Would you recommend either model? Explain.

12.25 For Example 12.5, analyze the residuals and comment.

12.26 Consider Example 12.6. Thoroughly analyze the residuals.

12.27 For a 2^4 factorial to be run in four blocks of four:
   a. Determine the blocks associated with the defining relation $I = ABD = BCD = AC$.
   b. Show that the defining equations $L_1 = x_1 + x_2 + x_4$ and $L_2 = x_2 + x_3 + x_4$ yield the same blocks as the defining equations $L_2 = x_2 + x_3 + x_4$ and $L_3 = x_1 + x_3$.

12.28 Design a 2^3 factorial in two blocks of four that is replicated three times with AB confounded in the first replicate and ABC confounded in the second and third replicates. Prepare a data layout and a partial ANOVA summary. Comment.

12.29 A study of the effects of plastic (virgin and regrind), screw speed (slow and fast), mold temperature (low and high), and mold pressure (low and high) on the flexibility of a molded headlamp component is to be conducted. Only eight treatment combinations can be run under similar conditions over a short period of time. Design an experiment, including a reasonable number of replications, that will provide good tests on the main effects and two-way interactions. Show an analysis layout for which three- and four-way interactions are assumed to be negligible over the ranges of the factors.

12.30 A 3^4 factorial is to be run in nine blocks of nine treatment combinations each with AC^2D and ABC confounded in blocks.
   a. Show that $I = ABC = AC^2D = AB^2D^2 = BC^2D^2$ is the defining relation.
      Hint: If α and β are confounded, so are $\alpha\beta$ and $\alpha\beta^2$.
   b. Determine defining equations for this experiment.
      Hint: The treatment combinations are 0000, 0001, ..., 2222. Only two defining equations are required, since each can attain one of three values (0, 1, or 2), giving $3 \times 3 = 9$ pairs of values.
   c. Determine the 9 blocks.

12.31 A 3^4 factorial is to be run in three blocks of 27 treatment combinations each.
   a. Using $I = AB^2CD$ as the defining relation, write the defining equation.
   b. Determine the principal block.
   c. Using the principal block, determine the other two blocks.

Chapter 13

 FRACTORIAL REPLICATION

13.1 INTRODUCTION

As the number of factors to be considered in a factorial experiment increases, the number of treatment combinations increases very rapidly. This can be seen with a 2^f factorial where $f = 5$ requires 32 experiments for one replication, $f = 6$ requires 64, $f = 7$ requires 128, and so on. Along with this increase in the amount of experimentation comes an increase in the number of high-order interactions. Some of these high-order interactions may be used as error, since those above second order (three-way) would be difficult to explain if found significant. Table 13.1 gives some idea of the number of main effects, first-order, second-order, . . . , interactions that can be recovered if a complete 2^f factorial can be run.

Considering $f = 7$, there will be 7 df for the seven main effects, 21 df for the 21 first-order interactions, and 35 df for the second-order interactions, leaving

$$35 + 21 + 7 + 1 = 64 \text{ df}$$

for an error estimate, assuming no blocking and no interactions above second order. Even if this experiment were confounded in blocks, there is still a large number of degrees of freedom for the error estimate. In such cases, it may not be economical to run a whole replicate of 128 observations. Nearly as much information can be gleaned from half as many observations. When only a fraction of a replicate is run, the design is called a *fractional replication* or a *fractional factorial*. A replicate that utilizes $(1/2)^k$ of the treatment combinations for a 2^f factorial is often called a 2^{f-k} *fractional factorial*, where k denotes an integer between 1 and f.

TABLE 13.1
Buildup of 2^f Factorial Effects

		Main	Order of Interaction						
f	2^f	Effect	1st	2nd	3rd	4th	5th	6th	7th
5	34	5	10	10	5	1			
6	64	6	15	20	15	6	1		
7	128	7	21	35	35	21	7	1	
8	256	8	28	56	70	56	28	8	1

For running a fractional replication, the methods of Chapter 12 are used to determine a confounding scheme such that the number of treatment combinations in a block is within the economic range of the experimenter. If, for example, the experimenter can run 60 or 70 experiments and is interested in seven factors, each at two levels, a half-replicate of a 2^7 factorial (i.e., a 2^{7-1} fractional factorial) can be used. The complete $2^7 = 128$ experiment is laid out in two blocks of 64 by confounding some high-order interaction. Then only one of these two blocks is run: a coin is tossed to decide which block will be run.

13.2 ALIASES

To see how a fractional replication is run, consider a simple case. Three factors are of interest, each at two levels. The experimenter cannot afford $2^3 = 8$ experiments, but will, however, settle for four. This suggests a half-replicate of a 2^3 factorial (i.e., a 2^{3-1} fractional factorial). Suppose ABC is confounded with blocks and

$$I = ABC$$

is the defining relation. The two blocks are then

| block 1 | (1) | ab | bc | ac |
| block 2 | a | b | c | abc |

from Example 12.2. A coin is flipped and the decision is made to run only block 2. What information can be gleaned from block 2, and what information is lost when only half of the experiment is run?

Referring to Table 9.5, which shows the proper coefficients ($+1$ or -1) for the treatment combinations in a 2^3 factorial to give the effects desired, we prepare another orthogonal table (Table 13.2) that contains only those treatment combinations in block 2. Notice that the ABC column has all plus signs, which is a result of confounding blocks with ABC. If the table containing only those treatment combinations in block 1 were prepared, the ABC column would contain only minus signs.

From Table 13.2, we see that the contrasts associated with A and BC are identical. That is,

TABLE 13.2
Orthogonal Table for a Half-Replicate of a 2^3 Factorial Experiment

Treatment Combination	Linear Contrast Coefficients							Total
	A	B	C	AB	AC	BC	ABC	
a	$+1$	-1	-1	-1	-1	$+1$	$+1$	T_1
b	-1	$+1$	-1	-1	$+1$	-1	$+1$	T_2
c	-1	-1	$+1$	$+1$	-1	-1	$+1$	T_3
abc	$+1$	$+1$	$+1$	$+1$	$+1$	$+1$	$+1$	T_4

$$C_A = T_1 - T_2 - T_3 + T_4 = C_{BC}$$

so we cannot distinguish between A and BC in block 2. Two or more effects for which the absolute values of the linear expressions of their contrasts are identical are called *aliases*. We cannot tell them apart. Since the B and AC columns of Table 13.2 are identical, as are the C and AB columns, this design also aliases B with AC and C with AB. Because of this confounding when only a fraction of the experiment is run, we must check the aliases and be reasonably sure they are not both present if such a design is to be of value.

A partial ANOVA summary for the half-replication of Table 13.2 would be as shown in Table 13.3. This would hardly be a practical experiment unless the experimenter is sure that no first-order interactions exist and has some external source of error to use in testing A, B, and C. The real advantage of such fractionating will be seen on designs with larger f values.

If block 1 were run for the experiment instead of block 2, orthogonal Table 13.4 could be used. In this case, $A = -BC$, $B = -AC$, and $C = -AB$, since $C_A = -T_5 + T_6 + T_7 - T_8 = -C_{BC}$, $C_B = -T_5 + T_6 - T_7 + T_8 = -C_{AC}$, and $C_C = -T_5 - T_6 + T_7 + T_8 = -C_{AB}$. The definition given earlier still holds when one effect is the alias of another if the absolute values of the linear expressions for the two contrasts in the treatment totals are equal. Therefore, the sums of squares of aliases are equal.

A quick way to find the aliases of an effect in a fractional replication of a 2^f factorial experiment is to multiply the effect by the terms in the defining relation: I is treated like an identity for multiplication, as is the square of any letter, and exponents are determined using modulo 2 arithmetic. The results will be aliases of the effect used as a multiplier. This simple rule works for any fractional replication of a 2^f factorial. It works also with a slight modification for a 3^f factorial run as a fractional replication.

TABLE 13.3
Partial ANOVA Summary for a
Half-Replication of a 2^3 Factorial

Source	df
A (or BC)	1
B (or AC)	1
C (or AB)	1
Total	3

TABLE 13.4
Orthogonal Table for a Second Half-Replicate of a 2^3 Factorial Experiment

Treatment Combination	Linear Contrast Coefficients							Total
	A	B	C	AB	AC	BC	ABC	
(1)	-1	-1	-1	$+1$	$+1$	$+1$	-1	T_5
ab	$+1$	$+1$	-1	$+1$	-1	-1	-1	T_6
ac	$+1$	-1	$+1$	-1	$+1$	-1	-1	T_7
bc	-1	$+1$	$+1$	-1	-1	$+1$	-1	T_8

■ **Example 13.1 (Aliases for a Half-Replicate of a 2^3 Factorial Experiment)**

For a half-replicate of a 2^3 factorial experiment, the defining relation is $I = ABC$. Beginning with this relation and multiplying by factors or interactions not previously listed gives the equations

$$I = ABC$$

$$A = A(I) = A(ABC) = A^2BC = BC$$

$$B = B(I) = B(ABC) = AB^2C = AC$$

$$C = C(I) = C(ABC) = ABC^2 = AB$$

Thus, A is aliased with BC, B is aliased with AC, and C is aliased with AB.

Using Computer Software to Design an Experiment

Computer programs such as JMP and Minitab can be used to determine the design matrix, defining relation, and alias structure for a fractional factorial. To do so with Minitab, select **Stat ▶ DOE ▶ Fractional Factorial . . .** and enter the number of factors, blocks, and runs in the appropriate boxes. Options can be selected that have Minitab output the alias structure and data matrix. When using JMP, select **Tables ▶ Design Experiment . . .** and choose **2-level Design** from the **Choose Design Type** menu.

Minitab outputs for a 2^{3-1} fractional factorial are given in Table 13.5. The design consists of the same treatment combinations as those in Table 13.2. The treatment combinations were output in random order, so we should run the experiment in the order a, c, abc, b.

TABLE 13.5
Minitab Design for a 2^{3-1} Fractional Factorial

```
Factorial Design

Fractional Factorial Design
Factors: 3              Design: 3, 4        Resolution: III
Runs: 4                 Replicates: 1       Fraction: 1/2
Blocks: none            Center points: 0
*** NOTE *** Some main effects are confounded with two-way interactions
Design Generators: C = AB
Defining Relation: I = ABC

Alias Structure
I + ABC
A + BC
B + AC
C + AB

Data Matrix (randomized)
  Run  A  B  C
    1  +  -  -
    2  -  -  +
    3  +  +  +
    4  -  +  -
```

13.3 2^f FRACTIONAL REPLICATIONS

As an example, consider the problem suggested in the introduction. An experimenter wishes to study the effect of seven factors, each at two levels, but cannot afford to run all 128 experiments and so will settle for 64, or a half-replicate of a 2^7 factorial. If the highest order interaction is confounded with blocks, the defining relation is

$$I = ABCDEFG$$

The two blocks are found by placing (1) and all pairs, quadruples, and sextuples of the seven letters in one block and the single letters, triples, quintuples, and one septuple of the seven letters in the other block. One of the two blocks is chosen at random and run.

Before carrying out this experiment, the experimenter should check on the aliases. Multiplying the defining relation by A, we find that

$$A = A(I) = A(ABCDEFG) = BCDEFG$$

So, A is aliased with $BCDEFG$, and all main effects are likewise aliased with fifth-order interactions. Multiplying the defining relation by AB gives

$$AB = AB(I) = AB(ABCDEFG) = CDEFG$$

So, AB is aliased with $CDEFG$, a fourth-order interaction, and all first-order interactions are aliased with fourth-order interactions. Further, a second-order interaction such as ABC is aliased with a third-order interaction as

$$ABC = ABC(I) = ABC(ABCDEFG) = DEFG$$

If the second- and third-order interactions are taken as error, the analysis will be as in Table 13.6.

This is a very practical design because, if all higher order interactions are assumed to be zero, there are good tests on all main effects and first-order interactions. The degrees of freedom for each test would be 1 and 35. If the experimenter has reason to suspect a second-order interaction, it could be omitted from the error estimate and there would still

TABLE 13.6
A Half Replication of a 2^7 Factorial

Source	df	
Main effects $A, B, \ldots, G$ (or fifth order)	1 each for 7	
First-order interaction AB, AC (or fourth order)	1 each for 21	
Second-order interaction ABC, ABD (or third order)	1 each for 35	use as error
Total	63	

be sufficient degrees of freedom for the error estimate. The analysis of such an experiment follows the methods given in Chapter 9.

If this same experimenter is further restricted and can afford to run only 32 experiments, a one-fourth replication of a 2^7 factorial might be tried. Here 3 df must be confounded with blocks. If the two fourth-order interactions $ABCDE$ and $CDEFG$ are confounded with blocks, the third-order interaction $(ABCDE)(CDEFG) = ABFG$ is automatically confounded. Thus, the defining relation is

$$I = ABCDE = CDEFG = ABFG$$

which confounds 3 df with the four blocks. The principal block, which consists of all treatment combinations sharing an even number of letters with both $ABCDE$ and $CDEFG$, is

(1)	ab	cd	ce	de	fg	acf	acg
adf	adg	aef	aeg	bcf	bcg	bdf	bdg
bef	beg	abcd	abce	abde	abfg	cdfg	cefg
defg	acdef	acdeg	bcdef	bcdeg	abcdfg	abcefg	abdefg

In this design, if only one of the four blocks of 32 observations is run, each effect has three aliases. These are as follows:

$A = BCDE = ACDEFG = BFG$	$CD = ABE = EFG = ABCDFG$
$B = ACDE = BCDEFG = AFG$	$CE = ABD = DFG = ABCEFG$
$C = ABDE = DEFG = ABCFG$	$CF = ABDEF = DEG = ABCG$
$D = ABCE = CEFG = ABDFG$	$CG = ABDEG = DEF = ABCF$
$E = ABCD = CDFG = ABEFG$	$DE = ABC = CFG = ABDEFG$
$F = ABCDEF = CDEG = ABG$	$DF = ABCEF = CEG = ABDG$
$G = ABCDEG = CDEF = ABF$	$DG = ABCEG = CEF = ABDF$
$AB = CDE = ABCDEFG = FG$	$EF = ABCDF = CDG = ABEG$
$AC = BDE = ADEFG = BCFG$	$EG = ABCDG = CDF = ABEF$
$AD = BCE = ACEFG = BDFG$	$ACF = BDEF = ADEG = BCG$
$AE = BCD = ACDFG = BEFG$	$ACG = BDEG = ADEF = BCF$
$AF = BCDEF = ACDEG = BG$	$ADF = BCEF = ACEG = BDG$
$AG = BCDEG = ACDEF = BF$	$ADG = BCEG = ACEF = BDF$
$BC = ADE = BDEFG = ACFG$	$AEG = BCDG = ACDF = BEF$
$BD = ACE = BCEFG = ADFG$	$BEG = ACDG = BCDF = AEF$
$BE = ACD = BCDFG = AEFG$	

This is quite a formidable list of aliases; but when only one block of 32 is run, there are 31 degrees of freedom within the block. The preceding list accounts for these 31 df. If, in this design, all second-order (three-way) and higher interactions can be considered to be negligible, the main effects are all clear of first-order interactions. The first-order interactions AB, AF, and AG are aliased with FG, BG, and BF, respectively. If the choice of factors A, B, F, and G can be made with the assurance that FG, BG, and BF are negligible, one would have clear information on the interactions AB, AF, and AG. If that choice can be made with the assurance that they are either negligible or quite small, those six interactions (with 3 df) could be pooled for error. The remaining 15 first-order interactions are all clear of other interactions except second order or higher. There are also 6 degrees of freedom left over for error involving only second-order interactions or higher. An analysis might be that shown in Table 13.7.

Tests can be made with 1 and 6 df or 1 and 9 df if the last two lines of Table 13.7 can be pooled for error. This is a fairly good design when it is necessary to run only a one-fourth replication of a 2^7 factorial.

One could also screen for active effects by preparing a normal probability plot of the estimates of the 31 effects $A, B, \ldots, AEF$, and BEG. The procedure is the same as that in Section 9.6.

TABLE 13.7
A One-Fourth Replication of a 2^7 Factorial

Source	df
Main effects $A, B, \ldots, G$	1 each for 7
First-order interaction $AC, AD, \ldots$	1 each for 15
AB (or FG), AF (or BG), AG (or BF), $\ldots$	1 each for 3
Second-order interaction or higher ($ACF, \ldots$)	1 each for 6
Total	31

■ **Example 13.2 (A $2^{5\text{-}1}$ Fractional Factorial)**[1]

In an experiment on the total yield of oil per batch of peanuts, Kilgo [17] considered the following five factors as possible sources of significant variation: A, CO_2 pressure; B, CO_2 temperature; C, peanut moisture; D, CO_2 flow rate; and E, peanut particle size. If each factor were set at two levels near the extremes of reasonable operating conditions, $2^5 = 32$ experiments would be required to run one complete replication. To reduce experimental costs, $ABCDE$ was confounded with blocks and the principal block was run as a half-replicate of this factorial. The resulting design and alias structure are given in Table 13.8.

When $ABCDE$ is confounded with blocks, the treatment combinations sharing 0, 2, or 4 letters with $ABCDE$ form the principal block {(1), ab, ac, ad, ae, bc, bd, be, cd, ce, de, $abcd$, $abce$, $abde$, $acde$, $bcde$}. This block contains the treatment combinations for which $ABCDE$ is -1. Thus, the defining relation for this half-replicate is $I = -ABCDE$, as reported in Table 13.8. By multiplication, one obtains the alias structure summarized in that table, where $A - BCDE$ indicates that A is aliased with $-BCDE$. Notice that each main effect is confounded with a four-way interaction, and each two-way interaction is confounded with a three-way interaction.

Experimental results, obtained in random order, were (1) $= 63$, $ae = 21$, $be = 36$, $ab = 99$, $ce = 24$, $ac = 66$, $bc = 71$, $abce = 54$, $de = 23$, $ad = 74$, $bd = 80$, $abde = 33$, $cd = 63$, $acde = 21$, $bcde = 44$, and $abcd = 96$. Entering these in a

---

[1]Used by courtesy of Marcel Dekker, Inc.

Minitab worksheet, selecting **Stat ▸ DOE ▸ Fit Factorial Model . . .** , and entering the model

$$Y_{ijklm} = \mu + A_i + B_j + AB_{ij} + C_k + AC_{ik} + BC_{jk} + D_l + AD_{il} + BD_{jl}$$
$$+ CD_{kl} + E_m + AE_{im} + BE_{jm} + CE_{km} + DE_{lm} + \varepsilon_{ijklm}$$

we can produce the outputs in Table 13.9.

Each effect in Table 13.9 is obtained by calculating the contrast and then dividing by $0.5 \times 1 \times 2^{5-1} = 8$, since data are for a half-replicate. This is easily seen by considering AB. Coefficients for the AB contrast are $1, -1, -1, 1, 1, -1, -1, 1, 1, -1, -1, 1, 1, -1, -1$, and 1. Thus,

$$C_{AB} = (1)(63) + (-1)(21) + (-1)(36) + (1)(99) + (1)(24) + (-1)(66)$$
$$+ (-1)(71) + (1)(54) + (1)(23) + (-1)(74) + (-1)(80) + (1)(33)$$
$$+ (1)(63) + (-1)(21) + (-1)(44) + (1)(96) = 42$$

and

$$AB_{eff} = 42/8 = 5.25$$

TABLE 13.8

Minitab Alias Structure and Data Matrix for Example 13.2

```
Factorial Design

Fractional Factorial Design

Factors:     5          Design:      5, 16        Resolution: V
Runs:        16         Replicates:   1           Fraction: 1/2, number 1
Blocks:   none          Center points:   0

Design Generators:  E = -ABCD

Defining Relation:  I = -ABCDE

Alias Structure              Data Matrix
I  - ABCDE           Run A B C D E
A  - BCDE             1  - - - - -
B  - ACDE             2  + - - - +
C  - ABDE             3  - + - - +
D  - ABCE             4  + + - - -
E  - ABCD             5  - - + - +
AB - CDE              6  + - + - -
AC - BDE              7  - + + - -
AD - BCE              8  + + + - +
AE - BCD              9  - - - + +
BC - ADE             10  + - - + -
BD - ACE             11  - + - + -
BE - ACD             12  + + - + +
CD - ABE             13  - - + + -
CE - ABD             14  + - + + +
DE - ABC             15  - + + + +
                     16  + + + + -
```

A normal probability plot of the effects (Figure 13.1) is also included when the **Fit Factorial Model . . .** routine in Minitab is selected. In that plot, all estimated effects except those for B and E fall reasonably close to a line through $(0, 0)$. Assuming that the effects of $ACDE$ and $ABCD$ are negligible, we conclude that B and E significantly affect the variability in the yields.

The average yields at the low and high levels of B are 44.375 and 64.125%, respectively; those for E are 76.5 and 32.0%. It appears that higher average yields can be attained when the CO_2 temperature is near the high level (reported to be 95°C) and the peanut particle size is near the low level (reported to be 1.28 mm). Cost can be considered when levels of the other factors are selected.

TABLE 13.9
Estimated Effects for Example 13.2

```
Estimated Effects and Coefficients for y
Term              Effect                          Coef
Constant                                         54.25
A                   7.50                           3.75
B                  19.75                           9.87
C                   1.25                           0.62
D                  -0.00                          -0.00
E                 -44.50                         -22.25
A*B                 5.25                           2.62
A*C                 1.25                           0.62
A*D                -4.00                          -2.00
A*E                -7.00                          -3.50
B*C                 3.00                           1.50
B*D                -1.75                          -0.87
B*E                -0.25                          -0.13
C*D                 2.25                           1.13
C*E                 6.25                           3.12
D*E                -3.50                          -1.75
```

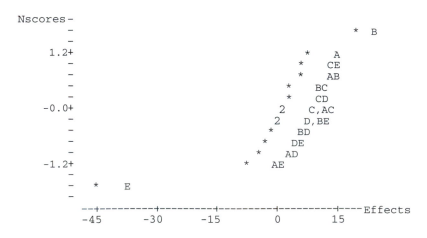

Figure 13.1   Minitab normal probability plot of effects: Example 13.2.

13.3.1 The Yates Method with 2^{f-k} Fractional Factorials

The contrasts required to calculate the effects in Table 13.9 can be obtained by a Yates method as shown in Table 13.10. Values in the *cycle* 1 column are obtained by treating blanks in the Sample Total column as zeros. Notice that numerical values for an effect and its alias are the same in absolute value, since the linear contrasts for each are negatives of one another.

TABLE 13.10
The Yates Method on the Example 13.2 Data

Treatment Combination	Sample Total	cycle 1	cycle 2	cycle 3	cycle 4	cycle 5
(1)	63	63	162	299	612	868 = grand total
a		99	137	313	256	$60 = c_A$
b		66	154	135	58	$158 = c_B$
ab	99	71	159	121	2	$42 = c_{AB}$
c		74	57	31	80	$10 = c_C$
ac	66	80	78	27	78	$10 = c_{AC}$
bc	71	63	56	15	30	$24 = c_{BC}$
abc		96	65	-13	12	$28 = c_{ABC}$
d		21	36	41	-20	$0 = c_D$
ad	74	36	-5	39	30	$-32 = c_{AD}$
bd	80	24	-6	45	-2	$-14 = c_{BD}$
abd		54	33	33	12	$-50 = c_{ABD}$
cd	63	23	-15	25	-4	$18 = c_{AD}$
acd		33	30	5	28	$2 = c_{ACD}$
bcd		21	10	21	14	$56 = c_{BCD}$
$abcd$	96	44	-23	-9	14	$356 = c_{ABCD}$
e		-63	36	-25	14	$-356 = c_E$
ae	21	99	5	5	-14	$-56 = c_{AE}$
be	36	66	6	21	-4	$-2 = c_{BE}$
abe		-71	33	9	-28	$-18 = c_{ABE}$
ce	24	74	15	-41	-2	$50 = c_{CE}$
ace		-80	30	39	-12	$14 = c_{ACE}$
bce		-63	10	45	-20	$32 = c_{BCE}$
$abce$	54	96	23	-33	-30	$0 = c_{ABCE}$
de	23	21	162	-31	30	$-28 = c_{DE}$
ade		-36	-137	27	-12	$-24 = c_{ADE}$
bde		-24	-154	15	80	$-10 = c_{BDE}$
$abde$	33	54	159	13	-78	$-10 = c_{ABDE}$
cde		-23	-57	-299	58	$-42 = c_{ADE}$
$acde$	21	33	78	313	-2	$-158 = c_{ACDE}$
$bcde$	44	21	56	135	612	$-60 = c_{BCDE}$
$abcde$		-44	-65	-121	-256	$-868 = c_{ABCDE}$

13.3.2 Fractional Replicates in Blocks

Sometimes a fractional factorial involves so many treatment combinations that it is necessary to run the fractional replicate in blocks. Suppose, for example, that the experimenters had had to run the half-replicate of Example 13.2 in two blocks of eight units each. The five-factor interaction was confounded with blocks to obtain the half-replicate, but what information is lost when that half-replicate is also blocked? If we have reason to assume that the effect of AB is negligible, we can confound AB with the two blocks from the half-replicate by selecting the eight treatment combinations that share zero or two letters as one block and the remaining eight treatment combinations as the other. Thus, our experiment becomes

block 1	(1)	ab	cd	abcd	ce	abce	de	abde
block 2	ac	bc	ad	bd	ae	be	acde	bcde

with the blocks run in random order and treatments within blocks randomized. AB and its alias $-CDE$ are confounded with these blocks, whereas the remaining alias structure is unchanged.

In addition to being necessary for the reasons just given, running a large fractional replicate in smaller blocks may also serve as a damage control strategy. When mistakes occur with one or more runs within smaller blocks, only the blocks affected by the error must be repeated. If the external conditions change between the initial run of a block and its rerun, the blocking eliminates the effect of those changes.

13.4 PLACKETT–BURMAN DESIGNS

Fractional replicates of 2^f factorials are most useful when there are many variables to be considered and the experimenter wants to identify the one or two factors that have the greatest effect on the response variable. The design of such an experiment is called a *screening design*. In 1946 R. L. Plackett and J. P. Burman [27] developed classes of screening designs that may be used to extract the most important main effects if no interactions are present.

Plackett–Burman designs have been devised for 2^f fractional factorials that enable experimenters to study up to $N - 1$ main effects with only N runs, where N is a multiple of 4. A brief but excellent discussion of these designs is given by Nelson [22].

If N is a power of 2, these designs are like our 2^f fractional factorials. If N is a multiple of 4 and not a power of 2, Plackett–Burman designs indicate the levels of each factor to be used in each run for retrieving main effects. Often one may choose a design with a higher f than an $N - 1$ factorial and use the extra factors as "dummy" factors. The interactions associated with those factors can be combined and used as an error term.

Six of the Plackett–Burman designs for 2^f fractional replicates are summarized in Table 13.11. In each case, main effects are confounded with two-factor and higher interactions. The sequence of $+$ and $-$ signs to the right of the value of N describes the first row of an orthogonal table for the design, with $+1$ and -1 associated with $+$ and $-$, respectively. The second row is found by shifting the first row one place to the right and using the last entry of row 1 as the first entry of row 2. This procedure continues until $N - 1$ rows have been generated. Each entry of the Nth row is set to -1.

The N = 8 Plackett–Burman Design

The Plackett–Burman design using eight runs or treatment combinations is illustrated in Nelson's paper [27]. According to Table 13.11, the first run (row of the orthogonal table) is

TABLE 13.11
Six of the Plackett–Burman Screening Designs

$N = 8$	$+ + + - + - -$
$N = 12$	$+ + - + + + - - - + -$
$N = 16$	$+ + + + - + - + + - - + - - -$
$N = 20$	$+ + - - + + + + - + - + - - - - + + -$
$N = 24$	$+ + + + + - + - + + - - + + - - + - + - - - -$
$N = 32$	$- - - + - + - + + + - + + - - - + + + + + - - + + - + - - +$

$+1, +1, +1, -1, +1, -1, -1$. Shifting that row one position to the right and moving the last entry of row 1 (i.e., -1) to the first position of row 2 gives $-1, +1, +1, +1, -1, +1, -1$ as the second row. Continuing in this fashion produces the first seven rows of Table 13.12. As with all Plackett–Burman designs, each entry of the Nth row is -1. Randomization should be used to determine the order of the experimental runs.

TABLE 13.12
An $N = 8$ Plackett–Burman Design with Seven Factors

Run	tc	Factors							y
		A	B	C	D	E	F	G	
1	abce	+1	+1	+1	−1	+1	−1	−1	—
2	bcdf	−1	+1	+1	+1	−1	+1	−1	—
3	cdeg	−1	−1	+1	+1	+1	−1	+1	—
4	adef	+1	−1	−1	+1	+1	+1	−1	—
5	befg	−1	+1	−1	−1	+1	+1	+1	—
6	acfg	+1	−1	+1	−1	−1	+1	+1	—
7	abdg	+1	+1	−1	+1	−1	−1	+1	—
8	(1)	−1	−1	−1	−1	−1	−1	−1	—
Two-factor aliases		$-BF$	$-AF$	$-AD$	$-AC$	$-AG$	$-AB$	$-AE$	
		$-CD$	$-CG$	$-BG$	$-BE$	$-BD$	$-CE$	$-BC$	
		$-EG$	$-DE$	$-EF$	$-FG$	$-CF$	$-DG$	$-DF$	
Three-factor aliases*		BCE	ACE	ABE	ABG	ABC	ACG	ABD	
		BDG	ADG	AFG	AEF	ADF	ADE	ACF	
		CFG	CDF	BDF	BCF	BFG	BCD	BEF	
		DEF	EFG	DEG	CEG	CDG	BEG	CDE	

*ABF, ACD, AEG, BCG, BDE, CEF, and DFG are confounded with blocks.

■ **Example 13.3 (An $N = 8$ Plackett–Burman Design Experiment with Four Factors)**

In an experiment on slicing yield from pork bellies, four factors were considered as possible sources of significant variation: A, press effect; B, molding temperature; C, slicing temperature; and D, side of the hog. Only eight bellies could be handled at one time, so the experimenters decided to use the $N = 8$ Plackett–Burman design described in Table 13.12. Since there are only four main effects, the last three columns of the table provide information about some of the interactions. By multiplication of columns,

TABLE 13.13
Orthogonal Table with Data for Example 13.3

Run	tc	A	B	C	D	E = −BD	F = −AB	G = −BC	y
						Factors			
3	abc	+1	+1	+1	−1	+1	−1	−1	83.06
1	bcd	−1	+1	+1	+1	−1	+1	−1	83.16
6	cd	−1	−1	+1	+1	+1	−1	+1	92.21
8	ad	+1	−1	−1	+1	+1	+1	−1	98.26
2	be	−1	+1	−1	−1	+1	+1	+1	84.77
4	ac	+1	−1	+1	−1	−1	+1	+1	88.32
5	abd	+1	+1	−1	+1	−1	−1	+1	93.19
7	(1)	−1	−1	−1	−1	−1	−1	−1	97.18
Value of contrast		5.51	−31.78	−26.63	13.49	3.55	−11.13	−3.17	720.15
Estimated effect		1.38	−7.95	−6.66	3.37	2.78	0.79	0.89	
Alias		−CD	−ABCD	−AD	−AC	ABC	BCD	ABD	

we find that columns E, F, and G of Table 13.12 contain the contrast coefficients for $-BD$, $-AB$, and $-BC$, respectively, as indicated in Table 13.13.

The aliases of the column effects in Table 13.13 can also be found by multiplication. For example, the products of corresponding entries in columns C and D are the negatives of the corresponding entries in column A. Thus, A is aliased with $-CD$. By proceeding in a similar fashion, one can obtain the other aliases noted in Table 13.13.

The effect of ACD is not among the aliases and factors listed in Table 13.13; if it were added to that table, all entries of an ACD column would be -1. Thus, $I = -ACD$ is the defining contrast for this design and ACD is confounded in blocks.

The yields (in percentages) obtained by the experimenters are included in the y column of Table 13.13. The observed value of the contrast associated with A (or $-CD$) is

$$c_A = (1)(83.06) + (-1)(83.16) + (-1)(92.21) + (1)(98.26) + (-1)(84.77)$$
$$+ (1)(88.32) + (1)(93.19) + (-1)(97.18) = 5.51$$

Since this is a half-replicate of a 2^4 factorial, the estimated effect of A is found by dividing c_A by $0.5 \times 1 \times 2^{4-1} = 4$. Thus,

$$A_{eff} = 5.51/4 = 1.3775 \approx 1.38$$

Proceeding in the same manner, we obtain the seven estimated effects of Table 13.13. These effects and the alias structure can also be obtained using the **Fit Factorial Model . . .** procedure in Minitab, as shown in Table 13.14.

A normal probability plot of the estimated effects (Figure 13.2) is included when Minitab's **Fit Factorial Model . . .** procedure is selected. In that plot, all estimated effects except those for B and C fall reasonably close to a line. Since B is aliased with $-ABCD$, we assume that $ABCD$ is negligible and conclude that B (molding temperature) significantly affects the variability in yield. However, C (slicing temperature) is aliased with $-AD$, so C and/or AD would seem to significantly affect the variability in yield.

At a later date, experimenters were able to run a complete 2^4 factorial to further investigate the effects of A, B, C, and D. That experiment showed that B, C, D, and AD were significant, so we see how a screening design can extract the most important factors.

TABLE 13.14
Minitab Summary of Estimated Effects and Aliases: Example 13.3

```
Fractional Factorial Fit
Estimated Effects and Coefficients for y                  Alias Structure

Term            Effect          Coef                      I - A*C*D
Constant                        90.019                    A - C*D
A                1.378           0.689                     B - A*B*C*D
B               -7.948          -3.974                     C - A*D
C               -6.663          -3.331                     D - A*C
D                3.373           1.686                     A*B - B*C*D
A*B              2.782           1.391                     B*C - A*B*D
B*C              0.793           0.396                     B*D - A*B*C
B*D              0.888           0.444
```

```
    Nscores-
           -                                                        *  D
           =
     1.0+
           -                                                 *      AB
           -
           -                                          *      A
           -
    -0.0+                                           *      BD
           -
           -                                     *      BC
           -
           -             *    C
    -1.0+
           -
           - *     B
           --+---------+---------+---------+---------+---------+--- Effects
         -7.5      -5.0      -2.5       0.0       2.5
```

Figure 13.2   Minitab normal probability plot of effects: Example 13.3.

The Geometric Plackett–Burman Designs

There are two categories of Plackett–Burman designs—geometric and nongeometric. The *geometric designs* with factors at two levels each contain run sizes of $2^3 = 8$, $2^4 = 16$, $2^5 = 32$, and so forth. When the design tables associated with these designs are used with k factors, where $k \leq N - 1$ and $2^k > N$, the resulting design is a fractional factorial. In Example 13.3, $k = 4$, $N = 8$, and the design is a 2^{4-1} fractional factorial.

Consider the $N = 8$ Plackett–Burman design. When a factor, say A, shows a very large effect, we could just be observing the effect of a very large BF, CD, or EG interaction. To determine whether this is the case, we can multiply each column of the design matrix by -1 and conduct a second experiment using the runs for this second design, which is called a *reflected design*. An analysis of the combined results (the 16 runs of the original and reflected designs) will give an estimate of the effect of A that is free of two-factor interactions.

The Nongeometric Plackett–Burman Designs

Nongeometric Plackett–Burman designs with factors at two levels each contain run sizes of $3 \times 4 = 12, 5 \times 4 = 20, 6 \times 4 = 24, 7 \times 4 = 28, 9 \times 4 = 36$, and so forth. Such designs are not fractional factorials.

The alias structures of these nongeometric designs are much more complex than those of the geometric designs. When a main effect is assigned to each column of the design table, each main effect is partially confounded with every two-factor interaction that does not involve that main effect. Thus, an active two-factor interaction may bias the estimates of a number of the main effects. For this reason, the results of such a design should be interpreted with extreme caution.

13.5 DESIGN RESOLUTION

In Table 13.5, the Minitab summary for Example 13.1 includes the statement **Resolution: III** for the 2^{3-1} fractional factorial with defining relation $I = ABC$, whereas the summary for the 2^{5-1} fractional factorial of Example 13.2 (Table 13.8) includes the statement **Resolution: V**. In general, the design of a 2^{f-k} fractional design is of *resolution R* if no p-factor effect (main or interaction) is aliased with another effect containing fewer than $R - p$ factors. Roman numeral subscripts are usually used to denote the resolution of a 2^{f-k} fractional factorial. Thus, the design of Example 13.1 is a 2^{3-1}_{III} design and that of Example 13.2 is a 2^{5-1}_{V} design.

The defining relation is of particular value when it is necessary to determine the resolution of a 2^{f-k} fractional design. The resolution is the length of the shortest sequence of letters (other than I) in the defining relation. The most common resolutions are III, IV, and V.

Resolution III Designs

A 2^{f-k} fractional factorial design is of *resolution III* when no main effect is aliased with another main effect, but each main effect is aliased with a two-factor interaction. Thus, in the 2^{3-1}_{III} design of Example 13.1, no main effect ($p = 1$) is aliased with another effect with fewer than $3 - 1 = 2$ factors—that is, with other main effects. This is easily seen because the defining contrast $I = ABC$ confounds ABC with blocks and determines the aliases $A = BC, B = AC$, and $C = AB$.

■ **Example 13.4 (Design Resolution of a Plackett–Burman N = 8 Design with Seven Factors)**

The Plackett–Burman design with seven factors (Table 13.12) is a 2^{7-4} fractional factorial. That is, it is a one-sixteenth replicate of a 2^7 factorial. As Table 13.12

indicates, each main effect is aliased with three two-factor interactions. Therefore, we have a 2_{III}^{7-4} design.

To view this another way, we shall determine the defining contrast. The product of columns A and B gives $F = -AB$; the product of columns A and C gives $D = -AC$; the product of columns B and C gives $G = -BC$; and the product of columns A, B, and C gives $E = ABC$. When the contrasts $-ABF$, $-ACD$, $-BCG$, and $ABCE$ are used to generate 16 blocks of eight for the 2^7 factorial, the defining relation is found to be

$$I = -ABF = -ACD = -BCG = ABCE = BCDF$$
$$= ACFG = -CEF = ABDG = -BDE = -AEF$$
$$= -DFG = ADEF = BEFG = CDEG = ABCDEFG$$

Since the shortest sequence of letters (apart from I) is of length 3, we have a resolution III design.

Resolution IV Designs

A 2^{f-k} fractional factorial design is of *resolution IV* when no main effect is aliased with another main effect or a two-factor interaction, but some two-factor interactions are aliased with other two-factor interactions. The 2^{7-4} fractional factorial considered in Section 13.3 is of resolution IV, since all main effects are confounded with three-factor and higher interactions but AB, AF, and AG are confounded with FG, BG, and FB, respectively. This is also evident from the defining relation $I = ABCDE = CDEFG = ABFG$, since the shortest sequence of letters (apart from I) in that relation is of length 4.

■ **Example 13.5 (A 2_{IV}^{4-1} Design)**

For a half-replicate of a 2^4 factorial with factors A, B, C, and D, one usually confounds the four-way interaction with blocks, citing $I = ABCD$ as the defining relation. Here $A = A(I) = A(ABCD) = BCD$ and so on, so that all main effects are aliased with three-way effects. But $AB = AB(I) = (AB)(ABCD) = CD$, and so on, so that all two-way interactions are aliased with other two-way interactions. Thus, the design has resolution IV. Also note that the block generator $ABCD$ consists of four letters, indicating that the design has resolution IV.

This is not a bad design if there is no interest in two-way or higher interactions. For formal tests, two-way and three-way effects are usually lumped into an error term with 3 degrees of freedom.

Resolution III designs can be used to obtain resolution IV designs if the orthogonal array of the resolution III design is combined with that of the reflected design. When this is done with a 2_{III}^{f-k} design, the resulting orthogonal array can be used to obtain a 2_{IV}^{f+1-k} design.

■ **Example 13.6 (Obtaining a 2_{IV}^{4-1} Design from a 2_{III}^{3-1} Design and Its Reflection)**

Consider a 2^3 factorial situation with factors A, B, and C. We shall use $I = ABC$ to obtain two half-replicates (blocks): $\{a, b, c, abc\}$ and $\{(1), ab, bc, ac\}$. Table 13.15 is obtained when the orthogonal tables for these replicates (Tables 13.2 and 13.4) are combined. The first three columns of the first four rows in that table form the design matrix for a 2_{III}^{3-1} fractional factorial, whereas the first three columns of the last four rows form the reflected design matrix for that factorial. To obtain a design for a 2_{IV}^{4-1} fractional factorial from the combined designs, we only need to assign the fourth factor (D) to the column associated with blocks in the original designs—that is, the seventh column of contrast coefficients. The resulting design is for a half-replicate of a 2^4 factorial.

   The products of corresponding entries in columns A, B, C, and D of Table 13.15 all equal $+1$. Thus, $ABCD$ is confounded with blocks and $I = ABCD$ for this new design. This informs us that the new design has resolution IV.

TABLE 13.15
An Eight-Run Orthogonal Table for a 2_{IV}^{4-1} Fractional Factorial

Treatment Combination	Linear Contrast Coefficients						D	Total
	A	B	C				D	Total
ad	+1	−1	−1	−1	−1	+1	+1	T_1
bd	−1	+1	−1	−1	+1	−1	+1	T_2
cd	−1	−1	+1	+1	−1	−1	+1	T_3
abcd	+1	+1	+1	+1	+1	+1	+1	T_4
(1)	−1	−1	−1	+1	+1	+1	−1	T_5
ab	+1	+1	−1	+1	−1	−1	−1	T_6
ac	+1	−1	+1	−1	+1	−1	−1	T_7
bc	−1	+1	+1	−1	−1	+1	−1	T_8

Resolution V Designs

A 2^{f-k} fractional factorial design is of *resolution V* when no main effect is aliased with another main effect or a two-factor interaction, and two-factor interactions are aliased with three-factor interactions or higher. Resolution V designs are quite good because one can study all main effects and two-way effects clear of each other. If all interactions of three factors and higher are negligible, these designs permit one to obtain unique estimates of the effects of each factor and two-way interaction.

■ **Example 13.7 (A 2_V^{5-1} Fractional Factorial)**

The 2^{5-1} fractional factorial experiment of Example 13.2 has defining contrast $I = -ABCDE$, indicating that $ABCDE$ is confounded with blocks. Since the sequence

$ABCDE$ contains five letters, the experiment has a resolution V design, as indicated by the Minitab summary of Table 13.8.

Now $A = A(I) = A(-ABCDE) = -BCDE$, so A is aliased with $-BCDE$. Further, $AB = AB(I) = (AB)(ABCDE) = CDE$, indicating that AB is aliased with CDE. Proceeding in like manner, we find that each main effect is aliased with a four-way effect and each two-way effect is aliased with a three-way effect. The complete alias structure is given in the Minitab summary of Table 13.8.

Design Resolutions Greater than V

Section 13.3 opened with a discussion of a half-replicate of a 2^7 factorial. Since the block generator $ABCDEFG$ has seven letters, the design is that of a 2^{7-1}_{VII} fractional factorial—a resolution VII design. Designs such as this, with higher resolutions, are considered to be better, but we often have to settle for designs of resolution III, IV, or V.

Determining the Minimum f for a "Good" Experiment

Another use of the resolution concept is to determine the minimum value of f for a "good" 2^{f-k} fractional factorial. "Good" here usually means a design resolution of IV, V, or higher. To construct a design with resolution R when $k = 1, 2, 3, 4, \ldots$ for one-half, one-fourth, one-eighth, one-sixteenth, $\ldots$ fractional factorials, respectively, we must choose the number of factors to satisfy the inequality

$$f \geq \frac{R(2^k - 1)}{2^{k-1}} \tag{13.1}$$

■ **Example 13.8 [Using Inequality (13.1) to Design Experiments]**

a. Suppose a half-replicate ($k = 1$) of a 2^f factorial is to be designed. From Inequality (13.1), $f \geq R(2 - 1)/2^0 = R$. Thus, the minimum number of factors for half-replicates of resolution III, IV, or V is 3, 4, and 5, respectively.

b. Now consider a one-fourth replicate ($k = 2$) of a 2^f factorial. To achieve the desired resolution, f must satisfy $f \geq R(4 - 1)/2 = 1.5R$.
  1. For a resolution III design, $f \geq (1.5)(3) = 4.5$; so the minimum number of factors is five and the minimum number of experimental runs is $2^5/4 = 8$.
  2. For a resolution IV design, $f \geq (1.5)(4) = 6$; so the minimum number of factors is six and the minimum number of experimental runs is $2^6/4 = 16$.
  3. For a resolution V design, $f \geq (1.5)(5) = 7.5$; so the minimum number of factors is eight and the minimum number of experimental runs is $2^8/4 = 64$.

c. Finally, consider a one-eighth replicate ($k = 3$), which requires that f satisfy the inequality $f \geq R(8 - 1)/4 = 1.75R$.
  1. For a resolution III design, $f \geq (1.75)(3) = 5.25$; so the minimum number of factors is six and the minimum number of experimental runs is $2^6/8 = 8$.

2. For a resolution IV design, $f \geq (1.75)(4) = 7$; so the minimum number of factors is seven and the minimum number of experimental runs is $2^7/8 = 16$.
3. For a resolution V design, $f \geq (1.75)(5) = 8.75$; so the minimum number of factors is nine and the minimum number of experimental runs is $2^9/8 = 64$.

13.6 3^{f-k} FRACTIONAL FACTORIALS

The concepts of the preceding sections may easily be extended to 1/3, 1/9, 1/27, . . . , $1/3^k$ fractions of a 3^f factorial experiment. Usually, 3^{f-k} is used to denote a $1/3^k$ factional replicate of a 3^f factorial.

A 3^{2-1} Fractional Factorial

The simplest 3^f factorial is a 3^2, requiring nine experiments for a complete factorial. If only a fraction of this can be run, we might consider a one-third replication. As we will see, such a design produces a poor experiment, but it serves to illustrate the basics.

The two-factor interaction can be partitioned into two components, AB and AB^2, each of which has 2 degrees of freedom. Confounding AB^2 with blocks gives the defining relation $I = AB^2$ and the defining equation

$$L = x_1 + 2x_2, \ \mathrm{mod} \ 3$$

From Section 12.5, this gives the following three blocks:

block 1 $L = 0$	block 2 $L = 1$	block 3 $L = 2$
00	10	20
12	22	02
21	01	11

Suppose only one of these blocks is run. We can use the defining relation to determine the aliases, but we must remember to use arithmetic modulo 3 on the exponents and bear in mind that the exponent on the first element of an interaction component is never left greater than 1. Since

$$A = A(I) = A(AB^2) = A^2B^2 = (A^2B^2)^2 = (A^2)^2(B^2)^2 = AB$$

and

$$B = B(I) = B(AB^2) = A \cdot B \cdot B^2 = A$$

A, B, and the AB component of the two-factor interaction are aliases. The experiment is poor not only because an estimate of the effect of A cannot be made free of B, but also because no degrees of freedom are left for error. However, the procedure used to obtain this design shows that only slight modification of the procedures in earlier sections is required to determine the aliases in a 3^{f-k} fractional factorial.

A 3^{3-1} Fractional Factorial

Consider a 3^3 factorial in three blocks of nine treatment combinations each. The effects can be broken down into thirteen 2-df effects: $A, B, C, AB, AB^2, AC, AC^2, BC, BC^2, ABC, AB^2C, ABC^2$, and AB^2C^2. Confounding AB^2C with blocks gives

$$I = AB^2C$$

as the defining contrast and

$$L = x_1 + 2x_2 + x_3, \text{mod } 3$$

as the defining equation. Using the methods of Chapter 12, the block composition is

block 1	000	011	110	121	102	212	220	022	201	$L = 0$
block 2	100	111	210	221	202	012	020	122	011	$L = 1$
block 3	200	211	010	021	002	112	120	222	101	$L = 2$

If only one of these blocks is run, the aliases of A are

$$A = A(AB^2C) = A^2B^2C = (A^2B^2C)^2 = ABC^2$$

and

$$A = A(AB^2C)^2 = A(A^2BC^2) = BC^2.$$

Likewise, the aliases of B are

$$B = B(AB^2C) = AC$$

and

$$B = B(AB^2C)^2 = B(A^2BC^2) = A^2B^2C^2 = (A^2B^2C^2)^2 = ABC$$

and the aliases of C are

$$C = C(AB^2C) = AB^2C^2$$

and

$$C = C(AB^2C)^2 = C(A^2BC^2) = A^2B = (A^2B)^2 = AB^2$$

Finally, the aliases of AB are

$$AB = (AB)(AB^2C) = A^2C = (A^2C)^2 = AC^2$$

and

$$AB = (AB)(AB^2C)^2 = (AB)(A^2BC^2) = B^2C^2 = (B^2C^2)^2 = BC$$

A partial ANOVA summary is given in Table 13.16. The design would be somewhat practical if we could consider all interactions negligible and be content with 2 df for error.

TABLE 13.16
Partial ANOVA Summary for a 3^{3-1} Fraction

Source	df
A (or ABC^2 or BC^2)	2
B (or AC or ABC)	2
C (or AB^2 or AB^2C^2)	2
AB (or AC^2 or BC)	2
Total	8

■ **Example 13.9 (A One-Third Replicate of a 3^3 Factorial)**

Torque after preheat in pounds was to be measured on some rubber material with the following factors: A, temperature at 145, 155, and 165°C; B, mix I, II, and III; and C, laboratory 1, 2, and 3. The $3^3 = 27$ experiments seemed too costly, so a one-third replicate was run. Blocks were confounded with AB^2C using $L = x_1 + 2x_2 + x_3$, mod 3. The numerical results for the block

$$L = 1 \quad \boxed{100 \quad 111 \quad 210 \quad 221 \quad 202 \quad 012 \quad 020 \quad 122 \quad 001}$$

are as given in Table 13.17 and the analysis as in Table 13.18. Because of the tiny error term, all main effects are highly significant, even with only 2 degrees of freedom for error.

In this case, the computer analysis is based on the model

$$Y_{ijk} = \mu + A_i + B_j + C_k + \varepsilon_{ijk}$$

which does not account for a particular alias structure. If the results are to be interpreted properly, the alias structure must be determined manually.

TABLE 13.17
Rubber Torque Data for Example 13.9

	Temperature, A (°C)								
	145 Mix, B			155 Mix, B			165 Mix, B		
Lab, C	I	II	III	I	II	III	I	II	III
1			16.8	11.2				9.9	
2	15.8				14.4				17.8
3		17.1				20.5	15.7		

TABLE 13.18
Minitab ANOVA for Example 13.9

```
General Linear Model

Factor      Levels    Values
A               3         1    2    3
B               3         1    2    3
C               3         1    2    3

Analysis of Variance for y

Source       DF    Seq SS     Adj SS     Adj MS        F        P
A             2     6.660      6.660      3.330      52.58    0.019
B             2    38.127     38.127     19.063     301.00    0.003
C             2    40.807     40.807     20.403     322.16    0.003
Error         2     0.127      0.127      0.063
Total         8    85.720
```

Design Resolution

The concept of design resolution extends to fractional replicates of 3^f factorials. For the 3^{3-1} fractional factorial of Example 13.9, A is aliased with BC^2, where BC^2 is a component of the effect of the interaction between B and C. However, main effects are free of each other. Thus, this design has resolution III. That is, Example 13.9 is based on a 3^{3-1}_{III} design.

Comparison of a 3^{3-1} Fractional Replicate and a Latin Square

It might be instructive to examine somewhat further the fractional factorial design of Example 13.9. Begin by constructing a 3×3 table for the nine AB treatment combinations, as shown in Table 13.19. In each cell of that table, list the level of C used in Example 13.9. Careful consideration of the results reveals that Table 13.19 is none other than a Latin square!

In a sense our designs have now come full circle, from a Latin square as two restrictions on a single-factor experiment (Chapter 4) to a Latin square as a one-third replication of a 3^3 factorial. The designs come out the same, but they result from entirely different objectives. In a single-factor experiment the general model

$$Y_{ij} = \mu + \tau_j + \varepsilon_{ij}$$

is partitioned by refining the error term due to restrictions on the randomization to give

TABLE 13.19
The One-Third Replicate Used in Example 13.9 in Terms of Factor C

Factor B	Factor A		
	0	1	2
0	1	0	2
1	2	1	0
2	0	2	1

$$Y_{ijk} = \mu + \tau_i + \beta_j + \gamma_k + \varepsilon'_{ij}$$

where the block and position effects are taken from the original error term ε_{ij}.

In the one-third replication of a 3^3 factorial, three factors of interest comprise the treatments, giving

$$Y_{ijk} = \mu + A_i + B_j + C_k + \varepsilon_{ijk}$$

where A, B, and C are taken from the treatment effect. The two models look alike but are derived from experiments of different types. In both cases we assume that there is no interaction among the factors. This is more reasonable when the factors represent only randomization restrictions such as blocks and positions than when each factor is of vital concern to the experimenter.

By choosing each of the other two blocks in the confounding of AB^2C ($L = 0$ or $L = 2$), two other Latin squares are obtained. If another confounding scheme is chosen, such as ABC^2, ABC, or AB^2C^2, more and different Latin squares may be generated. This shows how it is possible to select a Latin square at random for a particular design.

The need to ensure that no interaction exists before these designs are used on a 3^3 factorial cannot be overstressed. It is not enough to say that there is no interest in the interactions, because, unfortunately, the interactions are badly confounded with main effects as aliases.

13.7 SAS PROGRAMS

The type III sums of squares obtained with the **GLM** procedure can be used to analyze fractional factorials. The methods considered in earlier chapters are used to prepare a command file.

We start by reviewing the data for a 2^{4-1} fractional factorial with ACD confounded with blocks given in Table 13.13. That study (Example 13.3), entailed slicing yield from pork bellies, where A, B, C, and D were press effect, molding temperature, slicing temperature, and side of hog, respectively. The alias structure also is given in Table 13.13. For illustration, we will assume that the interactions are negligible, pooling all interactions that are not aliased with main effects as error. This gives the mathematical model

$$Y_{ijklm} = \mu + A_i + B_j + C_k + D_l + \varepsilon_{m(ijkl)}$$

which can be expressed in a SAS command file as

```
MODEL Y = A B C D;
```

We include this statement in the command file of Table 13.20 to produce the outputs in Figure 13.3, where only the type III sums of squares are shown, to conserve space. The results agree with those obtained in Example 13.3, where a probability plot of the estimated effects is used for analysis.

By default, the **GLM** procedure includes type I and type III sums of squares in the outputs. To conserve space, the type I sums of squares are not included in Figure 13.3 (the two sums of squares are identical for these data).

TABLE 13.20
SAS Command File for Example 13.3

```
OPTIONS LINESIZE=80;
DATA YIELD;
INPUT A B C D Y @@;
CARDS;
2 2 2 1 83.06 1 2 2 2 83.16
1 1 2 2 92.21 2 1 1 2 98.26
1 2 1 1 84.77 2 1 2 1 88.32
2 2 1 2 93.19 1 1 1 1 97.18
;
PROC GLM;
CLASS A B C D;
MODEL Y = A B C D;
```

```
                  GENERAL LINEAR MODELS PROCEDURE
DEPENDENT VARIABLE: Y
                              SUM OF          MEAN
SOURCE            DF          SQUARES         SQUARE      F VALUE
MODEL             4       241.64585000    60.41146250       9.89
ERROR             3        18.31603750     6.10534583     PR > F
CORRECTED TOTAL   7       259.96188750                    0.0448

   R-SQUARE                   ROOT MSE         Y MEAN
   0.929543                 2.47089980    90.01875000

SOURCE            DF       TYPE III SS       F VALUE     PR > F
A                 1         3.79401250          0.62     0.4880
B                 1       126.32551250         20.69     0.0199
C                 1        88.77781250         14.54     0.0317
D                 1        22.74751250          3.73     0.1491
```

Figure 13.3 SAS output for Example 13.3.

13.8 SUMMARY

Continuing the summary at the end of Chapter 12 under 2.b.ii gives:

Experiment	Design	Analysis
II. Two or more factors A. Factorial (crossed)		
	1. Completely randomized	1.
	. . .	. . .
	2. Randomized block	2.
	a. Complete	a.
	. . .	. . .
	b. Incomplete, confounding:	b.

i. Main effect—split plot

    · · ·

ii. Interactions in 2^f and 3^f

  (1) Several replications

$$Y_{ijkq} = \mu + R_i + \beta_j + R\beta_{ij}$$
$$+ A_k + B_q$$
$$+ AB_{kq} + \varepsilon_{ijkq}, \ldots$$

  (2) One replication only

$$Y_{ijk} = \mu + \beta_i + A_j$$
$$+ B_k + AB_{jk}, \ldots$$

  (3) Fractional replication—aliases

    a. in 2^f

    b. in 3^f; $f = 3$: Latin square

$$Y_{ijk} = \mu + A_i + B_j$$
$$+ AB_{ij}, \ldots$$

i. Split-plot ANOVA

ii.

  (1) Factorial ANOVA with replications R_i and blocks β_j or confounded interation

  (2) Factorial ANOVA with blocks β_j or confounded interaction

  (3) Factorial ANOVA with aliases or aliased interaction

13.9 FURTHER READING

Box, G. E. P., W. G. Hunter, and J. S. Hunter, *Statistics for Experimenters: An Introduction to Design, Data Analysis, and Model Building.* New York: John Wiley & Sons, 1978. See Chapter 12: "Fractional Factorial Designs at Two Levels."

Barrentine, L. B., "Illustration of Confounding in Plackett–Burman Designs," *Quality Engineering*, Vol. 9, No. 1, 1996, pp. 11–20.

Bisgaard, S., "Blocking Generators for Small 2^{k-p} Designs," *Journal of Quality Technology*, Vol. 25, No. 1, January 1993, pp. 28–35.

Bisgaard, S., "A Method for Identifying Defining Contrasts for 2^{k-p} Experiments," *Journal of Quality Technology*, Vol. 26, No. 4, October 1994, pp. 288–296.

Draper, N. R., and I. Guttman, "Two-Level Factorial and Fractional Factorial Designs in Blocks of Size Two," *Journal of Quality Technology*, Vol. 29, No. 1, January 1997, pp. 71–75.

Mason, R. L., R. F. Gunst, and J. L. Hess, *Statistical Design and Analysis of Experiments with Applications to Engineering and Science.* New York: John Wiley & Sons, 1989. See pages 177–181 for a discussion of Plackett–Burman screening designs.

Montgomery, D. C., *Design and Analysis of Experiments*, 4th ed. New York: John Wiley & Sons, 1997. See Chapter 9: "Two–Level Fractional Factorial Designs."

Montgomery, D. C., and G. C. Runger, "Foldovers of 2^{k-p} Resolution IV Experimental Designs," *Journal of Quality Technology*, Vol. 28, No. 4, October 1996, pp. 446–450.

Schneider, H., W. J. Kasperski, and L. Weissfeld, "Finding Significant Effects for Unreplicated Fractional Factorials Using the n Smallest Contrasts," *Journal of Quality Technology*, Vol. 25, No. 1, January 1993, pp. 18–27.

PROBLEMS

13.1 Assuming that not all of Problem 12.1 can be run, determine the aliases for a half-replication. Analyze the data for the principal block only.

13.2 Assuming that only a one-fourth replication can be run in Problem 12.6, determine the aliases.

13.3 For Problem 12.8, use a one-third replication and determine the aliases.

13.4   Run an analysis on the block in Problem 13.3 that contains treatment combination 022.

13.5   What would be the aliases in Problem 12.9 if a half-replication were run?

13.6   Determine the aliases in a one-fourth replication of Problem 12.11.

13.7   Determine the aliases in Problem 12.13 where a one-third replication is run.

13.8   If a one-eighth replication of Problem 12.17 is run, what are the aliases? Comment on the ANOVA table.

13.9   What are the aliases and the ANOVA outline if only one block of Problem 12.18 is run?

13.10   If data are taken from one furnace only in Problem 12.21, what are the aliases and the ANOVA outline?

13.11   What are the numerical results and conclusions from just one furnace of Problem 12.22?

13.12   In an experiment on slicing yield from pork bellies, four factors were considered as possible sources of significant variation: A, press effect (normal, high); B, molding temperature ($23, 40°C$); C, slicing temperature ($23, 40°C$); and D, side of the hog (left, right). Only eight bellies could be handled at one time, so the experimenters decided to confound $ABCD$ with blocks and run a half-replicate.
   a. Data for the principal block were $(1) = 95.29$, $ab = 86.58$, $ac = 88.70$, $ad = 96.45$, $bc = 86.79$, $bd = 89.38$, $cd = 90.35$, and $abcd = 89.57$. Analyze these data by constructing an orthogonal table, estimating the effects, and preparing a normal quantile plot of those estimates. Determine the alias structure and comment about your results.
   b. At a later date, the second block was run with results as follows: $a = 89.26$, $b = 88.08$, $c = 85.95$, $d = 90.87$, $abc = 85.68$, $abd = 93.16$, $acd = 95.25$, and $bcd = 83.65$. Analyze these data by constructing an orthogonal table, estimating the effects, and preparing a normal quantile plot of those estimates. Determine the alias structure and comment about your results.

13.13   Take the results of Problems 13.12a and 13.12b and combine into the full factorial in two blocks of eight. Analyze the results and comment.

13.14   a. A four-factor experiment is to be run with each factor at two levels only. However, only eight treatment combinations can be run in each block. If one wishes to confound the ABD interactiion, what are the block compositions?
   b. Outline the ANOVA for just one run of this experiment. Consider all factors fixed and note what you might use as an error term.
   c. If only one block of this experiment can be run, explain how this restriction complicates the experiment and how the ANOVA table must be altered.

13.15   Consider a 2^5 factorial experiment with all fixed levels to be run in four blocks of eight treatment comtinations each. This whole experiment can be replicated in a total of three replications. The following interactions are to be confounded in each replication:

Replication I—$ABC, CDE, ABDE$

Replication II—$BCDE, ACD, ABE$

Replication III—$ABCD, BCE, ADE$

a. Determine the block composition for the principal block in replication I.
b. If only a one-fourth replication were run of replication II, what would be the aliases of BC in this block?
c. Determine the number of degrees of freedom for the replication by three-way interactions in this problem.

13.16 Consider an experiment designed to study the effect of six factors—A, B, C, D, E, and F—each at two levels.
a. Determine at least six entries in the principal block if this experiment is to be run in two blocks of 32 and the highest order interaction is to be confounded.
b. If only one of the two blocks can be run, what are the aliases of the effects A, BC, and ABC?
c. Outline the ANOVA table for part b and indicate what effects might be tested and how they would be tested.

13.17 In determining which of four factors A, B, C, or D might affect the underpuffing of cereal in a plant, the following one-fourth replicate was run.

		A_1		A_2	
		B_1	B_2	B_1	B_2
C_1	D_1			96.6	125.7
	D_2	43.5	22.4		
C_2	D_1	14.1	9.5		
	D_2			28.8	52.5

a. Determine which interaction was confounded in this design.
b. Analyze these data and comment on the results and on the design.

13.18 A company manufacturing engines was concerned about emission characteristics. To try to minimize several undesirable emission variables, the company proposed to build and operate experimental engines varying five factors: throat diameter to be set at three levels, ignition system at three levels, temperature at three levels, velocity of the jet stream at three levels, and timing system at three levels. Since the proposed design would require 243 experiments, it was decided to run a one-third replicate. Set up a confounding scheme for this problem and indicate a few terms in the principal block. Also show an ANOVA outline for the running of one of the three blocks, indicating some aliases. Comment on the design.

13.19 In studying the surface finish of steel, five factors were varied each at two levels: B, slab width; C, percent sulfur; D, tundish temperature; E, casting speed; and F, mold level. The following data were recorded on the number of longitudinal face cracks observed.

Tundish temperature D (°F)	Casting speed E (L/min)	Mold Level F (in.)		< 60		> 62	
				< 0.015	> 0.017	< 0.015	> 0.017
>2845	>59	>10			2	3	
		5–8		0			0
	<54	>10		5			0
		5–8			6	0	
<2840	>59	>10		6			3
		5–8			2	0	
	<54	>10			7	0	
		5–8		1			0

Determine what interacton is confounded here to run this half-replicate. Outline its ANOVA and comment on the design.

13.20 Using the data of Problem 13.19, do the analysis and state your conclusions.

13.21 For Problem 13.6, determine the resolution, explain how you found it, and express the experiment in resolution form.

13.22 For Problem 13.7, determine the resolution, explain how you found it, and express the experiment in resolution form.

13.23 Determine the resolution in Problem 13.8 and comment on your results.

13.24 Consider Problem 13.19. Explain how you might work this problem as a Plackett–Burman design where $N = 8$.

13.25 Using the data of the complete factorial in Problem 9.6, try to check the results using the Plackett–Burman design where $N = 8$. (Use only the first reading in each cell and add some error as you see fit.)

13.26[2] At the time of the experiment of Example 13.2, data were also taken on the amount of oil that dissolved in the carbon dioxide used to extract oil from peanuts. The amount Y, known as the *solubility* of the oil, is measured in milligrams of oil per liter of CO_2. Using the same factors and treatment combinations as those in Example 13.2, Kilgo [17] reports the following results: $(1) = 29.2$, $ae = 23.0$, $be = 37.0$, $ab = 139.7$, $ce = 23.3$, $ac = 38.3$, $bc = 42.6$, $abce = 141.4$, $de = 22.4$, $ad = 37.2$, $bd = 31.3$, $abde = 48.6$, $cd = 22.9$, $acde = 36.2$, $bcde = 33.6$, and $abcd = 172.6$.
  a. Determine the estimated effects and prepare a normal probability plot.
  b. Identify any factors that appear to have strong effects on the solubility. Assuming that higher solubility is desirable, make recommendations.
  c. Do you notice any unusual patterns in the normal probability plot of the estimated effects? If so, suggest possible reasons for such a pattern and discuss how those may affect your conclusions in part b.

13.27 Determine the alias structure for a 2^{5-2} fractional factorial and determine the effects confounded with blocks. What is the resolution of this design?

13.28 Determine the alias structure for an $N = 8$ Plackett–Burman designed experiment with five factors, where the treatment combinations are determined by the first five columns of Table 13.12.

13.29 The defining relation for a one-fourth replicate of six factors is $I = ABCD = ABEF = CDEF$.
  a. Write the defining equations and determine the principal block.
  b. Prepare an orthogonal table for the principal block. Include aliases.
  c. Suppose the principal block cannot be run in a completely random manner, but must be run in four blocks of four treatment combinations each. Which four blocks are determined when ACE and ADE are used as block generators? What other effects are confounded with these blocks?
  d. Would you recommend the design in part c? Explain.

13.30 Determine the Latin squares determined by the other two blocks ($L = 0$ and $L = 2$) for the confounding scheme of Example 13.9.

13.31 A 3^3 factorial is to be run in three blocks of nine treatment combinations each by confounding ABC^2 with blocks.

---

[2] Used by courtesy of Marcel Dekker, Inc.

a. Determine the defining relation and defining equation.
b. Use the defining equation to determine the three blocks.
c. Determine the alias structure for a one-third replicate of this factorial.
d. Prepare a partial ANOVA summary for the one-third replicate.
e. Determine the Latin square associated with the block $L = 1$.

13.32 Five factors (A, resin type; B, hardener type; C, exposure time; D, developing time; and E, cure rate), each at two levels, were studied to determine their effects on the percent of solids (Y) in a chemical process. The experiment was set up as a completely randomized factorial experiment. Smaller values of Y are better.

a. Data for the principal block determined by the defining contrast $I = ABCDE$ follow. Conduct a complete analysis.

tc	A	B	C	D	E	Data				
(1)	1	1	1	1	1	19.0	18.8	19.6	18.3	17.9
ab	2	2	1	1	1	16.6	16.4	17.2	16.8	16.6
ac	2	1	2	1	1	13.3	14.6	14.8	15.0	14.8
ad	2	1	1	2	1	14.0	14.1	13.6	13.1	12.0
ae	2	1	1	1	2	12.6	12.9	13.7	13.1	13.6
bc	1	2	2	1	1	16.9	18.2	21.3	20.8	21.0
bd	1	2	1	2	1	20.2	19.6	19.8	20.1	21.2
be	1	2	1	1	2	20.0	21.0	22.3	18.4	20.6
cd	1	1	2	2	1	20.5	19.6	21.0	21.0	21.4
ce	1	1	2	1	2	24.3	23.2	20.2	24.1	23.8
de	1	1	1	2	2	28.6	24.8	24.9	26.8	27.7
abcd	2	2	2	2	1	16.3	16.9	17.7	17.3	17.1
abce	2	2	2	1	2	17.7	17.5	16.9	16.4	16.8
abde	2	2	1	2	2	15.3	15.2	16.6	13.5	14.9
acde	2	1	2	2	2	13.4	14.2	14.2	14.0	13.3
bcde	1	2	2	2	2	21.6	20.7	20.9	21.5	21.2

b. Data for the other half-replicate of this experiment follow. Conduct a complete analysis.

tc	A	B	C	D	E	Data				
a	2	1	1	1	1	12.7	12.8	13.7	12.3	13.4
b	1	2	1	1	1	19.5	18.3	21.5	21.3	22.1
c	1	1	2	1	1	22.3	18.6	24.3	22.5	22.3
abc	2	2	2	1	1	18.6	17.3	18.4	17.9	18.1
d	1	1	1	2	1	28.8	28.0	29.1	24.7	24.3
abd	2	2	1	2	1	15.5	16.6	16.7	16.9	19.7
acd	2	1	2	2	1	12.5	13.1	13.3	13.8	13.2
bcd	1	2	2	2	1	20.9	20.7	20.2	20.3	20.1
e	1	1	1	1	2	18.4	19.5	18.1	21.2	19.5
abe	2	2	1	1	2	16.5	15.4	15.9	17.6	16.5
ace	2	1	2	1	2	14.7	14.8	15.3	14.4	15.0
ade	2	1	1	2	2	12.1	12.4	11.4	12.3	12.9
bce	1	2	2	1	2	19.0	20.2	21.9	21.4	21.5
bde	1	2	1	2	2	20.0	19.3	18.9	19.8	20.7
cde	1	1	2	2	2	20.3	21.0	22.6	19.6	19.8
abcde	2	2	2	2	2	16.5	17.3	16.6	17.1	16.5

c. Compare the results of parts a and b.
d. Assuming that the experiment was run in two half-replicates, analyze further.
e. The data in parts a and b were obtained for a completely randomized 2^5 factorial. Combine those data and analyze appropriately. Comment.

Chapter 14

THE TAGUCHI APPROACH TO THE DESIGN OF EXPERIMENTS

14.1 INTRODUCTION

Genichi Taguchi [32] has made several significant contributions to industrial statistics problems. These include:

1. An emphasis on minimizing variation about a target value.
2. An emphasis on the use of designed experiments to design a product whose performance is insensitive to variation among components and environmental conditions.
3. The use of orthogonal arrays and their associated linear graphs for designing experiments.
4. The use of loss functions to assess the economic impact of product variation.
5. An emphasis on the adage that "Quality must be engineered in, not added on."

The Taguchi approach to the engineering design process is a three-stage model consisting of system design, parameter design, and tolerance design. *System design* is the work of those who concentrate on the technical aspects of product or process development. Scientific and engineering knowledge are used to select materials, determine manufacturing equipment needs, and set tentative specifications on product parameters and process factors. This design stage is often far removed from production facilities.

The use of designed experiments to design a product whose performance is insensitive to variation among components and environmental conditions is the major thrust of *parameter design*. The goal is to determine the best operating conditions for the manufacture of the items developed at the system design stage so that they perform as intended under a variety of conditions that change with customer use.

Sometimes work at the parameter design stage fails to achieve sufficient reduction in variation, with the result that the intended product design is not realized. Such difficulties are studied at the *tolerance design* stage, and their resolution calls for additional expenditures to obtain better grade materials, more expensive components, and new equipment. Whereas production costs are often reduced and quality increased at the parameter design stage, increases in quality at the tolerance design stage usually incur cost increases.

In this chapter, we center our attention on Taguchi's experimental designs as used at the parameter design stage. We believe that to understand and use Taguchi's methods properly, one must study experimental design through fractional factorial experiments as in Chapter 13.

14.2 THE $L_4(2^3)$ ORTHOGONAL ARRAY

To acquaint the reader with the basic Taguchi design procedures, we first consider his $L_4(2^3)$ orthogonal array. The notation $L_4(2^3)$ means that we can use the corresponding array to design an experiment requiring four observations and at most three factors, each at two levels. All cases use the orthogonal array design of Table 14.1, where "1" represents the low level of a factor and "2" the high level. (In a more familiar notation, the low level would be -1 and high level $+1$, as shown in Table 14.2.)

For Table 14.2, note that column 3 is the negative of the products of corresponding entries in columns 1 and 2. If columns 1 and 2 are used as the design matrix for a two-factor factorial with factors A and B, then column 3 gives the coefficients for the contrast $-AB$ or the interaction.

One can also use columns 1 and 2 of Table 14.1 to find column 3 of that table. To do so, first add corresponding entries in columns 1 and 2. If a sum is even, enter 1 in the corresponding position of row 3; otherwise, enter 2.

Triangular Table for L$_4$(2^3)

Taguchi developed triangular tables to help experimenters determine the columns associated with the two-way interaction effects of those main effects assigned to other columns. These

TABLE 14.1
Taguchi's $L_4(2^3)$ Orthogonal Array

Run	Column 1	2	3
1	1	1	1
2	1	2	2
3	2	1	2
4	2	2	1

TABLE 14.2
$L_4(2^3)$ Orthogonal Array Using +1 and −1

Column	1	2	3
1	-1	-1	-1
2	-1	$+1$	$+1$
3	$+1$	-1	$+1$
4	$+1$	$+1$	-1

tables show how each column can be obtained from two other columns. Table 14.3 shows the triangular table for $L_4(2^3)$. From the main diagonal (the one with numbers in parentheses), $(1) \times (2)$ gives column 3; $(2) \times (3)$ gives column 1; and $(1) \times (3)$ gives column 2, if entries of the orthogonal table are -1's and $+1$'s.

Linear Graph for $L_4(2^3)$

To aid in the design of experiments Taguchi developed for each orthogonal table a linear graph made up of three parts: dots, line segments, and numbers. Each dot represents a main effect, a line segment connecting two dots represents the negative of the effect of the two-way interaction of the factors associated with those dots, and the number assigned to a dot or line segment indicates the column that describes the contrast for the corresponding effect.

The linear graph for $L_4(2^3)$ is given in Figure 14.1. Suppose a half-replicate of a 2^3 factorial experiment is to be designed. If the design matrix is determined by assigning A, B, and C to columns 1, 2, and 3, respectively, the 3 assigned to the line segment connecting the dots labeled 1 and 2 indicates that C is aliased with $-AB$. Clear information is available on C only if AB is negligible.

Three Designs

We can use $L_4(2^3)$ to design and analyze three cases. The simplest is a single factor A with two observations at each level. If column 1 is used to define the levels of A, calculate SS_A using column 1 and then find the error sum of squares by subtracting SS_A from the total sum of squares.

The second case is a 2^2 factorial. If A and B are assigned to columns 1 and 2, respectively, use those columns to calculate SS_A and SS_B, and use column 3 to calculate SS_{AB}. If there is more than one observation per treatment combination, the error sum of squares can be obtained by subtracting the sum of these three sums of squares from the total sum of squares.

The third case is a 2^{3-1} fractional factorial. If A and B are assigned to columns 1 and 2, respectively, then C must be assigned to column 3. From the triangular table or the linear

TABLE 14.3
Triangular Table for $L_4(2^3)$

Column	1	2	3
	(1)	3	2
		(2)	1
			(3)

Figure 14.1   Linear graph for $L_4(2^3)$.

graph, we know that C is confounded with $-AB$. The triangular table also indicates that A is confounded with $-BC$ and B is confounded with $-AC$. Further, ABC is confounded in blocks. So, this is merely the principal block for a 2^3 factorial in two blocks of four. The defining relation is $I = ABC$.

Sums of Squares

When $L_4(2^3)$ is expressed as in Table 14.2, one can use the methods of earlier chapters to calculate the observed contrasts and sums of squares. Alternatively, one can use Table 14.1 and calculate the sum of squares associated with a column using the formula

$$\text{SS}_{\text{column}} = \frac{(\text{sum of all values at level 1})^2}{\text{values at level 1}} + \frac{(\text{sum of all values at level 2})^2}{\text{values at level 2}}$$
$$- \frac{(\text{sum of all values})^2}{\text{number of values}}$$

This is equivalent to

$$\text{SS}_{\text{column}} = \frac{(\text{sum of all values at level 2} - \text{sum of all values at level 1})^2}{\text{number of values}} \quad (14.1)$$

which is the formula used by Taguchi. For larger tables, it is more economical to use a computer package, as illustrated in earlier chapters.

■ Example 14.1 [Using $L_4(2^3)$ to Design a 2^{3-1} Fractional Factorial]

In Example 9.3 we considered a 2^3 factorial experiment involving A (tool type: 1 or 2), B (bevel angle: 15° or 30°), and C (type of cut: continuous or interrupted). Suppose conditions dictate that four replications of four treatment combinations can be run in random order, but the other four treatment combinations cannot be run until later. Also suppose that prior experience indicates that the effect of the AB interaction is negligible. Assigning A, B, and C to columns 1, 2, and 3, respectively of Table 14.2 gives the design in Table 14.4. For illustration, the totals are from Example 9.3.

TABLE 14.4
$L_4(2^3)$ Orthogonal Array for Example 14.1

tc	Column $1 = A$	$2 = B$	$3 = C$	Total
(1)	−1	−1	−1	2
bc	−1	+1	+1	−2
ac	+1	−1	+1	−17
ab	+1	+1	−1	13
High total: H	−4	11	−19	
Low total: L	0	−15	15	
$H − L$	−4	26	−34	
Sum of squares	1	42.25	72.25	

Using the manual method of the Taguchi literature to determine SS_A, first add the totals associated with ac and ab (the high levels of A). This gives $-17 + 13 = -4$, as indicated in Table 14.4. Then, add the totals associated with (1) and bc (the low levels of A) to obtain $2 + -2 = 0$. This gives a difference of $(-4) - 0 = -4$. There are four observations per treatment combination for a total of $4 \times 4 = 16$ observations. Using this information and Equation (14.1), we find

$$SS_A = \frac{(-4)^2}{16} = 1$$

The other sums of squares in Table 14.3 are obtained in like manner.

The sum of the sums of squares associated with main effects and their aliases is $1 + 42.25 + 72.25 = 115.50$. The reader may show that the (corrected) total sum of squares is 229.00. Thus, the error sum of squares is $229.00 - 115.50 = 113.50$.

Plots of Factor Means

To improve process centering when one is analyzing results from orthogonal tables, Taguchi recommends the use of individual plots of the means (sometimes called *marginal means plots*) at each level of a factor to aid in determining the best choice of levels. When such plots are used without supporting evidence or when interaction is present, however, erroneous conclusions may result.

■ **Example 14.2 (Marginal Means Plots for Example 14.1)**

Again consider the study of power requirements as discussed in Example 14.1. The sample means for the Table 14.4 data are summarized in Table 14.5. Plotting $\bar{y}$ versus the factor level gives the marginal means plots of Figure 14.2. These plots suggest that either level of A with the low level of B and the high level of C would minimize the average power requirement.

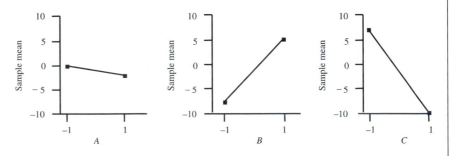

Figure 14.2   Marginal means plots for data in Table 14.4.

TABLE 14.5
Sample Means for Data in Table 14.4

Level	Factor A	Factor B	C
−1	0	−7.5	7.5
+1	−2	5.5	−9.5

Percent Contribution

Practitioners of Taguchi methods often consider the percent contribution of a factor in lieu of an F test. Factors with an extremely small percent contribution to the overall variability are usually pooled with the error term. If A denotes a factor with degrees of freedom ν_A, the *percent contribution for A* is

$$P_A = \frac{SS_A - (MS_{error})\,\nu_A}{SS_{total}} \times 100 \tag{14.2}$$

Use $P_A = 0$ when Equation (14.2) produces a negative value.

■ **Example 14.3 (Percent Contribution for Factors in Example 14.1)**

In Example 14.1, the sums of squares for A, B, and C are 1.0, 42.5, and 72.5, respectively, each with 1 df. Since there are 12 degrees of freedom for error and the error sum of squares is 113.5, $MS_{error} = 113.5/12 \approx 9.5$. Further, $SS_{total} = 229.0$. Using Equation (14.2), we write

$$P_A = \frac{1.0 - 9.5}{229} \times 100 \approx 0.0,$$

$$P_B = \frac{42.25 - 9.5}{229} \times 100 \approx 14.3,$$

and

$$P_C = \frac{72.25 - 9.5}{229} \times 100 \approx 27.4$$

These results indicate that within the ranges of this experiment, A (tool type) and its alias $-BC$ have little effect on the variability among power means, B (bevel angle) and its alias $-AC$ have a mild effect, and C (type of cut) has a moderate effect. Since the percent contribution for tool type is so small, Taguchi would remove A from the model (i.e., pool A with error). Assuming that all interactions are negligible and noting that the observed average power requirements are smaller at the low level of B (15° bevel angle) and the high level of C (interrupted cut), these results seem to indicate that the average power requirement can be minimized by using either tool type with a 15° bevel angle and an interrupted cut. This conclusion agrees with that arrived at earlier.

The Confirmation Experiment

Conclusions such as the one obtained in Example 14.2 are rather intuitive. When is a contribution of a certain percentage large enough to be considered important and another small enough to allow us to pool the associated effect with error? To assist in the evulation of borderline cases, Taguchi insists that a second experiment be conducted to reproduce the conclusions, if possible, or to observe whether the suggested experimental setup produces the anticipated results. Such experiments are called *confirmation experiments*.

14.3 OUTER ARRAYS

Taguchi emphasizes the use of statistical methods to produce products that are robust to outside influence. A product design is *robust* when factors beyond the control of the producer (called *noise factors*) have little effect on the intended product performance. Extending the concept of blocking, Taguchi recommends that a design matrix for the noise factors, called an *outer array*, be crossed with the design matrix for the controllable factors, called an *inner array*, to render the design more robust. Each experimental run from the inner array is then tested across the runs from the outer array. We illustrate with a simple example.

A metal stamping process places a hole near the edge of a stamped bracket. The target (or nominal) spacing from the hole to the edge of the bracket is 0.40 inch. To study how well the target is being met, a project team decides to consider roller height, A, material supplier, B, and feeder adjustment, C, as controllable factors with amount of oil, D, material thickness, E, and material hardness, F, as noise factors. The crossed array, with distances recorded in hundredths of an inch, is given in Figure 14.3.

Consider the row for run 4 of the inner array in Figure 14.3. Here, the controllable factors A and B are set at their high levels and C is set at its low level. The four observed values in this row were obtained for this treatment combination—one each at (1), ef, df, and de for the noise factors D, E, and F. That is, when the six factors are considered, these four observations are $ab = 41$, $abef = 52$, $abdf = 46$, and $abde = 42$. The remaining 12 observations were obtained in a similar manner.

				Outer Array				
						Run		Factor
				1	2	3	4	
	Inner Array			1	1	2	2	D
	Factor			1	2	1	2	E
Run	A	B	C	1	2	2	1	F
1	1	1	1	37	38	36	37	
2	1	2	2	35	39	40	33	
3	2	1	2	45	44	44	46	
4	2	2	1	41	52	46	42	

Figure 14.3   Inner and outer array layout with data for hole-to-edge distance study.

Analysis When the Outer Array Is Used for Blocking

One assumes that noise factors are selected because they are very difficult to control. Thus, once an experimental setup has been made for a treatment combination of the outer array, all treatment combinations of the inner array are usually run. If the data in Figure 14.3 were obtained under such conditions, we would have four replicates of a 2^{3-1} fractional factorial. We are assuming that the blocking order is randomly determined, as is the experimental order within each block. Thus, if the first block is determined by ef, the inner array treatment combinations (1), bc, ac, and ab are obtained in random order. Combining the six factors, we might arrive at the ordering $bcef$, $abef$, $acef$, and ef.

We use the model

$$Y_{ijklm} = \mu + \beta_i + A_j + B_k + C_l + \varepsilon_{m(ijkl)}$$

to obtain the ANOVA summary of Table 14.6. Suppose interactions are negligible. It appears that within the factor ranges of this experiment, only roller height significantly affects the hole spacing distance. The observed average distances at levels 1 and 2 of A are 36.875 and 45.000, respectively. Neither is near the target of 40. Further exploration of the effect of roller height at levels 1, 2, and some intermediate levels may lead to an average distance closer to that target value.

A better procedure might be to use the other replicate of the 2^{3-1} fractional factorial in two of the blocks from the outer array. This approach would help us break down the alias structure and give information about A, B, and C that is clear of any two-way interactions between those factors. When the experiment is run according to the design in Figure 14.3, A, B, and C are aliased with $-BC$, $-AC$, and $-AB$, respectively.

Analysis When the Data Are Obtained in Random Order

If the data in Figure 14.3 were obtained in random order, we could combine the arrays to obtain the array for a 2^{6-2} fractional factorial, as shown in Table 14.7. The defining relation is

$$I = ABC = DEF = ABCDEF$$

so the design is of resolution 3. Further, A, B, C, D, E, and F are confounded with the two-factor interactions $-BC$, $-AC$, $-AB$, $-EF$, $-DF$, and $-DE$, respectively. The reader is asked to conduct a complete analysis of these data in Problem 14.9.

TABLE 14.6
ANOVA Summary for Data in Figure 14.3

Source	df	SS	MS	F	p value
Block	3	39.1875	13.0628	1.61	0.2548
A	1	264.0625	264.0625	32.53	0.0003
B	1	0.0625	0.0625	0.01	0.9320
C	1	0.5625	0.5625	0.07	0.6983
Error	9	73.9375	8.1181	6.24	
Totals	15	376.9375			

TABLE 14.7
Orthogonal Table for a 2^{6-2} Fractional Factorial from Figure 14.3

tc	A	B	C	D	E	F	y
(1)	1	1	1	1	1	1	37
bc	1	2	2	1	1	1	35
ac	2	1	2	1	1	1	45
ab	2	2	1	1	1	1	41
ef	1	1	1	1	2	2	38
bcef	1	2	2	1	2	2	39
acef	2	1	2	1	2	2	44
abef	2	2	1	1	2	2	52
df	1	1	1	2	1	2	36
bcdf	1	2	2	2	1	2	40
acdf	2	1	2	2	1	2	44
abdf	2	2	1	2	1	2	46
de	1	1	1	2	2	1	37
bcde	1	2	2	2	2	1	33
acde	2	1	2	2	2	1	46
abde	2	2	1	2	2	1	42

In this case, the Taguchi design when randomization is possible is equivalent to one of the standard fractional factorials. Further, we cannot design a 2^{6-2} fractional factorial with resolution greater than III. However, the inner and outer array procedure will sometimes result in larger experiments of lower resolution than might be obtained by means of the methods of Chapter 13. One should always remember to ascertain the nature of the information that can be obtained from a given experiment *before* spending the time and money to conduct that experiment.

14.4 SIGNAL-TO-NOISE RATIO

For Taguchi, no deviation of a response variable from its intended (target) value is good. He views our task as one of producing products (or services) at optimal levels with minimal variation in the measured characteristics. Taguchi recommends the use of a statistic called the *signal-to-noise ratio*, denoted *S/N*, as an aid in reducing variation. Several ratios have been developed for characteristics of different types. The three most commonly used are as follows:

Smaller is better:

$$S/N_{\text{SB}} = -10\log_{10}\left[\frac{\left(\sum_{i=1}^{n} Y_i^2\right)}{n}\right] \tag{14.3}$$

Nominal is best:

$$S/N_{\text{NB}} = -10\log_{10}\left[S^2\right] \tag{14.4}$$

Larger is better:

$$S/N_{\text{LB}} = -10\log_{10}\left[\frac{1}{n}\sum_{i=1}^{n}\left(\frac{1}{Y_i^2}\right)\right] \qquad (14.5)$$

where $n \geq 2$ is the number of observations in a sample of size n and S^2 is the variance of that sample. These ratios have been defined so that *larger values are better*. A signal-to-noise ratio is calculated for each treatment combination. In general, Taguchi claims that the treatment combination with the largest true average signal-to-noise ratio produces the least variation in the response variable. This claim has been questioned, as noted in the following paragraph.

Taguchi includes two signal-to-noise ratios for use when the target mean is considered best. Hunter [14] recommends using only that defined in Equation (14.4), since the other is highly sensitive to shifts in the mean. The smaller-is-better and larger-is-better ratios are also highly sensitive to shifts in the mean, so the nominal-is-best ratio of Equation (14.4) is often used for all three cases.

■ Example 14.4 (Signal-to-Noise Ratios for Figure 14.3)

Consider again Figure 14.3. A sample of four distances is associated with each row of the inner array. Those samples are summarized in Table 14.8. The observed sample variance for {37, 38, 36, 37} is $s^2 = 2/3 \approx 0.67$. Using the nominal-is-best ratio defined in Equation (14.4), we find that the signal-to-noise ratio for this sample is

$$S/N_{NB} = -10\log_{10}(0.67) \approx 1.74$$

Similar calculations give the other ratios in Table 14.8.

TABLE 14.8
Signal-to-Noise Ratios for Inner Array of Figure 14.3

Row	A	B	C	Sample	Variance	S/N_{NB}
1	1	1	1	{37, 38, 36, 37}	0.67	1.74
2	1	2	2	{35, 39, 40, 33}	10.92	−10.38
3	2	1	2	{45, 44, 44, 46}	0.92	0.36
4	2	2	1	{41, 52, 46, 42}	24.92	−13.97

Analysis of Variance Using S/N Ratios

Taguchi recommends the use of the signal-to-noise ratio as a response variable to identify factors that can reduce variation. Active factors are identified by means of a standard analysis of variance approach. Taguchi notes that a compromise must be made when the levels of factors that improve centering do not agree with those that improve variability. Such compromise is left to analysts who have knowledge of the system. A confirmation experiment is recommended to verify the experimental results.

To illustrate, consider the design matrix of Table 14.8. Using the signal-to-noise ratios of that table as response values, we obtain Table 14.9. Even though we have no degrees of freedom for error, it is evident that B (supplier) has the greatest effect on the S/N ratio. In fact, if A and C are pooled with error, B is significant at the 0.02 significance level.

Notice that the largest signal-to-noise ratios of Table 14.8 occur for the treatment combinations (1) and ac. This may indicate that material from supplier 1 has the least effect on the variability in the measured distance. Recalling that A (roller height) has the greatest effect on centering, it appears that further experimentation involving roller height and material from supplier 1 is warranted.

TABLE 14.9
ANOVA Summary Using S/N Ratios of Table 14.8

Source	df	SS	MS	F	p value
A	1	6.2250	6.2250	•	•
B	1	175.1652	175.1652	•	•
C	1	1.1990	1.1990	•	•
Error	0	•	•	•	
Totals	3	182.5892			

14.5 THE $L_8(2^7)$ ORTHOGONAL ARRAY

Designs based on the $L_8(2^7)$ orthogonal array require only eight observations to examine up to seven factors, each at two levels. This table, depicted in Table 14.10, is equivalent to that for the $N = 8$ Plackett–Burman design with seven factors (i.e., Table 13.11). It can be used to design 2_{III}^{7-4}, 2_{III}^{6-3}, 2_{III}^{5-2}, 2_{IV}^{4-1}, and 2^3 factorial experiments.

TABLE 14.10
Taguchi's $L_8(2^7)$ Orthogonal Array

Run	Column						
	1	2	3	4	5	6	7
1	1	1	1	1	1	1	1
2	1	1	1	2	2	2	2
3	1	2	2	1	1	2	2
4	1	2	2	2	2	1	1
5	2	1	2	1	2	1	2
6	2	1	2	2	1	2	1
7	2	2	1	1	2	2	1
8	2	2	1	2	1	1	2

To see that the $L_8(2^7)$ orthogonal array is equivalent to Table 13.11, assign factors A, B, F, E, G, D, and C to columns 1, 2, 3, 4, 5, 6, and 7, respectively. The eight experimental runs under this assignment are (1), *cdeg*, *bcdf*, *befg*, *acfg*, *adef*, *abdg*, and *abce*, which form the same block as that of Table 13.11. This is the principal block of a 2^{7-4} fractional factorial, with each main effect aliased with a two-way effect. This design, as with all Taguchi designs for which each column is assigned a main effect, is of resolution III.

Triangular Table for $L_8(2^7)$

Suppose A, B, C, D, E, F, and G are assigned (in that order) to columns 1 through 7 of $L_8(2^7)$. The two-factor effects that are aliased with main effects can be determined using Table 14.11. If (2) and (4) are chosen on the main diagonal, the intersection of the corresponding row and column, respectively, contains 6. But B, D, and F are assigned to columns 2, 4, and 6, respectively, so $-BD$ is aliased with F. If BD is active, any information about F will be "'mixed up'" with information about BD. Further, $-AG$ and $-CE$ are also aliased with F.

Taguchi is often criticized for failing to mention confounding and aliases. Triangular tables such as Table 14.11 are used to determine the column to which a potentially active interaction should be assigned. In other words, Taguchi assumes that the experimenter is wise enough to know whether two factors are likely to interact. Such determinations are possible in many cases, but surprises do occur!

Linear Graphs for $L_8(2^7)$

The linear graphs associated with $L_8(2^7)$, given in Figure 14.4, represent two of the more common experimental situations and can be modified for other situations. The numbers on the linear graphs represent columns in $L_8(2^7)$, solid dots indicate the main effects, and a line segment connecting two solid dots represents the interaction between the main effects assigned to those dots. If the experimenter has reason to believe that an interaction effect is negligible (or can be ignored), a main effect can be assigned to the column associated with a line segment.

TABLE 14.11
Two-Factor Interactions for Columns in an $L_8(2^7)$ Array

Column	1	2	3	4	5	6	7
	(1)	3	2	5	4	7	6
		(2)	1	6	7	4	5
			(3)	7	6	5	4
				(4)	1	2	3
					(5)	3	2
						(6)	1
							(7)

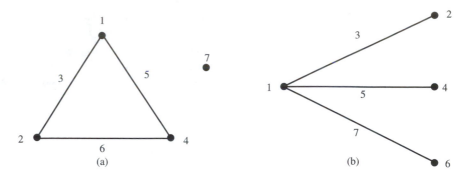

Figure 14.4   Taguchi's linear graphs for $L_8(2^7)$.

■ Example 14.5 (Four Factors with Two-Way Interactions Among Only Three)

Figure 14.4a is the suggested linear graph for an experimental design involving four factors (A, B, C, and D) and the two-factor interactions among three of those factors. If, for example, the only two-factor interactions are AB, AC, and BC, the assignments could be as shown in Figure 14.5. The design matrix consists of columns 1, 2, 4, and 7 only. If -1 and $+1$ are substituted for 1 and 2, respectively, the entries in columns 3, 5, and 6 of $L_8(2^7)$ determine the coefficients of the linear contrasts for the effects of $-AB$, $-AC$, and $-BC$, respectively. This design is that of a 2^{4-1} fractional factorial of resolution IV with defining contrast $I = ABCD$.

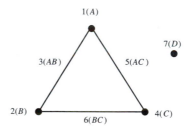

Figure 14.5   Linear graph for Example 14.5.

■ Example 14.6 (Four Factors, One of Which Interacts with the Other Three)

Figure 14.4b is the suggested linear graph for an experimental design involving four factors (A, B, C, and D), one of which interacts with each of the others. If, for example, A interacts with each of B, C, and D, the assignments could be as shown in Figure 14.6.

Here the assumption is that all other two-factor interactions are negligible. The design matrix consists of columns 1, 2, 4, and 6 only. If -1 and $+1$ are substituted for 1 and 2, respectively, the entries in columns 3, 5, and 7 of $L_8(2^7)$ determine the coefficients of the linear contrasts for the effects of $-AB$, $-AC$, and $-AD$, respectively. This design is that of a 2^{4-1} fractional factorial of resolution III with defining contrast $I = -BCD$.

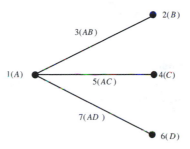

Figure 14.6   Linear graph for Example 14.6.

■ **Example 14.7 (Six Factors, Only Two of Which Interact)**

Suppose we have six factors (A, B, C, D, E, and F) at two levels each. If we wish to include one interaction, say AB, we can use Figure 14.4b with A, B, and AB assigned to columns 1, 4, and 5, respectively. The remaining factors are then assigned to unused columns. If (say) C, D, E, and F were assigned to columns 2, 3, 6, and 7, respectively, Taguchi would suggest that the line segments associated with columns 3 and 7 be replaced with dots to give the linear graph in Figure 14.7. Here, he assumes that the experimenter will be aware that AB is the only active two-factor interaction. The design is that for a 2_{III}^{6-3} fractional factorial, since the two-factor interactions assumed to equal zero are confounded with main effects.

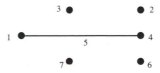

Figure 14.7   Linear graph for Example 14.7.

■ **Example 14.8 (Five Factors, One of Which Interacts with Two Others)**

Suppose we have five factors (A, B, C, D, and E) at two levels each. If A interacts with each of (say) B and C, and all other interactions are assumed zero, we can use Figure 14.4a with A, B, and C assigned to columns 1, 2, and 4, respectively. The effects of AB and AC are then estimated using columns 3 and 5, respectively. If D is assigned to column 6, then E must be assigned to column 7. The design matrix consists of columns 1, 2, 4, 6, and 7. Since A is aliased with $-DE$, the design is that for a 2_{III}^{5-2} fractional factorial.

Since a main effect has been assigned to column 6, we replace the associated line with a dot and obtain the linear graph of Figure 14.8. This graph is based on the assumption, commom to all Taguchi's designs, that only the two-factor interactions associated with its line segments are active.

■ **Example 14.9 (Data Analysis Using the Design of Example 14.8)**

The factors A (casting speed), B (slab width), C (percent sulfur), D (tundish tempera-ture), and E (mold level) are to be considered at two levels each in a study of the surface finish of steel. The experimenters decide that the only active two-factor interactions are AB and AC. Using Figure 14.8 to design a one-fourth replicate of this 2^5 factorial, they obtain the Taguchi array of Table 14.12. The shaded columns form the design matrix and the data are from those in Problem 13.19, with the factor labels changed for convenience.

Using the model

$$Y_{ijklm} = \mu + A_i + B_j + AB_{ij} + C_k + AC_{ik} + D_l + E_m + \varepsilon_{1(ijklm)}$$

we obtain the sums of squares in Table 14.13.

TABLE 14.12
Taguchi Array for Example 14.9

	Columns							
	A	B	AB	C	AC	D	E	
tc	1	2	3	4	5	6	7	y
(1)	1	1	1	1	1	1	1	1
cde	1	1	1	2	2	2	2	6
bde	1	2	2	1	1	2	2	0
bc	1	2	2	2	2	1	1	0
ae	2	1	2	1	2	1	2	6
acd	2	1	2	2	1	2	1	2
abd	2	2	1	1	2	2	1	3
abce	2	2	1	2	1	1	2	3

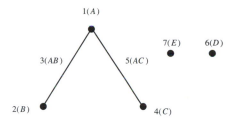

Figure 14.8   Linear graph for Examples 14.8 and 14.9.

TABLE 14.13
ANOVA Summary for Example 14.9

Source or Factor	df	SS	MS	F	p value
A (casting speed)	1	6.125	6.125	•	•
B (slab width)	1	10.125	10.125	•	•
AB	1	3.125	3.125	•	•
C (% sulfur)	1	0.125	0.125	•	•
AC	1	10.125	10.125	•	•
D (tundish temp.)	1	0.125	0.125	•	•
E (mold level)	1	10.125	10.125	•	•
Error	0	•	•		
Totals	7	39.875	•		

Taguchi uses a pooling strategy to obtain an error term. The resulting F tests are then used as reference values. In this case, C and D are pooled to give Table 14.14. Even with only 2 degrees of freedom for error, all remaining factors and interactions are significant for any $\alpha \geq 0.0377$. Since such a procedure has been shown to give biased results, Taguchi uses a confirmation experiment to validate the conclusions.

Taguchi also relies heavily on the percent contribution when making decisions regarding the importance of an effect. Values in the Contribution (%) column of Table 14.14 were obtained by means of Equation (14.2). It appears that B (slab width), AC (casting speed × % sulfur), and E (mold level) have the greatest effect on the surface finish.

TABLE 14.14
ANOVA Summary for Example 14.9 After Pooling

Source	df	SS	MS	F	p value	Contribution (%)
A	1	6.125	6.125	49.00	0.0198	15
B	1	10.125	10.125	81.00	0.0121	25
AB	1	3.125	3.125	25.00	0.0377	8
AC	1	10.125	10.125	81.00	0.0121	25
E	1	10.125	10.125	81.00	0.0121	25
Error	2	0.250	0.125			
Totals	7	39.875				

In Problem 13.20, where we used a half-replicate to investigate this same problem, slab width, mold level, and the interaction between slab width and casting speed were identified as significant. The Taguchi method agrees well on the two main effects and uses only 8 observations, versus 16 in Problem 13.20. The interaction effects fail to agree, but this may be expected when only half the data are run.

More examples could be used to show the many applications of the $L_8(2^7)$ orthogonal array. We hope those included here are adequate to provide the reader with a good overview of the basic Taguchi techniques. Now we present a classic experiment that has appeared in nearly all Taguchi papers and seminars, in which seven main effects are examined with only eight observations.

■ **Example 14.10 (A Classic Taguchi Example)**

A Japanese company was concerned with excessive variability in the dimensions of tiles that it produced. Seven factors were identified as possible sources of serious variability. These were labeled A through G, where

$$
\begin{aligned}
A &= \text{amount of limestone (5 or 1\%)} \\
B &= \text{coarseness of additive (existing or finer)} \\
C &= \text{amount of agalmatolite (43 or 53\%)} \\
D &= \text{type of agalmatolite (existing or new)} \\
E &= \text{quantity of charging material (1300 or 1200 kg)} \\
F &= \text{amount of waste return (0 or 4\%)} \\
G &= \text{amount of feldspar (0 or 5\%)}
\end{aligned}
$$

Here a full factorial would require $2^7 = 128$ experiments. Assuming *no* interactions present, the experimenters decided to use the experimental design described by the $L_8(2^7)$ orthogonal array with A, B, C, D, E, F, and G assigned to columns 1, 2, 3, 4, 5, 6, and 7, respectively. A random sample of 100 tiles was taken from each of the eight experimental runs. The response (Y) was defined as the number of defective tiles in the sample, and the results are shown in Table 14.15.

When the sums of squares in Table 14.15 are considered, it seems evident that factors A, F, and G are the largest contributors to the variation in the dimensions of the tiles. Further, we note that B and C seem to have little effect on that variation.

Now consider the percent defective at each factor level. Those percentages are summarized in Table 14.16. Since 400 tiles were inspected at each level of a factor, the percent defective is the total number of defectives at a particular level divided by 4.

Since the goal is to reduce the number of defective tiles, the percentages in Table 14.16 can be used to select the "best" level of each factor. Relative to the other factors, B and C have such small sums of squares that Taguchi recommends choosing the processing level for these by considering cost and other characteristics. In each case, the low level is less expensive. For the other five factors, the suggested level is the one that produces the smaller number of defectives at the time of the experimental runs. Thus, the suggestion is to process the tiles using E high (1200 kg of charging material), G high (5% feldspar), and the other five factors at their low levels. This combination was discovered by running a one-sixteenth replicate of a 2^7 factorial. That is, the design is a 2_{III}^{7-4}.

TABLE 14.15
Taguchi Array for Example 14.10

tc	A 1	B 2	C 3	Factor D Column 4	E 5	F 6	G 7	y
(1)	1	1	1	1	1	1	1	16
defg	1	1	1	2	2	2	2	17
bcfg	1	2	2	1	1	2	2	12
bcde	1	2	2	2	2	1	1	6
aceg	2	1	2	1	2	1	2	6
acdf	2	1	2	2	1	2	1	68
abef	2	2	1	1	2	2	1	42
abdg	2	2	1	2	1	1	2	26
SS	1035	55	10	210	325	903	630	3168

TABLE 14.16
Percent Defective Data for Factors in Table 14.15

Level	A	B	C	Factor D	E	F	G
1	12.75	26.75	25.25	19.00	30.50	13.50	33.00
2	35.50	21.50	23.00	29.25	17.75	34.75	15.25

14.6 THE $L_{16}(2^{15})$ ORTHOGONAL ARRAY

Taguchi proposes an orthogonal array for examining 15 effects with only 16 experiments, denoted $L_{16}(2^{15})$, as seen in Table 14.17. The design of this array is comparable to the $N = 16$ Plackett–Burman design of Table 13.11.

As with any orthogonal array, a design matrix is identified once the main effects have been assigned to columns. Table 14.18 can be used to determine the two-factor interactions that are aliased with a main effect and which columns can be used to estimate a two-factor effect.

Taguchi also prepared the six types of linear graph shown in Figure 14.9 to aid in designing experiments based on $L_{16}(2^{15})$. He makes the following recommendations for their application:

TABLE 14.17
Taguchi's $L_{16}(2^{15})$ Orthogonal Array

Experiment	|	1	2	3	4	5	6	7	8	9	10	11	12	13	14	15
																Columns
1	|	1	1	1	1	1	1	1	1	1	1	1	1	1	1	1
2	|	1	1	1	1	1	1	1	2	2	2	2	2	2	2	2
3	|	1	1	1	2	2	2	2	1	1	1	1	2	2	2	2
4	|	1	1	1	2	2	2	2	2	2	2	2	1	1	1	1
5	|	1	2	2	1	1	2	2	1	1	2	2	1	1	2	2
6	|	1	2	2	1	1	2	2	2	2	1	1	2	2	1	1
7	|	1	2	2	2	2	1	1	1	1	2	2	2	2	1	1
8	|	1	2	2	2	2	1	1	2	2	1	1	1	1	2	2
9	|	2	1	2	1	2	1	2	1	2	1	2	1	2	1	2
10	|	2	1	2	1	2	1	2	2	1	2	1	2	1	2	1
11	|	2	1	2	2	1	2	1	1	2	1	2	2	1	2	1
12	|	2	1	2	2	1	2	1	2	1	2	1	1	2	1	2
13	|	2	2	1	1	2	2	1	1	2	2	1	1	2	2	1
14	|	2	2	1	1	2	2	1	2	1	1	2	2	1	1	2
15	|	2	2	1	2	1	1	2	1	2	2	1	2	1	1	2
16	|	2	2	1	2	1	1	2	2	1	1	2	1	2	2	1

TABLE 14.18
Interactions Between Two Columns for $L_{16}(2^{15})$

Column	|	1	2	3	4	5	6	7	8	9	10	11	12	13	14	15
	|	(1)	3	2	5	4	7	6	9	8	11	10	13	12	15	14
	|		(2)	1	6	7	4	5	10	11	8	9	14	15	12	13
	|			(3)	7	6	5	4	11	10	9	8	15	14	13	12
	|				(4)	1	2	3	12	13	14	15	8	9	10	11
	|					(5)	3	2	13	12	15	14	9	8	11	10
	|						(6)	1	14	15	12	13	10	11	8	9
	|							(7)	15	14	13	12	11	10	9	8
	|								(8)	1	2	3	4	5	6	7
	|									(9)	3	2	5	4	7	6
	|										(10)	1	6	7	4	5
	|											(11)	7	6	5	4
	|												(12)	1	2	3
	|													(13)	3	2
	|														(14)	1

Type 1: The interactions between main effects are equally important.

Type 2: One factor is very important: the interactions between this factor and other factors are required, and two other interactions are also needed.

Type 3: Two groups of interactions, occurring at two stages of a process, are required.

Type 4: The interactions between one factor and all others are important.

Type 5: Can be used to prepare Latin , Graeco–Latin, and hyper-Graeco–Latin squares.

Type 6: Convenient for pseudofactor designs.

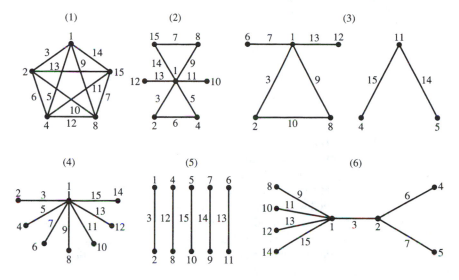

Figure 14.9   Linear graphs for $L_{16}(2^{15})$.

Thus, for example, Taguchi's type 1 linear graph might be used to examine 15 factors if no interactions were believed present, producing a design of resolution III. At the other extreme, it could be used for examining 5 factors (5 vertices of the pentagon) and all 10 two-way interactions ($_5C_2 = 10$, as sides and diagonals of the pentagon). The latter assignment would produce a 2_V^{5-1} design.

Examples of the use of $L_{16}(2^{15})$ with 15 factors are available but are not included because it is unlikely that one would try to handle 15 factors in one experiment. Are all 15 equally important? Probably not! One should then ascertain from the team designing the experiment which are the five or six most important factors. If each member of the team listed first four or five most important factors, the combined list would probably include only six or seven as most important. Then a smaller design could be used. After all, a 2^{15} = 32,768 run in only $16 = 2^4$ experiments is a 2^{15-11} fractional factorial, or a 1/2048th replicate of a 2^{15} factorial.

14.7 THE $L_9(3^4)$ ORTHOGONAL ARRAY

We now consider the orthogonal array denoted $L_9(3^4)$, which can be used to design experiments involving four factors at three levels each (fewer than four factors if some interactions are examined) and only nine treatment combinations. Here the interactions are of the form AB and AB^2 as discussed in Chapter 10, so that each factor and component of interaction has 2 degrees of freedom. The array can be used to design experiments involving:

1. Four factors, say A, B, C, and D, each at three levels. This would require $3^4 = 81$ experiments for a full factorial. To do this with nine experiments calls for a one-ninth

replicate of a 3^4 factorial with main effects aliased with components of two-factor interactions. That is, the design is that of a 3_{III}^{4-2} fractional factorial.

2. Three factors and one interaction, say A, B, AB, and C, where AB is only part of the $A \times B$ interaction.

3. Two factors (say) A and B with both parts of the interaction AB and AB^2. This design is a complete 3^2 factorial.

4. One factor, say A, with pseudofactors for error. This design is a one-way ANOVA.

The orthogonal array $L_9(3^4)$ is displayed in Table 14.19, where 1 is the lowest level, 2 the middle level, and 3 the highest level, corresponding to -1, 0, and $+1$, respectively, in the notation of Chapter 10.

The linear graph for $L_9(3^4)$ is given in Figure 14.10. Suppose that two factors, A and B, at three levels each, are assigned to columns 1 and 2. Figure 14.10 informs us that the sums of squares obtained using columns 3 and 4 should be added to give the sum of squares for the $A \times B$ interaction. The sums of squares associated with columns 3 and 4 are the AB and AB^2 components, respectively, of the $A \times B$ interaction.

Taguchi-type experiments often use $L_9(3^4)$ as an inner array and $L_8(2^7)$ as an outer array. To investigate variability, the signal-to-noise ratios are calculated for the nine treatment combinations of $L_9(3^4)$ and used as responses for a standard analysis of variance. For each factor in the inner array, the average signal-to-noise level is plotted against the factor levels to select the "best" level of each factor. Recall that larger S/N ratios are better.

Sums of Squares Associated with $L_9(3^4)$

The sum of squares associated with a given column of $L_9(3^4)$ can be calculated manually using

TABLE 14.19
Taguchi's $L_9(3^4)$ Array

Experiment	Column			
	1	2	3	4
1	1	1	1	1
2	1	2	2	2
3	1	3	3	3
4	2	1	2	3
5	2	2	3	1
6	2	3	1	2
7	3	1	3	2
8	3	2	1	3
9	3	3	2	1

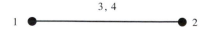

Figure 14.10   Linear graph for $L_9(3^4)$.

$$\text{SS}_{\text{column}} = \frac{(\text{total at level 1})^2}{\text{observations at level 1}} + \frac{(\text{total at level 2})^2}{\text{observations at level 2}} + \frac{(\text{total at level 3})^2}{\text{observations at level 3}}$$
$$- \frac{(\text{grand total})^2}{\text{total number of observations}}$$

When only one observation is recorded per treatment combination, the sum of the four column sums of squares is the (corrected) total sum of squares. When more than one observation is recorded per treatment combination, the total sum of squares is calculated in the usual manner.

■ **Example 14.11 (One Factor at Three Levels with Three Observations per Level)**

A study of leakage in milliamperes of capacitors from three vendors produced the following data:

Vendor 1:	7.3	8.0	8.1
Vendor 2:	10.7	10.2	10.2
Vendor 3:	10.5	10.1	10.8

Using Taguchi's $L_9(3^4)$ array with A (vendor) assigned to column 1 and error E as a pseudofactor, we obtain Table 14.20. The design matrix is determined by column 1. The sums of squares associated with columns 2, 3, and 4 are components of the error sum of squares.

From column 1 the totals are $7.3 + 8.0 + 8.1 = 23.4$, $10.7 + 10.2 + 10.2 = 31.1$, and $10.5 + 10.1 + 10.8 = 31.4$ for levels 1, 2, and 3, respectively, of A. Summing these totals, we find the grand total to be $23.4 + 31.1 + 31.4 = 85.9$. It follows that

$$\text{SS}_A = \frac{(23.4)^2}{3} + \frac{(31.1)^2}{3} + \frac{(31.4)^2}{3} - \frac{(85.9)^2}{9} \approx 13.71$$

TABLE 14.20
Taguchi Array for Example 14.11

			Factor		
	A	E	AE	AE^2	
			Column		
Experiment	1	2	3	4	y
1	1	1	1	1	7.3
2	1	2	2	2	8.0
3	1	3	3	3	8.1
4	2	1	2	3	10.7
5	2	2	3	1	10.2
6	2	3	1	2	10.2
7	3	1	3	2	10.5
8	3	2	1	3	10.1
9	3	3	2	1	10.8

From column 2 the totals are $7.3 + 10.7 + 10.5 = 28.5$, $8.0 + 10.2 + 10.1 = 28.3$, and $8.1 + 10.2 + 10.8 = 29.1$. This gives

$$SS_E = \frac{(28.5)^2}{3} + \frac{(28.3)^2}{3} + \frac{(29.1)^2}{3} - \frac{(85.9)^2}{9} \approx 0.12$$

as one component of the error sum of squares. Showing that the other two components of error are $SS_{AE} \approx 0.62$ and $SS_{AE^2} \approx 0.06$ is left to the reader. These give

$$SS_{error} = SS_E + SS_{AE} + SS_{AE^2} \approx 0.12 + 0.62 + 0.06 = 0.80$$

Notice that $SS_{total} = 7.3^2 + \cdots + 10.8^2 - (85.9^2/9) \approx 14.50$ and $SS_A + SS_{error} \approx 14.51$. The difference is due to round-off error.

■ **Example 14.12 (Two Factors at Three Levels and Their Interaction)**

The hypothetical data for a complete 3^2 factorial, first considered in Section 10.2, are included in Table 14.21. If A and B are assigned to columns 1 and 2, respectively, we can use columns 3 and 4 to obtain the sums of squares for the two interaction components. Notice that columns 1 and 2 determine the design matrix.

From column 3, the totals are $1 + -1 + 1 = 1$, $0 + -2 + 2 = 0$, and $2 + 4 + 3 = 9$ for levels 1, 2, and 3, respectively, of the AB component of the A by B interaction. Thus,

$$SS_{AB} = \frac{1^2 + 0^2 + 9^2}{3} - \frac{10^2}{9} = \frac{146}{9}$$

Likewise,

$$SS_{AB^2} = \frac{7^2 + 2^2 + 1^2}{3} - \frac{10^2}{9} = \frac{62}{9}$$

TABLE 14.21
Taguchi Array for Example 14.12

Treatment Combination	A	B	AB	AB²	
			Factor		
			Column		
	1	2	3	4	y
00	1	1	1	1	1
01	1	2	2	2	0
02	1	3	3	3	2
10	2	1	2	3	-2
11	2	2	3	1	4
12	2	3	1	2	-1
20	3	1	3	2	3
21	3	2	1	3	1
22	3	3	2	1	2

Thus, the interaction sum of squares is

$$SS_{interaction} = SS_{AB} + SS_{AB^2} = \frac{146 + 62}{9} = 23.111$$

to three decimal places. This result agrees with that in Table 10.2.

There is little to be gained by using Taguchi's orthogonal arrays to analyze data such as those in Examples 14.11 and 14.12. Those examples were presented to illustrate basic principles on familiar data sets. When the number of factors exceeds two, and few or no interactions can be assumed, the merit of a Taguchi approach becomes apparent.

■ **Example 14.13 (Three Factors and a Partial Interaction)**

In Example 10.4, we considered a 3^3 factorial with three replications. In this example, we will analyze only the data associated with the treatment combinations defined in $L_9(3^4)$. That is, we will use $L_9(3^4)$ to analyze data for a 3^{3-1} fractional factorial.

Defining factors A, B, and C as day, operator, and concentration, respectively, we assign A to column 1 and B to column 2, which leaves two possibilities for the assignment of C. If C is assigned to column 4, we will have C aliased with the AB^2 component of the $A \times B$ interaction. Column 3 can then be used to calculate a sum of squares for the AB component of that interaction. In this case, it makes sense to assume that C interacts with neither A nor B, so we have reason to believe that we can obtain full information about A and B. These assignments, the treatment combinations, and the relevant data are presented in Table 14.22. Columns 1, 2, and 4 make up the design matrix.

TABLE 14.22
Taguchi Array for Example 14.13: Data From Table 10.16

Treatment Combination	A	B	AB	C	Observations			Total
	Column							
	1	2	3	4				
000	1	1	1	1	1.0	1.2	1.7	3.9
011	1	2	2	2	3.2	3.7	3.5	10.4
022	1	3	3	3	7.2	6.5	6.7	20.4
102	2	1	2	3	6.5	6.0	6.2	18.7
110	2	2	3	1	1.0	0.0	0.0	1.0
121	2	3	1	2	3.7	4.0	4.2	11.9
201	3	1	3	2	4.5	5.0	4.7	14.2
212	3	2	1	3	7.5	6.0	6.0	19.5
220	3	3	2	1	0.5	1.0	1.7	3.2

Totals for levels 1, 2, and 3 of A are $3.9 + 10.4 + 20.4 = 34.7$, $18.7 + 1.0 + 11.9 = 31.6$, and $14.2 + 19.5 + 3.2 = 36.9$, respectively, each based on nine observations. Thus, the grand total, based on 27 observations, is $34.7 + 31.6 + 36.9 = 103.2$. So,

$$SS_A = \frac{(34.7)^2 + (31.6)^2 + (36.9)^2}{9} - \frac{(103.2)^2}{27} \approx 1.5756$$

Verification of the remaining sums of squares, summarized in Figure 14.11, is left to the reader.

Assuming that the AB^2 interaction component is negligible, we see from Figure 14.11 that C accounts for almost all of the variability in the average yield. The observed average yields at levels 1, 2, and 3 are 0.90, 4.06, and 6.51, respectively. It appears that level 3 of C should be used for maximum yield.

In this problem, the experimenters would be interested in maximizing the yield. Suppose they are also concerned with minimizing the variability in the yields, as Taguchi would be. As recommended by Hunter [14], we will use Equation (14.4) to calculate the S/N ratio at each level of each factor. The sample variances and corresponding S/N ratios are summarized in Table 14.23. Plots are given in Figure 14.12.

Recall that larger signal-to-noise ratios are better. Since we have no control over days and there seems to be little difference among operators, it appears that variability can be minimized by using level 3 of factor C. This is most fortunate, since that level also seems to maximize the average yield.

The use of signal-to-noise ratios in the manner presented here is criticized by many. Advocates argue that they have obtained good results. Critics claim that more efficient (and accurate) methods exist and that even the inferior one-factor-at-a-time procedure will produce good results when used with long-neglected processes. Here, our intent is to make the reader aware of the general procedure and philosophy.

Source	DF	Sum of Squares	Mean Square	F Ratio
Model	8	146.8667	18.3583	84.2985
Error	18	3.9200	0.2178	Prob>F
C Total	26	150.786		0.0000

Source	DF	Sum of Squares	F Ratio	Prob>F
A	2	1.5756	3.6173	0.0478
B	2	2.1356	4.9031	0.0200
C	2	142.4156	326.9745	0.0000
AB	2	0.7400	1.6990	0.2109

Figure 14.11   JMP outputs for Example 14.13.

TABLE 14.23
S/N Ratios for Factors in Example 14.13

	Level in Factor A		
	1	**2**	**3**
s^2	4.8378	6.7586	6.0550
S/N	−7.6625	−8.2986	−7.8211

	Level in Factor B		
	1	**2**	**3**
s^2	4.8611	7.4175	6.3028
S/N	−6.8674	−8.7026	−7.995

	Level in Factor C		
	1	**2**	**3**
s^2	0.3975	0.3528	0.2961
S/N	4.0066	4.5247	5.2856

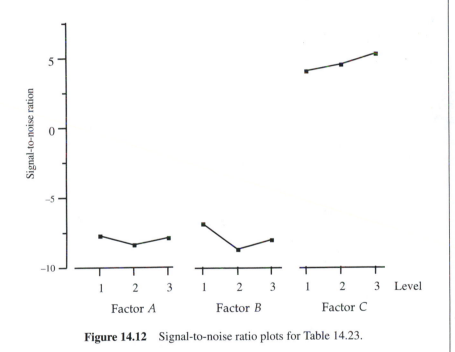

Figure 14.12   Signal-to-noise ratio plots for Table 14.23.

We should note that the experimental design used in Example 14.13 is not unique to Taguchi. If $I = AB^2C^2$ is used as a defining contrast for a 3^{3-1} fractional factorial, then by the methods of Chapter 10 we can write

$$A = A(AB^2C^2) = A^2B^2C^2 = ABC$$

and

$$A = A(AB^2C^2)^2 = BC$$

So, A is aliased with the ABC and BC components of interaction. Proceeding in a similar manner, we can show that B is aliased with ABC^2, C is aliased with AB^2 and AB^2C, and AB is aliased with AC and BC^2. This is precisely the design used in our example.

14.8 SOME OTHER TAGUCHI DESIGNS

So far, the designs have been for either a 2^f of a 3^f factorial experiment. Taguchi also has some orthogonal arrays for 2^f3^g. One is the $L_{12}(2^{11})$ of Table 14.24, where the factorial could be 2^23^1 for 12 treatment combinations. This array can be used to handle up to 11 factors at two levels each with only 12 observations and is equivalent to the array associated with the nongeometric $N = 12$ Plackett–Burman design of Table 13.11.

TABLE 14.24
Taguchi's $L_{12}(2^{11})$ Orthogonal Array

Number	1	2	3	4	5	6	7	8	9	10	11
1	1	1	1	1	1	1	1	1	1	1	1
2	1	1	1	1	1	2	2	2	2	2	2
3	1	1	2	2	2	1	1	1	2	2	2
4	1	2	1	2	2	1	2	2	1	1	2
5	1	2	2	1	2	2	1	2	1	2	1
6	1	2	2	2	1	2	2	1	2	1	1
7	2	1	2	2	1	1	2	2	1	2	1
8	2	1	2	1	2	2	2	1	1	1	2
9	2	1	1	2	2	2	1	2	2	1	1
10	2	2	2	1	1	1	1	2	2	1	2
11	2	2	1	2	1	2	1	1	1	2	2
12	2	2	1	1	2	1	2	1	2	2	1
Group	1	2									

Taguchi claims that $L_{12}(2^{11})$ is a specially designed array with interactions distributed rather uniformly to columns. He discourages its use in the analysis of interactions, but highly recommends it for screening designs involving 11 main effects. We should note, however, that the alias structure for such a design is quite complex in that every main effect is partially aliased with every two-factor interaction that does not involve that main effect (e.g., the nine main effects C through K are partially aliased with AB), and each main effect is partially aliased with 45 two-factor interactions.

Taguchi also highly recommends $L_{18}(2^1 \times 3^7)$ of Table 14.25 for screening experiments. Columns 1 and 2 can be used to design a 2×3 factorial with three observations per treatment combination. As such, complete information can be obtained for both main effects and their interaction. As with $L_{12}(2^{11})$, the alias structure for this array is quite complex. For example, the interaction between two three-level factors is partially aliased with all other three-level factors. A large effect due to one of these interactions could lead the analyst astray.

TABLE 14.25
Taguchi's $L_{18}(2^1 \times 3^7)$ Orthogonal Array

Number	1	2	3	4	5	6	7	8
1	1	1	1	1	1	1	1	1
2	1	1	2	2	2	2	2	2
3	1	1	3	3	3	3	3	3
4	1	2	1	1	2	2	3	3
5	1	2	2	2	3	3	1	1
6	1	2	3	3	1	1	2	2
7	1	3	1	2	1	3	2	3
8	1	3	2	3	2	1	3	1
9	1	3	3	1	3	2	1	2
10	2	1	1	3	3	2	2	1
11	2	1	2	1	1	3	3	2
12	2	1	3	2	2	1	1	3
13	2	2	1	2	3	1	3	2
14	2	2	2	3	1	2	1	3
15	2	2	3	1	2	3	2	1
16	2	3	1	3	2	3	1	2
17	2	3	2	1	3	1	2	3
18	2	3	3	2	1	2	3	1
Group	1	2			3			

14.9 Summary

Several observations can be made concerning the use of Taguchi's experimental designs.

1. They present a procedure for examining many factors in an experiment using a very small number of observations.
2. In using such procedures one must usually assume that no interactions are present or, at least, be able to identify before the experiment some interactions and include them in the design. The assumption of no interactions must, of course, resemble the true state of affairs.
3. Orthogonal arrays are not unique to Taguchi. Neither were they discovered by Taguchi. As noted earlier, many are equivalent to the orthogonal arrays associated with the Plackett–Burman designs. See Box, Bisgaard, and Fung [6] for further information.
4. Taguchi does not include the defining contrasts for the designs associated with his orthogonal arrays. However, one can use his triangular tables and linear graphs for two-level experiments to determine some of the two-factor interactions that are aliased with main effects.
5. Use of the Taguchi approach tends to require considerable time in discussion before the experiment, since it is necessary to assure the experimenter that no

interactions exist or that some that may exist can be worked into the design. Of course, when resources are limited and only a fractional factorial can be run, that time would be spent in developing a good experiment using the methods of Chapter 13.

6. Taguchi's arrays, when used on the maximum number of factors, are all of resolution III. This means that all main effects are independent of other main effects but are confounded with two–way interactions. Such designs are seldom recommended because of their low resolution number.

7. Although it is true that any fractional factorial must confound some interactions, the usual designs attempt to confound only three-way or higher interactions and allow one to safely assess main effects and two-way interactions free of each other. Again, an investigating team needs to carefully consider what aliases and confoundings exist *before* a fractional factorial is run.

8. Taguchi-type designs should be considered only as a screening type of experiment. As with the Plackett–Burman designs, they will usually find the most important main effects.

9. To really find out what factors and interactions are important, one should run either a complete factorial or a fractional factorial of resolution V or higher.

10. Taguchi advocates some unusual methods of analysis. The role of interactions is minimized. Marginal means plots are often used when an interaction plot would provide better information. Signal-to-noise ratios that are highly sensitive to changes in centering are often used to study variability. In many cases a fractional factorial that requires fewer experimental runs could be substituted for crossed designs that involve inner and outer arrays.

11. If using Taguchi methods does nothing more than make experimenters aware of the need for a good experimental design, this innovative researcher has accomplished an important objective.

12. Taguchi's philosophy that design of experiment techniques should be used to design products, processes, and services that are robust to outside sources of variability and that minimizing variability about a target value is more important than merely meeting specifications should be supported. If those goals can be achieved by using experimental methods that are simpler and more efficient than the methods proposed by Taguchi, we should strive to discover and use them.

14.10 FURTHER READING

Bisgaard, S., "A Method for Identifying Defining Contrasts for 2^{k-p} Experiments," *Journal of Quality Technology*, Vol. 25, No. 1, January 1993, pp. 28–35.

Bullington, K. E., J. N. Hool, and S. Maghsoodloo, "A Simple Method for Obtaining Resolution IV Designs for Use with Taguchi's Orthogonal Arrays," *Journal of Quality Technology*, Vol. 22, No. 4, October 1990, pp. 260–264.

Byrne, D. M., and S. Taguchi, "The Taguchi Approach to Parameter Design," *Quality Progress*, December 1987, pp. 19–26.

Kacker, R. N., and K. Tsui, "Interaction Graphs: Graphical Aids for Planning Experiments," *Journal of Quality Technology*, Vol. 22, No. 1, January 1990, pp. 1–14.

Montgomery, D. C., *Design and Analysis of Experiments*, 4th ed. New York: John Wiley & Sons, 1997. See pages 622–641 for a discussion of Taguchi methods.

Ross, P. J., *Taguchi Techniques for Quality Engineering: Loss Function, Orthogonal Experiments, Parameter and Tolerance Design*. New York: McGraw-Hill Book Company, 1988.

Ryan, T. P., "Taguchi's Approach to Experimental Design: Some Concerns," *Quality Progress*, May 1988, pp. 34–36.

Tribus, M., and G. Szonyi, "An Alternative View of the Taguchi Approach," *Quality Progress*, May 1989, pp. 46–52.

PROBLEMS

14.1 Consider the first observation only in each cell of Table 1.2. Analyze these data using a Taguchi $L_4(2^3)$ orthogonal array. Compare the results with those of Examples 14.1 through 14.3.

14.2 Using a Taguchi $L_8(2^7)$ orthogonal array, analyze the data of Problem 9.17. Compare your results with the text answers.

14.3 Use the cell totals in Example 10.1 to solve this problem with an $L_9(3^4)$ orthogonal array. Compare your results with those of the example.

14.4 Problem 13.12a is handled as a half-replicate of a 2^4 factorial. Analyze these data using an $L_8(2^7)$ orthogonal array. Include three interactions that you designate. Compare your results with those of the problem.

14.5 Using an $L_9(3^4)$ orthogonal array, analyze the data of Example 13.9. Compare your results those of the example.

14.6 Using a Taguchi approach, analyze the data of Problem 13.17. Include three two-factor interactions in your model.

14.7 Prove Equation (14.1).

14.8 For Example 14.1, show that SS_B, SS_C, and SS_{total} are as claimed.

14.9 Assuming that three-factor and higher interactions are negligible, analyze the data in Table 14.7 by preparing a normal quantile plot of the estimated effects. Include comments about aliased effects and suggestions for further study.

14.10 An experiment is to be set up using one $L_8(2^7)$ orthogonal array as an inner array and a second $L_8(2^7)$ orthogonal array as an outer array. For the inner array, the experimenters decide to use four factors (A, B, C, D) and three interactions (AB, AC, and AD). The outer array will contain only three noise factors (F, G, and H). Only one observation is to be obtained for each cell of the crossed array.
a. Sketch the resulting layout.

b. Assuming that the outer array is used to determine experimental blocks, write a mathematical model. How might the experiment be improved?

c. Comment about the nature of the experiment if the $8 \times 8 = 64$ observations are obtained in completely random order. Include a mathematical model. If possible, describe a more efficient design.

14.11 Using Taguchi's method of calculating sums of squares, verify the sums of squares associated with Table 14.22.

14.12 When designing an experiment involving the main effects A, B, C, D, and E at two levels each, the experimenters agreed that the only interactions likely to be active were AB and AC. Columns 1, 2, 4, 6, and 7 of $L_8(2^7)$ were used as a design matrix, which yielded the coded data in the following array. The shaded columns form the design matrix and the indicated treatment combinations form the principal block for the resulting 2^{5-2} fractional factorial.

	Factor and Column							**Data**	
	A	**B**	**AB**	**C**	**AC**	**D**	**E**		
tc	1	2	3	4	5	6	7	y_1	y_2
(1)	1	1	1	1	1	1	1	30	26
cde	1	1	1	2	2	2	2	32	27
bde	1	2	2	1	1	2	2	52	44
bc	1	2	2	2	2	1	1	27	29
ae	2	1	2	1	2	1	2	37	35
acd	2	1	2	2	1	2	1	1	6
abd	2	2	1	1	2	2	1	2	3
abce	2	2	1	2	1	1	2	21	12

a. Determine the mathematical model.

b. Using either the manual methods of Taguchi or computer software, prepare an ANOVA table. Include the percent contributions for each main effect and interaction. Interpret. In this case, smaller means are better.

c. Determine the alias structure, defining relation, and design resolution for this design.

14.13 An experiment was designed using $L_9(3^4)$ as an inner array with four main effects (A, B, C, and D) at three levels each and three columns of $L_4(2^3)$ as an outer array with the noise factors E, F, and G. The resulting coded data are included in the following layout. In this case, larger average values are better.

a. Assuming that the outer array was used for blocking, write a mathematical model, prepare an ANOVA summary for the average responses, and interpret.

b. Assuming that the 36 observations were obtained in completely random order, write a mathematical model, prepare an ANOVA summary for the average responses, and interpret. At which levels should each factor be set to maximize the average response?

c. Using the S/N ratios as the responses for the nine treatment combinations of the inner array, prepare an ANOVA summary and interpret.

d. Calculate *S/N* ratios for each level of each main effect. Prepare a plot of *S/N* versus the factor level for each main effect. Interpret.

e. Based on the results of parts b–d, prepare a short summary of your findings and recommendations.

					Outer Array				
						Run			
					1	2	3	4	Factor
	Inner Array				1	1	2	2	E
		Factor			1	2	1	2	F
Run	A	B	C	D	1	2	2	1	G
1	1	1	1	1	13.0	17.3	17.0	17.4	
2	1	2	2	2	12.4	17.0	17.2	21.6	
3	1	3	3	3	13.7	13.0	15.6	20.7	
4	2	1	2	3	15.7	16.0	16.3	20.6	
5	2	2	3	1	17.1	22.5	18.8	24.9	
6	2	3	1	2	13.6	17.2	17.0	19.9	
7	3	1	3	2	13.8	21.0	16.0	21.7	
8	3	2	1	3	11.6	14.2	17.0	20.6	
9	3	3	2	1	13.5	14.7	20.1	20.0	

14.14 Experimenters wish to design an experiment to study six factors (A, B, C, D, E, and F), each at two levels. They have good reason to believe that only B and D interact. Funds and time are limited, so they agree that only eight experimental runs are feasible.

a. Using the linear graph of Figure 14.4b, design an experiment with B and D assigned to columns 1 and 2, respectively, of $L_8(2^7)$. Are any two-factor interactions confounded with main effects? If so, will this pose a problem? Comment.

b. Determine the defining relation and complete alias structure for the design in part a. Comment.

c. The experimental design of part a is a one-eighth replicate of a 2^6 factorial of resolution III. Improve on that design by designing a one-eighth replicate that is of resolution IV. Give the defining relation and complete alias structure. Prepare an orthogonal table like $L_8(2^7)$ with aliases and treatment combinations included.

14.15 In the manufacture of integrated circuits, monitor wafers, coated with one of three films, are used to verify the etch rate of the process. A project team wishes to determine which of three types of film (A, B, or C) is most sensitive to changes in the process. The three films were etched at constant pressure and power, with the flow rates of three gases (D, E, and F) varied according to a 2^3 factorial used as an outer array. The order in which the eight combinations of gas flow rates were used was randomized, and two wafers were etched for each of the three types of film at each combination. This experimental design produced the following data and layout. Two observations were identified as outliers. Further study revealed processing problems due to the experimental setups, so the observations were deleted from the data set.

Outer Array

Run

	1	2	3	4	5	6	7	8	Gas
	1	1	1	1	2	2	2	2	D
	1	1	2	2	1	1	2	2	E
Film	1	2	1	2	1	2	1	2	F
A	999	1197	1080	1186	931	1003	1051	1023	
	1031	1219	1089	1224	992	1071	1059	1041	
B	251	328	290	324	248	222	230	269	
		328	315	361	251	247	249	281	
C	4599	5962	4464	5111	4910	5792	4682	5424	
	4671	6003	4549	5137	5041	5833	4923		

Write a mathematical model, prepare an ANOVA summary for the average responses, and interpret.

Chapter 15

REGRESSION

15.1 INTRODUCTION

In Figure 3.1 (After ANOVA, what?), we noted several procedures that might follow an analysis of variance when the effects of factors were significant. However, we have not yet treated cases in which the factor levels are quantitative. Figure 3.1 suggests two procedures to follow when significant effects are present: the method of orthogonal polynomials when the factor levels are equispaced, and general methods of curve fitting or regression.

Some experimenters prefer to treat all analyses as regression models because even the ANOVA procedures considered to this point can be cast into a regression model. This is evidenced by the term GLM (general linear model) of most computer packages. If any one of the independent variables (X's) is quantitative and its effect is statistically significant, one probably should use regression analysis in an attempt to determine a mathematical model suitable for predicting the magnitude of the response variable from values of the independent variables. This regression procedure will be examined for Y as a function of one X, where the relationship may be linear, quadratic, or a higher order polynomial. The case of multiple regression, where Y is a function of several X's, will also be considered. Our search for an adequate prediction equation relies on computer software because most real problems are too difficult to handle otherwise.

Some Notation

Suppose an experiment is conducted for which n_j observations are made when $X = x_j$. Throughout this chapter we will let y_{ij} denote the ith value of Y at x_j. The average of the sample of n_j observations is denoted $\bar{y}_{.j}$, and the predicted value of Y is denoted y'_{x_j} with the subscripts omitted as the context permits.

Before data are collected, the ith value of Y at $X = x_j$ is a random variable and, as such, will be denoted Y_{ij}. Likewise, $\bar{Y}_{.j}$ denotes the sample mean that will result once the sample data have become available, and $Y'_{X=x_j}$ denotes the value of Y that will result.

15.2 LINEAR REGRESSION

To review linear regression and become familiar with the notation used, consider an experiment designed to determine the effect of the amount of drug dosage X (in milligrams)

403

on a person's reaction time Y (in milliseconds) to a given stimulus. When 15 subjects were randomly assigned one of five dosages of drug—0.5, 1.0, 1.5, 2.0, or 2.5 mg—with three subjects assigned to a given drug dosage, the reaction times of Table 15.1 were recorded.

From Table 15.2, the ANOVA results for the data in Table 15.1, it is obvious that dosage has a highly significant effect on reaction time. But since reaction time is a quantitative factor, the next question might well be: How does reaction time vary with drug dosage? Can one find a functional relationship between these two variables that might allow the prediction of reaction time from drug dosage?

As a first step, we construct Figure 15.1, a scattergram of all 15 (x, y) pairs of points, where the $\times$'s denote the observed average, $\bar{y}_{.j}$, at each x_j. The plot shows that Y increases with increases in X.

A reasonable second step is to try a straight-line "fit" as a first approximation for predicting Y from X. Such a line is shown in Figure 15.1. One can see how close the line comes to $\bar{y}_{.j}$ for each x_j. Note that for a particular value (or level) of X, the ith value of Y can be partitioned into three parts:

1. y'_{x_j}, which is the predicted value of Y at $X = x_j$.
2. $\bar{y}_{.j} - y'_{x_j}$, which is the amount by which the sample average for Y at $X = x_j$ deviates from its predicted value (referred to as the *departure from linear regression*).

TABLE 15.1
Reaction Time Data

		Dosage			
	0.5	1.0	1.5	2.0	2.5
Time	26	28	28	32	38
	28	26	30	33	39
	29	30	31	31	38

TABLE 15.2
Minitab ANOVA for Reaction Time

One-Way Analysis of Variance

Analysis of Variance on Time

Source	DF	SS	MS	F	p
Dosage	4	229.73	57.43	28.72	0.000
Error	10	20.00	2.00		
Total	14	249.73			

```
                                   Individual 95% CIs For Mean
                                   Based on Pooled StDev
Level      N      Mean     StDev   -------+----------+----------+----------+
   1       3    27.667     1.528   (---*----)
   2       3    28.000     2.000   (----*----)
   3       3    29.667     1.528      (---*----)
   4       3    32.000     1.000           (----*----)
   5       3    38.333     0.577                            (----*---)
                                   ------+----------+---------+---------+
Pooled StDev =     1.414             28.0       32.0      36.0      40.0
```

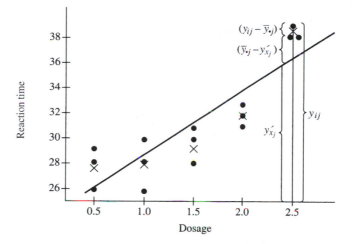

Figure 15.1 Reaction time versus drug dosage.

3. $y_{ij} - \bar{y}_{.j}$, which is the amount by which the *i*th value of Y at $X = x_j$ varies from the mean of the sample of n_j observations.

15.2.1 The Simple Linear Model

In the analysis of variance model we have

$$Y_{ij} = \mu + (\mu_j - \mu) + (Y_{ij} - \mu_j) \tag{15.1}$$

and now we try to predict μ_j from some function based on X_j. If this predicted mean is labeled $\mu_{Y|X}$, the model can be expanded to read

$$Y_{ij} = \mu + (\mu_{Y|X} - \mu) + (\mu_j - \mu_{Y|X}) + (Y_{ij} - \mu_j) \tag{15.2}$$

where $\mu_{Y|X}$ is the true predicted mean based on X and $\mu_j - \mu_{Y|X}$ is the amount by which the true mean μ_j departs from its predicted value.

For the sample model, Equation (15.2) becomes

$$Y_{ij} = \bar{Y}_{..} + (Y'_X - \bar{Y}_{..}) + (\bar{Y}_{.j} - Y'_X) + (Y_{ij} - \bar{Y}_{.j}) \tag{15.3}$$

If the second and third terms on the right of this equation are combined, we have the ANOVA sample model. Thus, we attempt here to partition the mean of a given treatment into two parts: that which can be predicted by regression on X, and the amount by which the mean shows a departure from such a regression model. The purpose, of course, is to find a model for predicting the response means for which the departures from these means are very small. By choosing a polynomial of high enough degree, one can find a model that actually goes through all the Y means. However, it is hoped that an adequate lower degree model can be found whose departures are small and insignificant compared with the error of individual Y's around their means.

The Straight-Line Model

In Equation (15.3) Y'_X represents the predicted Y (or average Y) for a given X for any assumed model. The straight-line model is given by

$$Y'_X = B_0 + B_1 X \tag{15.4}$$

where B_0 is the Y intercept and B_1 is the slope. When the true average Y and the corresponding level of X are linearly related, B_0 and B_1 are estimators of the intercept and slope, respectively, of the true line of the means. In many cases, of course, such a linear relation is only approximate but adequate.

15.2.2 The Least Squares Line

The usual method for determining estimators of parameters in a simple linear model is the method of *least squares*. The estimators in Equation (15.4) are determined from a random sample in a way that minimizes the sum of squares of the deviations of each Y_{ij} from its predicted value Y'_X. [One could also find estimators such that the sum of the squares of the departures from regression (i.e., $\bar{Y}_{.j} - Y'_X$) is minimized, but this leads to the same results.] In this case,

$$Y_{ij} - Y'_X = Y_{ij} - B_0 - B_1 X_j$$

and

$$\text{SS}_{\text{deviations}} = \sum_i \sum_j (Y_{ij} - Y'_X)^2 = \sum_i \sum_j (Y_{ij} - B_0 - B_1 X_j)^2$$

Because the summations of X's and Y's can be found from a given set of data, $\text{SS}_{\text{deviations}}$ is a function of B_0 and B_1. To minimize $\text{SS}_{\text{deviations}}$ then, one differentiates $\text{SS}_{\text{deviations}}$ and sets the two partial derivatives to zero. This gives

$$\frac{\partial (\text{SS}_{\text{deviations}})}{\partial B_0} = 2 \sum_i \sum_j (Y_{ij} - B_0 - B_1 X_j)(-1) = 0$$

$$\frac{\partial (\text{SS}_{\text{deviations}})}{\partial B_1} = 2 \sum_i \sum_j (Y_{ij} - B_0 - B_1 X_j)(-X_j) = 0$$

Dividing through by -2 and simplifying gives the system of equations

$$\left. \begin{aligned} \sum_i \sum_j Y_{ij} &= B_0 N + B_1 \sum_j n_j X_j \\[2ex] \sum_i \sum_j X_j Y_{ij} &= B_0 \sum_j n_j X_j + B_1 \sum_j n_j X_j^2 \end{aligned} \right\} \tag{15.5}$$

with N the total sample size and n_j the sample size at the jth level of X. These equations are often called the *least squares normal equations*, which can now be solved to obtain expressions for the random variables B_0 and B_1. The sample values obtained for a specific sample of (x, y) pairs are denoted b_0 and b_1. These are obtained by solving the system

$$\left.\begin{array}{l}\sum_i \sum_j y_{ij} = b_0 N + b_1 \sum_j n_j x_j \\[2ex] \sum_i \sum_j x_j y_{ij} = b_0 \sum_j n_j x_j + b_1 \sum_j n_j x_j^2 \end{array}\right\} \qquad (15.6)$$

and the resulting sample equation is

$$y_x' = b_0 + b_1 x \qquad (15.7)$$

■ **Example 15.1   (Least Squares Line for Reaction Time Data)**

The analysis of variance for the reaction time data of Table 15.1 revealed a significant time effect. We will now begin a search for an adequate model of reaction time as a function of dosage. For convenience, the data have been reproduced with totals in Table 15.3.

From Table 15.3, we write $\sum_i \sum_j y_{ij} = \sum_j T_{.j} = 83 + 84 + 89 + 96 + 115 = 467$; $n_j = 3$ for $j = 1, 2, 3, 4, 5$; and $N = 3 \times 5 = 15$. Further,

$$\sum_j x_j = 0.5 + 1.0 + 1.5 + 2.0 + 2.5 = 7.5$$

and

$$\sum_j x_j^2 = (0.5)^2 + (1.0)^2 + (2.0)^2 + (2.5)^2 = 13.75$$

To find the cross product, note that since $\sum_i y_{ij} = T_{.j}$, $\sum_i \sum_j x_i y_{ij} = \sum_j x_j T_{.j}$. Thus,

$$\sum_i \sum_j x_i y_{ij} = (0.5)(83) + (1.0)(84) + (1.5)(89) + (2.0)(96) + (2.5)(115) = 738.5$$

Substituting in Equation (15.6) gives

$$467.0 = 15.0 b_0 + 22.50 b_1$$
$$738.5 = 22.5 b_0 + 41.25 b_1$$

TABLE 15.3
Reaction Time Data with Totals

	Dosage				
	0.5	1.0	1.5	2.0	2.5
Time	26	28	28	32	38
	28	26	30	33	39
	29	30	31	31	38
$T_{.j}$	83	84	89	96	115

To solve the system given here, one might multiply the first equation by 1.5, giving

$$700.5 = 22.5b_0 + 33.75b_1$$

$$738.5 = 22.5b_0 + 41.25b_1$$

Subtracting and solving for b_1, we have

$$b_1 = \frac{700.5 - 738.5}{33.75 - 41.25} = \frac{-38.0}{-7.5} = \frac{76}{15}$$

and from the original first equation,

$$b_0 = \frac{467 - (22.5)(76/15)}{15} = \frac{353}{15}$$

So, the linear model for this set of sample data is

$$y'_x = \frac{353}{15} + \frac{76}{15}x \approx 23.53 + 5.07x$$

The sample equation for a least squares line is seldom obtained by solving the system of normal equations directly. Even moderately priced calculators often contain a simple linear regression routine. Regression modules are also included in statistical computing programs. For example, when the Table 15.3 data are analyzed using the **Regression** module in Minitab we obtain Figure 15.2. The test for a linear effect (a t test on milligrams or an F test on regression) indicates that inclusion of an x term in our model is reasonable ($p = 0.000$).

A SAS Program

The least squares line obtained manually in Example 15.1 and included in the Minitab summary of Figure 15.2 can also be obtained using the **GLM** (or **REG**) procedure in SAS.

```
Regression Analysis

The regression equation is
Time = 23.5 + 5.07 dose

Predictor        Coef       Stdev      t-ratio          p
Constant       23.533       1.270        18.53      0.000
dose            5.067       0.766         6.61      0.000

s = 2.098     R-sq = 77.1%       R-sq(adj) = 75.3%

Analysis of Variance

SOURCE           DF          SS          MS          F          p
Regression        1      192.53      192.53      43.76      0.000
Error            13       57.20        4.40
Total            14      249.73
```

Figure 15.2 Minitab regression analysis for reaction time study.

An appropriate command file is given in Table 15.4. Since we are estimating a functional relationship between the two variables, a class statement is not used. Except for rounding, the slope and intercept included in the outputs of Figure 15.3 are the same as those obtained manually and using Minitab.

A Formula for B_0 and B_1

If the normal equations of Equation (15.5) are solved simultaneously, we have

$$B_0 = \bar{Y} - B_1\bar{X} \tag{15.8}$$

and

$$B_1 = \frac{\sum_i \sum_j X_i Y_{ij} - \frac{(\sum_i \sum_j X_i)(\sum_i \sum_j Y_{ij})}{N}}{\sum_i \sum_j X_i^2 - \frac{(\sum_i \sum_j X_j)^2}{N}} = \frac{\sum_i \sum_j (X_i - \bar{X})(Y_{ij} - \bar{Y})}{\sum_i \sum_j (X_i - \bar{X})^2} \tag{15.9}$$

TABLE 15.4
SAS Command File for Example 15.1

```
OPTIONS LINESIZE=80;
DATA REACTION;
INPUT DOSE TIME @@;
CARDS;
0.5 26 0.5 28 0.5 29 1.0 28 1.0 26 1.0 30 1.5 28 1.5 30 1.5 31
2.0 32 2.0 33 2.0 31 2.5 38 2.5 39 2.5 38
;
PROC GLM;
MODEL TIME = DOSE;
```

```
                    GENERAL LINEAR MODELS PROCEDURE

DEPENDENT VARIABLE: TIME

                                SUM OF           MEAN
SOURCE                  DF      SQUARES          SQUARE        F VALUE
MODEL                    1   192.53333333    192.53333333       43.76
ERROR                   13    57.20000000      4.40000000       PR > F
CORRECTED TOTAL         14   249.73333333                       0.0001

R-SQUARE              C.V.      ROOT MSE       TIME MEAN
0.770956           6.7375    2.09761770     31.13333333

SOURCE                  DF     TYPE I SS       F VALUE         PR > F
DOSE                     1   192.53333333       43.76          0.0001

SOURCE                  DF    TYPE III SS      F VALUE         PR > F
DOSE                     1   192.53333333       43.76          0.0001

                                T FOR H0:     PR > |T|    STD ERROR OF
PARAMETER           ESTIMATE   PARAMETER=0                    ESTIMATE
INTERCEPT         23.53333333       18.53        0.0001     1.27017059
DOSE               5.06666667        6.61        0.0001     0.76594169
```

Figure 15.3 SAS output for Example 15.1.

Denoting the numerator and denominator of the last fraction in Equation (15.9) by SP_{XY} and SS_X, respectively, we have

$$B_1 = \frac{\text{SP}_{XY}}{\text{SS}_X} \tag{15.10}$$

■ **Example 15.2 [Using Equations (15.8) and (15.9) with Reaction Time Data]**

From Example 15.1, we know that $\sum_i \sum_j y_{ij} = 467$, $\sum_i \sum_j x_j = 3(7.5) = 22.5$, $\sum_i \sum_j x_j^2 = 3(13.75) = 41.25$, $N = 15$, and $\sum_i \sum_j x_i y_{ij} = 738.5$ for the reaction time data of Table 15.1. Using these with the sample equivalents of Equations (15.8) and (15.9), we find

$$b_1 = \frac{\sum_i \sum_j x_i y_{ij} - \dfrac{\left(\sum_i \sum_j x_i\right)\left(\sum_i \sum_j y_{ij}\right)}{N}}{\sum_i \sum_j x_i^2 - \dfrac{\left(\sum_i \sum_j x_j\right)^2}{N}} = \frac{738.5 - \dfrac{(22.5)(467)}{15}}{41.25 - \dfrac{(22.5)^2}{15}} = \frac{76}{15}$$

and

$$b_0 = \bar{y} - b_1 \bar{x} = \frac{467}{15} - \frac{76}{15} \times \frac{22.5}{15} = \frac{353}{15}$$

as before.

15.2.3 Departure from the Linear Model: A Lack of Fit Test

We noted earlier that the line of Figure 15.1 does not provide a perfect fit to the data. In fact, that figure indicates that a nonlinear effect is present. To see how well the empirical model

$$y_x' = \frac{353}{15} + \frac{76}{15}x$$

predicts the observed value of Y from a given value of X, we construct Table 15.5. This table shows the departures from linear regression. For example, when the observed value of X is 1.0, the predicted value of Y is $y' = 353/15 + (76/15)(1) = 143/5$, or 28.6. But the average value of Y when X is 1.0 is $84/3 = 28$. In this instance, the departure from linear regression is

$$\bar{y} - y' = 28.0 - 28.6 = -0.6 = -\tfrac{3}{5}$$

Note that the sum of the departures adds to zero and the observed sum of squares of departures from linear regression is

$$\text{SS}_{\text{departures}} = \sum_j n_j (\bar{y}_{.j} - y_{x_j}')^2 = 3\left(\frac{2791}{225}\right) = \frac{2791}{75} \approx 37.21$$

as each departure of a mean must be weighted with three observed values for each value of X.

TABLE 15.5
Departures from Linear Regression

x_j	$\bar{y}._j$	y'_x	$\bar{y}._j - y'_x$	$(\bar{y}._j - y'_x)^2$
0.5	83/3	391/15	24/15	576/225
1.0	84/3	429/15	−9/15	81/225
1.5	89/3	467/15	−22/15	484/225
2.0	96/3	505/15	−25/15	625/225
2.5	115/3	543/15	32/15	1025/225
Totals			0	2791/225

From Equation (15.3), we write

$$(Y'_X - \bar{Y}..) + (\bar{Y}._j - Y'_X) = \bar{Y}._j - \bar{Y}..$$

where the right-most term is the deviation of a treatment mean from the overall mean in an ANOVA model. Squaring both sides and summing over i and j gives

$$\sum_i \sum_j (Y'_X - \bar{Y}..)^2 + \sum_i \sum_j (\bar{Y}._j - Y'_X)^2 = \sum_i \sum_j (\bar{Y}._j - \bar{Y}..)^2 \qquad (15.11)$$

with the third sum of squares the treatment sum of squares for the ANOVA model. When the sum of squares of departures from linear regression is known, the ANOVA Table 15.2 can be expanded to give Table 15.6. This table shows a highly significant linear effect but also a significant departure from the linear model (p value $= 0.012$). This indicates that a higher order model may be required to adequately predict the reaction time from drug dosage.

Notice that the sum of $SS_{departure}$ and SS_{error} in Table 15.6 is the error sum of squares in Figures 15.2 and 15.3. Thus, the regression error has been decomposed into the error you get from ANOVA (often called *pure error*) and a portion due to (nonlinear) treatment effects.

Many statistical computing programs include such an analysis, called a lack of fit test, as an option but to use this option requires multiple observations on at least one x. For example, Minitab includes

TABLE 15.6
ANOVA on Reaction Time with Linear Regression

Source	df	SS	MS	F	p value
Between dosages*	4	229.73			
Linear	1	192.53	192.53	96.3	0.000
Departure from linear	3	37.21	12.40	6.2	0.012
Error	10	20.00	2.00		
Totals	14	249.73			

*$SS_{linear} + SS_{departure} \neq SS_{between}$ due to round-off error.

Pure error test $-\text{F} = 6.20$ $\text{P} = 0.0119$ $\text{DF(pure error)} = 10$

with the outputs of Figure 15.2 when a test for departure from the linear model is requested.

Because there is a strong linear effect in this example and, in some problems, the linear model is adequate to explain most of the variation in the Y variable, we now consider some useful statistics that can be determined from Table 15.6.

Coefficient of Determination

The proportion of the total sum of squares that can be accounted for by linear regression is sometimes called the *coefficient of determination* and denoted r^2. From Table 15.6, the coefficient of determination for the reaction time data is found to be

$$r^2 = \frac{\text{SS}_{\text{linear}}}{\text{SS}_{\text{total}}} = \frac{192.53}{249.73} = 0.7712$$

Thus, linear regression will account for about 77% of the variation seen in the reaction time Y.

Recall that the positive square root of the coefficient of determination with the sign of b_1, affixed, denoted r, is the *Pearson product–moment correlation coefficient*. In this case, $r = (0.7712)^{0.5} = 0.88$.

Eta Squared

The ratio of the sum of squares between means to the total sum of squares is called *eta squared* and is denoted η^2. This ratio gives the maximum amount of the total variation that could be accounted for by a curve or model that passes through all the mean Y's for each X. Here

$$\eta^2 = \frac{\text{SS}_{\text{between}}}{\text{SS}_{\text{total}}} = \frac{229.73}{249.73} = 0.9199$$

Thus, approximately 92% of the variation in reaction time can be accounted for by a model through all the means for each of the five dosage levels.

Standard Error of Estimate

Another statistic used with a linear model is the standard error of estimate, denoted $S_{Y.X}$. It is the standard deviation of the deviations of the Y_{ij}'s from their predicted values

$$Y_{ij} - Y'_X = (Y_{ij} - \bar{Y}_{.j}) + (\bar{Y}_{.j} - Y'_X)$$

This expression adds the error about the mean and the departure of the mean from the predicted curve to obtain

$$S_{Y.X} = \sqrt{\frac{\text{SS}_{\text{departure}} + \text{SS}_{\text{error}}}{N - 2}}$$

and is appropriate only if the departure is nonsignificant. This statistic is often used to set confidence limits around a line of best fit when linear regression is appropriate.

15.2.4 The Use of Equispaced Levels

To this point, no use has been made of the fact that the dosages of the reaction time study are equispaced. In such cases, one can code the X_j's by considering their overall mean $\bar{X}$ and the width c of the interval between two consecutive values. If we choose

$$U_j = \frac{X_j - \bar{X}}{c} \tag{15.12}$$

the simple linear model becomes

$$Y'_U = B'_0 + B'_1 U \tag{15.13}$$

whose least squares normal equations are given by

$$\left.\begin{array}{l} \displaystyle\sum_i \sum_j Y_{ij} = B'_0 N + B'_1 N \sum_j U_j \\[2em] \displaystyle\sum_i \sum_j U_j Y_{ij} = B'_0 \sum_j n_j U_j + B'_1 \sum_j n_j U_j^2 \end{array}\right\} \tag{15.14}$$

Upon equispacing the values of X, we find that the corresponding values of U sum to zero. Knowing this, we can give the intercept and slope of the least squares line for a particular sample by

$$\left.\begin{array}{l} b'_0 = \dfrac{\sum_i \sum_j y_{ij}}{N} = \bar{y} \\[2em] b'_1 = \dfrac{\sum_i \sum_j u_j y_{ij}}{\sum_j n_j u_j^2} = \dfrac{\sum_j u_j T_{.j}}{\sum_j n_j u_j^2} \end{array}\right\} \tag{15.15}$$

Equation (15.15) is relatively easy to solve. It also has the advantage that the numerator of the expression for the slope is a contrast in the treatment totals. Thus, the observed sum of squares due to linear regression can be found by means of

$$SS_{\text{linear}} = \frac{\left(\sum_j u_j T_{.j}\right)^2}{\sum_j n_j u_j^2} \tag{15.16}$$

■ **Example 15.3 (Another Look at the Least Squares Line for Reaction Time Data)**

For the reaction time data of Table 15.1, the levels of X are 0.5, 1.0, 1.5, 2.0, and 2.5, with three observations per level. Thus, $\bar{x} = (0.5 + 1.0 + 1.5 + 2.0 + 2.5)/5 = 1.5$. Since the width of the interval between any two consecutive levels is 0.5, the lowest level of U is $u_1 = (0.5 - 1.5)/0.5 = -2$. Likewise, $u_2 = -1, u_3 = 0, u_4 = 1$, and $u_5 = 2$. From Table 15.3, the sample totals are $T_{.1} = 83, T_{.2} = 84, T_{.3} = 89, T_{.4} = 96$, and $T_{.5} = 115$. Summing these totals gives $\bar{y} = 467/15$. From Equation (15.15),

$$b'_0 = \bar{y} = \frac{467}{15}$$

and

$$b_1' = \frac{\sum_j u_j T_{\cdot j}}{\sum_j n_j u_j^2} = \frac{(-2)(83) + (-1)(84) + (0)(89) + (1)(96) + (2)(115)}{3[(-2)^2 + (-1)^2 + (0)^2 + (1)^2 + (2)^2]}$$

$$= \frac{76}{30} = \frac{38}{15}$$

So, the straight-line model is

$$y_u' = \frac{467}{15} + \frac{38}{15}u$$

To see that this is the same model, substitute $(x - 1.5)/0.5$ for u to obtain

$$y_x' = \frac{467}{15} + \frac{38}{15} \times \frac{x - 1.5}{0.5} = \frac{467}{15} - \frac{114}{15} + \frac{76}{15}x = \frac{353}{15} + \frac{76}{15}x$$

as before. Note also that we can use Equation (15.16) to obtain

$$SS_{\text{linear}} = \frac{(76)^2}{30} = \frac{2888}{15} = 192.53$$

as shown in Table 15.6.

15.3 CURVILINEAR REGRESSION

When a least squares line is not sufficient to explain all the significant variation in the means, the next logical step is to consider the second-degree (or quadratic) least squares polynomial:

$$y_x' = b_0 + b_1 x + b_2 x^2 \tag{15.17}$$

In fact, careful consideration of the scattergram in Figure 15.1 combined with the significant departure from linear for the reaction time data leads us to believe that such a model will provide a better description of the relationship between dosage and reaction time than that of the least squares straight line.

Formulas for the estimates b_0, b_1, and b_2 are obtained by setting the three partial derivatives for Equation (15.17) to zero and solving the resulting system of normal equations. In Problem 15.37, the reader is asked to show that these equations are

$$\left. \begin{array}{l} \sum_i \sum_j y_{ij} = b_0 N + b_1 \sum_j n_j x_j + b_2 \sum_j n_j x_j^2 \\[2mm] \sum_i \sum_j x_j y_{ij} = b_0 \sum_j n_j x_j + b_1 \sum_j n_j x_j^2 + b_2 \sum_j n_j x_j^3 \\[2mm] \sum_i \sum_j x_j^2 y_{ij} = b_0 \sum_j n_j x_j^2 + b_1 \sum_j n_j x_j^3 + b_2 \sum_j n_j x_j^4 \end{array} \right\} \tag{15.18}$$

The standard Gaussian elimination procedures can be used to solve these equations.

When the levels of X are equispaced, as is the case with the reaction time data, solution of Equation (15.18) is simplified by using the coding of Equation (15.12). This gives

$$y_u' = b_0' + b_1'u + b_2'u^2 \tag{15.19}$$

$$\left. \begin{aligned} \sum_i \sum_j y_{ij} &= b_0'N + b_1' \sum_j n_j u_j + b_2' \sum_j n_j u_j^2 \\ \sum_i \sum_j u_j y_{ij} &= b_0' \sum_j n_j u_j + b_1' \sum_j n_j u_j^2 + b_2' \sum_j n_j u_j^3 \\ \sum_i \sum_j u_j^2 y_{ij} &= b_0' \sum_j n_j u_j^2 + b_1' \sum_j n_j u_j^3 + b_2 \sum_j n_j u_j^4 \end{aligned} \right\} \tag{15.20}$$

Because of the choice of the u_j's, each term in Equation (15.20) that involves odd powers of the u_j's equals zero. So, the equations become

$$\left. \begin{aligned} \sum_i \sum_j y_{ij} &= b_0'N + b_2' \sum_j n_j u_j^2 \\ \sum_i \sum_j u_j y_{ij} &= b_1' \sum_j n_j u_j^2 \\ \sum_i \sum_j u_j^2 y_{ij} &= b_0' \sum_j n_j u_j^2 + b_2' \sum_j n_j u_j^4 \end{aligned} \right\} \tag{15.21}$$

■ **Example 15.4 (Least Squares Parabola for Reaction Time Data)**

Consider again the reaction time data of Table 15.1. There are three observations at each of the five dosage levels so $N = 3 \times 5 = 15$ and $n_j = 3$ for $j = 1, 2, 3, 4,$ and 5. From Table 15.3, $T_{.1} = 83$, $T_{.2} = 84$, $T_{.3} = 89$, $T_{.4} = 96$, and $T_{.5} = 115$. So,

$$\sum_i \sum_j y_{ij} = \sum_j T_{.j} = 83 + 84 + 89 + 96 + 115 = 467$$

From Example 15.3, $u_1 = -2$, $u_2 = -1$, $u_3 = 0$, $u_4 = 1$, and $u_5 = 2$. Thus,

$$\sum_i \sum_j u_j y_{ij} = \sum_j u_j T_{.j} = (-2)(83) + (-1)(84) + (0)(89) + (1)(96)$$

$$+ (2)(115) = 76$$

$$\sum_i \sum_j u_j^2 y_{ij} = \sum_j u_j^2 T_{.j} = (-2)^2(83) + (-1)^2(84) + (0)^2(89) + (1)^2(96)$$

$$+ (2)^2(115) = 972$$

$$\sum_j n_j u_j^2 = (3)[(-2)^2 + (-1)^2 + (0)^2 + (1)^2 + (2)^2] = 3(10) = 30$$

$$\sum_j n_j u_j^4 = (3)[(-2)^4 + (-1)^4 + (0)^4 + (1)^4 + (2)^4] = 3(34) = 102$$

Substituting in Equation (15.21) gives the normal equations

$$
\begin{cases}
467 = 15b_0' + 30b_2' \\
76 = 30b_1' \\
972 = 30b_0' + 102b_2'
\end{cases}
$$

Solving by Gaussian elimination gives

$$
b_0' = \frac{3079}{105}, \ b_1' = \frac{38}{15}, \ \text{and} \ b_2' = \frac{19}{21}
$$

The quadratic model in terms of U is

$$
y_u' = \frac{3079}{105} + \frac{38}{15}u + \frac{19}{21}u^2
$$

But $u = (x - 1.5)/0.5 = 2x - 3$, so the quadratic model in terms of X is

$$
\begin{aligned}
y_x' &= \frac{3079}{105} + \frac{38}{15}(2x - 3) + \frac{19}{21}(2x - 3)^2 \\
&= \frac{1}{105}(3136 - 608x + 380x^2) \\
&\approx 29.87 - 5.79x + 3.62x^2
\end{aligned}
$$

Least squares polynomials are usually obtained with the aid of a statistics package. Some of these programs not only calculate the sample model but include the graph of the fitted polynomial and other information useful to the analyst. In Figure 15.4 we have included some of the information produced by JMP for the Table 15.1 data. Notice how well the parabola fits the scattergram. Also notice that except for rounding, the least squares equation (in terms of x) is the same as that obtained in Example 15.4.

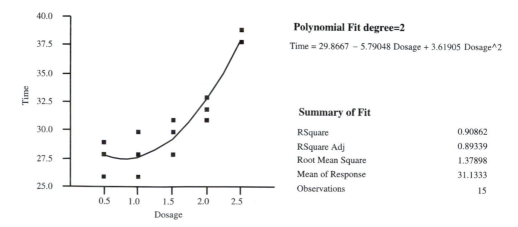

Figure 15.4   JMP quadratic fit for reaction time data.

A SAS Program

A SAS command file that produces a least squares line for the reaction time data was given in Table 15.4. We can have SAS calculate a least squares polynomial for these data by defining a new variable, the square of the dosage, and including that variable in the model statement. This is illustrated in Table 15.7. Except for rounding, the resulting model (see Figure 15.5) is the same as that obtained manually and using JMP.

15.3.1 Departure from the Quadratic Model: A Lack of Fit Test

The parabola of Figure 15.4 fits the reaction time data better than the line of Figure 15.1. This is evidenced by our visual perception as well as by the observation that the coefficient of determination for the quadratic model ($R^2 = 0.908626$) is markedly larger than that for the

TABLE 15.7
SAS Command File for Example 15.4

```
OPTIONS LINESIZE=80;
DATA REACTION;
INPUT DOSE TIME @@;
CARDS;
0.5 26 0.5 28 0.5 29 1.0 28 1.0 26 1.0 30 1.5 28 1.5 30 1.5 31
2.0 32 2.0 33 2.0 31 2.5 38 2.5 39 2.5 38
;
DOSE2 = DOSE*DOSE;
PROC GLM;
MODEL TIME = DOSE DOSE2;
```

GENERAL LINEAR MODELS PROCEDURE

DEPENDENT VARIABLE: TIME

SOURCE	DF	SUM OF SQUARES	MEAN SQUARE	F VALUE
MODEL	2	226.91428571	113.45714286	59.66
ERROR	12	22.81904762	1.90158730	PR > F
CORRECTED TOTAL	14	249.73333333		0.0001

R-SQUARE	C.V.	ROOT MSE	TIME MEAN	
0.908626	4.4293	1.37898053	31.13333333	

SOURCE	DF	TYPE I SS	F VALUE	PR > F
DOSE	1	192.53333333	101.25	0.0001
DOSE2	1	34.38095238	18.08	0.0011

SOURCE	DF	TYPE III SS	F VALUE	PR > F
DOSE	1	9.41339445	4.95	0.0460
DOSE2	1	34.38095238	18.08	0.0011

PARAMETER	ESTIMATE	T FOR H0: PARAMETER=0	PR > |T|	STD ERROR OF ESTIMATE
INTERCEPT	29.86666667	17.49	0.0001	1.70756177
DOSE	-5.79047619	-2.22	0.0460	2.60255133
DOSE2	3.61904762	4.25	0.0011	0.85112526

Figure 15.5   SAS output for Example 15.4.

linear model ($r^2 = 0.770956$). We now turn our attention to a formal test of the "goodness" of the fit of the quadratic model. Before proceeding, recall that eta squared is 0.9199 for these data, indicating that the variabilty in reaction times accounted for by our quadratic model is near the maximum amount that can be accounted for by a curve passing through the five sample means.

From Example 15.3, the quadratic model for the reaction time data is

$$y'_u = \frac{3079}{105} + \frac{38}{15}u + \frac{19}{21}u^2 = \frac{1}{105}(3079 + 266u + 95u^2)$$

when expressed in terms of U values or

$$y'_x = \frac{1}{105}(3136 - 608x + 380x^2)$$

when expressed in terms of X values. Either can be used to see how well the model predicts the Y's from the X's. Table 15.8 shows the predicted Y's and the departure of the means from the quadratic model.

Since each departure of a mean must be weighted with three observed values for each value of X, we can write the sum of the squares of the departure from the quadratic from Table 15.8 as follows:

$$SS_{departures} = \sum_j n_j(\bar{y}_{.j} - y'_{x_j})^2 = 3\left(\frac{10,360}{11,025}\right) = \frac{296}{105} \approx 2.82$$

This step can be added to refine Table 15.6 further and give Table 15.9. The departure from the quadratic is not significant at the 5% level ($p = 0.519$), so it is appropriate to stop with the quadratic for predicting reaction time from dosage.

Note that the error and nonsignificant departure term could be pooled, giving the error term in the SAS summary of Figure 15.5. Using this error term, the *standard error of estimate* for the quadratic model is

$$s_{Y.X,X^2} = \sqrt{\frac{20.00 + 2.82}{10 + 2}} = 1.38$$

which might be used to set confidence limits on the Y's around the quadratic curve.

TABLE 15.8
Departures from Quadratic

x_j	u_j	$\bar{y}_{.j}$	y'_x	$\bar{y}_{.j} - y'_x$	$(\bar{y}_{.j} - y'_x)^2$
0.5	−2	83/3	2927/105	−22/105	484/11,025
1.0	−1	84/3	2908/105	32/105	1,024/11,025
1.5	0	89/3	3079/105	36/105	1,296/11,025
2.0	1	96/3	3440/105	−80/105	6,400/11,025
2.5	2	115/3	3991/105	34/105	1,156/11,025
Totals				0	10,360/11,025

TABLE 15.9
ANOVA on Reaction Time with Quadratic Regression

Source	df	SS	MS	F	p value
Between dosages	4	229.73			
Linear	1	192.53	192.53	96.3	0.000
Quadratic	1	34.38	34.38	17.2	0.002
Departure from model	2	2.82	1.41	0.7	0.519
Error	10	20.00	2.00		
Totals	14	249.73			

15.3.2 Two Factors: One Qualitative and One Quantitative

We now consider how to determine the effects of depth (D) and position (P) in a tank on the concentration (Y) of a cleaning solution. Concentrations are measured at three depths from the surface of the tank: 0, 15, and 30 inches. At each depth, measurements are taken at five different lateral positions in the tank, as depicted in Figure 15.6. These are considered as five qualitative positions, although some orientation measure probably might be made on them. At each depth and position, two observations are taken. This is then a 5×3 factorial with two replications per cell (total of 30 observations). The data, collected in random order, are summarized in Table 15.10.

Analyzing the data as if both factors are qualitative, we have the model

$$Y_{ijk} = \mu + D_i + P_j + DP_{ij} + \varepsilon_{k(ij)}$$

with $i = 1, 2, 3$; $j = 1, 2, 3, 4, 5$; and $k = 1, 2$. As Figure 15.7 indicates, only the depth effect is significant.

Even though the interaction between depth and position is not significant, there may be an interaction between the linear effect of depth and positions or between the quadratic effect of depth and positions. To check for such interactions, we will fit a full model using the **GLM** procedure in SAS with the **MODEL** statement

```
MODEL CONC=DEPTH DEPTH*DEPTH POS DEPTH*POS DEPTH*DEPTH*POS;
```

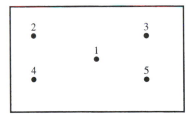

Figure 15.6   Positions in tank at each depth.

TABLE 15.10
Cleaning Solution Concentration Data

Position P_j	Depth from Top of Tank D_i (in.)		
	0	15	30
1	5.90	5.90	5.94
	5.91	5.89	5.80
2	5.90	5.89	5.75
	5.91	5.89	5.83
3	5.94	5.91	5.86
	5.90	5.91	5.83
4	5.93	5.94	5.83
	5.91	5.90	5.89
5	5.90	5.94	5.83
	5.87	5.90	5.86

```
Analysis of Variance

Source      DF      Sum of Squares    Mean Square      F Ratio
Model       14              0.0390        0.0028        2.2135
Error       15              0.0189        0.0013      Prob > F
C Total     29              0.0579                      0.0694

Source      DF      Sum of Squares       F Ratio      Prob > F
depth        2              0.0282       11.1772        0.0011
pos          4              0.0050        1.0013        0.4374
depth*pos    8              0.0058        0.5787        0.7802
```

Figure 15.7   JMP ANOVA for Table 15.10 data.

Here POS is a classification variable. Coding the data by subtracting 5.90 and multiplying by 100 gives Figure 15.8.

Only the depth factor (linear and quadratic) from the SAS analysis of Figure 15.8 needs to be considered, since the other factors are not significant. [**Note:** The Type I sums of squares are used.] The plot of the sample means versus depth (Figure 15.9) confirms the appropriateness of a quadratic model.

Using JMP with the Table 15.10 data to fit a second-degree polynomial gives Figure 15.10. From that figure, the least squares equation is

$$y' = 5.9070 + 0.0022d - 0.0001d^2 \tag{15.22}$$

which accounts for about 49% of the variation in concentration. The large error variance leads us to believe that the experimenters failed to consider at least one factor that has a significant effect on that variability. Further, the scattergram seems to indicate that the concentration readings at a depth of 30 inches vary more than those at the other depths.

GENERAL LINEAR MODELS PROCEDURE

DEPENDENT VARIABLE: CONC

SOURCE	DF	SUM OF SQUARES	MEAN SQUARE	F VALUE
MODEL	14	390.46666667	27.89047619	2.21
ERROR	15	189.00000000	12.60000000	PR > F
CORRECTED TOTAL	29	579.46666667		0.0694

R-SQUARE	C.V.	ROOT MSE	CONC MEAN
0.673838	242.0214	3.54964787	-1.46666667

SOURCE	DF	TYPE I SS	F VALUE	PR > F
DEPTH	1	211.25000000	16.77	0.0010
DEPTH*DEPTH	1	70.41666667	5.59	0.0320
POS	4	50.46666667	1.00	0.4374
DEPTH*POS	4	41.50000000	0.82	0.5303
DEPTH*DEPTH*POS	4	16.83333333	0.33	0.8508

SOURCE	DF	TYPE III SS	F VALUE	PR > F
DEPTH	1	16.25000000	1.29	0.2739
DEPTH*DEPTH	1	70.41666667	5.59	0.0320
POS	4	16.60000000	0.33	0.8539
DEPTH*POS	4	14.73076923	0.29	0.8794
DEPTH*DEPTH*POS	4	16.83333333	0.33	0.8508

Figure 15.8 SAS output for concentration data.

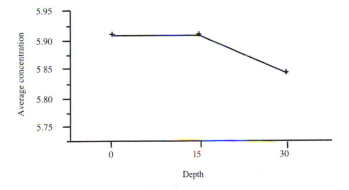

Figure 15.9 Means plot for concentration data.

Two Additional Comments

In the concentration study, the interaction effect was not significant. Had there been a significant interaction, we would have had to find a least squares equation in depth for each position. Had position been significant with no interaction, Equation (15.22) could have been used, with a constant added for each position.

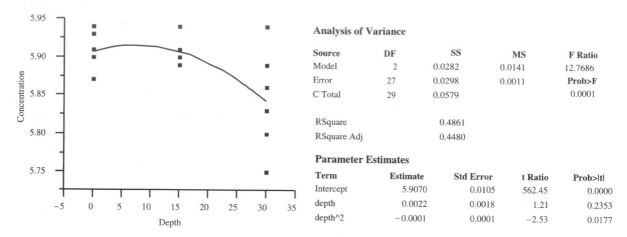

Analysis of Variance

Source	DF	SS	MS	F Ratio
Model	2	0.0282	0.0141	12.7686
Error	27	0.0298	0.0011	**Prob>F**
C Total	29	0.0579		0.0001

RSquare		0.4861
RSquare Adj		0.4480

Parameter Estimates

Term	Estimate	Std Error	t Ratio	Prob>ltl
Intercept	5.9070	0.0105	562.45	0.0000
depth	0.0022	0.0018	1.21	0.2353
depth^2	−0.0001	0.0001	−2.53	0.0177

Figure 15.10   JMP outputs for concentration study.

15.4 ORTHOGONAL POLYNOMIALS

In Section 10.1 it was shown that a linear and quadratic effect can be extracted by proper use of coefficients in orthogonal contrasts when the levels of X are equispaced. This concept can be extended to cubic, quartic, and so on, contrasts as the number of levels (k) of X increases. Statistical Table F gives the proper coefficients for such contrasts for $k = 3$ through 10. The use of these coefficients will provide orthogonal contrasts that can be treated independently of each other and from which a polynomial of the highest order whose coefficients are significant will provide an adequate equation to use for predicting the average value of Y at each level of X.

Begin by obtaining the treatment and error sums of squares for the analysis of variance model. Then calculate the numerical value of a contrast in treatment totals, starting with the linear contrast. Next determine the sum of squares associated with that contrast and divide that sum of squares by the error mean square. Under the usual ANOVA assumptions, this ratio is the observed value of an F with $v_1 = 1$ and $v_2 = N - (k+1)$. If the corresponding F test indicates significance, repeat the procedure for a second-degree polynomial. Continue this procedure to determine the highest order polynomial to consider as an adequate prediction equation. Finally, use appropriate computer software to find such a least squares polynomial.

We illustrate with the reaction time data of Example 15.1. From Table 15.3, the treatment totals are $T_1 = 83$, $T_2 = 84$, $T_3 = 89$, $T_4 = 96$, and $T_5 = 115$. Also, the error mean square for a one-way analysis of variance is 2.00 and $v_2 = 10$ from Table 15.2. Because five levels of dosage were considered, we use the $k = 5$ block of Statistical Table F to determine the contrast coefficients. Each contrast value is found by means of

$$C = \xi_1 T_{.1} + \xi_2 T_{.2} + \xi_3 T_{.3} + \xi_4 T_{.4} + \xi_5 T_{.5}$$

with ξ_j the value in the j column of the appropriate row. The associated sum of squares is found by means of

$$SS = \frac{c}{3 \sum_{j=1}^{5} (\xi_j)^2}$$

since $n = 3$ observations were obtained at each treatment level. Notice that the value of $\sum_{j=1}^{k} (\xi_j)^2$ is also included in Statistical Table F.

For the linear effect, we use the linear row of the $k = 5$ block to find

$$c_{\text{linear}} = (-2)(T_{.1}) + (-1)(T_{.2}) + (0)(T_{.3}) + (1)(T_{.4}) + (2)(T_{.5})$$
$$= (-2)(83) + (-1)(84) + (0)(89) + (1)(96) + (2)(115) = 76$$

and

$$SS_{\text{linear}} = \frac{(c_{\text{linear}})^2}{3 \sum_{j=1}^{5} (\xi_j)^2} = \frac{(76)^2}{3(10)} = \frac{2888}{15} = 192.53$$

Thus, $f = 192.53/2 = 96.3$ and $P(F_{1,10} \geq 96.3) = 0.000$, indicating that the linear effect is significant for any reasonable value of α. This agrees with Table 15.9.

Proceeding in the same manner, we use the quadratic row of the $k = 5$ block to find

$$c_{\text{quadratic}} = (2)(83) + (-1)(84) + (-2)(89) + (-1)(96) + (2)(115) = 38$$

and

$$SS_{\text{quadratic}} = \frac{(38)^2}{3(14)} = \frac{722}{21} = 34.38$$

Thus, $f = 34.38/2 = 17.2$ and $P(F_{1,10} \geq 17.2) = 0.002$, indicating that the quadratic effect is significant for any reasonable value of α, which also agrees with Table 15.9.

Likewise,

$$c_{\text{cubic}} = (-1)(83) + (2)(84) + (0)(89) + (-2)(96) + (1)(115) = 8$$

and

$$SS_{\text{cubic}} = \frac{(8)^2}{3(10)} = \frac{32}{15} = 2.13$$

Thus, $f = 2.13/2 = 1.1$ and $P(F_{1,10} \geq 1.1) = 0.325$, indicating that the cubic effect may be negligible.

Since the test for a cubic effect has such a large p value ($p = 0.325$) but the linear and quadratic effects test highly significant (p values of 0.000 and 0.002, respectively), we would use a quadratic as an adequate equation. This conclusion agrees with that in Section 15.3.1.

We are now prepared to use a statistical computing program to determine the least squares quadratic associated with our data. Recall that, in Section 15.3, such a polynomial was obtained manually, using JMP, and using SAS.

15.5 MULTIPLE REGRESSION

We now consider a more general situation in which Y may be a function of several independent variables $X_1, X_2, \ldots, X_k$ with no restrictions on the settings of these k independent variables. In fact, in most such multiple regression situations the X variables have already acted and we simply record their values along with those of the dependent variable Y. This is, of course, *ex-post-facto* research, as opposed to experimental research in which one manipulates the X's and observes the effect on Y.

In practice there are many studies of this type. For example, one may wish to predict the surface finish of steel from dropout temperature and back-zone temperature. Here Y is a function of two recorded temperatures X_1 and X_2. Or, it may be of interest to predict college grade-point average (GPA) in the freshman year for students whose input data include rank in high school, high school Regents' average, SAT (Scholastic Aptitude Test) verbal score, and SAT mathematics score. Here Y, the freshman year GPA, is to be predicted from four independent variables: X_1, high school rank; X_2, high school Regent's average; X_3, SAT verbal; and X_4, SAT mathematical.

As in the preceding sections a mathematical model is written and the coefficients in the model are determined from the observed sample data by the method of least squares, making the sum of squares of deviations from this model a minimum.

To predict the value of a dependent variable (Y) from its regression on several independent variables ($X_1, X_2, \ldots, X_k$), the linear population model is given as follows:

$$Y = \beta_0 + \beta_1 X_1 + \beta_2 X_2 + \cdots + \beta_k X_k \tag{15.23}$$

where the β's are the true coefficients to be used to weight the observed X's. In practice, a random sample of size N is chosen and the values of all variables ($Y, X_1, X_2, \ldots, X_k$) are recorded for each item in the sample. The corresponding sample model is

$$Y' = B_0 + B_1 X_1 + B_2 X_2 + \cdots + B_k X_k \tag{15.24}$$

where the B's are estimators of the β's. When $k = 1$, this gives the straight-line model of Section 15.2.

After sampling, sample estimates of the β's are calculated, giving

$$y' = b_0 + b_1 x_1 + b_2 x_2 + \cdots + b_k x_k \tag{15.25}$$

Since each observed y can be expressed as

$$y_i = y_i' + e_i$$

with

$$i = 1, 2, \ldots, N$$

where the e's are called *residuals*, the b_i's are determined by minimizing $\sum e_i^2$. Differentiating and setting each partial derivative to zero gives the following least squares equations:

$$\left.\begin{array}{l} \sum y = b_0 N + b_1 \sum x_1 + b_2 \sum x_2 + \cdots + b_k \sum x_k \\[2mm] \sum x_1 y = b_0 \sum x_1 + b_1 \sum x_1^2 + b_2 \sum x_1 x_2 + \cdots + b_k \sum x_1 x_k \\[2mm] \sum x_2 y = b_0 \sum x_2 + b_1 \sum x_2 x_1 + b_2 \sum x_2^2 + \cdots + b_k \sum x_2 x_k \\[2mm] \quad\vdots \\[2mm] \sum x_k y = b_0 \sum x_k + b_1 \sum x_k x_1 + b_2 \sum x_k x_2 + \cdots + b_k \sum x_k^2 \end{array}\right\} \quad (15.26)$$

This set of $k + 1$ equations in $k + 1$ unknowns $(b_0, b_2, \ldots, b_k)$ can be solved for the b's. Solving these equations is tedious if more than two or three independent variables are involved, so a specialized computer program is usually used to obtain a solution.

Even though many independent variables may be used, simpler models are more appealing and easier to interpret. Thus, we try to identify the smallest subset of the independent variables that will provide an adequate model. To this end, many practitioners compare the values of the adjusted R-squares of the different models. If N observations are made and $k + 1$ parameters are estimated, *adjusted R-square* is given by

$$R_{\text{adj}}^2 = \frac{(N-1)R^2 - k}{N - 1 - k} \tag{15.27}$$

with $R^2 = SS_{\text{model}}/SS_{\text{total}}$ the coefficient of multiple determination for the regression model. If there is little difference between the adjusted R-squares for two models, we have evidence that supports using the simpler model. Before any model is recommended, of course, its adequacy should be thoroughly assessed, along with the adequacy of the regression assumptions.

To illustrate the procedure, we consider an experiment involving 24 samples of a steel alloy. For each sample, a chemical analysis is made and the percentages of five specific chemical elements $(X_1, X_2, X_3, X_4, X_5)$ are recorded. Placing each sample under stress, the percent elongation (Y) is then determined. The resulting data are given in Table 15.11.

In Figure 15.11 we give Minitab output obtained using the **Best Subsets** module. Minitab searched through the $2^5 = 32$ models that could be developed from our five independent variables and reported the adjusted R-squares for the best two models in one, two, three, and four variables. That statistic is also reported for the full model in five variables. The best single predictor of Y is X_2, since the symbol $\times$ is below x_2 in the line that contains the first 1 in the variables column (Vars). Likewise, the best two-variable model contains X_2 and X_5, since the symbol $\times$ is below both x_2 and x_5 in the line that contains the first 2 in the variables column. Proceeding in this fashion, we find that the best three-variable model contains X_2, X_3, and X_5, whereas the best four-variable model contains X_2, X_3, X_4, and X_5.

Suppose we decide that an appropriate (tentative) model should contain only X_2 and X_5. For this analysis we would use the **Regression** module in Minitab, obtaining the outputs in Figure 15.12. Notice that an unusually large residual has been detected. Problem 15.38 asks the reader to conduct a residual analysis for this situation.

TABLE 15.11
Percent Elongation Data

Item	y	x_1	x_2	x_3	x_4	x_5
1	11.3	0.50	1.3	0.4	3.4	0.010
2	10.0	0.47	1.2	0.3	3.6	0.012
3	9.8	0.48	3.1	0.7	4.3	0.000
4	8.8	0.54	2.6	0.7	4.0	0.022
5	7.8	0.45	2.8	0.7	4.2	0.000
6	7.4	0.41	3.2	0.7	4.7	0.000
7	6.7	0.62	3.0	0.6	4.7	0.026
8	6.3	0.53	4.1	0.9	4.6	0.035
9	6.3	0.57	3.7	0.8	4.6	0.000
10	6.3	0.67	2.7	0.6	4.8	0.013
11	6.0	0.54	3.1	0.7	4.2	0.000
12	6.0	0.42	3.1	0.7	4.4	0.000
13	5.8	0.33	2.6	0.6	4.7	0.008
14	5.5	0.51	3.9	0.9	4.4	0.000
15	5.5	0.54	3.1	0.7	4.2	0.000
16	4.7	0.48	4.0	1.1	3.7	0.024
17	4.1	0.38	3.3	0.8	4.1	0.000
18	4.1	0.39	3.2	0.7	4.6	0.016
19	3.9	0.60	2.9	0.7	4.3	0.025
20	3.5	0.54	3.2	0.7	4.9	0.022
21	3.1	0.33	2.9	2.9	1.0	0.059
22	1.6	0.40	3.2	3.2	1.0	0.059
23	1.1	0.64	2.5	0.7	3.8	0.018
24	0.6	0.34	5.0	1.3	3.9	0.044

```
Best Subsets Regression
Response is y

                     Adj.                        x x x x x
Vars    R-sq         R-sq        C-p          s   1 2 3 4 5
1       31.4         28.3        5.4      2.3077   X
1       28.3         25.1        6.5      2.3590     X
2       49.2         44.4        0.8      2.0328   X       X
2       47.3         42.3        1.5      2.0703   X X
3       50.6         43.1        2.3      2.0552   X X     X
3       49.7         42.2        2.6      2.0727   X   X X
4       51.1         40.8        4.1      2.0975   X X X X
4       50.9         40.6        4.1      2.1007   X X X   X
5       51.3         37.8        6.0      2.1496   X X X X X
```

Figure 15.11   Minitab summary for elongation models.

It is interesting to note that observation 23 yields an unusually large residual for every model listed in Figure 15.11. In Problem 15.39, the reader is asked to conduct an analysis of the data in Table 15.11 after observation 23 has been removed.

The procedure illustrated here is descriptive and exploratory. Once a short list of possible models has been developed using a program such as Minitab, further analysis may indicate that one or more of these *may* serve as an adequate model. However, we

```
Regression Analysis

The regression equation is
y = 11.7 - 1.62 x2 - 63.2 x5

Predictor       Coef      Stdev    t-ratio          p
Constant       11.695     1.690       6.92      0.000
x2             -1.6200    0.5448     -2.97      0.007
x5             -63.18     23.30      -2.71      0.013

s = 2.033    R-sq = 49.2%    R-sq(adj) = 44.4%

Analysis of Variance
SOURCE          DF        SS         MS          F          p
Regression       2     84.071     42.035      10.17      0.001
Error           21     86.774      4.132
Total           23    170.845

SOURCE          DF      SEQ SS
x2               1      53.680
x5               1      30.391

Unusual Observations
Obs.            x2         y        Fit   Stdev.Fit   Residual   St.Resid
 23           2.50     1.100      6.508       0.524     -5.408     -2.75R

R denotes an obs. with a large st. resid.
```

Figure 15.12   Minitab output for the tentative two-factor elongation model.

cannot formally test or estimate using such tentative models. Rather, a new study should be conducted to assess the appropriateness of the proposed model.

Further Comments Concerning the Elongation Study

Residuals for a regression analysis are easily obtained when statistical computing programs are used for the analysis. From the Minitab output of Figure 15.12, we observed that one residual was suspiciously large. Further analysis of the residuals is requested in Problem 15.38. For now, however, consider the values of the residuals (obtained using Minitab) as given in Table 15.12. Note that the residuals (or errors of estimate) seem to be nonrandom inasmuch as, in general, they are positive for large values of Y and negative for small values of Y. This may indicate a need to consider some higher order terms such as X_2^2, X_5^2, or $X_2 X_5$. Only modest changes in the model statement of the computer program will effect such a refinement.

Alternatives to All-Subsets Regression

The procedure illustrated by means of the elongation data is an all-subsets regression analysis. In such an analysis, each possible subset model involving the variables under consideration is fitted to the sample data and assessed based on a given criterion (e.g., the coefficient of determination, the adjusted R-square value). Other commonly used procedures include stepwise regression and two special cases—forward selection and backward elimination. However, these procedures may miss some models considered by the

TABLE 15.12
Deviations from Regression on Elongation Data

y	y'	e	y	y'	e
11.3	8.95704	2.34296	5.8	6.97740	−1.17740
10.0	8.99268	1.00732	5.5	5.37684	0.12316
9.8	6.67284	3.12716	5.5	6.67284	−1.17284
8.8	6.09290	2.70710	4.7	3.69854	1.00146
7.8	7.15884	0.64116	4.1	6.34884	−2.24884
7.4	6.51084	0.88916	4.1	5.49997	−1.39997
6.7	5.19218	1.50782	3.9	5.41736	−1.51736
6.3	2.84157	3.45843	3.5	5.12090	−1.62090
6.3	5.70084	0.59916	3.1	3.01655	0.08345
6.3	6.49951	−0.19951	1.6	2.78327	−1.18327
6.0	6.67284	−0.67284	1.1	6.50761	−5.40761
6.0	6.67284	−0.67284	0.6	0.81496	−0.21496

all-subsets procedure. Thus, use of all-subsets regression is recommended when adequate computing facilities are available.

15.5.1 A SAS Program

The **RSQUARE** procedure in SAS is an all-subsets regression that can be used like the Minitab **Best Subsets** module. Since, however, stepwise regression is often used in determinations of the variables to be included in a regression model, we now illustrate the use of the **STEPWISE** procedure in SAS with the **MAXR** option. This technique is considered to be almost as good as an all-subsets regression.

A command file for the analysis of the elongation data of Table 15.11 is given in Table 15.13. Notice that the **MAXR** option is requested by placing the symbol / at the end of the model statement and following that symbol with the name of the option to be used. Other options include **F** for forward selection, **B** for backward elimination, and **STEPWISE** for stepwise regression.

The outputs from the command file of Table 15.13 are given in Figure 15.13. Notice that the "best" two-factor model

TABLE 15.13
SAS Command File for Elongation Study

```
DATA ELONG;
INPUT Y X1 X2 X3 X4 X5;
CARDS;
11.3 0.50 1.3 0.4 3.4 0.010
10.0 0.47 1.2 0.3 3.6 0.012
            :
 0.6 0.34 5.0 1.3 3.9 0.044
;
PROC STEPWISE;
MODEL Y = X1 X2 X3 X4 X5/MAXR;
```

```
MAXIMUM R-SQUARE IMPROVEMENT FOR DEPENDENT VARIABLE Y

STEP 1    VARIABLE X2 ENTERED    R SQUARE = 0.31420181   C(P) = 5.35523227
                        DF    SUM OF SQUARES   MEAN SQUARE        F    PROB>F
REGRESSION               1       53.67980887   53.67980887    10.08    0.0044
ERROR                   22      117.16519113    5.32569051
TOTAL                   23      170.84500000
                    B VALUE       STD ERROR    TYPE II SS         F    PROB>F
INTERCEPT        11.57748846
X2               -1.92211293      0.60542644   53.67980887    10.08    0.0044
THE ABOVE MODEL IS THE BEST 1 VARIABLE MODEL FOUND.

STEP 2    VARIABLE X5 ENTERED    R SQUARE = 0.49208798   C(P) = 0.77845055
                        DF    SUM OF SQUARES   MEAN SQUARE        F    PROB>F
REGRESSION               2       84.07077035   42.03538517    10.17    0.0008
ERROR                   21       86.77422965    4.13210617
TOTAL                   23      170.84500000
                    B VALUE       STD ERROR    TYPE II SS         F    PROB>F
INTERCEPT        11.69482586
X2               -1.61999580      0.54479609   36.53686509     8.84    0.0073
X5              -63.17916865     23.29632535   30.39096148     7.35    0.0131
THE ABOVE MODEL IS THE BEST 2 VARIABLE MODEL FOUND.

STEP 3    VARIABLE X3 ENTERED    R SQUARE = 0.50554119   C(P) = 2.28106031
DF                SUM OF SQUARES    MEAN SQUARE        F    PROB>F
REGRESSION               3       86.36918442   28.78972814     6.82    0.0024
ERROR                   20       84.47581558    4.22379078
TOTAL                   23      170.84500000
                    B VALUE       STD ERROR    TYPE II SS         F    PROB>F
INTERCEPT        11.82239196
X2               -1.54994707      0.55893256   32.48009675     7.69    0.0117
X3               -0.74853877      1.01473201    2.29841407     0.54    0.4693
X5              -42.22569494     36.89984445    5.53104177     1.31    0.2660
THE ABOVE MODEL IS THE BEST 3 VARIABLE MODEL FOUND.

STEP 4    VARIABLE X4 ENTERED    R SQUARE = 0.51071485   C(P) = 4.08978045
                        DF    SUM OF SQUARES   MEAN SQUARE        F    PROB>F
REGRESSION               4       87.25307854   21.81326963     4.96    0.0066
ERROR                   19       83.59192146    4.39957481
TOTAL                   23      170.84500000
                    B VALUE       STD ERROR    TYPE II SS         F    PROB>F
INTERCEPT        14.31358314
X2               -1.22427578      0.92375797    7.72774578     1.76    0.2008
X3               -1.66044527      2.28291111    2.32745790     0.53    0.4759
X4               -0.65610814      1.46379635    0.88389412     0.20    0.6591
X5              -41.70034653     37.80329379    5.87923996     1.34    0.2620
THE ABOVE MODEL IS THE BEST 4 VARIABLE MODEL FOUND.

STEP 5    VARIABLE X1 ENTERED    R SQUARE = 0.51314320   C(P) = 6.0000000
                        DF    SUM OF SQUARES   MEAN SQUARE        F    PROB>F
REGRESSION               5       87.66794928   17.53358986     3.79    0.0160
ERROR                   18       83.17705072    4.62094726
TOTAL                   23      170.84500000
                    B VALUE       STD ERROR    TYPE II SS         F    PROB>F
INTERCEPT        14.91722299
X1               -1.57654684      5.26157760    0.41487075     0.09    0.7679
X2               -1.27616946      0.96242424    8.12483590     1.76    0.2014
X3               -1.69396517      2.34231361    2.41685102     0.52    0.4788
X4               -0.57829883      1.52248084    0.66670232     0.14    0.7085
X5              -43.14470468     39.67043862    4.97078416     1.08    0.3134
THE ABOVE MODEL IS THE BEST 5 VARIABLE MODEL FOUND.
```

Figure 15.13 SAS analysis of the elongation data.

$$\hat{y} \approx 11.695 - 1.620x_2 - 63.179x_5$$

is the same as that obtained using Minitab.

15.5.2 Testing for Significance of the Coefficient of Determination

The proper test for checking the significance of a given R^2 when the model contains k factors is

$$F_{k,N-k-1} = \frac{R^2/k}{(1 - R^2)/(N - k - 1)} \tag{15.28}$$

This is the statistic for the test of H_0: The coefficients of the predictor variables in the model are all zero. The test is valid when data are collected to check the appropriateness of a specific model.

When exploratory procedures are applied in the selection of a "best" model, F tests are often used as guides in deciding the next step. To decide whether any more variables are needed after a subset of r variables has been found significant, for example, one can use

$$F_{k-r,N-k-1} = \frac{(R_k^2 - R_r^2)/(k - r)}{(1 - R_k^2)/(N - k - 1)} \tag{15.29}$$

where R_k^2 is the coefficient of multiple determination for the model containing the entire set of k independent variables and R_r^2 is that coefficient for a subset of r of the k independent variables. Usually, no more variables are added to the model when the p value of the test is large.

■ **Example 15.5 (Model Tests for the Elongation Study)**

From Figure 15.13, the coefficients of multiple determination are $R_5^2 = 0.51314320$ for the five-variable model and $R_2^2 = 0.49208798$ for the model containing only X_2 and X_5.

To test the five-variable model, we use Equation (15.28) and find

$$F_{5,18} = \frac{0.51314320/5}{(1 - 0.51314320)/18} = 3.79$$

as in ANOVA. If the original intent was to use the five-variable model, we conclude that at least one of the five independent variables in the model has a nonzero coefficient.

When deciding whether to add variables other than X_2 and X_5 to the linear model, we use Equation (15.29) and find

$$F_{3,18} = \frac{(0.51314320 - 0.49208798)/3}{(1 - 0.51314320)/18} = 0.2695$$

Since $P(F_{3,18} > 0.2695) = 0.8465$, one may decide not to go beyond X_2 and X_5.

15.6 SUMMARY

The examples of this chapter give the basic adequate equations. The methods may be augmented to include terms in multiple variables that are more complex. For example, with three variables (X_1, X_2, and X_3) one might create new variables such as $X_4 = X_1^2$, $X_5 = X_1 X_2^2$, and $X_6 = X_3^3$, allowing for prediction surfaces that go beyond the rectilinear model of Equation (15.24). So there is much more that can be done when the X's are quantitative.

The following practical steps summarize some of the procedures discussed in this chapter.

CASE I. Polynomial regression when X is set at equispaced levels.

1. Plot a rough scattergram of y versus x to get an idea of the underlying polynomial.
2. Determine the treatment totals for each level of X.
3. Use the table of orthogonal polynomials to test for the highest degree polynomial for an adequate prediction equation. This step may not be needed with certain computer programs.
4. Use appropriate computer software to find this adequate equation.
5. Note the size of R^2 to see how much of the variation in Y is associated with the proposed linear model and whether this is worthwhile (even though statistically significant) from a practical point of view. Include a residual analysis to determine the appropriateness of the model.
6. If the model is deemed appropriate, note the standard error of estimate (square root of MS_{error}) and place confidence limits around the predictions, if desirable.

CASE II. Polynomial regression when the levels of X are not equispaced—usually an ex-post-facto situation.

1. Plot a rough scattergram to help in estimating the highest degree that might be used.
2. Run a computer program to find this highest degree equation and note if all B's are statistically significant at some desired level.
3. If some B's are not significant, try a polynomial of one degree less and check the B's.
4. Continue as in step 3 until an adequate prediction model is found.
5. Based on the value of R^2 and a thorough residual analysis, decide whether the tentative regression equation should be recommended for use. A second experiment may be necessary to check the adequacy of the proposed model.
   Caution: When writing the model for a given polynomial, be sure to include as variables all lower powers of X as well.

CASE III.   Multiple regression.

1. For each experimental unit, record the values of Y and each X (X_1 to X_k) to be studied. Some X's may be powers or products of powers of other X's.
2. If computer resources permit, use an all-subsets regression procedure to find a tentative regression equation. Otherwise, use stepwise regression.
3. Based on the value of R^2 and a thorough residual analysis, decide whether the tentative regression equation should be recommended for use. A second experiment may be necessary to check the adequacy of the proposed model.

A word of caution pertaining to the planning of a regression study is advisable. Since computers can deal with so many X's, experimenters often feel compelled to include all X values that enter their minds. However, it is difficult to believe that all X's would be deemed important in their effect on Y by those knowledgeable in the field to be investigated. Thus, the use of multiple regression should not substitute for careful "soaking" in a problem and searching out of past studies and theoretical considerations before including a given X in a multiple regression study. Sometimes simple scattergrams of Y versus a given X will suggest whether they may be related, and the scattergram may also indicate an association that is nonlinear. This latter may indicate a need for a term such as X^2. The literature in the area may also suggest that some X is logarithmically related to Y and that one should use $\log(X)$ as the independent variable instead of X. Careful consideration of all these notions before the experiment is essential to good experimentation and increases the prospect of meaningful prediction.

Once a data set has been used to develop a least squares equation, a careful analysis of the residuals may indicate inadequacies that may be eliminated by choosing a different model or by transforming the data in an appropriate fashion. This step is often neglected in the haste to "solve" the problem and move on to the next problem.

15.7 FURTHER READING

Graybill, F. A., and H. K. Iyer, *Regression Analysis: Concepts and Applications*. Belmont, CA: Duxbury Press, 1994.

Hamilton, L. C., *Modern Data Analysis: A First Course in Applied Statistics*. Pacific Grove, CA: Brooks/Cole Publishing Company, 1990. Chapters 14, 15, and 16 are devoted to regression analysis.

Mason, R. L.; R. F. Gunst, and J. L. Hess. *Statistical Design and Analysis of Experiments with Applications to Engineering and Science*. New York: John Wiley & Sons, 1989. See "Part IV: Data Fitting" for a detailed discussion of regression models.

Neter, J., W. Wasserman, and M. H. Kutner. *Applied Linear Statistical Models: Regression, Analysis of Variance, and Experimental Design,* Homewood, IL: Richard D. Irwin, 1983.

PROBLEMS

15.1   For the following data on X and Y, plot a scattergram and determine the least squares straight line of best fit.

			x		
	3	4	5	6	7
y	7	8	10	11	10
	8	8	9	9	10
	9	9	9	10	9

15.2 For Problem 15.1 present your results in an ANOVA table and comment on these results.

15.3 From your table in Problem 15.2 find r^2, η^2, and the standard error of estimate.

15.4 Use the method of orthogonal polynomials on the data of Problem 15.1 and find the best fitting polynomial.

15.5 An experiment to determine the effect of planting rate (X) in thousands of plants per acre on yield (Y) in bushels per acre of corn gave the following results.

Planting Rate	Yield
12	130.5, 129.6, 129.9
16	142.5, 140.3, 143.4
20	145.1, 144.8, 144.1
24	147.8, 146.6, 148.4
28	134.8, 135.1, 136.7

Plot a scattergram of these data and comment on the degree polynomial that might prove appropriate for predicting yield from planting rate.

15.6 From the data of Problem 15.5, use orthogonal polynomials and determine what degree polynomial is appropriate here.

15.7 Find the equation of the polynomial in Problem 15.6.

15.8 From an ANOVA table of the results of Problem 15.6, find r^2, R^2, and η^2, and comment on what these statistics tell you.

15.9 Seasonal indexes of hog prices taken in two different years for five months gave the following results.

			Month		
Year	1	2	3	4	5
1	93.9	94.3	101.2	96.7	107.5
2	95.4	96.5	94.7	97.1	107.2

a. Sketch a scatterplot of these data and comment on what type of curve might be used to predict hog prices when the month is known.
b. Do a complete analysis of these data and find the equation of the best fitting curve for predicting hog price from month.

15.10 A bakery is interested in the effect of six baking temperatures on the quality of its cakes. The bakers agree on a measure of cake quality, Y. A large batch of cake mix is prepared and 36 cakes are poured. The baking temperature is assigned at random to the cakes such that six cakes are baked at each temperature. Results are as follows:

Temperature (°F)					
175	185	195	205	215	225
22	4	10	22	32	20
10	21	22	14	27	36
18	18	18	21	22	33
26	28	32	25	37	33
21	21	28	26	27	20
21	28	25	25	31	25

Do a complete analysis of these data and write the "best" fitting model for predicting cake quality from temperature.

15.11 Compute statistics that will indicate how good your fit is in Problem 15.10 and comment. Include a residual analysis.

15.12 Here are some data on the effect of age on retention of information.

Ages:	6–14	15–23	24–32	33–41
Total score of 10 people, $T_{.j}$:	80	92	103	90

Source	df	SS		MS
Between ages	3	26.675		8.892
Linear	1		8.405	8.405
Quadratic	1		15.625	15.625
Cubic	1		2.645	2.645
Error	36	72.000		2.000
Total	39	98.675		

a. Determine the significance of each prediction equation and state which degree equation (linear, quadratic, or cubic) will adequately predict retention score from age.
b. Write the equation for part a.
c. Find the departure from regresson of the mean retention score of a 37-year-old based on your answer to part b.

15.13 A study of the effect of chronological age on history achievement scores gave the following results.

History Achievement Score, Coded by $(Y - 18)/15$	Chronological Age				
	8	9	10	11	12
5			1	6	5
4		1	7	4	5
3		6	2		
2	3	2			
1	6	1			
0	1				

Do a complete analysis of justifying your curve of best fit.

15.14 For a study of the effect of the distance a road sign stands from the edge of the road on the amount by which a driver swerves from the edge of the road, five distances were chosen. Then the five distances were set at random, and four cars were observed at each set distance. The results gave

$$\eta^2 = 0.70$$

$$r^2 = 0.61$$

a. Test whether a linear function will "fit" these data. (You may assume a total SS of 10,000 if you wish, but this is not necessary.)
b. In fitting the proper curve, what are you assuming about the four readings at each specified distance?

15.15 Thickness of a film is studied to determine the effect of two types of resin and three gate settings on the thickness.

Resin Type	Gate Setting (mm)		
	2	4	6
1	1.5	2.5	3.6
	1.3	2.5	3.8
2	1.4	2.6	3.7
	1.3	2.4	3.6

Do an analysis of variance of the results above, following the methods of Chapter 5.

15.16 For Problem 15.15, extract a linear and a quadratic effect of gate setting and test for significance. Also partition the interaction and test its components.

15.17 Write any appropriate prediction equation or equations from the data of Problem 15.15.

15.18 The experiment of Problem 15.15 is extended to include a third factor, weight fraction at three levels, as indicated by the accompanying tabulation. Compile an ANOVA table for this three-factor experiment as in Chapter 5.

Gate Setting (mm)	Resin Type					
	1			2		
	Weight Fraction			Weight Fraction		
	0.20	0.25	0.30	0.20	0.25	0.30
2	1.6	1.5	1.5	1.5	1.4	1.6
	1.5	1.3	1.3	1.4	1.3	1.4
4	2.7	2.5	2.4	2.4	2.6	2.2
	2.7	2.5	2.3	2.3	2.4	2.1
6	3.9	3.6	3.5	4.0	3.7	3.4
	4.0	3.8	3.4	4.0	3.6	3.3

15.19 Outline a further breakdown of the ANOVA table in Problem 15.18 based on how the independent variables are set. Show the proper degrees of freedom.

15.20 For the significant ($\alpha = 0.05$) effect in Problem 15.18 complete the ANOVA breakdown suggested in Problem 15.19.

15.21 Graph any significant effects found in Problem 15.20 and discuss.

15.22 Data on the effect of knife-edge radius R in inches and feedroll force F in pounds per inch on the energy necessary to cut 1-inch lengths of alfalfa are given as follows:

| Feedroll Force F_j (lb/in.) | Knife-Edge Radius, R_i (in.) | | | | | | | $T_{\cdot j}$ |
	0.000		0.005		0.010		0.015	
5	29		98		44		84	
	30		128		81		100	
	20		67		77		63	
		79		293		202		247	821
10	22		35		53		103	
	26		80		93		90	
	16		29		59		98	
		64		144		205		291	704
15	18		49		58		80	
	17		68		103		91	
	11		61		128		77	
		46		178		289		248	761

20	38		68		87		86		
	31		74		116		113		
	21		47		90		81		
		90		189		293		280	852

$T_{i..}$	279	804	989	1066	$T_{...} = 3138$

Write the mathematical model for this experiment and do an ANOVA on the two factors and their interaction.

15.23 Outline a further ANOVA based on Problem 15.22 with single-degree-of-freedom terms.

15.24 Do an analysis for each of the terms in Problem 15.23.

15.25 Plot some results of Problem 15.22 and try to argue that they confirm some results found in Problem 15.24.

15.26 The sales volume Y of a product in thousands of dollars, price X_1 per unit in dollars, and advertising expense X_2 in hundreds of dollars were recorded for $n = 8$ cases. The results are as follows.

				Case				
	1	2	3	4	5	6	7	8
y	10.1	6.5	5.1	11.0	9.9	14.7	4.8	12.2
x_1	1.3	1.9	1.7	1.5	1.6	1.2	1.6	1.4
x_2	8.8	7.1	5.5	13.8	18.5	9.8	6.4	10.2

a. Find the means and standard deviations of each variable and the intercorrelation matrix.
b. On the basis of the discussion in Section 15.2, find the regression equation Y from X_1 alone.

15.27 In studying the percent conversion (Y) in a chemical process as a function of time in hours (X_1) and average fusion temperature in degrees Celsius (X_2) the following data were collected.

y	x_1	x_2
62.7	3	297.5
76.2	3	322.5
80.8	3	347.5
80.8	6	297.5
89.2	6	322.5
78.6	6	347.5
90.1	9	297.5
88.0	9	322.5
76.1	9	347.5

Analyze these data using multiple linear regression. Reexamine the data and comment on any peculiarities noted.

15.28 A management consulting firm attempted to predict annual salary of executives from the executives' years of experience (X_1), years of education (X_2), sex (X_3), and number of employees supervised (X_4). A sample of 25 executives gave an average salary of $79,700 with a standard deviation of $1300. From a computer program the following statistics were recorded.

$$r^2_{Y4} = 0.42$$

$$R^2_{Y.24} = 0.78$$

$$R^2_{Y.124} = 0.90$$

$$R^2_{Y.1234} = 0.95$$

a. Explain how you would interpret these statistics.
b. Test whether one could stop after variables 2 and 4 have been entered into the equation.
c. If variables 1, 2, and 4 were used in the prediction equation, what would the limits on the salaries expected for a predicted salary have to be in order to be correct about 95% of the time?

15.29 Consider a prediction situation in which some dependent variable Y is to be predicted from four independent variables X_1, X_2, X_3, and X_4 and you have a stepwise regression printout on this problem involving 25 observations. Explain briefly each of the following with respect to this problem and its printout.
a. How does the computer decide which independent variable to enter first into a regression equation?
b. How does it decide which variable to enter next in the equation?
c. How do you decide when to stop adding variables based on the printout?
d. Do you need to add any more variables if $R^2_{Y.1234} = 0.7$ and $R^2_{Y.24} = 0.6$?

15.30 A study was made to determine the effect of four variables on the grade-point average of 55 students who attended a junior college after graduation from high school. These variables were high school rank (X_1), SAT score (X_2), IQ (X_3), and age in months (X_4). From the computer printout the following statistics were recorded.

$$R^2_{Y.1234} = 0.45 \qquad R^2_{Y.12} = 0.33$$

$$R^2_{Y.123} = 0.42 \qquad r^2_{Y1} = 0.19$$

Use this information to test for the significance of each added variable and determine whether an ordered subset of these four will make a satisfactory prediction. If you wish, you may assume a total of sum of squares in this study of 10,000 units. Also explain how you would predict with approximate 2 : 1 odds the maximum GPA expected for an individual student based on his or her scores of the variables in the appropriate equation.

15.31 In a study of several variables that might affect freshman GPA (Y) a sample of 55 students reported their scores on the college entrance arithmetic test (X_1), and the analogies test on an Ohio battery (X_2), as well as their high school average (X_3) and their interest score on an interest inventory (X_4). Results of this study are given in the following (oversimplified) ANOVA table.

Source*	df	SS	MS
Due to X_2	1	1360	1360
Due to X_3/X_2	1	480	480
Due to $X_1/X_2, X_3$	1	80	80
Due to $X_4/X_1, X_2, X_3$	1	80	80
Error	50	2000	40
Total	54	4000	

* The symbol/means "given."

Assuming a stepwise procedure was used so that variables were entered in their order of importance, answer the following.

a. Determine $R_{Y.1234}$ and explain its meaning in this problem.
b. Determine which of the four variables are sufficient to predict GPA (Y) and show that your choice is sufficient.
c. If a regression equation based on your answer to part b is used to predict student A's GPA based on test scores, how close do you think this prediction will be to the actual GPA? Justify your answer and note any assumptions you are making.

15.32 In an attempt to predict the average value per acre of farm land in Iowa in 1920, the following data were collected.

$$Y = \text{average value in dollars per acre}$$
$$X_1 = \text{average corn yield in bushels per acre for 10 preceding years}$$
$$X_2 = \text{percentage of farmland in small grain}$$
$$X_3 = \text{percentage of farmland in corn}$$

y	x_1	x_2	x_3	y	x_1	x_2	x_3
87	40	11	14	193	41	13	28
133	36	13	30	203	38	24	31
174	34	19	30	279	38	31	35
385	41	33	39	179	24	16	26
363	39	25	33	244	45	19	34
274	42	23	34	165	34	20	30
235	40	22	37	257	40	30	38
104	31	9	20	252	41	22	35
141	36	13	27	280	42	21	41
208	34	17	40	167	35	16	23
115	30	18	19	168	33	18	24
271	40	23	31	115	36	18	21
163	37	14	25				

Use multiple linear regression to analyze these data, and comment on the results.

15.33 In a study involving handicapped students researchers attempted to predict GPA from two demographic variables, sex and ethnicity, and two independent measured variables, interview score and contact hours used in counseling and/or tutoring. A dummy variable for sex was taken as $1 =$ male, $2 =$ female. For ethnicity: $1 =$ African-American, $2 =$ Hispanic, and $3 =$ white, non-Hispanic. Do a complete analysis of the data below and justify the regression equation that will adequately "fit" the data.

Students	Ethnic	Sex	Interview	Hours	GPA
1	1	2	11.0	4.0	5.50
2	1	2	10.0	5.0	4.10
3	1	2	12.0	73.0	5.00
4	1	2	11.5	68.0	4.22
5	1	2	10.8	82.0	5.00
6	1	1	12.5	72.5	5.00
7	1	1	9.5	64.0	4.60
8	1	1	9.5	78.0	4.25
9	1	1	8.0	64.0	4.00
10	1	1	7.5	13.0	2.00
11	2	2	9.0	37.0	4.25
12	2	2	8.2	4.0	4.00
13	2	2	10.7	38.5	4.61
14	2	2	8.5	3.0	2.93
15	2	2	12.5	10.5	5.50
16	2	1	12.0	80.0	4.77
17	2	1	12.2	6.0	5.00
18	2	1	7.0	6.5	3.25
19	2	1	8.6	22.0	2.66
20	2	1	8.3	28.5	3.37
21	3	2	10.9	12.0	5.00
22	3	2	9.0	9.0	4.00
23	3	2	10.0	5.0	5.00
24	3	2	7.2	12.0	3.87
25	3	2	8.5	4.0	3.00
26	3	1	10.0	8.0	4.77
27	3	1	8.5	8.0	5.00
28	3	1	10.0	22.0	5.08
29	3	1	11.4	61.5	5.57
30	3	1	11.9	37.0	6.00

15.34 In a study on the effect of several variables on posttest achievement scores of 48 biology students in a rural high school, the following table was presented. It included variables that had a significant effect on achievement at the 10% level of significance.

Step	Variable Entered	Multiple R	R^2	R^2 Change
1	Figural elaboration pretest	0.46587	0.21704	0.21704
2	Achievement pretest	0.83022	0.68927	0.47223
3	Piagetian pretest	0.86536	0.74885	0.05958
4	Attitudes pretest	0.90606	0.82094	0.07209

Discuss these results and test whether or not the last two variables were necessary.

15.35 To study the effect of planting rate on yield in an agricultural problem, points were taken with the following results.

x:	12	12	16	16	16	20	20
y:	130.5	129.6	142.5	140.3	143.4	144.8	144.1

x:	23	23	23	23	28	28
y:	145.1	147.8	146.6	148.4	134.8	135.1

a. Plot a scattergram and note what type of curve you might expect of Y versus X.
b. Taking $X^2 = X_2$ and $X = X_1$, use multiple regression to find the equation of the best fitting quadratic.
c. Compare this quadratic with the best linear fit.

15.36 The printout of a computer program yielded the following statistics.

$$B_0 = 140.27 \qquad r_{y1}^2 = 0.0907$$
$$B_1 = 1.24 \qquad R_{y.12}^2 = 0.3285$$
$$B_2 = 2.65 \qquad R_{y.123}^2 = 0.4211$$
$$B_3 = -0.30 \qquad R_{y.1234}^2 = 0.4548$$
$$B_4 = -1.27 \qquad S_y = 16.0$$
$$N = 45$$

a. Explain the meaning of the R squares given above.
b. Test and determine which regression model is adequate for predicting Y.
c. Compute the standard error of estimate for your model in part b.
d. Explain whether the program was a stepwise (step-up) program.

15.37 Derive Equation (15.18)—the system of normal equations for a least squares parabola.

15.38 Conduct a residual analysis for the least squares fit $y' = 11.695 - 1.62x_2 - 63.18x_5$ for the elongation study of Section 15.5.

15.39 For the elongation data of Table 15.11, remove item 23 and conduct a complete regression analysis on the remaining 23 observations. Include a residual analysis.

Chapter 16

MISCELLANEOUS TOPICS

16.1 INTRODUCTION

Several techniques have been developed, based on the general design principles presented in the first 15 chapters of this book and on other well-known statistical methods. These approaches are not discussed in detail because they are presented very well in the references. It is hoped that a background in experimental design as given in the first 15 chapters will make it possible for the experimenter to read and understand the references.

The methods to be discussed are covariance analysis, response surface experimentation, evolutionary operation, analysis of attribute data, incomplete block design, and Youden squares.

16.2 COVARIANCE ANAYLSIS

Philosophy

Occasionally, when a study is made of the effect of one or more factors on some response variable, say Y, there is another variable, or variables, that vary along with Y. It is often not possible to control this other variable (or variables) at a constant level throughout the experiment, but the variable can be measured along with the response variable. This variable is referred to as a *concomitant variable X* because it "runs along with" the response variable Y. Thus, to assess the effect of the treatments on Y, one should first attempt to remove the effects of this concomitant variable X. This technique of removing the effect of X (or several X's) on Y and then analyzing the residuals for the effect of the treatments on Y is called *covariance analysis*.

Several examples of the use of this technique might be cited. If one wishes to study the effect on student achievement of several teaching methods, it is customary to use the difference between a pretest and a posttest score as the response variable. It may be, however, that the gain Y is also affected by the student's pretest score X as gain, and pretest scores may be correlated. Covariance analysis will provide for an adjustment in the gains due to differing pretest scores, assuming some type of regression of gain on pretest scores. The advantage of using this technique is that one is not limited to running an experiment only on pupils who have approximately the same pretest scores or on matching pupils with the same pretest scores, and randomly assigning them to control and experimental groups. Covariance analysis has the effect of providing "handicaps" as if each student had the same pretest scores while actually letting the pretest scores vary along with the gains.

This technique has often been used to adjust weight gains in animals by their original weights to assess the effect of certain feed treatments on these gains. Many industrial examples might be cited such as the one given below in which original weight of a bracket may affect the weight of plating applied to the bracket by different methods.

Snedecor and Cochran [30] give several examples of the use of covariance analysis. Ostle [25] is an excellent reference on the extension of covariance analysis to many different experimental designs. The *Biometrics* article of 1957 [6] devotes most of its pages to a discussion of many of the fine points of covariance analysis. The problem discussed in Section 16.2.1 illustrates the technique for the case of a single-factor experiment run in a completely randomized design. Because the covariance technique represents a marriage between regression and the analysis of variance, we will use the methods of Chapters 3 and 15.

16.2.1 Covariance Analysis by Example

Several steel brackets were sent to three different vendors to be zinc plated. The chief concern in this process is the thickness of the zinc plating, and it must be determined whether there was any difference in thickness in the plates offered by the three vendors. Data on this thickness in hundred thousandths of an inch (inch $\times 10^{-5}$) for four brackets plated by the three vendors are given in Table 16.1.

Assuming a single-factor experiment and a completely randomized design, an analysis of variance model for this experiment would be

$$Y_{ij} = \mu + \tau_j + \varepsilon_{ij} \tag{16.1}$$

where Y_{ij} is the observed thickness of the ith bracket from the jth vendor, μ is a common effect, τ_j represents the vendor effect, and ε_{ij} represents the random error. Considering vendors as a fixed factor and random errors as normally and independently distributed with mean zero and common variance σ_ε^2, an ANOVA table was complied (Figure 16.1).

TABLE 16.1
Thickness (10^{-5} in.) from Three Vendors

| | Vendor | |
A	B	C
40	25	27
38	32	24
30	13	20
47	35	13

```
One-Way Analysis of Variance
Analysis of Variance on Y
Source      DF        SS        MS         F        p
Vendor       2       665.2     332.6     5.51     0.027
Error        9       543.5      60.4
Total       11      1208.7
                                          Individual 95% CIs For Mean
                                          Based on Pooled StDev
 Level       N       Mean      StDev      --------+--------+--------+--------
   1         4      38.750     6.994                       (-------*--------)
   2         4      26.250     9.777            (-------*--------)
   3         4      21.000     6.055      (-------*--------)
                                          --------+--------+--------+--------
Pooled StDef =      7.771                     20       30       40
```

Figure 16.1   Minitab ANOVA on plating thickness.

The F statistic equals 5.51 and is significant at the 5% significance level with 2 and 9 df. One might conclude, therefore, that there is a real difference in average plating thickness among these three vendors, and steps should be taken to select the most desirable vendor.

During a discussion of these results, it was pointed out that some of the difference among products of the three vendors might be due to unequal thickness in the brackets before plating. In fact, there might be a correlation between the thickness of the brackets before plating and the thickness of the plating. To see whether such an idea is at all reasonable, a scatterplot was made on the thickness of the bracket before plating X and the thickness of the zinc plating Y. Results are shown in Figure 16.2.

A glance at this scattergram would lead one to suspect that there is a positive correlation between the thickness of the bracket and the plating thickness. Because another variable, X, can be measured on each piece and may be related to the variable of interest Y, it may be desirable, when comparing vendors, to remove the linear effect of X on Y by running a covariance analysis. Such an analysis can be used to remove the linear or higher order relationship of one or more independent variables from a dependent variable when the effect of a factor on the dependent variable needs to be assessed.

Covariance

Since a positive correlation is suspected between X and Y above, a regression model might be written as

$$Y_{ij} = \mu + \beta(X_{ij} - \bar{X}) + \varepsilon_{ij} \tag{16.2}$$

where β is the true linear regression coefficient (or slope) between Y and X over all the data; $\bar{X}$ is the mean of the X values. In such a model it is assumed that a measure of variable X can be made on each unit along with a corresponding measure of Y. For a covariance analysis, the models of Equations (16.1) and (16.2) can be combined to give the covariance model

$$Y_{ij} = \mu + \beta(X_{ij} - \bar{X}) + \tau_j + \varepsilon_{ij} \tag{16.3}$$

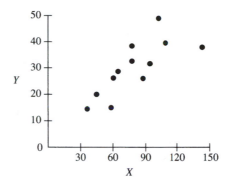

Figure 16.2   Scattergram of bracket thickness X versus plating thickness Y.

In this model the error term ε_{ij} should be smaller than in the first model, Equation (16.1), because of the removal of the effect of the covariate X_{ij}.

To determine how to adjust the analysis of variance to provide for the removal of this covariate, consider the regression model, Equation (16.2), in deviation form

$$y = bx + e \tag{16.4}$$

where x and y are now deviations from their respective means ($x = X_{ij} - \bar{X}$, $y = Y_{ij} - \bar{Y}$) and b is the sample slope. By the method of least squares it will be recalled that

$$b = \frac{\sum xy}{\sum x^2}$$

where summations are over all points. The sum of the squares of the errors of estimate, which have been minimized, is now, from Equation (16.4),

$$e = y - bx$$
$$\sum e^2 = \sum (y - bx)^2$$
$$= \sum y^2 - 2b \sum xy + b^2 \sum x^2$$

Substituting for b, we write

$$b = \frac{\sum xy}{\sum x^2}$$

and

$$\sum e^2 = \sum y^2 - 2 \frac{\sum xy}{\sum x^2} \cdot \sum xy + \left(\frac{\sum xy}{\sum x^2}\right)^2 \sum x^2$$
$$= \sum y^2 - \frac{(\sum xy)^2}{\sum x^2} \tag{16.5}$$

In this expression the term $(\sum xy)^2 / \sum x^2$ is the amount of reduction in the sum of squares of the Y variable due to its linear regression on x. This term is then the sum of squares due to linear regression. It involves the sum of the cross products of the two variables $(\sum xy)$ and the sum of squares of the x variable alone $(\sum x^2)$. If a term of this type is subtracted from the sum of squares for the dependent variable y, the result will be a corrected or adjusted sum of squares for y.

The sums of squares and cross products can be computed on all the data (total), within each vendor or between vendors. In each case it will be helpful to recall that the sums of squares and cross products in terms of the original data are

$$\sum x^2 = \sum X^2 - \frac{(\sum X)^2}{N}$$

$$\sum y^2 = \sum Y^2 - \frac{(\sum Y)^2}{N}$$

$$\sum xy = \sum XY - \frac{(\sum X)(\sum Y)}{N}$$

Data on both variables appear with the appropriate totals in Table 16.2.

From Table 16.2 the sums of squares and cross products for the total data are

$$T_{xx} = (110)^2 + (75)^2 + \cdots + (59)^2 - \frac{(944)^2}{12} = 9240.7$$

$$T_{yy} = (40)^2 + (38)^2 + \cdots + (13)^2 - \frac{(344)^2}{12} = 1208.7$$

$$T_{xy} = (110)(40) + (75)(38) + \cdots + (59)(13) - \frac{(944)(344)}{12} = 2332.7$$

where T, V, and E are used to denote sum of squares and cross products for totals, vendors, and error, respectively. Next, the between-vendors sums of squares and cross products are computed by vendor totals as in an ANOVA:

$$V_{xx} = \frac{(375)^2 + (313)^2 + (256)^2}{4} - \frac{(944)^2}{12} = 1771.2$$

TABLE 16.2
Bracket Thickness X and Plating Thickness Y (in. $\times$ 10^{-5}) from Three Vendors

| | Vendor | | | | | | | |
| | A | | B | | C | | Total | |
	X	Y	X	Y	X	Y	X	Y
	110	40	60	25	62	27		
	75	38	75	32	90	24		
	93	30	38	13	45	20		
	97	47	140	35	59	13		
Totals	375	155	313	105	256	84	944	344

$$V_{yy} = \frac{(155)^2 + (105)^2 + (84)^2}{4} - \frac{(344)^2}{12} = 665.2$$

$$V_{xy} = \frac{(375)(155) + (313)(105) + (256)(84)}{4} - \frac{(944)(344)}{12} = 1062.2$$

By subtraction, the error sums of squares are

$$E_{xx} = T_{xx} - V_{xx} = 9240.7 - 1771.2 = 7469.5$$

$$E_{yy} = T_{yy} - V_{yy} = 1208.7 - 665.2 = 543.5$$

$$E_{xy} = T_{xy} - V_{xy} = 2332.7 - 1062.2 = 1270.5$$

This information may be used to adjust the sums of squares of the dependent variable Y for regression on X. On the totals, the adjusted sum of squares is then

$$\text{adjusted} \sum y^2 = T_{yy} - \frac{T_{xy}^2}{T_{xx}} \quad \text{as in Equation (16.5)}$$

$$= 1208.7 - \frac{(2332.7)^2}{9240.7} = 619.8$$

and the adjusted sum of squares within vendors is

$$\text{adjusted} \sum y^2 = E_{yy} - \frac{(E_{xy})^2}{E_{xx}}$$

$$= 543.5 - \frac{(1270.5)^2}{7469.5} = 327.4$$

Then, by subtraction, the adjusted sum of squares between vendors is

$$619.8 - 327.4 = 292.4$$

These results are usually displayed as in Tabel 16.3.

With the adjusted sums of squares, the new F statistic is

$$F = \frac{146.2}{40.9} = 3.57$$

TABLE 16.3
Analysis of Covariance for Bracket Data

Source	df	SS and Products $\sum y^2$	$\sum xy$	$\sum x^2$	Adjusted $\sum y^2$	df	MS
Between vendors	2	665.2	1062.2	1771.2	—	—	—
Within vendors	9	543.5	1270.5	7469.5	372.4	8	40.9
Totals	11	1208.7	2332.7	9240.7	619.8	10	
Between vendors					292.4	2	146.2

with 2 and 8 df. This is now not significant at the 5% significance level. That is, after the effect of bracket thickness has been removed, the average plating thickness on the brackets no longer differs from vendor to vendor.

Table 16.3 should have a word or two of explanation. The degrees of freedom on the adjusted sums of squares are reduced by 2 instead of 1 because estimates of both the mean and the slope are necessary in their computation. Adjustments are made here on the totals and the within-sums-of-squares rather than on the between sum of squares, since we are interested in making the adjustment based on an overall slope of Y on X and a within-group average slope of Y on X, as explained in more detail later.

A Computer Analysis

The adjusted sums of squares displayed in Table 16.3 are readily obtained by means of statistical software. Selecting **Stat ▶ ANOVA ▶ Analysis of Covariance** in Minitab after the data of Table 16.2 have been entered produces the output in Figure 16.3. The entries in the Vendor row of the ADJ SS column and the Error row of that column are, respectively, the between-vendors and within-vendors adjusted sums of squares of Table 16.3.

Three Types of Regression

In this analysis three different types of regression can be identified: the "overall" regression of all the Y's on all the X's, the within-vendor regression, and the regression of the three Y means on the three X means. This last regression could be quite different from the other two and is of little interest, since we are attempting to adjust the Y's and the X's within the vendors. An estimate of this "average within vendor" slope is

$$b = \frac{\sum xy}{\sum x^2} = \frac{E_{xy}}{E_{xx}} = \frac{1270.5}{7469.5} = 0.17$$

If this slope or regression coefficient is used to adjust the observed Y means for the effect of X on Y, one finds

$$\text{adjusted } \bar{Y}_{.j} = \bar{Y}_{.j} - b(\bar{X}_{.j} - \bar{X}_{..})$$

```
Analysis of Covariance (Orthogonal Designs)

Factor          Levels Values
Vendor             3            1            2            3

Analysis of Covariance for y
Source          DF       ADJ SS         MS         F          P
Covariates       1       216.10     216.10       5.28      0.051
Vendor           2       292.42     146.21       3.57      0.078
Error            8       327.40      40.92
Total           11      1208.67

Covariate        Coeff        Stdev    t-value          P
x               0.1701       0.0740      2.298      0.051
```

Figure 16.3 Minitab covariate analysis for Table 16.2

For vendor A,

$$\text{adjusted } \bar{Y}_{.1} = \frac{155}{4} - (0.17)\left(\frac{375}{4} - \frac{944}{12}\right)$$

$$= 38.75 - (0.17)(93.75 - 78.67) = 36.19$$

For vendor B,

$$\text{adjusted } \bar{Y}_{.2} = 26.25 - (0.17)(78.25 - 78.67) = 26.31$$

For vendor C,

$$\text{adjusted } \bar{Y}_{.3} = 21.00 - (0.17)(64.00 - 78.67) = 23.49$$

A comparison of the last column above with the column of unadjusted means just to the right of the equal signs shows that these adjusted means are closer together than the unadjusted means, which is confirmed in the preceding analysis. This can be seen graphically by using different symbols to plot each vendor's data and "sliding" the means along lines parallel to this regression slope (Figure 16.4).

At the left-hand ordinate in Figure 16.4, the three unadjusted Y means are marked off. By moving these three means along lines parallel to the general slope ($b = 0.17$), we can compare the three means (on the $\bar{X}_{..}$ ordinate) after adjustment. This shows graphically the results of the covariance analysis. By examining each group of observations for a given vendor, we see that all three slopes within vendors are in the same direction and of similar magnitude. By considering all 12 points, we can visualize the overall slope. On the other hand, if the three means (indicated by the capital letters) are considered, the slope of these $\bar{Y}$'s on their corresponding $\bar{X}$'s is much steeper. Hence adjustments are made on the basis of the overall slope and the "average" or "pooled" within-groups slope.

16.2.2 The Assumptions of Covariance Analysis

One of the problems in covariance analysis is that many assumptions are made in its application. These should be examined in some detail.

First, there are the usual assumptions of analysis of variance on the dependent variable Y: an additive model, normally and independently distributed error, and homogeneity of

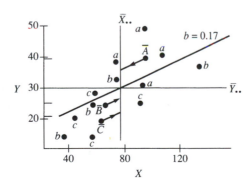

Figure 16.4 Plating thickness Y versus bracket thickness X plotted by vendors: a, b, and c, vendors A, B, and C, respectively; $\bar{A}$, $\bar{B}$, and $\bar{C}$, means for vendors A, B, and C, respectively.

variance within the groups. This last assumption is often tested with Barlett's or a similar test. In this problem, since the ranges of the three vendor readings are 17, 22, and 14, the assumption seems tenable.

In covariance analysis it is further assumed that the regression is linear, that the slope is not zero (covariance was necessary), that the regression coefficients within each group are homogeneous (so that the "average" or "pooled" within-groups regression can be used on all groups), and finally that the independent variable X is not affected by the treatments given to the groups.

Checking the Assumptions

The linearity assumption may be tested if the experimental design incorporates more than one Y observation for each X. Since this was not done here, a look at the scattergram will have to suffice for this linearity assumption.

To test the hypothesis that the true slope β in Equation (16.3) is zero, consider whether the reduction in the error sum of squares is significant compared with the error sum of squares. By an F test, the ratio would be

$$F_{1,N-k-1} = \frac{E_{xy}^2/E_{xx}}{\text{adjusted } E_{yy}/(N-k-1)}$$

$$F_{1,8} = \frac{216.1}{327.4/8} = \frac{216.1}{40.9} = 5.28$$

Since the 5% F is 5.32, this result is nearly significant and certainly casts considerable doubt on the hypothesis that $\beta = 0$. This test on the covariate is included in the Minitab summary of Figure 16.3 as both an F test and a t test. Since the results after covariance adjustment were not significant compared with the results before adjustment, it seems reasonable to conclude that covariance was necessary.

To test the hypothesis that all three regression coefficients are equal, let us compute each sample coefficient and adjust the sum of squares within each group by its own regression coefficient.

For vendor A,

$$b_A = \frac{\sum xy}{\sum x^2} = \frac{\sum XY - (\sum X)(\sum Y)/n}{\sum X^2 - (\sum X)^2/n}$$

Within group A,

$$b_A = \frac{14,599 - 58,125/4}{35,783 - 140,625/4} = 0.108$$

$$\text{adjusted } \sum y^2 = \sum y^2 - \frac{(\sum xy)^2}{\sum x^2} = 146.75 - 7.32 = 139.43$$

For vendor B,

$$b_B = \frac{9294 - 32,865/4}{30,269 - 24,492.25/4} = 0.187$$

$$\text{adjusted } \sum y^2 = 286.75 - 201.72 = 85.03$$

For vendor C,

$$b_C = \frac{5501 - 21{,}504/4}{17{,}450 - 65{,}536/4} = 0.117$$

$$\text{adjusted } \sum y^2 = 110.00 - 14.66 = 95.34$$

Summarizing, we obtain Table 16.4.

Adjusting each within-vendor sum of squares by its own regression coefficient reduces the degrees of freedom to 2 per vendor. Next the new adjusted sums of squares are added, and Table 16.4 shows a within-vendor sum of squares of 319.80 based on 6 df. When the within-vendor sum of squares was adjusted by a "pooled" within-vendor regression, Table 16.3 showed an adjusted sum of squares within vendors of 327.4 based on 8 df. If there were significant differences in regression within the three vendors, this would show as a difference in these two figures: $327.4 - 319.8 = 7.6$. An F test for this hypothesis would then be

$$F_{k-1,N-2k} = \frac{\left[\begin{array}{l} \text{adjusted } \sum y^2 \text{ (based on pooled within-groups regression)} \\ -\text{adjusted } \sum y^2 \text{ (based on regressions within each group)} \end{array} \right] \Big/ (k-1)}{[\text{adjusted } \sum y^2 \text{ (within each group)}]/(N-2k)}$$

and

$$F_{2,6} = \frac{(327.4 - 319.8)/2}{319.8/6} = \frac{3.8}{53.3} = <1$$

hence nonsignificant, and we conclude that the three regression coefficients are homogeneous.

The final assumption—namely, that the vendors do not affect the covariate X—is tenable here on practical grounds, since the vendor has nothing to do with the bracket thickness before plating. However, in some applications this assumption needs to be checked by an F test on the X variable. For our data from Table 16.3, such an F test would give

$$F_{2,9} = \frac{1771.2/2}{7469.5/9} = \frac{885.6}{829.9} = 1.07$$

which is obviously nonsignificant.

TABLE 16.4
Adjusted Sums of Squares Within Each Vendor

Within Vendor	df	$\sum y^2$	b	Adjusted $\sum y^2$	df
A	3	146.75	0.108	139.43	2
B	3	286.75	0.187	85.03	2
C	3	110.00	0.117	95.34	2
Totals	9	543.50		319.80	6

Some Comments

As the foregoing discussion indicates, satisfying all the assumptions of covariance analysis can be problematic. Indeed, such difficulties may cause many people to avoid covariance when it is applicable if they suspect that another measured variable or variables might affect the dependent variable.

The techniques illustrated in this section may be extended to handle several covariates, nonlinear regression, and several factors and interactions of interest. In handling randomized blocks, Latin squares, or experiments with several factors, Ostle [25] points out that the proper technique is to add the sum of squares and sum of cross products of the term of interest to the sum of squares and sum of cross products of the error, adjust this total, then adjust the error and determine the adjusted treatment effect by subtraction.

When several covariates are involved, many questions arise concerning how many covariates can be handled, how many really affect the dependent variable, how a significant subset can be found, and so on. All these questions apply in any multiple regression problem, and some discussion of these and other problems can be found in [5].

16.2.3 Covariance Analysis Using SAS

The bracket study of this section can be easily analyzed by means of SAS. A command file is given in Table 16.5. The **SOLUTION** option is included in the **MODEL** statement to have SAS include the solution to the normal equations with the other outputs. The statement

```
LSMEANS VENDOR/STDERR PDIFF;
```

has SAS calculate the adjusted means. The option **STDERR** has the standard errors of those means calculated, and the option **PDIFF** has SAS output the p values for the tests of equality of pairs of factor level means.

The outputs for the Table 16.5 command file are given in Figure 16.5. The type I sum of squares for vendor is the same as the vendor sum of squares included in the Minitab ANOVA of Figure 16.1. The type III sum of squares for vendor is the adjusted sum of squares for vendor included in Figure 16.3 and Table 16.3. As with the Minitab summary in Figure 16.3, SAS outputs both F and t results for the test that the true slope β is zero.

TABLE 16.5
SAS Command File for Covariate Analysis of Table 16.2

```
DATA BRACKET;
INPUT VENDOR $ X Y @@;
CARDS;
A 110  40 A 75 38 A 93 30 A  97 47
B  60  25 B 75 32 B 38 13 B 140 25
C  62  27 C 90 24 C 45 20 C  59 13
;
PROC GLM;
CLASS VENDOR;
MODEL Y=VENDOR X/SOLUTION;
LSMEANS VENDOR/STDERR PDIFF;
```

```
                              COVARIANCE EXAMPLE
                        GENERAL LINEAR MODELS PROCEDURE
```

DEPENDENT VARIABLE: Y

SOURCE	DF	SUM OF SQUARES	MEAN SQUARE	F VALUE	PR > F	R-SQUARE	C.V.
MODEL	3	881.26817949	293.75605983	7.18	0.00117	0.729124	22.3160
ERROR	8	327.39848718	40.92481090		ROOT MSE		Y MEAN
CORRECTED TOTAL	11	1208.66666667			6.39725026		28.66666667

SOURCE	DF	TYPE I SS	F VALUE	PR > F	DF	TYPE III SS	F VALUE	PR > F
VENDOR	2	665.16666667	8.13	0.0118	2	292.42172781	3.57	0.0778
X	1	216.10151282	5.28	0.0506	1	216.10151282	5.28	0.0506

PARAMETER		ESTIMATE	T FOR H0: PARAMETER=0	PR > |T|	STD ERROR OF ESTIMATE
INTERCEPT		10.11413080 B	1.77	0.1148	5.71601862
VENDOR	A	12.68977174 B	2.52	0.0357	5.03106293
	B	2.82619319 B	0.61	0.5598	4.64488634
	C	0.00000000 B	.	.	.
X		0.17009171	2.30	0.0506	0.07401974

```
                              COVARIANCE EXAMPLE
                        GENERAL LINEAR MODELS PROCEDURE
                           LEAST SQUARES MEANS
```

VENDOR	Y LSMEAN	STD ERR LSMEAN	PROB > |T| H0:LSMEAN=0	H0: LSMEAN(I)=LSMEAN(J)		
				PROB > |T|		
				I/J	1	2	3
A	36.1844501	3.3878748	0.0001	1	.	0.0675	0.0357
B	26.3208715	3.1987738	0.0001	2	0.0675	.	0.5598
C	23.4946784	3.3778366	0.0001	3	0.0357	0.5598	.

NOTE: TO ENSURE OVERALL PROTECTION LEVEL, ONLY PROBABILITIES ASSOCIATED WITH PRE-PLANNED COMPARISONS SHOULD BE USED.

Figure 16.5 SAS covariance analysis for Table 16.2.

For either test, the p value is 0.0506, indicating that inclusion of the covariate in the model is reasonable.

The procedure illustrated can be extended to include two or more factors. The analysis is quite straightforward when a computer package such as SAS is used. This is illustrated as follows.

■ **Example 16.1 (Covariance Analysis Involving Two Factors and One Covariate)**

Consider a study in which two factors (A and B) and one covariate (X) are to be included in the model. Letting Y denote the response, we obtain the data in Table 16.6.

A covariance model for this situation is

$$Y_{ijk} = \mu + \beta(X_{ijk} - \bar{X}) + A_j + B_k + AB_{jk} + \varepsilon_{ijk}$$

This can be included in a SAS command file using the statement

```
MODEL Y=A|B X/SOLUTION;
```

The **SOLUTION** option is included to have the solution of the normal equations output with the other statistics. A complete command file is given in Table 16.7.

TABLE 16.6
Data for Example 16.1

A	B	X	Y	A	B	X	Y
1	1	22	5	1	2	30	8
1	1	19	3	1	2	25	5
1	1	20	4	1	2	35	8
2	1	27	3	2	2	40	9
2	1	30	4	2	2	29	7
2	1	32	6	2	2	30	6

TABLE 16.7
SAS Command File for Example 16.1

```
DATA EXAMPLE;
INPUT A B X Y @@;
CARDS;
1 1 22 5 1 2 30 8 1 1 19 3 1 2 25 5
1 1 20 4 1 2 35 8 2 1 27 3 2 2 40 9
2 1 30 4 2 2 29 7 2 1 32 6 2 2 30 6
;
PROC GLM;
CLASS A B;
MODEL Y=A|B X/SOLUTION;
LSMEANS A B/STDERR PDIFF;
```

Outputs for Table 16.7 are given in Figure 16.6. Using the type III sums of squares, we find that A has a significant effect on the response ($p = 0.0480$). The test of $H_0 : \beta = 0$ versus $H_1 : \beta \neq 0$ has a p value of 0.0033, indicating that inclusion of the covariate is reasonable. From the Least Squares Means portion of Figure 16.6, we note that there is sufficient evidence to conclude that the true average response at level 1 of A exceeds that at level 2 of A.

16.3 RESPONSE SURFACE EXPERIMENTATION

Philosophy

The concept of a response surface involves a dependent variable Y, called the response variable, and several independent or controlled variables, $X_1, X_2, \ldots, X_k$. If all these variables are assumed to be measurable, the response surface can be expressed as follows:

$$Y = f(X_1, X_2, \ldots, X_k)$$

For the case of two independent variables such as temperature X_1 and time X_2, the yield Y of a chemical process can be expressed as follows:

$$Y = f(X_1, X_2)$$

This surface can be plotted in three dimensions, with X_1 on the abscissa, X_2 on the ordinate, and Y perpendicular to the $X_1 X_2$ plane. If the values of X_1 and X_2 that yield the same Y are connected, we can picture the surface with a series of equal-yield lines, or contours. These are similar to the contours of equal height on topographic maps and the isobars on weather maps. Figure 16.7 presents some typical response surfaces. Hunter [13] is an excellent reference for understanding response surface experimentation.

The Twofold Problem

The response surface experimentation entails a twofold problem: to determine, on the basis of one experiment, where to move in the next experiment toward the optimal point on the underlying response surface, and, having located the optimum or near optimum of the surface, to determine the equation of the response surface in an area near this optimum point.

One method of experimentation that seeks the optimal point of the response surface might be the traditional one-factor-at-a-time method. As shown in Figure 16.7a, if X_2 is fixed and X_1 is varied, we find the X_1 optimal (or near optimal) value of response Y at the fixed value of X_2. Having found this X_1 value, we can now run experiments at this fixed X_1, and the X_2 for optimal response can be found. In the case of the mound in Figure 16.7a, this method would lead eventually to the peak of the mound or near it. When applied to a surface such as the rising ridge in Figure 16.7c, however, the same method fails to lead to the maximum point on the response surface. In experimental work the type of response surface is usually unknown. Thus a better method is necessary if the optimum set of conditions is to be found for any surface.

COVARIANCE EXAMPLE
GENERAL LINEAR MODELS PROCEDURE

DEPENDENT VARIABLE: Y

SOURCE	DF	SUM OF SQUARES	MEAN SQUARE	F VALUE	PR > F	R-SQUARE	C.V.
MODEL	4	40.01336478	10.00334119	15.05	0.0015	0.895822	14.3881
ERROR	7	4.65330189	0.66475741			ROOT MSE	Y MEAN
CORRECTED TOTAL	11	44.66666667				0.81532657	5.6666667

SOURCE	DF	TYPE I SS	F VALUE	PR > F	DF	TYPE III SS	F VALUE	PR > F
A	1	0.33333333	0.50	0.5018	1	3.80390312	5.72	0.0480
B	1	27.00000000	40.62	0.0004	1	1.75390980	2.64	0.1483
A*B	1	0.00000000	0.00	1.0000	1	2.22532394	3.35	0.1100
X	1	12.68003145	19.07	0.0033	1	12.68003145	19.07	0.0033

PARAMETER		ESTIMATE		T FOR H0: PARAMETER=0	PR > |T|	STD ERROR OF ESTIMATE
INTERCEPT		-2.55110063		-1.10	0.3063	2.31163732
A	1	0.56525157	B	0.81	0.4439	0.69678035
	2	0.00000000	B	.	.	.
B	1	-2.00157233	B	-2.84	0.0249	0.70386960
	2	0.00000000	B	.	.	.
A*B	1 1	1.89701258	B	1.83	0.1100	1.03682434
	1 2	0.00000000	B	.	.	.
	2 1	0.00000000	B	.	.	.
	2 2	0.00000000	B	.	.	.
X		0.29952830		4.37	0.0033	0.06858187

LEAST SQUARES MEANS

A	Y LSMEAN	STD ERR LSMEAN	PROB > |T| H0:LSMEAN=0	PROB > |T| H0: LSMEAN1=LSMEAN2
1	6.42354560	0.39434573	0.0001	0.0480
2	4.90978774	0.39434573	0.0001	

B	Y LSMEAN	STD ERR LSMEAN	PROB > |T| H0:LSMEAN=0	PROB > |T| H0: LSMEAN1=LSMEAN2
1	5.14013365	0.40059122	0.0001	0.1483
2	6.19319969	0.40059122	0.0001	

Figure 16.6   SAS covariance analysis for Table 16.6.

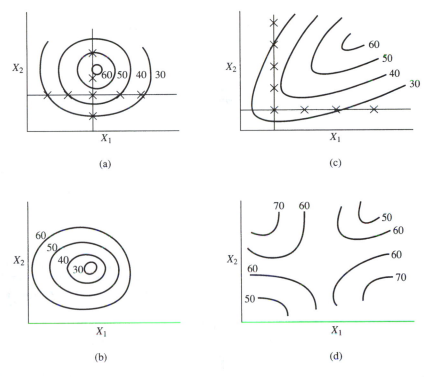

Figure 16.7   Some typical response surfaces in two dimensions: (a) mound, (b) depression, (c) rising ridge, and (d) saddle.

The method developed by those who have worked in this area is called the *path of steepest ascent* method. The idea here is to run a simple experiment over a small area of the response surface where, for all practical purposes, the surface may be regarded as a plane. We then determine the equation of this plane and from it the direction we should take from this experiment in order to move toward the optimum of the surface. The next experiment should be in a direction in which we hope to scale the height the fastest; hence the designation *path of steepest ascent*. This technique does not determine how far away from the original experiment succeeding sequential experiments should be run, but it does indicate the direction along which the next experiment should be performed. A simple example will illustrate this method.

To determine the equation of the response surface, several special experimental designs have been developed that attempt to approximate this equation using the smallest number of experiments possible. In two dimensions the simplest surface is a plane given by

$$Y = B_0 X_0 + B_1 X_1 + B_2 X_2 + \varepsilon \qquad (16.6)$$

where Y is the observed response, X_0 is taken as unity, and estimates of the B's are to be determined by the method of least squares, which minimizes the sum of the squares of the errors ε. Such an equation is referred to as a first-order equation, since the power on each independent variable is unity.

If there is some evidence that the surface is not planar, a second-order equation in two dimensions may be a more suitable model:

$$Y = B_0 X_0 + B_1 X_1 + B_2 X_2 + B_{11} X_1^2 + B_{12} X_1 X_2 + B_{22} X_2^2 + \varepsilon \qquad (16.7)$$

Here the $X_1 X_2$ term represents an interaction between the two variables X_1 and X_2.

If there are three independent or controlled variables, the first-order equation is again a plane or hyperplane,

$$Y = B_0 X_0 + B_1 X_1 + B_2 X_2 + B_3 X_3 + \varepsilon \qquad (16.8)$$

and the second-order equation is

$$Y = B_0 X_0 + B_1 X_1 + B_2 X_2 + B_3 X_3 + B_{11} X_1^2 + B_{22} X_2^2$$
$$+ B_{33} X_3^2 + B_{12} X_1 X_2 + B_{13} X_1 X_3 + B_{23} X_2 X_3 + \varepsilon \qquad (16.9)$$

As the complexity of the surface increases, more coefficients must be estimated, and the number of experimental points must necessarily increase. Several very clever designs have been developed that minimize the amount of work necessary to estimate these response surface equations.

To determine the coefficients for these more complex surfaces and to interpret their geometric nature, both multiple regression techniques and the methods of solid analytical geometry are used. The example that follows explores only the simplest type of surface. The cited references give many more complex examples.

■ EXAMPLE 16.2

Consider an example in which an experimenter is seeking the proper values for both concentration of filler to epoxy resin X_1 and position in the mold X_2 to minimize the abrasion on a plastic die. This abrasion or wear is measured as a decrease in thickness of the material after 10,000 cycles of abrasion. Since the maximum thickness is being sought, the first experiment should attempt to discover the direction in which succeeding experiments should be run to approach this maximum by the steepest path. Assuming the surface to be a plane in a small area, the first experiment will be used to determine the equation of this plane. The response surface is then

$$Y = B_0 X_0 + B_1 X_1 + B_2 X_2 + \varepsilon$$

Since there are three parameters to be estimated, B_0, B_1, and B_2, at least three experimental points must be taken to estimate these coefficients. Such a design might be an equilateral triangle, but since there are two factors, X_1 and X_2, each can be set at two levels and a 2^2 factorial may be used. Two concentrations were chosen, 0.5:1 and 1:1 (ratio of filler to resin), and two positions, 1 inch and 2 inches from a reference point, with responses Y, as shown in Figure 16.8; the thickness of the material is given in ten-thousandths of an inch.

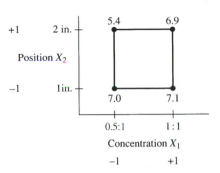

Figure 16.8 A 2^2 factorial example on response surface.

To determine the equation of the best fitting plane for these four points, consider the error in the prediction equation

$$\varepsilon = Y - B_0 X_0 - B_1 X_1 - B_2 X_2$$

The sum of the squares for this error is

$$\sum \varepsilon^2 = \sum (Y - B_0 X_0 - B_1 X_1 - B_2 X_2)^2$$

where the summation is over all points given in the design. To find the B's, we differentiate this expression with respect to each parameter and set these three expressions equal to zero. This provides three least squares normal equations, which can be solved for the best estimates of the B's. These estimates are designated as b's.

Differentiating the expression above gives

$$\frac{\partial (\sum \varepsilon^2)}{\partial B_0} = -2 \sum (Y - B_0 X_0 - B_1 X_1 - B_2 X_2) X_0 = 0$$

$$\frac{\partial (\sum \varepsilon^2)}{\partial B_1} = -2 \sum (Y - B_0 X_0 - B_1 X_1 - B_2 X_2) X_1 = 0$$

$$\frac{\partial (\sum \varepsilon^2)}{\partial B_2} = -2 \sum (Y - B_0 X_0 - B_1 X_1 - B_2 X_2) X_2 = 0$$

and from these results we get

$$\sum X_0 Y = b_0 \sum X_0^2 + b_1 \sum X_0 X_1 + b_2 \sum X_0 X_2$$

$$\sum X_1 Y = b_0 \sum X_0 X_1 + b_1 \sum X_1^2 + b_2 \sum X_1 X_2$$

$$\sum X_2 Y = b_0 \sum X_0 X_2 + b_1 \sum X_1 X_2 + b_2 \sum X_2^2 \qquad (16.10)$$

as the least squares normal equations.

By a proper choice of experimental variables, it is possible to reduce these equations considerably for simple solution. More complex models can be solved best by means of matrix algebra. For the 2^2 factorial, the following coding scheme simplifies the solution of Equation (16.10). Set

$$X_1 = 4C - 3$$

$$X_2 = 2P - 3$$

where C = concentration, P = position; X_0 is always taken as unity. For the experimental variables X_1 and X_2, the responses can be recorded as in Table 16.8. These experimental variables are also indicated on Figure 16.8.

An examination of the data as presented in Table 16.8 shows that X_0, X_1, and X_2 are all orthogonal to each other as

$$\sum X_1 = \sum X_2 = 0 \quad \text{and} \quad \sum X_1 X_2 = 0$$

Hence the least squares normal equations become

$$\sum X_0 Y = b_0 n + b_1 \cdot 0 + b_2 \cdot 0$$

$$\sum X_1 Y = b_0 \cdot 0 + b_1 \cdot \sum X_1^2 + b_2 \cdot 0 \tag{16.11}$$

$$\sum X_2 Y = b_0 \cdot 0 + b_1 \cdot 0 + b_2 \sum X_2^2$$

Solving gives

$$b_0 = \frac{\sum X_0 Y}{n}$$

$$b_1 = \frac{\sum X_1 Y}{\sum X_1^2} \tag{16.12}$$

$$b_2 = \frac{\sum X_2 Y}{\sum X_2^2}$$

For this problem

$$b_0 = \frac{26.4}{4} = 6.60$$

$$b_1 = \frac{1.6}{4} = 0.40$$

$$b_2 = \frac{-1.8}{4} = -0.45$$

and the response surface can be approximated as

$$\hat{Y} = 6.60 + 0.40X_1 - 0.45X_2$$

To determine the sum of squares due to each of these terms in the model, we can use the sum of squares due to b_i's:

$$SS_{b_i} = b_i \cdot \sum X_i Y$$

Here

$$SS_{b_0} = (6.60)(26.4) = 174.24$$

$$SS_{b_1} = (0.40)(1.6) = 0.64$$

$$SS_{b_2} = (-0.45)(-1.8) = 0.81$$

TABLE 16.8
Orthogonal Layout for Figure 16.8

Y	X_0	X_1	X_2
7.0	1	−1	−1
5.4	1	−1	1
7.1	1	1	−1
6.9	1	1	1

TABLE 16.9
ANOVA for 2^2 Factorial Response Surface Example

Source	df	SS
b_0	1	174.24
b_1	1	0.64
b_2	1	0.81
Residual	1	0.49
Totals	4	176.18

each carrying 1 df. Table 16.9, the ANOVA table, shows all four degrees of freedom (since the b_0 term represents the degree of freedom usually associated with the mean, i.e., the correction term in many examples). The total sum of squares is $\sum Y^2$ of the responses and the residual is what is left over. With this 1-df residual, no good test is available on the significance of each term in the model, nor is there any way to assess the adequacy of the planar model to describe the surface.

To decide on the direction for the next experiment, plot contours of equal response using the equation of the plane determined above:

$$\hat{Y} = 6.60 + 0.40X_1 - 0.45X_2$$

Solve for X_2,

$$X_2 = \frac{6.60 - \hat{Y} + 0.40X_1}{0.45}$$

If $\hat{Y} = 5.5$,

$$X_2 = \frac{1.10 + 0.40X_1}{0.45}$$

when

$$X_1 = -1 \qquad X_2 = 1.56$$
$$X_1 = 1 \qquad X_2 = 3.33$$

If $\hat{Y} = 6.0$,

$$X_2 = \frac{0.60 + 0.40X_1}{0.45}$$

when

$$X_1 = -1 \qquad X_2 = 0.44$$
$$X_1 = 1 \qquad X_2 = 2.22$$

If $\hat{Y} = 6.5$,

$$X_2 = \frac{0.10 + 0.40X_1}{0.45}$$

when

$$X_1 = -1 \qquad X_2 = -0.67$$
$$X_1 = 1 \qquad X_2 = 1.11$$

If $\hat{Y} = 7.0$,

$$X_2 = \frac{-0.40 + 0.40X_1}{0.45}$$

when

$$X_1 = -1 \qquad X_2 = -1.67$$
$$X_1 = 1 \qquad X_2 = 0$$

If $\hat{Y} = 7.5$,

$$X_2 = \frac{-0.90 + 0.40X_1}{0.45}$$

when

$$X_1 = -1 \qquad X_2 = -2.89$$
$$X_1 = 1 \qquad X_2 = -1.11$$

Plotting these five contours on the original diagram gives the pattern shown in Figure 16.9.

By moving in a direction normal to these contours and "up" the surface, we can anticipate larger values of response until the peak is reached. To decide on a possible set of conditions for the next experiment, consider the equation of a normal to these contours through the point (0, 0). The contours are

$$X_2 = \frac{6.60 - \hat{Y} + 0.40X_1}{0.45}$$

and their slope is $0.40/0.45 = 8/9$. The normal will have a slope $= -9/8$, and its equation is

$$X_2 - 0 = \frac{-9}{8}(X_1 - 0)$$

$$X_2 = \frac{-9}{8}X_1 \qquad \text{(see arrow in Figure 16.9)}$$

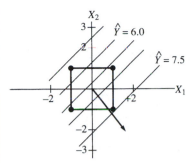

Figure 16.9 Contours on 2^2 factorial response surface example.

This technique will not tell how far to go in this direction, but we might run the next experiment with the center of the factorial at $(+1, -9/8)$. The four points would be

X_1	X_2
0	$-\frac{1}{8}$
0	$-2\frac{1}{8}$
2	$-\frac{1}{8}$
2	$-2\frac{1}{8}$

These points are expressed in terms of the experimental variables, but they must be decoded to see where to set the concentration and position. Using the coding,

$$X_1 = 4C - 3 \quad \text{or} \quad C = \frac{X_1 + 3}{4}$$

$$X_2 = 2P - 3 \quad \text{or} \quad P = \frac{X_2 + 3}{2}$$

we find the following new points:

X_1	X_2	C	P
0	$\frac{1}{8}$	$\frac{3}{4} : 1$	$1\frac{7}{16}$
0	$-2\frac{1}{8}$	$\frac{3}{4} : 1$	$\frac{7}{16}$
2	$-\frac{1}{8}$	$1\frac{1}{4} : 1$	$1\frac{7}{16}$
2	$-2\frac{1}{8}$	$1\frac{1}{4} : 1$	$\frac{7}{16}$

if these settings are possible. Responses may now be taken at these four points on a new 2^2 factorial, and after analysis we can again decide the direction of steepest ascent. This procedure continues until the optimum has been obtained.

The design above is sufficient to indicate the direction for subsequent experiments, but it does not provide a good measure of experimental error to test the significance of b_0, b_1, and b_2, nor is there any test of how well the plane approximates the surface. One way to improve on this design is to take two or more points at the center of the square. By replication at the same point, we can obtain an estimate of experimental error, and the average of the center-point responses will provide an estimate of "goodness of fit" of the plane. If the experiment is near the maximum response, this center point might be somewhat above the four surrounding points, which would indicate the need for a more complex model.

To see how this design would help, consider two observations of response at the center of the example above. Using the same responses at the vertices of the square, the results might be those shown in Figure 16.10.

The coefficients in the model are still estimated by Equation (16.12):

$$b_0 = \frac{\sum X_0 Y}{n} = \frac{39.8}{6} = 6.63$$

$$b_1 = \frac{\sum X_1 Y}{\sum X_1^2} = \frac{1.6}{4} = 0.40$$

$$b_2 = \frac{\sum X_2 Y}{\sum X_2^2} = \frac{-1.8}{4} = -0.45$$

and

$$SS_{b_0} = 6.63(39.8) = 263.87$$

$$SS_{b_1} = 0.40(1.6) = 0.64$$

$$SS_{b_2} = -0.45(-1.8) = 0.81$$

For the sum of squares of the error at (0, 0), we have

$$SS_e = (6.6)^2 + (6.8)^2 - \frac{(13.4)^2}{2} = 89.80 - 89.78 = 0.02$$

The analysis is shown in Table 16.10.

Y	X_0	X_1	X_2
7.0	1	−1	−1
5.4	1	−1	1
7.1	1	1	−1
6.9	1	1	1
6.6	1	0	0
6.8	1	0	0

Testing these effects against error gives

$$b_0: F_{1,1} = \frac{236.87}{0.02} = 11,843.5$$

$$b_1: F_{1,1} = \frac{0.64}{0.02} = 32$$

$$b_2: F_{1,1} = \frac{0.81}{0.02} = 40.5$$

$$\text{lack of fit: } F_{2,1} = \frac{0.32}{0.02} = 16$$

With such a small number of degrees of freedom, only b_0 shows significance at the 5% level. However, the tests on the b terms are larger than the test on lack of fit, which may indicate that the plane

$$\hat{Y} = 6.63 + 0.40X_1 - 0.45X_2$$

is a fair approximation to the surface where this first experiment was run.

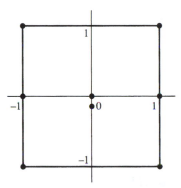

Figure 16.10   A 2^2 factorial with two points in center.

TABLE 16.10
ANOVA for 2^2+ Two Center Points

Source	df	SS	MS
Total	6	265.98	
b_0	1	263.87	236.87
b_1	1	0.64	0.64
b_2	1	0.81	0.81
Residual	3	0.66	—
Error	1	0.02	0.02
Lack of fit	2	0.64	0.32

Response Surface Analysis Using Minitab

If Minitab is available, selection of **Stat ▶ DOE ▶ Fit RS Model** will produce a response surface analysis of a data set. The outputs for the data in Table 16.8 are given in Figure 16.11. Using the entries in the Coef column, the fitted model is

$$\hat{Y} = 6.60 + 0.40X_1 - 0.45X_2$$

as before. The ANOVA results agree with those of Table 16.9. Notice, however, that the uncorrected total sum of squares is included in Table 16.9, whereas the Minitab ANOVA includes only the sum of squares associated with the linear model, that for the residual, and the corrected total sum of squares.

```
Response Surface Regression

Estimated Regression Coefficients for y

Term                 Coef      Stdev    t-ratio          p
Constant           6.6000     0.3500    18.857      0.034
x1                 0.4000     0.3500     1.143      0.458
x2                -0.4500     0.3500    -1.286      0.421

s = 0.7000    R-sq = 74.7%    R-sq(adj) = 24.2%

Analysis of Variance for y

Source            DF     Seq SS     Adj SS     Adj MS        F        P
Regression         2    1.45000    1.45000   0.725000     1.48    0.503
  Linear           2    1.45000    1.45000   0.725000     1.48    0.503
Residual Error     1    0.49000    0.49000   0.490000
Total              3    1.94000
```

Figure 16.11   Minitab response surface analysis of the Table 16.8 data.

More Complex Surfaces

For more factors—controlled variables—the model for the response variable is more complex, but several designs have been found useful in estimating the coefficients of these surfaces.

When three variables are involved, a first approximation is again a plane or hyperplane of the form

$$Y = B_0 X_0 + B_1 X_1 + B_2 X_2 + B_3 X_3 + \varepsilon$$

For this first-order surface, at least four points must be taken to estimate the four B's. Because three dimensions are involved, we might consider a 2^3 factorial. Since this design has eight experimental conditions at its vertices, the design often used is a half-replication of a 2^3 factorial with two or more points at the center of the cube. These six points (two at the center) are sufficient to estimate all four B's and to test for lack of fit of this plane to the surface in three dimensions. After this initial half-replication of the 2^3, the other half might be run, giving more information for a better fit.

If a plane is not a good fit in two dimensions, a second-order model might be tried. Such a response surface has the form

$$Y = B_0 X_0 + B_1 X_1 + B_2 X_2 + B_{11} X_1^2 + B_{12} X_1 X_2 + B_{22} X_2^2 + \varepsilon$$

Here six B's are to be estimated. The simplest design for this model, a pentagon (five points) plus center points, would yield six or more responses, and all six B's could be estimated.

In developing these designs Box and others found that the calculations can be simplified if the design can be rotated. A *rotatable design* is one that has equal predictability in all directions from the center, and the points are at a constant distance from the center. All first-order designs are rotatable, as the square and half-replication of the cube in the cases above. The simplest second-order design that is rotatable is the pentagon with a point at the center, as given above.

In three dimensions a second-order surface is given by

$$Y = B_0 X_0 + B_1 X_1 + B_2 X_2 + B_3 X_3 + B_{11} X_1^2 + B_{22} X_2^2$$
$$+ B_{33} X_3^2 + B_{12} X_1 X_2 + B_{13} X_1 X_3 + B_{23} X_2 X_3 + \varepsilon \qquad (16.13)$$

which has 10 unknown coefficients. The cube for a three-dimensional model has only eight points, so a special design has been developed, called a *central composite* design. It is a 2^3 factorial with points along each axis at a distance from the center equal to the distance to each vertex. This gives 15 points (i.e., 8 + 6 + a point at the center), which is adequate for estimating the B's in Equation (16.13). This design is pictured in Figure 16.12.

In attempting to determine the equation of the response surface, some concepts in solid analytical geometry are often helpful. If a second-order model in two dimensions is

$$Y = B_0 X_0 + B_1 X_1 + B_2 X_2 + B_{11} X_1^2 + B_{12} X_1 X_2 + B_{22} X_2^2 + \varepsilon$$

the shape of this surface can be determined by reducing this equation to what is called *canonical form:*

$$Y = B_{11}' X_1'^2 + B_{22}' X_2'^2$$

This is accomplished by a translation of axes to remove terms in X_1 and X_2, and then a rotation of axes to remove the $X_1 X_2$ term. From B_{11}' and B_{22}' we can determine whether the surface is a sphere, ellipsoid, paraboloid, hyperboloid, or other shape. In higher dimensions this becomes more complicated, but it can still be useful in describing the response surface.

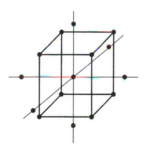

Figure 16.12   Central composite design.

16.4 EVOLUTIONARY OPERATION (EVOP)

Philosophy

Evolutionary operation is a method of process operation that has a built-in procedure to increase productivity. The technique was developed by Box and Draper [7]. Their book and

Barnett's article [3] should be read to understand how the method works. Many chemical companies have reported considerable success using EVOP.

The procedure consists of running a simple experiment, usually a factorial, within the range of operability of a process as it is currently running. It is assumed that the variables to be controlled are measurable and can be set within a short distance of the current settings without disturbing production quality. The idea is to gather data on a response variable, usually yield, at the various points of an experimental design. When one set of data has been taken at all the points, one *cycle* is said to have been completed. One cycle is usually not sufficient to detect any shift in the response, so a second cycle is taken. Cycles are taken until the effect of one or more control variables, their interactions, or a change in the mean shows up as significant compared with a measure of experimental error. This estimate of error is obtained from the cycle data, thus making the experiment self-contained. After a significant increase in yield has been detected, one *phase* is said to have been completed, and at this point the basic operating conditions usually can be changed in a direction that should improve the yield. Several cycles may be necessary before a shift can be detected. The objective here, as with response surfaces, is to move in the direction of an optimum response. Response surface experimentation is primarily a laboratory or research technique; evolutionary operation is a production-line method.

To facilitate the EVOP procedure, a simple form has been developed to be used on the production line for each cycle of a 2^2 factorial with a point at the center. In the sections that follow, we run an example using these forms; the details of the form are developed later.

■ Example 16.3

To illustrate the EVOP procedure, consider a chemical process in which temperature and pressure are varied over short ranges, and the resulting chemical yield is recorded. Since two controlled variables affect the yield, a 2^2 factorial should indicate the effect of each factor as well as a possible interaction between them. If we take a point at the center of a 2^2 factorial, we can also check on a change in the mean (CIM) by comparing this point at the center with the four points around the vertices of the square. If the process happens to be straddling a maximum, the center point should eventually (after several cycles) lie significantly above the peripheral points at the vertices. The standard form for EVOP locates the five points in this design as indicated in Figure 16.13.

By comparing the responses (or average responses) at points 3 and 4 with those at 2 and 5, it may be possible to detect an effect of variable X_1. Likewise, by comparing responses at 3 and 5 with those at 2 and 4, we may assess the X_2 effect. Comparing the responses at 2 and 3 with those at 4 and 5 will indicate an interaction effect, and comparing the responses at 2, 3, 4, and 5 with those at 1 will indicate a change in the mean, if present. The forms shown in Table 6.11 are filled in with data to show their use, which is self-explanatory.

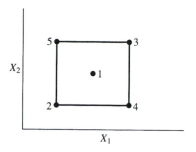

Figure 16.13   A 2^2 EVOP design.

TABLE 16.11
EVOP Worksheet: First Cycle

5		3			**Cycle: $n = 1$**		**Project: 424**
	1				**Response: Yield**		**Phase: 1**
2		4					**Date: 10/12**

Operating Conditions	Calculation of Averages					Calculation of Standard Deviation
	(1)	*(2)*	*(3)*	*(4)*	*(5)*	
(i) Previous cycle sum						Previous sum $s =$
(ii) Previous cycle average						Previous average $s =$
(iii) New observations	94.0	94.5	96.5	94.5	94.5	Range $=$
(iv) Differences [(ii) less (iii)]						New $s = $ range $\times f_{k,n} =$
(v) New sums	94.0	94.5	96.5	94.5	94.5	New sum $s =$
(vi) New averages $\hat{Y}_i$	94.0	94.5	96.5	94.5	94.5	New average $s = \frac{\text{new sum } s}{n-1}$

Calculation of Effects	Calculations Error Limits
Temperature effect $= 0.5(\bar{Y}_3 + \bar{Y}_4 - \bar{Y}_2 - \bar{Y}_5) = 1.00$	For new average $= \frac{2}{\sqrt{n}}\, s =$
Pressure effect $= 0.5(\bar{Y}_3 + \bar{Y}_5 - \bar{Y}_2 - \bar{Y}_4) = 1.00$	For new effects $= \frac{2}{\sqrt{n}}\, s =$
$T \times P$ interaction effect $= 0.5(\bar{Y}_2 + \bar{Y}_3 - \bar{Y}_4 - \bar{Y}_5) = 1.00$	For change in mean $= \frac{1.78}{\sqrt{n}}\, s =$
Change in mean effect $= 0.2(\bar{Y}_2 + \bar{Y}_3 + \bar{Y}_4 + \bar{Y}_5 - 4\bar{Y}_1)$ $= 0.80$	

Not much can be learned from the first cycle unless some separate estimate of standard deviation is available. The second cycle (Table 16.12) will begin to show the method for testing the effects.

TABLE 16.12
EVOP Worksheet: Second Cycle

5	3			Project: 424
[1]		Cycle: $n = 2$		Phase: 1
2	4	Response: Yield		Date: 10/12

	Calculation of Averages					Calculation of Standard Deviation
Operating Conditions	*(1)*	*(2)*	*(3)*	*(4)*	*(5)*	
(i) Previous cycle sum	94.5	94.5	96.5	94.5	94.5	Previous sum $s =$
(ii) Previous cycle average	94.0	94.5	96.5	94.5	94.5	Previous average $s =$
(iii) New observations	96.0	95.0	95.0	96.5	94.0	Range $= 3.5$
(iv) Differences [(ii) less (iii)]	−2.0	−0.5	1.5	−2.0	−0.5	New $s =$ range $\times f_{k,n}$
						$= 1.05$
(v) New sums	190.0	189.5	191.5	191.0	188.5	New sum $s = 1.05$
(vi) New averages $\bar{Y}_i$	95.0	94.7	95.7	95.5	94.2	New average $s = \frac{\text{new sum } s}{n-1}$
						$= 1.05$

Calculation of Effects	Calculations of Error Limits
Temperature effect $= 0.5(\bar{Y}_3 + \bar{Y}_4 - \bar{Y}_2 - \bar{Y}_5) = 1.15$	For new average $= \frac{2}{\sqrt{n}} s$ $= 1.48$
Pressure effect $= 0.5(\bar{Y}_3 + \bar{Y}_5 - \bar{Y}_2 - \bar{Y}_4) = -0.15$	For new effects $= \frac{2}{\sqrt{n}} s$ $= 1.48$
$T \times P$ interaction effect $= 0.5(\bar{Y}_2 + \bar{Y}_3 - \bar{Y}_4 - \bar{Y}_5)$ $= 0.35$	For change in mean $= \frac{1.78}{\sqrt{n}} s$ $= 1.32$
Change in mean effect $= 0.2(\bar{Y}_2 + \bar{Y}_3 + \bar{Y}_4 + \bar{Y}_5 - 4\bar{Y}_1)$ $= 0.02$	

Since, in the second cycle, none of the effects are numerically larger than their error limits, the true effect could easily be zero. In such a case, another cycle must be run. The only items that might need explaining are under the calculation of standard deviation. The range referred to here is the range of the differences (iv), and $f_{k,n}$ is found in a table where $k = 5$ for these five-point designs, and n is the cycle number. Part of such a table shows

$n =$	2	3	4	5	6	7	8
$f_{5,n} =$	0.30	0.35	0.37	0.38	0.39	0.40	0.40

The third cycle is shown in Table 16.13. Here the temperature effect is seen to be significant, with an increase in temperature giving a higher yield. Once a significant effect has been found, the first *phase* of the EVOP procedure has been completed. The average results at this point are usually displayed on an EVOP bulletin board as in Figure 16.14.

TABLE 16.13
EVOP Worksheet: Third Cycle

5	3
1	
2	4

Cycle: $n = 3$
Response: Yield

Project: 424
Phase: 1
Date: 10/12

	Calculation of Averages					Calculation of Standard Deviation
Operating Conditions	*(1)*	*(2)*	*(3)*	*(4)*	*(5)*	
(i) Previous cycle sum	190.0	189.5	191.5	191.0	188.5	Previous sum $s = 1.05$
(ii) Previous cycle average	95.0	94.7	95.7	95.5	94.2	Previous average $s = 1.05$
(iii) New observations	94.5	93.5	96.0	97.0	94.0	Range $= 2.7$
(iv) Differences [(ii) less (iii)]	0.5	1.2	-0.3	-1.5	0.2	New $s = $ range $\times f_{k,n}$ $= 0.95$
(v) New sums	284.5	283.0	287.2	286.5	282.7	New sum $s = 2.00$
(vi) New averages $\bar{Y}_i$	94.8	94.3	95.7	95.5	94.2	New average $s = \frac{\text{new sum } s}{n-1}$ $= 1.00$

Calculation of Effects	Calculations of Error Limits

Temperature effect $= 0.5(\bar{Y}_3 + \bar{Y}_4 - \bar{Y}_2 - \bar{Y}_5) = 1.35*$

Pressure effect $= 0.5(\bar{Y}_3 + \bar{Y}_5 - \bar{Y}_2 - \bar{Y}_4) = 0.05$

$T \times P$ interaction effect $= 0.5(\bar{Y}_2 + \bar{Y}_3 - \bar{Y}_4 - \bar{Y}_5)$
$= -0.15$

Change in mean effect $= 0.2(\bar{Y}_2 + \bar{Y}_3 + \bar{Y}_4 + \bar{Y}_5 - 4\bar{Y}_1)$
$= 0.10$

For new average $= \frac{2}{\sqrt{n}} s$
$= 1.16$

For new effects $= \frac{2}{\sqrt{n}} s$
$= 1.16$

For change in mean $= \frac{1.78}{\sqrt{n}} s$
$= 1.02$

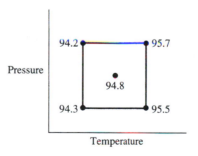

Figure 16.14   Final cycle of EVOP.

An EVOP committee usually reviews these data and decides whether to reset the operating conditions. If they change the operating conditions (point 1), EVOP is reinstated around this new point and the second phase is begun. EVOP is continued again until significant changes are detected. In fact, it goes on continually, seeking to optimize a process.

This is a very simple example with just two independent variables. The references should be consulted for more complex situations.

EVOP Form Rationale

Most of the steps in the EVOP form above are quite clear. It may be helpful to track down the source of some of the constants.

In estimating the standard deviation from the range of differences in step iv, consider a general expression for these differences

$$D_p = \frac{X_{p1} + X_{p2} + \cdots + X_{p,n-1}}{n - 1} - X_{p,n}$$

where D_p represents the difference at any point p of the design. X_{pi} represents the observation at point p in the ith cycle, and n is, of course, the number of cycles. Because the variance of a sum equals the sum of the variances for independent variables, and since the variance of a constant times a random variable equals the square of the constant times the variance of the random variable, we can write

$$\sigma_{Dp}^2 = \frac{1}{(n-1)^2}[\sigma_{xp1}^2 + \sigma_{xp2}^2 + \cdots + \sigma_{xp,n-1}^2] + \sigma_{xp,n}^2$$

Since all these x's represent the same population, their variances are all alike, and then we have

$$\sigma_D^2 = \frac{1}{(n-1)^2}[(n-1)\sigma_x^2] + \sigma_x^2 = \frac{n}{n-1}\sigma_x^2$$

and

$$\sigma_D = \sqrt{\frac{n}{n-1}}\sigma_x$$

The standard deviation of the population can then be wrtten in terms of the standard deviation of these differences

$$\sigma_x = \sqrt{\frac{n-1}{n}}\sigma_D$$

Now σ_D can be estimated from the range of these differences R_d. From the quality control field

$$\sigma_D = \frac{R_d}{d_2}$$

where d_2 depends on the number of differences in the range, which is 5 on this form. Here $d_2 = 2.326$ for samples of 5,

$$\sigma_D = \frac{R_d}{2.326}$$

and the standard deviation of the population is estimated by

$$\sigma_x = \sqrt{\frac{n-1}{n} \frac{R_d}{2.326}}$$

The quantity

$$\sqrt{\frac{n-1}{n} \frac{1}{2.326}}$$

is called $f_{k,n}$ in the EVOP form where $k = 5$.

Note that

$$f_{5,2} = \sqrt{\frac{1}{2} \frac{R_d}{2.326}} = 0.30 R_d$$

$$f_{5,6} = \sqrt{\frac{5}{6} \frac{R_d}{2.326}} = 0.39 R_d$$

which tallies with the table values given.

To determine the error limits for the effects, two standard deviation limits are used, since they represent the approximate 95% confidence limits on the parameter being estimated.

For any effect such as

$$E = \frac{\bar{Y}_3 + \bar{Y}_4 - \bar{Y}_2 - \bar{Y}_5}{2}$$

its variance would be

$$V_E = \frac{\sigma^2_{\bar{Y}_3} + \sigma^2_{\bar{Y}_4} + \sigma^2_{\bar{Y}_2} + \sigma^2_{\bar{Y}_5}}{4} = \frac{4\sigma^2_{\bar{Y}}}{4} = \sigma^2_{\bar{Y}} = \frac{\sigma^2_{\bar{Y}}}{n}$$

and two standard deviation limits on an effect would be

$$\pm 2 \frac{\sigma_Y}{\sqrt{n}} \quad \text{estimated by} \quad \pm 2 \frac{s}{\sqrt{n}}$$

For the change in mean effect

$$CIM = \frac{\bar{Y}_2 + \bar{Y}_3 + \bar{Y}_4 + \bar{Y}_5 - 4\bar{Y}_1}{5}$$

$$V_{\text{CIM}} = \frac{\sigma^2_{\bar{Y}_2} + \sigma^2_{\bar{Y}_3} + \sigma^2_{\bar{Y}_4} + \sigma^2_{\bar{Y}_5} + 16\sigma^2_{\bar{Y}_1}}{25}$$

$$= \frac{20}{25}\sigma^2_{\bar{Y}} = \frac{20}{25}\frac{\sigma^2_{\bar{Y}}}{n}$$

$$\sigma_{\text{CIM}} = \sqrt{\frac{4}{5}} \cdot \frac{\sigma_Y}{\sqrt{n}}$$

and two standard deviation limits would be

$$\pm 2 \cdot \sqrt{\frac{4}{5}} \frac{s}{\sqrt{n}} = \pm 1.78 \frac{s}{\sqrt{n}}$$

as given on the EVOP form.

■ Example 16.4

This EVOP procedure often has three factors, each at two levels. The $2^3 = 8$ experimental points are run as well as two points in the center of the cube giving 10 experimental points per cycle. Usually this design is run in two blocks, where the *ABC* interaction is confounded with blocks. The standard form for the location of the 10 points is shown in Figure 16.15.

The first five points (1–5) are run as the first block and the last five points (6–10) are run as the second block. Tables 16.14 and 16.15 illustrate this EVOP technique for the second cycle of an example on the mean surface factor for grinder slivers, where the factors are dropout temperature *A*, back-zone temperature *B*, and atmosphere *C*. The worksheets, based on the design principles of confounding a 2^3 factorial in two blocks of four observations, are self-explanatory. One could terminate this process with the second cycle as it shows dropout temperature *A*, back-zone temperature *B*, and their *AB* interaction to be significant in producing changes in the mean surface factor for grinder slivers.

TABLE 16.14
EVOP Worksheet: Block 1 in 2^3

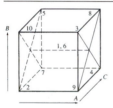

Cycle: $n = 2$ Project: 478
Response: Mean surface   Phase: 1
    factor for grinder slivers   Date: 11/6
Factors: *A* = Dropout temperature 2145 and 2165°F
         B = Back-zone temperature 2140 and 2160°F
         C = Atmosphere: Reducing and oxidizing

Operating Conditions (Block 1)	Calculation of Averages: Block 1					Calculation of Standard Deviation
	(1)	*(2)*	*(3)*	*(4)*	*(5)*	
(i) Previous cycle sum	3.9	0.0	1.6	9.8	4.3	Previous sum s = (all blocks)
(ii) Previous cycle sum	3.9	0.0	1.6	9.8	4.3	Previous average s =
(iii) New observations	5.6	4.0	6.7	15.0	3.4	Range = 4.3
(iv) Differences [(ii) less (iii)]	−1.7	−4.0	−5.1	−5.2	−0.9	New sum s = range $\times f_{k,n}$ = 1.29
(v) New sums	9.5	4.0	8.3	24.8	7.7	New sum s = 1.29 (all blocks)
(vi) New averages $\bar{Y}_i$	4.8	2.0	4.2	12.4	3.8	New average $s = \frac{\text{new sum } s}{2n-3}$ = 1.29

Calculation of Effects: Block 1

$A - BC = 0.5(\bar{Y}_3 + \bar{Y}_4 - \bar{Y}_2 - \bar{Y}_5) = 5.40$
$B - AC = 0.5(\bar{Y}_3 + \bar{Y}_5 - \bar{Y}_2 - \bar{Y}_4) = -3.20$
$C - AB = 0.5(\bar{Y}_4 + \bar{Y}_5 - \bar{Y}_2 - \bar{Y}_3) = 5.00$
Change in mean effect $= 0.2(\bar{Y}_2 + \bar{Y}_3 + \bar{Y}_4 + \bar{Y}_5 - 4\bar{Y}_1) = 0.64$

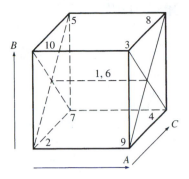

Figure 16.15   A 2^3 EVOP.

TABLE 16.15
EVOP Worksheet: Block 2 in 2^3

	Cycle: $n = 3$ Calculation of Averages—Block 2					Calculation of Standard Deviation
Operating Conditions *(Block 2)*	(6)	(7)	(8)	(9)	(10)	
(i) Previous cycle sum	4.2	4.4	2.8	10.1	0.0	Previous sum $s = 1.29$ (all blocks)
(ii) Previous cycle average	4.2	4.4	2.8	10.1	0.0	Previous average $s = 1.29$
(iii) New observations	3.2	3.4	6.6	15.0	1.9	Range $= 5.9$
(iv) Differences [(ii) less (iii)]	1.0	0.9	−3.8	−4.9	−1.9	New sum $s = $ range $\times f_{k,n}$ $= 1.77$
(v) New sums	7.4	7.7	9.4	25.1	1.9	New sum $s = 3.06$ (all blocks)
(vi) New averages $\bar{Y}_i$	3.7	3.8	4.7	12.6	1.0	New average $s = \frac{\text{new sum } s}{2n-2}$ $= 1.53$

Calculation of Effects: Block 2	Calculation of Error Limits
$A + BC = 0.5(\bar{Y}_8 + \bar{Y}_9 - \bar{Y}_7 - \bar{Y}_{10}) = 6.25$	For new averages $= \frac{2s}{\sqrt{n}}$ $= \pm 2.16$
$B + AC = 0.5(\bar{Y}_8 + \bar{Y}_{10} - \bar{Y}_7 - \bar{Y}_9) = -5.35$	For new effects $= \frac{1.42s}{\sqrt{n}}$ $= \pm 1.53$
$C + AB = 0.5(\bar{Y}_7 + \bar{Y}_8 - \bar{Y}_9 - \bar{Y}_{10}) = -2.55$	For change in mean $= \frac{1.26s}{\sqrt{n}}$ $= \pm 1.36$
Change in mean effect $= 0.2(\bar{Y}_7 + \bar{Y}_8 + \bar{Y}_9 + \bar{Y}_{10} - 4\bar{Y}_6) = 1.46$	

Calculation of Effects: Both Blocks

$A = 0.5[(A + BC) + (A - BC)] = 5.82^*$ $AB = 0.5[(C + AB) - (C - AB)] = -3.78^*$
$B = 0.5[(B + AC) + (B - AC)] = -4.28^*$ $AC = 0.5[(B + AC) - (B - AC)] = -1.08$
$C = 0.5[(C + AB) + (C - AB)] = 1.23$ $BC = 0.5[(A + BC) - (A - BC)] = -0.42$
Change in mean effect $= 0.5(\text{CIM}_1 + \text{CIM}_2) = 1.05$

* Significant at 5% level.

16.5 ANALYSIS OF ATTRIBUTE DATA

Philosophy

Two assumptions used in the application of the analysis of variance technique are (1) that the response variable is normally distributed and (2) that the variances of the experimental errors are equal throughout the experiment. In practice it is often necessary to deal with attribute data in which the response variable is either 0 or 1. In such cases one often records the number of occurrences of a particular phenomenon or the percentage of such occurrences.

It is well known that the number of occurrences per unit, such as defects per piece, errors per page, or customers per unit time, often follows a Poisson distribution where not only are such response variables nonnormal, but variances and means are equal. When proportions or percentages are used as the response variable, the data are binomial, and again variances are related to means and the basic assumptions of ANOVA do not hold. Some studies have shown that the lack of normality is not too serious in applying the ANOVA technique, but most statisticians recommend that a transformation be made on original data known to be nonnormal.

Another technique, called *factorial chi square,* has been very useful in treating attribute data in industrial problems. This technique is described by Batson [4], who gives several examples of its application to actual problems. Although this factorial chi-square technique may not be as precise as a regular ANOVA on transformed data, its simplicity makes it well worth consideration for many applied problems.

To illustrate these techniques, consider the following example from a consulting firm study.

■ **Example 16.5**

Interest centered on undesirable marks on steel samples from a grinding operation. The response variable was simply the occurrence of such marks. Four factors were believed to affect the marks: A, blade; B, centering; C, leveling; and D, speed. Each of these four factors was set at two levels, giving a 2^4 factorial experiment. Each of the 16 treatment combinations was run in a completely randomized order, and 20 steel samples were produced under each of the 16 experimental conditions. The number of damaged samples in each group of 20 is the recorded variable in Table 16.16.

Analysis Using Arcsin $\sqrt{p}$

If the responses are each divided by 20, the proportions of damaged samples can be determined. Since proportions follow a binomial distribution, an arc sine transformation is appropriate (see [10]). Taking the arcsin $\sqrt{p}$ as the response variable, and assuming that the three- and four-way interactions are negligible, we obtain the ANOVA of Figure 16.16.

TABLE 16.16
Damaged Steel Samples in 2^4 Experiment

		Factor A			
		0		1	
		Factor B		Factor B	
Factor C	Factor D	0	1	0	1
0	0	0	0	16	20
	1	0	0	10	20
1	0	0	0	10	14
	1	1	0	12	20

```
Analysis of Variance (Balanced Designs)
Factor      Type      Levels        Values
A           fixed       2             1            2
B           fixed       2             1            2
C           fixed       2             1            2
D           fixed       2             1            2
Analysis of Variance for TRNSFRMD
Source      DF         SS            MS            F          P
A            1      5.10992       5.10992       223.99      0.000
B            1      0.22896       0.22896        10.04      0.025
A*B          1      0.34959       0.34959        15.32      0.011
C            1      0.02068       0.02068         0.91      0.385
A*C          1      0.06582       0.06582         2.89      0.150
B*C          1      0.02132       0.02132         0.93      0.378
D            1      0.02132       0.02132         0.93      0.378
A*D          1      0.00111       0.00111         0.05      0.834
B*D          1      0.02068       0.02068         0.91      0.385
C*D          1      0.09418       0.09418         4.13      0.098
Error        5      0.11406       0.02281
Total       15      6.04765
```

Figure 16.16   Minitab ANOVA for transformed data of Example 16.5.

This analysis shows the highly significant blade size effect, which is obvious from a glance at Table 16.16. It also shows a significant centering effect and a blade size–centering interaction that is not as obvious from Table 16.16.

A Factorial Chi-Square Analysis

To apply the factorial chi-square technique to our steel sample experiment, we fill in Table 16.17. The plus and minus signs simply give the proper weights for the main effects and two-way interactions, and T is the value of the corresponding contrasts. The total number of samples in the total experiment, D, is 20 per cell or $20 \times 2^4 = 320$ samples. The chi-square (χ^2) value is determined by the formula

$$\chi^2_{[1]} = \frac{N^2}{(S)(F)} \times \frac{T^2}{D} \qquad (16.14)$$

where N is the total number of items in the whole experiment and S is the number of occurrences (successes) in the N; F is the number of nonoccurrences (or failures). In the example $N = 320$ samples, $S = 123$ with undesirable marks. For factor A, then

$$\chi^2_{[1]} = \frac{(320)^2}{(123)(197)}(45.75)$$

$$= (4.23)(45.75) = 193.52$$

The 4.23 is constant for these data. The resulting chi-square values are then compared with χ^2 with 1 df at the significance level desired. For the 5% level, $\chi^2_{[1]} = 3.84$ and again factors A, B, and AB are significant.

As seen in Table 16.17, the calculations for this technique are quite simple and the results are consistent with those using a more precise method. Extensions to problems other than 2^f factorials can be found in Batson [4].

TABLE 16.17
Factorial Chi-Square Worksheet for Table 16.16

Blade A				0								1												
Center B			0			1				0			1												
Level C		0		1		0		1		0		1		0		1									
Speed D	0	1	0	1	0	1	0	1	0	1	0	1	0	1	0	1	$\Sigma+$	$\Sigma-$	T	T^2	D	T^2/D	χ^2
---	---	---	---	---	---	---	---	---	---	---	---	---	---	---	---	---	---	---	---	---	---	---	---
$S/20$	0	0	0	1	0	0	0	0	16	10	10	12	20	20	14	20							
Main effects																							
Blade A	−	−	−	−	−	−	−	−	+	+	+	+	+	+	+	+	122	−1	121	14,641	320	45.75	193.52*
Center B	−	−	−	−	+	+	+	+	−	−	−	−	+	+	+	+	74	−49	25	625	320	1.95	8.25*
Level C	−	−	+	+	−	−	+	+	−	−	+	+	−	−	+	+	57	−66	−9	81	320	0.25	1.06
Speed D	−	+	−	+	−	+	−	+	−	+	−	+	−	+	−	+	63	−60	3	9	320	0.03	0.12
Interactions																							
AB	+	+	+	+	−	−	−	−	−	−	−	−	+	+	+	+	75	−48	27	729	320	2.28	9.65*
AC	+	+	−	−	+	+	−	−	−	−	+	+	−	−	+	+	56	−67	−9	81	320	0.25	1.06
AD	+	−	+	−	+	−	+	−	−	+	−	+	−	+	−	+	62	−61	1	1	320	0.01	0.04
BC	+	+	−	−	−	−	+	+	+	+	−	−	−	−	+	+	60	−63	−3	9	320	0.03	0.12
BD	+	−	+	−	−	+	−	+	+	−	+	−	−	+	−	+	66	−57	9	81	320	0.25	1.06
CD	+	−	−	+	+	−	−	+	+	−	−	+	+	−	−	+	69	−54	+15	225	320	0.70	2.96

16.6 RANDOMIZED INCOMPLETE BLOCKS: RESTRICTION ON EXPERIMENTATION

Method for Balanced Blocks

In some randomized block designs it is not possible to apply all treatments in every block. If there were, for example, six brands of tires to test, only four could be tried on a given car (not using the spare), and such a block having only four out of the six treatments, would be incomplete.

Take the problem of determining the effect of current flow of four treatments applied to the coils of TV tube filaments. Because each treatment application requires some time, it is not possible to run several observations of these treatments in one day. To have a randomized block design, we must take the days as blocks and run all four treatments in random order on each of several days. After checking, it is found that even four treatments cannot be completed in a day; three are the most that can be run. The question then is: Which treatments are to be run on the first day, which on the second, and so forth, if information is desired on all four treatments?

The solution to this problem is to use a balanced incomplete block design. An *incomplete block design* is simply one in which there are more treatments than can be put in a single block. A *balanced incomplete block design* is an incomplete block design in which every pair of treatments occurs the same number of times in the experiment. Tables of such designs may be found in Fisher and Yates [10]. The number of blocks necessary for balancing will depend on the number of treatments that can be run in a single block.

For our example there are four treatments, and only three treatments can be run in a block. The balanced design for this problem requires four blocks (days) as shown in Table 16.18.

In this design only treatments *A, C,* and *D* are run on the first day; *B, C,* and *D* on the second day, and so forth. Note that each pair of treatments, such as *AB,* occurs together twice in the experiment. In addition, *A* and *B* occur together on days 3 and 4; *C* and *D* occur together on days 1 and 2; and so on. As in randomized complete block designs, the order in which the three treatments are run on a given day is completely randomized.

The analysis of such a design is easier if some new notation is introduced. Let

$$b = \text{number of blocks in the experiment } (b = 4)$$

$$t = \text{number of treatments in the experment } (t = 4)$$

$$k = \text{number of treatments per block } (k = 3)$$

$$r = \text{number of replications of a given treatment throughout the experiment } (r = 3)$$

$$N = \text{total number of observations}$$

TABLE 16.18
Balanced Incomplete Block Design for TV Filament Example

Block (days)	Treatment				$T_{i.}$
	A	*B*	*C*	*D*	
1	2	—	20	7	29
2	—	32	14	3	49
3	4	13	31	—	48
4	0	23	—	11	34
$T_{.j}$	6	68	65	21	$160 = T_{..}$

$$= bk = tr(N = 12)$$

λ = number of times each pair of treatments appears together throughout the experiment

$$= r(k - 1)/(t - 1) \quad (\lambda = 2)$$

Table 16.18 gives current readings, coded by subtracting 513 milliamperes, and the block and treatment totals. The analysis for a balanced incomplete block design proceeds as follows:

1. Calculate the total sum of squares as usual:

$$SS_{total} = \sum_i \sum_j Y_{ij}^2 - \frac{T_{..}^2}{N}$$

$$= 3478 - \frac{(160)^2}{12} = 1344.67$$

2. Calculate the block sum of squares, ignoring treatments:

$$SS_{block} = \sum_{i=1}^{b} \frac{T_{i.}^2}{k} - \frac{T_{..}^2}{N}$$

$$= \frac{(29)^2 + (49)^2 + (48)^2 + (34)^2}{3} - \frac{(160)^2}{12} = 100.67$$

3. Calculate treatment effects, adjusting for blocks:

$$SS_{treatment} = \frac{\sum_{j=1}^{t} Q_j^2}{k\lambda t} \tag{16.15}$$

where

$$Q_j = kT_{.j} - \sum_i n_{ij} T_{i.} \tag{16.16}$$

where $n_{ij} = 1$ if treatment j appears in block i, and $n_{ij} = 0$ if treatment j does *not* appear in block i. Note that $\sum_i n_{ij} T_{i.}$ is merely the sum of all block totals that contain treatment j.

For the data given,

$$Q_1 = 3(6) - (29 + 48 + 34) = 18 - 111 = -93$$

$$Q_2 = 3(68) - 131 = 73$$

$$Q_3 = 3(65) - 126 = 69$$

$$Q_4 = 3(21) - 112 = \underline{-49}$$

$$0$$

Note that

$$\sum_{j=1}^{t} Q_j = 0$$

which is always true. Then

$$\text{SS}_{\text{treatment}} = \frac{(-93)^2 + (73)^2 + (69)^2 + (-49)^2}{3(2)4} = 880.83$$

4. Calculate the error sum of squares by subtraction

$$\text{SS}_{\text{error}} = \text{SS}_{\text{total}} - \text{SS}_{\text{block}} - \text{SS}_{\text{treatment}}$$

$$= 1344.67 - 100.67 - 880.83 = 363.17$$

Table 16.19 summarizes our data in an ANOVA table.

An F test gives $F_{3,5} = 293.61/72.63 = 4.04$, which is not significant at the 5% level (Statistical Table D).

In Table 16.19 the error degrees of freedom are determined by subtraction rather than as the product of the block and treatment degrees of freedom. However, this error degrees of freedom is seen to be the product of treatment and block degrees of freedom (9) if 4 is subtracted for the four missing values in the design.

In some incomplete block designs it may be desirable to test for a block effect. The mean square for blocks was not computed in Table 16.19 because it had not been adjusted for treatments. In the case of a *symmetrical balanced incomplete randomized block design*, where $b = t$, the block sum of squares may be adjusted in the same manner as the treatment sums of squares

$$Q_1' = r T_{i.} - \sum_j n_{ij} T_{.j}$$

$$Q_1' = 3(29) - \ 92 = -5$$

$$Q_2' = 3(49) - 154 = -7$$

$$Q_3' = 3(48) - 139 = +5$$

$$Q_4' = 3(34) - \ 95 = +7$$

$$\sum_i Q_i' = 0$$

TABLE 16.19
ANOVA for Incomplete Block Design Example

Source	df	SS	MS
Blocks (days)	3	100.67	—
Treatments (adjusted)	3	880.83	293.61
Error	5	363.17	72.63
Totals	11	1344.67	

$$SS_{block} = \sum_{i=1}^{b} (Q_i')^2 / r\lambda b = \frac{(-5)^2 + (-7)^2 + (5)^2 + (7)^2}{3(2)4} = 6.17$$

and

$$SS_{treatment\ (unadjusted)} = \frac{6^2 + 68^2 + 65^2 + 21^2}{3} - \frac{(160)^2}{12} = 975.34$$

The results of this adjustment and the one in treatments may now be summarized for this symmetrical case, as shown in Table 16.20.

In Table 16.20, the terms in parentheses are inserted only to show how the error term was computed for one adjusted effect and one unadjusted effect

$$SS_{error} = SS_{total} - SS_{treatment\ (adjusted)} - SS_{block}$$
$$= 1344.67 - 880.83 - 100.67 = 363.17$$

or

$$SS_{error} = SS_{total} - SS_{treatment} - SS_{block\ (adjusted)}$$
$$= 1344.67 - 975.34 - 6.17 = 363.17$$

Note also that the final sum of squares values used in Table 16.20 to get the mean square values do not add up to the total sum of squares. This is characteristic of a nonorthogonal design. The F test for blocks was not run because its value is obviously extremely small, which indicates no day-to-day effect on current flow.

General Regression Method

For nonsymmetrical or unbalanced designs, the general regression method may be a useful alternative. If contrasts are to be computed for an incomplete block design, it can be shown that the sum of squares for a contrast is given by

TABLE 16.20
ANOVA for Incomplete Block Design Example for Both Treatments and Blocks

Source	df	SS	MS
Blocks (adjusted)	3	6.17	2.06
Blocks	(3)	(100.67)	—
Treatments (adjusted)	3	880.83	293.61
Treatments	(3)	(975.34)	—
Error	5	363.17	72.63
Totals	11	1344.67	

$$SS_{C_m} = \frac{(C_m)^2}{(\sum_{j=1}^{t} c_{jm}^2)k\lambda t} \tag{16.17}$$

where contrasts C_m are made on the Q_j's rather than $T_{.j}$'s.

As an example consider the following orthogonal contrasts on the data of Table 16.18:

$$C_1 = Q_1 - Q_2 \qquad\qquad = -166$$
$$C_2 = Q_1 + Q_2 - 2Q_3 \qquad = -158$$
$$C_3 = Q_1 + Q_2 + Q_3 - 3Q_4 = 196$$

The corresponding sums of squares are

$$SS_{C_1} = \frac{(-166)^2}{(2)(3)(2)(4)} = 574.08$$

$$SS_{C_2} = \frac{(-158)^2}{(6)(3)(2)(4)} = 173.36$$

$$SS_{C_3} = \frac{(196)^2}{(12)(24)} \quad = \underline{133.39}$$

$$= 880.83$$

which checks with the adjusted sums of squares for treatments. Comparing each sum of squares with its 1 df against the error mean square in Table 16.19 indicates that contrast C_1 is the only one of the three that is significant at the 5% level of significance. This indicates a real difference in current flow between treatments A and B, even though treatments in general showed no significant difference in current flow.

Computer Analysis of Incomplete Block Experiments

Analysis of an incomplete block experiment is simplified by using the general linear model procedure in one of the better statistical computing programs. A Minitab summary for the filament data of Table 16.18 is presented in Figure 16.17. The results agree with those obtained manually. Notice that the sequential sums of squares (Seq SS) are the same as those given in Table 16.19, whereas the adjusted sums of squares (Adj SS) are the same as those in Table 16.20.

To obtain a SAS analysis of the filament data, we can use the command file of Table 16.21. Since the data are for an incomplete block experiment, **GLM** is the appropriate procedure. Unlike earlier studies, **ANOVA** will not provide us with the appropriate sums of squares for such experiments.

The outputs for Table 16.21 are given in Figure 16.18. Since this is a nonorthogonal design, the type III sums of squares (which are the same as Minitab's adjusted sums of squares) should be used to evaluate the effects of interest. The results are equivalent to those obtained manually and with Minitab.

General Linear Model

Factor	Levels	Values			
BLOCK	4	1	2	3	4
TRT	4	1	2	3	4

Analysis of Variance for Y

Source	DF	Seq SS	Adj SS	Adj MS	F	P
BLOCK	3	100.67	6.17	2.06	0.03	0.993
TRT	3	880.83	880.83	293.61	4.04	0.083
Error	5	363.17	363.17	72.63		
Total	11	1344.67				

Figure 16.17   Minitab ANOVA for incomplete block design.

TABLE 16.21
SAS Command File for TV Filament Example

```
OPTIONS LINESIZE=80;
DATA TV;
INPUT BLK TRT $ RDG @@;
CARDS;
1 A 2 1 C 20 1 D  7 2 B 32 2 C 14 2 D  3
3 A 4 3 B 13 3 C 31 4 A  0 4 B 23 4 D 11
;
PROC GLM;
CLASS BLK TRT;
MODEL TIME = BLK TRT;
```

GENERAL LINEAR MODELS PROCEDURE

DEPENDENT VARIABLE: RDG

SOURCE	DF	SUM OF SQUARES	MEAN SQUARE	F VALUE
MODEL	6	981.50000000	163.58333333	2.25
ERROR	5	363.16666667	72.63333333	PR > F
CORRECTED TOTAL	11	1344.66666667		0.1954

R-SQUARE	C.V.	ROOT MSE	RDG MEAN
0.729921	63.9189	8.52251919	13.33333333

SOURCE	DF	TYPE I SS	F VALUE	PR > F
BLK	3	100.66666667	0.46	0.7211
TRT	3	880.83333333	4.04	0.0834

SOURCE	DF	TYPE III SS	F VALUE	PR > F
BLK	3	6.16666667	0.03	0.9928
TRT	3	880.83333333	4.04	0.0834

Figure 16.18   SAS analysis for Table 16.18.

16.7 YOUDEN SQUARES

When the conditions for a Latin square are met except that only three treatments are possible (e.g., because in one block only three positions are available) and there are four blocks

TABLE 16.22
Youden Square Design

Block	Position 1	2	3
I	A	B	C
II	D	A	B
III	B	C	D
IV	C	D	A

altogether, the design is an incomplete Latin square. This design, called a *Youden square,* is illustrated in Table 16.22.

Note that the addition of a column (*D, C, A, B*) would make this a Latin square if another position were available. A situation calling for a Youden square might occur if four materials were to be tested on four machines but there were only three heads on each machine whose orientation might affect the results.

The analysis of a Youden square proceeds like the incomplete block analysis. Assuming hypothetical values for some measured variable Y_{ijk} where

$$Y_{ijk} = \mu + \beta_i + \tau_j + \gamma_k + \varepsilon_{ijk}$$

with

$$i = 1, \ldots, 4 \qquad j = 1, 2, \ldots 4 \qquad k = 1, 2, 3$$

we might have the data of Table 16.23.

In Table 16.23

$$t = b = 4$$
$$r = k = 3$$
$$\lambda = 2$$

Treatment totals of *A, B, C, D* are

TABLE 16.23
Youden Square Design Data

Block	Position 1	2	3	$T_{i..}$
I	$A = 2$	$B = 1$	$C = 0$	3
II	$D = -2$	$A = 2$	$B = 2$	2
III	$B = -1$	$C = -1$	$D = -3$	-5
IV	$C = 0$	$D = -4$	$A = 2$	-2
$T_{..k}$	-1	-2	+1	$-2 = T_{...}$

$$T_{.j} : 6, 2, -1, -9$$

From this we can write

$$SS_{total} = \sum_i \sum_j \sum_k Y_{ijk}^2 - \frac{T_{...}^2}{N} = 48 - \frac{(-2)^2}{12} = 47.67$$

Position effect may at first be ignored, since every position occurs once and only once in each block and once with each treatment, so that positions are orthogonal to blocks and treatments:

$$SS_{block \; (ignoring \; treatments)} = \frac{(3)^2 + (2)^2 + (-5)^2 + (-2)^2}{3} - \frac{(-2)^2}{12}$$

$$= \frac{42}{3} - \frac{1}{3} = \frac{41}{3} = 13.67$$

For treatment sum of squares adjusted for blocks we get

$$Q_1 = 3(+6) - 3 = 15$$
$$Q_2 = 3(2) - 0 = 6$$
$$Q_3 = 3(-1) - (-4) = 1$$
$$Q_4 = 3(-9) - (-5) = -22$$

and

$$\sum_i Q_i = 0$$

$$SS_{treatment} = \frac{(15)^2 + (6)^2 + (1)^2 + (-22)^2}{(4)(6)} = 31.08$$

$$SS_{position} = \frac{(-1)^2 + (-2)^2 + (1)^2}{4} - \frac{(-2)^2}{12} = 1.17$$

$$SS_{error} = SS_{total} - SS_{block} - SS_{treatment \; (adjusted)} - SS_{position}$$

$$= 47.67 - 13.67 - 31.08 - 1.17$$

$$= 1.75$$

The analysis appears in Table 16.24.

The position effect here is not significant, and it might be desirable to pool it with the error, getting

$$s_\varepsilon^2 = \frac{1.17 + 1.75}{2 + 3} = \frac{2.92}{5} = 0.58$$

as an estimate of error variance, with 5 df. Then the treatment effect is highly significant.

The block mean square is not given, since blocks must be adjusted by treatments if block effects are to be assessed. The procedure is the same as shown for incomplete block designs.

A Computer Analysis

If the **General Linear Model** procedure in Minitab is applied to the data in Table 16.23, the ANOVA of Figure 16.19 is obtained. The results, which agree with those of Table 16.24, indicate that the treatment effect is significant for any $\alpha \geq 0.021$.

```
General Linear Model

Factor      Levels      Values
BLOCK         4         1        2        3        4
POSITION      3         1        2        3
TRTMNT        4         1        2        3        4

Analysis of Variance for Y
Source        DF      Seq SS     Adj SS     Adj MS       F        P
BLOCK          3     13.6667     4.4167     1.4722     2.52     0.234
POSITION       2      1.1667     1.1667     0.5833     1.00     0.465
TRTMNT         3     31.0833    31.0833    10.3611    17.76     0.021
Error          3      1.7500     1.7500     0.5833
Total         11     47.6667
```

Figure 16.19   Minitab analysis of the Table 16.23 data.

TABLE 16.24
Youden Square ANOVA

Source	df	SS	MS
Treatment τ_j (adjusted)	3	31.08	10.36
Blocks β_i	3	13.67	—
Position γ_k	2	1.17	0.58
Error ε_{ijk}	3	1.75	0.58
Totals	11	47.67	

16.8 FURTHER READING

Box, G. E. P., W. G. Hunter, and J. S. Hunter, *Statistics for Experimenters: An Introduction to Design, Data Analysis, and Model Building.* New York: John Wiley & Sons, 1978. Chapter 11 contains a discussion of evolutionary operation and Chapter 15 is devoted to response surface methods.

Cornell, J. A., *How to Apply Response Surface Methodology.* Milwaukee, WI: American Society for Quality Control, 1984.

DeVor, R. E., T. Chang, and J. W. Sutherland, *Statistical Quality Design and Control: Contemporary Concepts and Methods.* New York: Macmillan Publishing Company, 1992. See pages 567–572 for a discussion of the response surface for 2^3 factorial experiment.

Gunter, B., "Statistically Designed Experiments. Part 2: The Universal Structure Underlying Experimentation," *Quality Progress,* February 1990, pp. 87–89. Design geometry and response surfaces are considered.

Gunter, B., "Statistically Designed Experiments. Part 5: Robust Process and Product Design and Related Matters," *Quality Progress,* August 1990, pp. 107–108. Ridge systems are considered.

Mason, R. L., R. F. Gunst, and J. L. Hess, *Statistical Design and Analysis of Experiments with Applications to Engineering and Science.* New York: John Wiley & Sons, 1989. Response surface designs and the analysis of covariance are considered in Chapters 11 and 19, respectively.

Montgomery, D. C., *Design and Analysis of Experiments,* 4th ed. New York: John Wiley & Sons, 1997. See Chapter 14 for discussions of response surface methods, evolutionary operation, and central composite designs.

Wadsworth, Jr., H. M., K. S. Stephens, and A. B. Godfrey, *Modern Methods for Quality Control and Improvement.* New York: John Wiley & Sons, 1986. See Chapter 17 for discussions of central composite designs and evolutionary operations.

PROBLEMS

16.1 One study reports on the breaking strength Y of samples of seven different starch films where the thickness of each film X was also recorded as a covariate. Results, rounded off considerably, are as follows:

Source	df	$\sum y^2$	$\sum xy$	$\sum x^2$	Adjusted $\sum y^2$	Adjusted df	Adjusted MS
Between starches		6,000,000	50,000	500			
Within starches	100						
Totals		8,000,000	60,000	600			

a. Complete this table and set up a test for the effect of starch film or breaking strength both before and after adjustment for thickness.

b. Write the mathematical model for this experiment, briefly describing each term.

c. State the assumptions implied in the tests of hypotheses made above and describe briefly how you would test their validity.

16.2 In a study of the effects of two different teaching methods, the gain score of each pupil in a public school test was used as the response variable. The pretest scores are also available. Run a covariance analysis on the data below to determine whether the new teaching method did show an improvement in average achievement as measured by this public school test.

	Method I			Method II			Method I			Method II	
Pupil	Pretest Score	Gain	Pupil	Pretest Score	Gain	Pupil	Pretest Score	Gain	Pupil	Pretest Score	Gain
1	28	5	1	24	7	15	25	7	15	43	−2
2	43	−8	2	40	9	16	43	1	16	34	2
3	36	4	3	41	4	17	34	3	17	44	7
4	35	1	4	33	4	18	40	5	18	43	4
5	31	4	5	45	5	19	36	4	19	28	3
6	34	1	6	41	−2	20	30	9	20	46	−1

7	33	4	7	41	6	21	35	−6	21	43	2
8	38	8	8	33	11	22	38	0	22	46	−4
9	39	1	9	41	−1	23	31	−14	23	21	4
10	44	−1	10	30	15	24	41	4			
11	36	−7	11	45	4	25	35	10			
12	21	2	12	50	2	26	27	10			
13	34	−1	13	42	9	27	45	1			
14	33	1	14	47	3						

16.3 Data below are for five individuals who have been subjected to four conditions or treatments represented by the group or lots 1–4. Y represents some measure of an individual supposedly affected by the variations in the treatments of the four lots. X represents another measure of an individual that may affect the value of Y even in the absence of the four treatments. Determine whether the Y means of the four groups differ significantly from each other after the effects of the X variable have been removed.

	Groups or Lots						
1		**2**		**3**		**4**	
X	Y	X	Y	X	Y	X	Y
29	22	15	30	16	12	5	23
20	22	9	32	31	8	25	25
14	20	1	26	26	13	16	28
21	24	6	25	35	25	10	26
6	12	19	37	12	7	24	23

16.4 Consider a two-way classified design with more than one observation per cell.
a. Show the F test for one of the main effects in a covariance analysis if both factors are fixed.
b. Show the F test for the interaction in a covariance analysis if one factor is fixed and the other one is random.

16.5 For Example 16.4, another response variable was measured—the mean surface factor for heat slivers. Results of the first three cycles at the 10 experimental points gave

Cycle	(1)	(2)	(3)	(4)	(5)	(6)	(7)	(8)	(9)	(10)
1	4.0	6.8	5.9	4.5	4.3	4.2	7.8	8.3	11.0	5.9
2	5.2	7.2	6.1	3.7	4.0	4.5	7.0	8.1	12.5	6.3
3	4.8	8.0	7.0	3.6	3.8	4.3	6.5	7.8	12.7	7.0

Set up EVOP worksheets for these data and comment on your results after three cycles.

16.6 Data on yield in pounds for varying concentration at levels 29, 30, and 31%, and varying power at levels 400, 450, and 500 watts, gave the following results in four cycles at the five points in a 2^2 design.

Cycle	(1)	(2)	(3)	(4)	(5)
1	477	472	411	476	372
2	469	452	430	468	453
3	465	396	375	468	292
4	451	469	363	432	460

Analyze by EVOP methods.

16.7 The response variable in the data below is the number of defective knife handles in samples of 72 knives in which the factors are machines, lumber grades, and replications. Make a suitable transformation of the data and present an analysis.

Lumber Grade	Replication	Machine 1	Machine 2
A	I	8	4
	II	4	6
B	I	9	3
	II	10	4

16.8 Analyze the data of Problem 16.7 using factorial chi square. Compare with the results of Problem 16.7.

16.9 Data on screen-color difference on a television tube measured in degrees Kelvin are to be compared for four operators. On a given day only three operators can be used in the experiment. A balanced incomplete block design gave results as follows:

Day	Operator A	Operator B	Operator C	Operator D
Monday	780	820	800	—
Tuesday	950	—	920	940
Wednesday	—	880	880	820
Thursday	840	780	—	820

Do a complete analysis of these data and discuss your findings with regard to differences between operators.

16.10 Run orthogonal contrasts on the operators in Problem 16.9.

16.11 An experiment was to be run on the wear resistance of a new experimental shoe leather compared with the standard leather in use by the Army. It was decided to equip several soldiers with one shoe of the experimental-type leather and the other shoe of standard leather and, after many weeks in the field, compare wear on the two types. Considering each soldier as a block, suggest a reasonable number of people to be used in this experiment and outline its possible analysis.

16.12 Three experimental types of leather were to be tested along with the standard in Problem 16.11, but it is obvious that only two of the four types can be tested on one soldier. Set up a balanced incomplete block design for this situation. Insert some arbitrary numerical values and complete an ANOVA table for your data.

16.13 For the values of Y in Table 16.16, let $P = Y/20$ and $X = \arcsin \sqrt{p}$. Use Yates's method to obtain the sums of squares associated with X. Compare your results with those of Figure 16.16.

SUMMARY AND SPECIAL PROBLEMS

Throughout this book the three phases of an experiment have been emphasized: the experiment, the design, and the analysis. Most chapters end with an outline of the designs considered thus far. It may prove useful to the reader to have a complete outline as a summary of the designs presented in Chapters 1–15.

This is not the only way such an outline could be constructed, but it represents an attempt to see each design as part of the overall picture of designed experiments. A look at the outline will reveal that one design is often used for entirely different experiments, which simply illustrates that much care must be taken to spell out the experiment and its design or method of randomization before the work is performed.

For each experiment, the chapter reference in this book is given.

After the summary several special problems are presented. Many do not fit any one pattern or design but may be of interest insofar as they represent the kinds of problems encountered in practice. These problems have no "pat" solutions, and the way they are set up may depend a great deal on assumptions made by the reader. They should serve to evoke some good discussion about designing an experiment to solve a given practical problem.

Experiment	Design	Analysis	Chapter Reference
I. Single factor			
	1. Completely randomized $Y_{ij} = \mu + \tau_j + \varepsilon_{ij}$	1. One-way ANOVA	Chapter 3
	2. Randomized block $Y_{ij} = \mu + \beta_i + \tau_j + \varepsilon_{ij}$	2.	
	a. complete	a. Two-way ANOVA	Chapter 4
	b. Incomplete, balanced	b. Special ANOVA	Chapter 16
	c. Incomplete, general	c. General regression method	
	3. Latin square $Y_{ijk} = \mu + \beta_i + \tau_j + \gamma_k + \varepsilon_{ijk}$	3.	
	a. Complete	a. Three-way ANOVA	Chapter 4
	b. Incomplete, Youden square	b. Special ANOVA (like 2b)	Chapter 16
	4. Graeco–Latin square $Y_{ijkm} = \mu + \beta_i + \tau_j + \gamma_k + \omega_m + \varepsilon_{ijkm}$	4. Four-way ANOVA	Chapter 4

Experiment	Design	Analysis	Chapter Reference

II. Two or more factors
  A. Factorial
    (crossed)			

	1. Completely randomized $Y_{ijk} = \mu + A_i + B_j$ $\quad + AB_{ij} + \varepsilon_{k(ij)}, \ldots$ for more factors	1.	
	a. General case	a. ANOVA with interactions	Chapter 5

	b. 2^f case	b. Yates method or general
		    ANOVA; use: (1),
		    $a, b, ab, \ldots$
			Chapter 9

	c. 3^f case	c. General ANOVA; use 00, 10,
		    20, 01, . . . and
		    $A \times B = AB + A\beta^2, \ldots$ for
		    interaction
			Chapter 10

	2. Randomized block	2.	
	a. Complete $Y_{ijk} = \mu + R_k + A_i + B_j$ $\quad + AB_{ij} + \varepsilon_{ijk}$	a. Factorial ANOVA with replication R_k	Chapter 8
	b. Incomplete, confounding:	b.	
	i. Main effect—split plot $Y_{ijk} = \mu + \underbrace{R_i + A_j + RA_{ij}}_{\text{whole plot}}$ $\underbrace{+ B_k + RB_{ik}}$ $\quad + AB_{jk} + RAB_{ijk}$ $\qquad\qquad\text{split plot}$	i. Split-plot ANOVA	Chapter 11
	ii. Interactions in 2^f and 3^f	ii.	Chapter 12

	(1) Several replications
	$Y_{ijkq} = \mu + R_i + \beta_j$
	$\quad + R\beta_{ij} + A_k$
	$\quad + B_q + AB_{kq}$
	$\quad + AB_{kq} + \varepsilon_{ijkq}$
		(1) Factorial ANOVA with
		        replications R_i and
		        blocks β_j or con-
		        founded interaction	

	(2) One replication only
	$Y_{ijk} = \mu + \beta_i + A_j$
	$\quad + B_k + AB_{jk}, \ldots$
		(2) Factorial ANOVA with
		        blocks β_i or
		        confounded interaction
			Chapter 12

	(3) Fractional replication—
	        aliases
	$Y_{ijk} = \mu + A_i + B_j$
	$\quad + AB_{ij}, \ldots$
	        (a) in 2^f
	        (b) in 3^f; $f = 3$: Latin
	           square
		(3) Factorial ANOVA with
		        aliases or aliased
		        interactions
			Chapter 13

	3. Latin square	3.	
	a. Complete $Y_{ijkm} = \mu + R_k + \gamma_m + A_i$ $\quad + B_j + AB_{ij} + \varepsilon_{ijkm}$	a. Factorial ANOVA with replications and positions	Chapter 8
B. Nested (hierarchical)			
	1. Completely randomized $Y_{ijk} = \mu + A_i +$ $\quad B_{j(i)} + \varepsilon_{k(ij)}$	1. Nested ANOVA	Chapter 7

Experiment	Design	Analysis	Chapter Reference
	2. Randomized block	2.	
	a. Complete $Y_{ijk} = \mu + R_k + A_i$ $+ B_{j(i)} + \varepsilon_{ijk}$	a. Nested ANOVA with blocks R_k	Chapters 4 and 7
	3. Latin square	3.	
	a. Complete $X_{ijkm} = \mu + R_k + \gamma_m$ $+ A_i + B_{j(i)}$ $+ \varepsilon_{ijkm}$	a. Nested ANOVA with blocks and positions	Chapters 4 and 7
C. Nested factorial			
	1. Completely randomized $Y_{ijkm} = \mu + A_i + B_{j(i)}$ $+ C_k + AC_{ik}$ $+ BC_{kj(i)} + \varepsilon_{m(ijk)}$	1. Nested-factorial ANOVA	Chapter 7
	2. Randomized block	2.	
	a. Complete $Y_{ijkm} = \mu + R_k + A_i$ $+ B_{j(i)} + C_m$ $+ AC_{im} + BC_{mj(i)}$ $+ \varepsilon_{ijkm}$	a. Nested- factorial ANOVA with blocks R_k	Chapters 4 and 7
	3. Latin square	3.	
	a. Complete $Y_{ijkmq} = \mu + R_k + \gamma_m$ $+ A_i + B_{j(i)}$ $+ C_q + AC_{iq}$ $+ BC_{qj(i)} + \varepsilon_{ijkmq}$	a. Nested-factorial ANOVA with blocks and positions	Chapters 4 and 7

SPECIAL PROBLEMS

1. Three different time periods are chosen in which to plant a certain vegetable: early May, late May, and early June. Plots are selected in early May and five different fertilizers are randomly tried in each plot. This is repeated in late May and again in early June. The criterion is the number of bushels per acre at harvest time. Assuming that r plots for each fertilizer can be used at each planting:
   a. Discuss the nature of this experiment.
   b. Recommend a value for r to make this an acceptable study. (Time periods and fertilizers are fixed.)

2. In the study of the factors affecting the surface finish of steel in Problem 13.19 there was a sixth factor—factor A (the flux type)—also set at two levels. It was possible to run all 64 (2^6) treatment combinations; the experimenter, however, wanted a complete replication of the experiment that was to be run. Thus, it was decided to do a half-replicate of the 2^6 and then replicate the same block at another time. It was also learned that first a flux type had to be chosen, and then a slab width, after which the other four factors had to be randomized. When these steps had been completed, the experimenter would run the other slab width and then repeat the experiment with the other flux type.

   Outline a data format for this problem and show what its ANOVA table might look like.

3. In Problem 13.18 regarding the 81 engines, it took about a year to gather all the data. When the data were submitted, they were found to be incomplete at the low-temperature level, so the experiment was really a one-third replication of a $3^4 \times 2$ factorial instead of the 3^5 factorial. How would you analyze the data if one whole level in your answer to this problem were missing?

4. In a food industry there was a strong feeling that viscosity readings on a slurry were not consistent. Some claimed that different laboratories gave different readings. Others said it was the fault of the instruments in these laboratories. And still others blamed the operators who read the instruments. To determine the main sources of variability, slurries from two products with vastly different viscosities were taken. Each sample of slurry was mixed well and then divided into four jars, one of which was sent to each of four laboratories for analysis. In each laboratory the contents of the jar was divided into eight samples, and each of two randomly chosen operators in that laboratory were to test two samples on each of two randomly chosen test instruments. Considering the laboratories as fixed, the two products as fixed, the operators as random, and the instruments as random, show a data layout for this problem. Show its model and its ANOVA outline, and discuss any concerns about the tests to be made.

5. A total of 256 seventh grade students (128 boys and 128 girls), representing both high and low socioeconomic status (SES), were subjected to a mathematics lesson that was believed to be either interesting or uninteresting under one of four reinforcement conditions: no knowledge of results, immediate verbal knowledge of results, delayed verbal knowledge of results, and delayed written knowledge of results. Task interest and type of reinforcement emerged as significant at the 1% significance level. The delayed verbal condition was significantly superior to the other three, which did not differ from each other. Also, low SES children learned the high-interest task as quickly as high SES children, and more quickly than they learned the low-interest task. On the basis of this information answer the following.
   a. What is the implied criterion Y?
   b. List each independent variable and state the levels of each.
   c. Set up an ANOVA table for this problem and indicate which results are significant.
   d. Show a data layout indicating the number of children in each treatment.
   e. For the significant results, set some hypothetical means to show possible interpretation of such results.

6. In a test for a significant effect of treatments on some measured variable Y, the resulting F test was almost, but not quite, significant at the 5% level (say, $F = 4.00$ with 3 and 8 df). Now the experimenter repeats the whole experiment on another set of independent observations under the same treatments, and again the F fails to reach the 5% level (say, $F = 4.01$ with 3 and 8 df). What conclusion—if any—would you reach with regard to the hypotheses of equal treatment effects? Briefly explain your reasoning.

7. Six different formulas or mixes of concrete are to be purchased from five competing suppliers. Tests of crushing strength are to be made on two blocks constructed according to each formula–supplier combination. One block will be poured, hardened, and tested for each combination before the second block is formed, thus making two replications of the whole experiment. Assuming mixes and suppliers fixed and replications random, set up a mathematical model for this situation, outline its ANOVA properties, and show what F tests are appropriate.

SUMMARY AND SPECIAL PROBLEMS

Throughout this book the three phases of an experiment have been emphasized: the experiment, the design, and the analysis. Most chapters end with an outline of the designs considered thus far. It may prove useful to the reader to have a complete outline as a summary of the designs presented in Chapters 1–15.

This is not the only way such an outline could be constructed, but it represents an attempt to see each design as part of the overall picture of designed experiments. A look at the outline will reveal that one design is often used for entirely different experiments, which simply illustrates that much care must be taken to spell out the experiment and its design or method of randomization before the work is performed.

For each experiment, the chapter reference in this book is given.

After the summary several special problems are presented. Many do not fit any one pattern or design but may be of interest insofar as they represent the kinds of problems encountered in practice. These problems have no "pat" solutions, and the way they are set up may depend a great deal on assumptions made by the reader. They should serve to evoke some good discussion about designing an experiment to solve a given practical problem.

Experiment	Design	Analysis	Chapter Reference
I. Single factor			
	1. Completely randomized $Y_{ij} = \mu + \tau_j + \varepsilon_{ij}$	1. One-way ANOVA	Chapter 3
	2. Randomized block $Y_{ij} = \mu + \beta_i + \tau_j + \varepsilon_{ij}$	2.	
	a. complete	a. Two-way ANOVA	Chapter 4
	b. Incomplete, balanced	b. Special ANOVA	Chapter 16
	c. Incomplete, general	c. General regression method	
	3. Latin square $Y_{ijk} = \mu + \beta_i + \tau_j + \gamma_k + \varepsilon_{ijk}$	3.	
	a. Complete	a. Three-way ANOVA	Chapter 4
	b. Incomplete, Youden square	b. Special ANOVA (like 2b)	Chapter 16
	4. Graeco–Latin square $Y_{ijkm} = \mu + \beta_i + \tau_j + \gamma_k + \omega_m + \varepsilon_{ijkm}$	4. Four-way ANOVA	Chapter 4

Experiment	Design	Analysis	Chapter Reference

II. Two or more factors
  A. Factorial
    (crossed)			

	1. Completely randomized $Y_{ijk} = \mu + A_i + B_j$ $\qquad + AB_{ij} + \varepsilon_{k(ij)}, \ldots$ for more factors	1.	
	a. General case	a. ANOVA with interactions	Chapter 5
	b. 2^f case	b. Yates method or general ANOVA; use: (1), $a, b, ab, \ldots$	Chapter 9
	c. 3^f case	c. General ANOVA; use 00, 10, 20, 01, $\ldots$ and $A \times B = AB + AB^2, \ldots$ for interaction	Chapter 10
	2. Randomized block	2.	
	a. Complete $Y_{ijk} = \mu + R_k + A_i + B_j$ $\qquad + AB_{ij} + \varepsilon_{ijk}$	a. Factorial ANOVA with replication R_k	Chapter 8
	b. Incomplete, confounding:	b.	
	i. Main effect—split plot $Y_{ijk} = \mu + \underbrace{R_i + A_j + RA_{ij}}_{\text{whole plot}}$ $\underbrace{\begin{array}{l} + B_k + RB_{ik} \\ + AB_{jk} + RAB_{ijk} \end{array}}_{\text{split plot}}$	i. Split-plot ANOVA	Chapter 11
	ii. Interactions in 2^f and 3^f	ii.	Chapter 12
	(1) Several replications $Y_{ijkq} = \mu + R_i + \beta_j$ $\qquad + R\beta_{ij} + A_k$ $\qquad + B_q + AB_{kq}$ $\qquad + AB_{kq} + \varepsilon_{ijkq}$	(1) Factorial ANOVA with replications R_i and blocks β_j or confounded interaction	
	(2) One replication only $Y_{ijk} = \mu + \beta_i + A_j$ $\qquad + B_k + AB_{jk}, \ldots$	(2) Factorial ANOVA with blocks β_i or confounded interaction	Chapter 12
	(3) Fractional replication— aliases $Y_{ijk} = \mu + A_i + B_j$ $\qquad + AB_{ij}, \ldots$	(3) Factorial ANOVA with aliases or aliased interactions	Chapter 13
	(a) in 2^f		
	(b) in 3^f; $f = 3$: Latin square		
	3. Latin square	3.	
	a. Complete $Y_{ijkm} = \mu + R_k + \gamma_m + A_i$ $\qquad + B_j + AB_{ij} + \varepsilon_{ijkm}$	a. Factorial ANOVA with replications and positions	Chapter 8
B. Nested (hierarchical)			
	1. Completely randomized $Y_{ijk} = \mu + A_i +$ $\qquad B_{j(i)} + \varepsilon_{k(ij)}$	1. Nested ANOVA	Chapter 7

8. Four chemical treatments, A, B, C, and D, are prepared in three concentrations (3 : 1, 2 : 1, and 1 : 1), and two samples of each combination are poured; then three determinations of pH are made from each sample. Considering treatments and concentrations fixed and samples and determinations random, set up the appropriate ANOVA table and comment on statistical tests to be made. Also explain in some detail how you would proceed to set 95% confidence limits on the average pH for treatment B, concentration 2 : 1.

9. A graduate student carried out an experiment involving six experimental conditions. From an available group of subjects, she assigned five men and five women at random to each of the six conditions, using 60 subjects in all. Upon completion of the experiment, she carried out the following analysis to test the differences among the condition means:

Source	df	
Between conditions	5	$F = \dfrac{MS_b}{MS_w}$ with 5 and 54 df
Within conditions	54	
Total	59	

At this point a statistical consultant was called in.
a. The consultant blew his top. Why?
b. You are now the consultant. Set up the analysis the student should use, with sources of variation, df, and F tests, and explain to her why she must use your analysis and why hers is incorrect.
c. As consultant you must point out the possibility that one of the significance tests in the appropriate analysis may affect the interpretation of the other. How? What should be done in such a case?

10. For the problem (Problem 9.6) on the effect of four factors on chemical yield, the original data had an unequal number of observations at each treatment combination. The numbers in each treatment combination below are the n's for each particular treatment. Explain how you might handle the analysis of this problem.

		A_1		A_2	
		B_1	B_2	B_1	B_2
C_1	D_1	2	2	7	5
	D_2	2	1	2	3
C_2	D_1	4	2	5	1
	D_2	1	1	2	2

GLOSSARY OF TERMS

Alias.   An effect in a fractionally replicated design that "looks like" or cannot be distinguished from another effect.

Alpha (α).   The size of the type I error, or the probability of rejecting a hypothesis when true.

Autocorrelation Coefficient.   A correlation between measures within the same sample in the order taken.

Beta (β).   The size of the type II error, or the probability of accepting a hypothesis when some alternative hypothesis is true.

Canonical Form.   A form of second-degree response surface equation that allows determination of the type of surface.

Completely Randomized Design.   A design in which all treatments are assigned to the experimental units in a completely random manner.

Confidence Limits.   Two values between which a parameter is said to lie with a specified degree of confidence.

Confounding.   An experimental arrangement in which certain effects cannot be distinguished from others. One such effect is usually blocks.

Consistent Estimator.   An estimator of a parameter whose values comes closer to the parameter as the sample size is increased.

Contrast.   A linear combination of treatment totals or averages in which the sum of the coefficient is zero.

Correlation Coefficient (Pearson r).   The square root of the proportion of total variation accounted for by simple linear regression, with the sign of the slope of the least squares line.

Correlation Index R.   the square root of the proportion of total variation accounted for by the regression equation of the degree being fitted to the data.

Covariance Analysis.   A technique for removing the effect of one or more variables on the response variable before assessing the effect of treatments on the response variable.

Critical Region.   A set of values of a test statistic in which the hypothesis under test is rejected.

Random Sample.   A sample in which each member of the population sampled has an equal chance of being selected in this sample.

Randomized Block Design.   An experimental design in which the treatment combinations are randomized within a block and several blocks are run.

Repeated-Measures Design.   A design in which measurements are taken on the experimental units more than once.

Research.   A systematic quest for undiscovered truth.

Residual Analysis.   Analysis of what remains of response variation after effects in the model have been removed.

Resolution.   A scheme for determining which effects and interactions can be assessed in a given fractional factorial experiment.

Rotatable Design.   An experimental design that has equal predictive power in all directions from a center point, and in which all experimental points are equidistant from this center point.

SAS (Statistical Analysis System).   A procedure used to analyze statistical data on a computer.

Split-Plot Design.   An experimental design in which the practical necessities of the order of experimentation lead to the confounding of a main effect with blocks.

Standard Error of Estimate.   The standard deviation of errors of estimate around a least squares fitted regression model.

Statistic u.   A measure computed from a sample.

Statistical Hypothesis H_0.   An assumption about a population being sampled.

Statistical Inference.   The act of inferring something about a population of measures from a sample of that population.

Statistics.   A tool for decision making in the light of uncertainty.

Steepest Ascent Method.   A method of sequential experiments that will direct the experimenter toward the optimum of a response surface.

Sum of Squares (SS).   The sum of the squares of deviations of a random variable from its mean,

$$SS = \sum_{i=1}^{n}(Y_i - \bar{Y})^2$$

Test of a Hypothesis.   A rule by which a hypothesis is accepted or rejected.

Test Statistic.   A statistic used to test a hypothesis.

Treatment Combination.   A given combination showing the levels of all factors to be run for that set of experimental conditions.

True Experiment.   A study in which certain independent variables are manipulated, their effect on one or more dependent variables is observed, and the levels of the independent variables are assigned at random to the experimental units of the study.

Unbiased Statistic.   A statistic whose expected value equals the parameter it is estimating.

Variance.   Of a sample:

$$s^2 = \frac{\sum_{i=1}^{n}(Y_i - \bar{Y})^2}{n - 1} = \frac{SS}{df}$$

Of a population:

$$\sigma^2 = E(Y - \mu)^2$$

Youden Square.   An incomplete Latin Square.

REFERENCES

1. Anderson, V. L., and R. A. McLean, *Design of Experiments: A Realistic Approach*. New York: Marcel Dekker, 1974.

2. Anderson, V. L., and R. A. McLean, "Restriction Errors: Another Dimension in Teaching Experimental Statistics," *The American Statistician*, November 1974.

3. Barnett, E. H., "Introduction to Evolutionary Operation," *Industrial and Engineering Chemistry*, Vol. 52, June 1960, p. 500.

4. Batson, H. C., "Applications of Factorial Analysis to Experiments in Chemistry," *National Convention Transactions of the American Society for Quality Control*, 1956, pp. 9–23.

5. *Biometrics*, Vol. 13, September 1957, pp. 261–405.

6. Box, G. E. P., S. Bisgaard, and C. A. Fung, "An Explanation and Critique of Taguchi's Contributions to Quality Engineering," *Quality and Reliability Engineering International*, Vol. 4, 1988, pp. 321–330.

7. Box, G. E. P., and N. R. Draper, *Evolutionary Operation: A Statistical Method for Process Improvement*. New York: John Wiley & Sons, 1969.

8. Box, G. E. P., W. G.. Hunter, and J. S. Hunter, *Statistics for Experimenters: An Introduction to Design, Data Analysis and Model Building*. New York: John Wiley & Sons, 1978.

9. Burr, I. W., *Statistical Quality Control Methods*. New York: Marcel Dekker, 1976.

10. Fisher, R. A., and F. Yates, *Statistical Tables for Biological, Agricultural, and Medical Research*, 4th ed. Edinburgh and London: Oliver & Boyd, 1953.

11. Hahn, G. J., and W. Q. Meeker, *Statistical Intervals: A Guide for Practitioners*. New York: John Wiley & Sons, 1991.

12. Hinkelmann, K., and O. Kempthorne, *Design and Analysis of Experiments: Volume 1: Introduction to Experimental Design*. New York: John Wiley & Sons, 1994.

13. Hunter, J. S., "Determination of Optimum Operating Conditions by Experimental Methods," *Industrial Quality Control*, December 1958–February 1959.

14. Hunter, J. S., "Signal-to-Noise Ratio Debated," *Quality Progress*, May 1987, pp. 7–9.

15. Kerlinger, F. N., *Foundations of Behavioral Research*, 2nd ed. New York: Holt, Rinehart & Winston, 1973.

16. Keuls, M., "The Use of the Studentized Range in Connection with an Analysis of Variance," *Euphytica*, Vol. 1, 1952, pp. 112–122.

17. Kilgo, M. B., "An Application of Fractional Factorial Experimental Designs," *Quality Engineering*, Vol. 1, 1988, pp. 19–23.

18. Leedy, P. D., *Practical Research: Planning and Design*. New York: Macmillan Publishing Company, 1974.

19. Lenth, R. V., "Quick and Easy Analysis of Unreplicated Factorials," *Technometrics*, Vol. 31, 1989, pp. 469–473.

20. McCall, C. H., Jr., "Linear Contrasts, Parts I, II, and III," *Industrial Quality Control*, July–September, 1960.

21. Miller, L. D., "An Investigation of the Machinability of Malleable Iron Using Ceramic Tools," MSIE thesis, Purdue University, 1959.

22. Nelson, L. S. "Extreme Screening Designs," *Journal of Quality Technology,* Vol. 14, April 1982.

23. Neter, J., and W. Wasserman, *Applied Linear Statistical Models: Regression, Analysis of Variance, and Experimental Design.* Homewood, IL: Richard D. Irwin, 1974.

24. Odeh, R. E., et al., eds., *Pocket Book of Statistical Tables.* New York: Marcel Dekker, 1977. Reprint by Books on Demand.

25. Ostle, B., and L. C. Malone, *Statistics in Research: Basic Concepts and Techniques for Research Work,* 4th ed. Ames, IA: Iowa State University Press, 1988.

26. Ostle, B., K. V. Turner Jr., C. R. Hicks, and G. W. McElrath, *Engineering Statistics: The Industrial Experience.* Belmont, CA: Duxbury Press, 1996.

27. Plackett, R. L., and J. P. Burman, "The Design of Optimum Multifactorial Experiments," *Biometrika,* Vol. 33, 1946, pp. 305–325.

28. Scheffé, H., "A Method for Judging All Contrasts in the Analysis of Variance," *Biometrics,* Vol. 40, June, 1953.

29. Shapiro, S. S., *How to Test Normality and Other Distributional Assumptions.* Milwaukee, WI: American Society for Quality, 1986.

30. Snedecor, G. W., and W. C. Cochran, *Statistical Methods,* 7th ed. Ames, IA: Iowa University Press, 1980.

31. Statistical Analysis System, *SAS User's Guide: Statistics Version 5 Edition.* SAS Institute, Inc., Cary, NC, 1985.

32. Taguchi, G., *Introduction to Quality Engineering.* Tokyo: Asian Productivity Organization, 1986.

33. Tukey, J. W., *Exploratory Data Analysis.* Reading, MA: Addison-Wesley Publishing Company, 1977.

34. Tanur, J. M., et al., eds., *Statistics: A Guide to the Unknown,* 2nd ed. San Francisco, CA: Holden-Day, 1978.

35. Winer, B. J., *Statistical Principles in Experimental Design,* 2nd ed. New York: McGraw-Hill Book Company, 1971.

36. Wortham, A. W., and T. E. Smith, *Practical Statistics in Experimental Design.* Dallas, TX: Dallas Publishing House, 1960.

37. Yates, F., *Design and Analysis of Factorial Experiments.* London: Imperial Bureau of Soil Sciences, 1937.

STATISTICAL TABLES

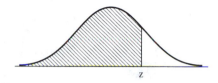

TABLE A
Areas Under the Standard Normal Curve (Proportion of Total Area Under the Curve from −∞ to Designated Z Value)

z	−0.09	−0.08	−0.07	−0.06	−0.05	−0.04	−0.03	−0.02	−0.01	0.00
−3.5	0.00017	0.00017	0.00018	0.00019	0.00019	0.00020	0.00021	0.00022	0.00022	0.00023
−3.4	0.00024	0.00025	0.00026	0.00027	0.00028	0.00029	0.00030	0.00031	0.00032	0.00034
−3.3	0.00035	0.00036	0.00038	0.00039	0.00040	0.00042	0.00043	0.00045	0.00047	0.00048
−3.2	0.00050	0.00052	0.00054	0.00056	0.00058	0.00060	0.00062	0.00064	0.00066	0.00069
−3.1	0.00071	0.00074	0.00076	0.00079	0.00082	0.00084	0.00087	0.00090	0.00094	0.00097
−3.0	0.00100	0.00104	0.00107	0.00111	0.00114	0.00118	0.00122	0.00126	0.00131	0.00135
−2.9	0.0014	0.0014	0.0015	0.0015	0.0016	0.0016	0.0017	0.0018	0.0018	0.0019
−2.8	0.0019	0.0020	0.0021	0.0021	0.0022	0.0023	0.0023	0.0024	0.0025	0.0026
−2.7	0.0026	0.0027	0.0028	0.0029	0.0030	0.0031	0.0032	0.0033	0.0034	0.0035
−2.6	0.0036	0.0037	0.0038	0.0039	0.0040	0.0041	0.0043	0.0044	0.0045	0.0047
−2.5	0.0048	0.0049	0.0051	0.0052	0.0054	0.0055	0.0057	0.0059	0.0060	0.0062
−2.4	0.0064	0.0066	0.0068	0.0069	0.0071	0.0073	0.0075	0.0078	0.0080	0.0082
−2.3	0.0084	0.0087	0.0089	0.0091	0.0094	0.0096	0.0099	0.0102	0.0104	0.0107
−2.2	0.0110	0.0113	0.0116	0.0119	0.0122	0.0125	0.0129	0.0132	0.0136	0.0139
−2.1	0.0143	0.0146	0.0150	0.0154	0.0158	0.0162	0.0166	0.0170	0.0174	0.0179
−2.0	0.0183	0.0188	0.0192	0.0197	0.0202	0.0207	0.0212	0.0217	0.0222	0.0228
−1.9	0.0233	0.0239	0.0244	0.0250	0.0256	0.0262	0.0268	0.0274	0.0281	0.0287
−1.8	0.0294	0.0301	0.0307	0.0314	0.0322	0.0329	0.0336	0.0344	0.0351	0.0359
−1.7	0.0367	0.0375	0.0384	0.0392	0.0401	0.0409	0.0418	0.0427	0.0436	0.0446
−1.6	0.0455	0.0465	0.0475	0.0485	0.0495	0.0505	0.0516	0.0526	0.0537	0.0548
−1.5	0.0559	0.0571	0.0582	0.0594	0.0606	0.0618	0.0630	0.0643	0.0655	0.0668
−1.4	0.0681	0.0694	0.0708	0.0721	0.0735	0.0749	0.0764	0.0778	0.0793	0.0808
−1.3	0.0823	0.0838	0.0853	0.0869	0.0885	0.0901	0.0918	0.0934	0.0951	0.0968
−1.2	0.0985	0.1003	0.1020	0.1038	0.1056	0.1075	0.1093	0.1112	0.1131	0.1151
−1.1	0.1170	0.1190	0.1210	0.1230	0.1251	0.1271	0.1292	0.1314	0.1335	0.1357
−1.0	0.1379	0.1401	0.1423	0.1446	0.1469	0.1492	0.1515	0.1539	0.1562	0.1587
−0.9	0.1611	0.1635	0.1660	0.1685	0.1711	0.1736	0.1762	0.1788	0.1814	0.1841
−0.8	0.1867	0.1894	0.1922	0.1949	0.1977	0.2005	0.2033	0.2061	0.2090	0.2119
−0.7	0.2148	0.2177	0.2206	0.2236	0.2266	0.2296	0.2327	0.2358	0.2389	0.2420
−0.6	0.2451	0.2483	0.2514	0.2546	0.2578	0.2611	0.2643	0.2676	0.2709	0.2743
−0.5	0.2776	0.2810	0.2843	0.2877	0.2912	0.2946	0.2981	0.3015	0.3050	0.3085
−0.4	0.3121	0.3156	0.3192	0.3228	0.3264	0.3300	0.3336	0.3372	0.3409	0.3446
−0.3	0.3483	0.3520	0.3557	0.3594	0.3632	0.3669	0.3707	0.3745	0.3783	0.3821
−0.2	0.3859	0.3897	0.3936	0.3974	0.4013	0.4052	0.4090	0.4129	0.4168	0.4207
−0.1	0.4247	0.4286	0.4325	0.4364	0.4404	0.4443	0.4483	0.4522	0.4562	0.4602
−0.0	0.4641	0.4681	0.4721	0.4761	0.4801	0.4840	0.4880	0.4920	0.4960	0.5000

TABLE A
Continued

z	0.00	0.01	0.02	0.03	0.04	0.05	0.06	0.07	0.08	0.09
0.0	0.5000	0.5040	0.5080	0.5120	0.5160	0.5199	0.5239	0.5279	0.5319	0.5359
0.1	0.5398	0.5438	0.5478	0.5517	0.5557	0.5596	0.5636	0.5675	0.5714	0.5753
0.2	0.5793	0.5832	0.5871	0.5910	0.5948	0.5987	0.6026	0.6064	0.6103	0.6141
0.3	0.6179	0.6217	0.6255	0.6293	0.6331	0.6368	0.6406	0.6443	0.6480	0.6517
0.4	0.6554	0.6591	0.6628	0.6664	0.6700	0.6736	0.6772	0.6808	0.6844	0.6879
0.5	0.6915	0.6950	0.6985	0.7019	0.7054	0.7088	0.7123	0.7157	0.7190	0.7224
0.6	0.7257	0.7291	0.7324	0.7357	0.7389	0.7422	0.7454	0.7486	0.7517	0.7549
0.7	0.7580	0.7611	0.7642	0.7673	0.7704	0.7734	0.7764	0.7794	0.7823	0.7852
0.8	0.7881	0.7910	0.7939	0.7967	0.7995	0.8023	0.8051	0.8078	0.8106	0.8133
0.9	0.8159	0.8186	0.8212	0.8238	0.8264	0.8289	0.8315	0.8340	0.8365	0.8389
1.0	0.8413	0.8438	0.8461	0.8485	0.8508	0.8531	0.8554	0.8577	0.8599	0.8621
1.1	0.8643	0.8665	0.8686	0.8708	0.8729	0.8749	0.8770	0.8790	0.8810	0.8830
1.2	0.8849	0.8869	0.8888	0.8907	0.8925	0.8944	0.8962	0.8980	0.8997	0.9015
1.3	0.9032	0.9049	0.9066	0.9082	0.9099	0.9115	0.9131	0.9147	0.9162	0.9177
1.4	0.9192	0.9207	0.9222	0.9236	0.9251	0.9265	0.9279	0.9292	0.9306	0.9319
1.5	0.9332	0.9345	0.9357	0.9370	0.9382	0.9394	0.9406	0.9418	0.9429	0.9441
1.6	0.9452	0.9463	0.9474	0.9484	0.9495	0.9505	0.9515	0.9525	0.9535	0.9545
1.7	0.9554	0.9564	0.9573	0.9582	0.9591	0.9599	0.9608	0.9616	0.9625	0.9633
1.8	0.9641	0.9649	0.9656	0.9664	0.9671	0.9678	0.9686	0.9693	0.9699	0.9706
1.9	0.9713	0.9719	0.9726	0.9732	0.9738	0.9744	0.9750	0.9756	0.9761	0.9767
2.0	0.9772	0.9778	0.9783	0.9788	0.9793	0.9798	0.9803	0.9808	0.9812	0.9817
2.1	0.9821	0.9826	0.9830	0.9834	0.9838	0.9842	0.9846	0.9850	0.9854	0.9857
2.2	0.9861	0.9864	0.9868	0.9871	0.9875	0.9878	0.9881	0.9884	0.9887	0.9890
2.3	0.9893	0.9896	0.9898	0.9901	0.9904	0.9906	0.9909	0.9911	0.9913	0.9916
2.4	0.9918	0.9920	0.9922	0.9925	0.9927	0.9929	0.9931	0.9932	0.9934	0.9936
2.5	0.9938	0.9940	0.9941	0.9943	0.9945	0.9946	0.9948	0.9949	0.9951	0.9952
2.6	0.9953	0.9955	0.9956	0.9957	0.9959	0.9960	0.9961	0.9962	0.9963	0.9964
2.7	0.9965	0.9966	0.9967	0.9968	0.9969	0.9970	0.9971	0.9972	0.9973	0.9974
2.8	0.9974	0.9975	0.9976	0.9977	0.9977	0.9978	0.9979	0.9979	0.9980	0.9981
2.9	0.9981	0.9982	0.9982	0.9983	0.9984	0.9984	0.9985	0.9985	0.9986	0.9986
3.0	0.99865	0.99869	0.99874	0.99878	0.99882	0.99886	0.99889	0.99893	0.99896	0.99900
3.1	0.99903	0.99906	0.99910	0.99913	0.99916	0.99918	0.99921	0.99924	0.99926	0.99929
3.2	0.99931	0.99934	0.99936	0.99938	0.99940	0.99942	0.99944	0.99946	0.99948	0.99950
3.3	0.99952	0.99953	0.99955	0.99957	0.99958	0.99960	0.99961	0.99962	0.99964	0.99965
3.4	0.99966	0.99968	0.99969	0.99970	0.99971	0.99972	0.99973	0.99974	0.99975	0.99976
3.5	0.99977	0.99978	0.99978	0.99979	0.99980	0.99981	0.99981	0.99982	0.99983	0.99983

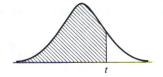

TABLE B
Student's *t* Distribution

df	Percentile Point						
	70	80	90	95	97.5	99	99.5
1	0.73	1.38	3.08	6.31	12.71	31.82	63.66
2	0.62	1.06	1.89	2.92	4.30	6.96	9.92
3	0.58	0.98	1.64	2.35	3.18	4.54	5.84
4	0.57	0.94	1.53	2.13	2.78	3.75	4.60
5	0.56	0.92	1.48	2.02	2.57	3.36	4.03
6	0.55	0.91	1.44	1.94	2.45	3.14	3.71
7	0.55	0.90	1.41	1.89	2.36	3.00	3.50
8	0.55	0.89	1.40	1.86	2.31	2.90	3.36
9	0.54	0.88	1.38	1.83	2.26	2.82	3.25
10	0.54	0.88	1.37	1.81	2.23	2.76	3.17
11	0.54	0.88	1.36	1.80	2.20	2.72	3.11
12	0.54	0.87	1.36	1.78	2.18	2.68	3.05
13	0.54	0.87	1.35	1.77	2.16	2.65	3.01
14	0.54	0.87	1.35	1.76	2.14	2.62	2.98
15	0.54	0.87	1.34	1.75	2.13	2.60	2.95
16	0.54	0.86	1.34	1.75	2.12	2.58	2.92
17	0.53	0.86	1.33	1.74	2.11	2.57	2.90
18	0.53	0.86	1.33	1.73	2.10	2.55	2.88
19	0.53	0.86	1.33	1.73	2.09	2.54	2.86
20	0.53	0.86	1.33	1.72	2.09	2.53	2.85
21	0.53	0.86	1.32	1.72	2.08	2.52	2.83
22	0.53	0.86	1.32	1.72	2.07	2.51	2.82
23	0.53	0.86	1.32	1.71	2.07	2.50	2.81
24	0.53	0.86	1.32	1.71	2.06	2.49	2.80
25	0.53	0.86	1.32	1.71	2.06	2.49	2.79
26	0.53	0.86	1.31	1.71	2.06	2.48	2.78
27	0.53	0.86	1.31	1.70	2.05	2.47	2.77
28	0.53	0.85	1.31	1.70	2.05	2.47	2.76
29	0.53	0.85	1.31	1.70	2.05	2.46	2.76
30	0.53	0.85	1.31	1.70	2.04	2.46	2.75
40	0.53	0.85	1.30	1.68	2.02	2.42	2.70
50	0.53	0.85	1.30	1.68	2.01	2.40	2.68
60	0.53	0.85	1.30	1.67	2.00	2.39	2.66
80	0.53	0.85	1.29	1.66	1.99	2.37	2.64
100	0.53	0.85	1.29	1.66	1.98	2.36	2.63
200	0.53	0.84	1.29	1.65	1.97	2.35	2.60
500	0.52	0.84	1.28	1.65	1.96	2.33	2.59
∞	0.52	0.84	1.28	1.64	1.96	2.33	2.58

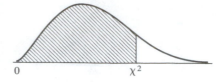

TABLE C
Cumulative Chi-Square Distribution
Entry is χ_p^2 where $F(\chi_p^2) = P(\chi^2 \leq \chi_p^2) = p$ and $\chi^2 \sim \chi_{(v)}^2$

v	0.005	0.01	0.025	0.05	0.10	0.90	0.95	0.975	0.99	0.995
1	0.000	0.000	0.001	0.004	0.016	2.706	3.841	5.024	6.635	7.879
2	0.010	0.020	0.051	0.103	0.211	4.605	5.991	7.378	9.210	10.597
3	0.072	0.115	0.216	0.352	0.584	6.251	7.815	9.348	11.345	12.838
4	0.207	0.297	0.484	0.711	1.064	7.779	9.488	11.143	13.277	14.860
5	0.412	0.554	0.831	1.145	1.610	9.236	11.070	12.832	15.086	16.750
6	0.676	0.872	1.237	1.635	2.204	10.645	12.592	14.449	16.812	18.548
7	0.989	1.239	1.690	2.167	2.833	12.017	14.067	16.013	18.475	20.278
8	1.344	1.646	2.180	2.733	3.490	13.362	15.507	17.535	20.090	21.955
9	1.735	2.088	2.700	3.325	4.168	14.684	16.919	19.023	21.666	23.589
10	2.156	2.558	3.247	3.940	4.865	15.987	18.307	20.483	23.209	25.188
11	2.603	3.053	3.816	4.575	5.578	17.275	19.675	21.920	24.725	26.757
12	3.074	3.571	4.404	5.226	6.304	18.549	21.026	23.337	26.217	28.299
13	3.565	4.107	5.009	5.892	7.042	19.812	22.362	24.736	27.688	29.819
14	4.075	4.660	5.629	6.571	7.790	21.064	23.685	26.119	29.141	31.319
15	4.601	5.229	6.262	7.261	8.547	22.307	24.996	27.488	30.578	32.801
16	5.142	5.812	6.908	7.962	9.312	23.542	26.296	28.845	32.000	34.267
17	5.697	6.408	7.564	8.672	10.085	24.769	27.587	30.191	33.409	35.718
18	6.265	7.015	8.231	9.390	10.865	25.989	28.869	31.526	34.805	37.156
19	6.844	7.633	8.907	10.117	11.651	27.204	30.143	32.852	36.191	38.582
20	7.434	8.260	9.591	10.851	12.443	28.412	31.410	34.170	37.566	39.997
21	8.034	8.897	10.283	11.591	13.240	29.615	32.671	35.479	38.932	41.401
22	8.643	9.542	10.982	12.338	14.042	30.813	33.924	36.781	40.289	42.796
23	9.260	10.196	11.689	13.090	14.848	32.007	35.172	38.076	41.638	44.181
24	9.886	10.856	12.401	13.848	15.659	33.196	36.415	39.364	42.980	45.558
25	10.520	11.524	13.120	14.611	16.473	34.382	37.652	40.647	44.314	46.928
26	11.160	12.198	13.844	15.379	17.292	35.563	38.885	41.923	45.642	48.290
27	11.808	12.878	14.573	16.151	18.114	36.741	40.113	43.195	46.963	49.645
28	12.461	13.565	15.308	16.928	18.939	37.916	41.337	44.461	48.278	50.993
29	13.121	14.257	16.047	17.708	19.768	39.087	42.557	45.722	49.588	52.336
30	13.787	14.953	16.791	18.493	20.599	40.256	43.773	46.979	50.892	53.672
35	17.192	18.509	20.596	22.465	24.797	46.059	49.801	53.203	57.342	60.274
40	20.707	22.164	24.433	26.509	29.051	51.805	55.759	59.342	63.691	66.766
45	24.314	25.901	28.365	30.612	33.350	57.505	61.658	65.411	69.956	73.166
50	27.991	29.707	32.357	34.764	37.689	63.167	67.505	71.420	76.154	79.490
55	31.734	33.570	36.398	38.958	42.060	68.796	73.311	77.380	82.292	85.749
60	35.535	37.485	40.482	43.188	46.459	74.397	79.082	83.298	88.379	91.952
70	43.275	45.442	48.758	51.739	55.329	85.527	90.531	95.023	100.425	104.215
80	51.172	53.540	57.153	60.392	64.278	96.578	101.879	106.629	112.329	116.321
90	59.196	61.754	65.647	69.126	73.291	107.565	113.145	118.136	124.116	128.299
100	67.328	70.065	74.222	77.930	82.358	118.498	124.342	129.561	135.807	140.169

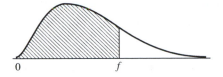

TABLE D
Cumulative F Distribution
Tabled value is F_p where $p = P(F \leq F_p)$ and $F \sim F_{(v_1, v_2)}$

v_2	p	1	2	3	4	5	6	7	8	9	10
							v_1				
1	0.005	0.00006	0.0051	0.018	0.032	0.044	0.054	0.062	0.068	0.073	0.078
	0.01	0.00025	0.010	0.029	0.047	0.062	0.073	0.082	0.089	0.095	0.100
	0.025	0.0015	0.026	0.057	0.082	0.100	0.113	0.124	0.132	0.139	0.144
	0.05	0.0062	0.054	0.099	0.130	0.151	0.167	0.179	0.188	0.195	0.201
	0.10	0.025	0.117	0.181	0.220	0.246	0.265	0.279	0.289	0.298	0.304
	0.90	39.9	49.5	53.6	55.8	57.2	58.2	58.9	59.4	59.9	60.2
	0.95	161	200	216	225	230	234	237	239	241	242
	0.975	648	800	864	900	922	937	948	957	963	969
	0.99	4050	5000	5400	5620	5760	5860	5930	5980	6020	6060
	0.995	16200	20000	21600	22500	23100	23400	23700	23900	24100	24200
2	0.005	0.00005	0.0050	0.020	0.038	0.055	0.069	0.081	0.091	0.099	0.106
	0.01	0.00020	0.010	0.032	0.056	0.075	0.092	0.105	0.116	0.125	0.132
	0.025	0.0013	0.026	0.062	0.094	0.119	0.138	0.153	0.165	0.175	0.183
	0.05	0.0050	0.053	0.105	0.144	0.173	0.194	0.211	0.224	0.235	0.244
	0.10	0.020	0.111	0.183	0.231	0.265	0.289	0.307	0.321	0.333	0.342
	0.90	8.53	9.00	9.16	9.24	9.29	9.33	9.35	9.37	9.38	9.39
	0.95	18.5	19.0	19.2	19.2	19.3	19.3	19.4	19.4	19.4	19.4
	0.975	38.5	39.0	39.2	39.2	39.3	39.3	39.4	39.4	39.4	39.4
	0.99	98.5	99.0	99.2	99.2	99.3	99.3	99.4	99.4	99.4	99.4
	0.995	198	199	199	199	199	199	199	199	199	199
3	0.005	0.00005	0.0050	0.021	0.041	0.060	0.077	0.092	0.104	0.115	0.124
	0.01	0.00019	0.010	0.034	0.060	0.083	0.102	0.118	0.132	0.143	0.153
	0.025	0.0012	0.026	0.065	0.100	0.129	0.152	0.170	0.185	0.197	0.207
	0.05	0.0046	0.052	0.108	0.152	0.185	0.210	0.230	0.246	0.259	0.270
	0.10	0.019	0.109	0.185	0.239	0.276	0.304	0.325	0.342	0.356	0.367
	0.90	5.54	5.46	5.39	5.34	5.31	5.28	5.27	5.25	5.24	5.23
	0.95	10.1	9.55	9.28	9.12	9.01	8.94	8.89	8.85	8.81	8.79
	0.975	17.4	16.0	15.4	15.1	14.9	14.7	14.6	14.5	14.5	14.4
	0.99	34.1	30.8	29.5	28.7	28.2	27.9	27.7	27.5	27.3	27.2
	0.995	55.6	49.8	47.5	46.2	45.4	44.8	44.4	44.1	43.9	43.7

TABLE D
(Continued)

v_2	p	|| v_1									
		11	12	15	20	24	30	40	60	120	∞
1	0.005	0.082	0.085	0.093	0.101	0.105	0.109	0.113	0.118	0.122	0.127
	0.01	0.104	0.107	0.115	0.124	0.128	0.132	0.137	0.141	0.146	0.151
	0.025	0.149	0.153	0.161	0.170	0.175	0.180	0.184	0.189	0.194	0.199
	0.05	0.207	0.211	0.220	0.230	0.235	0.240	0.245	0.250	0.255	0.261
	0.10	0.310	0.315	0.325	0.336	0.342	0.347	0.353	0.358	0.364	0.370
	0.90	60.5	60.7	61.2	61.7	62.0	62.3	62.5	62.8	63.1	63.3
	0.95	243	244	246	248	249	250	251	252	253	254
	0.975	973	977	985	993	997	1000	1010	1010	1010	1020
	0.99	6080	6110	6160	6210	6230	6260	6290	6310	6340	6370
	0.995	24300	24400	24600	24800	24900	25000	25100	25300	25400	25500
2	0.005	0.112	0.118	0.130	0.143	0.150	0.157	0.165	0.173	0.181	0.189
	0.01	0.139	0.144	0.157	0.171	0.178	0.186	0.193	0.201	0.209	0.217
	0.025	0.190	0.196	0.210	0.224	0.232	0.239	0.247	0.255	0.263	0.271
	0.05	0.251	0.257	0.272	0.286	0.294	0.302	0.309	0.317	0.326	0.334
	0.10	0.350	0.356	0.371	0.386	0.394	0.402	0.410	0.418	0.426	0.434
	0.90	9.40	9.41	9.42	9.44	9.45	9.46	9.47	9.47	9.48	9.49
	0.95	19.4	19.4	19.4	19.4	19.5	19.5	19.5	19.5	19.5	19.5
	0.975	39.4	39.4	39.4	39.4	39.5	39.5	39.5	39.5	39.5	39.5
	0.99	99.4	99.4	99.4	99.4	99.5	99.5	99.5	99.5	99.5	99.5
	0.995	199	199	199	199	199	199	199	199	199	200
3	0.005	0.132	0.138	0.154	0.172	0.181	0.191	0.201	0.211	0.222	0.234
	0.01	0.161	0.168	0.185	0.203	0.212	0.222	0.232	0.242	0.253	0.264
	0.025	0.216	0.224	0.241	0.259	0.269	0.279	0.289	0.299	0.310	0.321
	0.05	0.279	0.287	0.304	0.323	0.332	0.342	0.352	0.363	0.373	0.384
	0.10	0.376	0.384	0.402	0.420	0.430	0.439	0.449	0.459	0.469	0.480
	0.90	5.22	5.22	5.20	5.18	5.18	5.17	5.16	5.15	5.14	5.13
	0.95	8.76	8.74	8.70	8.66	8.63	8.62	8.59	8.57	8.55	8.53
	0.975	14.4	14.3	14.3	14.2	14.1	14.1	14.0	14.0	13.9	13.9
	0.99	27.1	27.1	26.9	26.7	26.6	26.5	26.4	26.3	26.2	26.1
	0.995	43.5	43.4	43.1	42.8	42.6	42.5	42.3	42.1	42.0	41.8

TABLE D
(Continued)

v_2	p	\multicolumn{10}{c}{v_1}
		1	**2**	**3**	**4**	**5**	**6**	**7**	**8**	**9**	**10**
4	0.005	0.00004	0.0050	0.022	0.043	0.064	0.083	0.100	0.114	0.126	0.137
	0.01	0.00018	0.010	0.035	0.063	0.088	0.109	0.127	0.143	0.156	0.167
	0.025	0.0011	0.026	0.066	0.104	0.135	0.161	0.181	0.198	0.212	0.224
	0.05	0.0044	0.052	0.110	0.157	0.193	0.221	0.243	0.261	0.275	0.288
	0.10	0.018	0.108	0.187	0.243	0.284	0.314	0.338	0.356	0.371	0.384
	0.90	4.54	4.32	4.19	4.11	4.05	4.01	3.98	3.95	3.94	3.92
	0.95	7.71	6.94	6.59	6.39	6.26	6.16	6.09	6.04	6.00	5.96
	0.975	12.2	10.6	9.98	9.60	9.36	9.20	9.07	8.98	8.90	8.84
	0.99	21.2	18.0	16.7	16.0	15.5	15.2	15.0	14.8	14.7	14.5
	0.995	31.3	26.3	24.3	23.2	22.5	22.0	21.6	21.4	21.1	21.0
5	0.005	0.00004	0.0050	0.022	0.045	0.067	0.087	0.105	0.120	0.134	0.146
	0.01	0.00017	0.010	0.035	0.064	0.091	0.114	0.134	0.151	0.165	0.177
	0.025	0.0011	0.025	0.067	0.107	0.140	0.167	0.189	0.208	0.223	0.236
	0.05	0.0043	0.052	0.111	0.160	0.198	0.228	0.252	0.271	0.287	0.301
	0.10	0.017	0.108	0.188	0.247	0.290	0.322	0.347	0.367	0.383	0.397
	0.90	4.06	3.78	3.62	3.52	3.45	3.40	3.37	3.34	3.32	3.30
	0.95	6.61	5.79	5.41	5.19	5.05	4.95	4.88	4.82	4.77	4.74
	0.975	10.0	8.43	7.76	7.39	7.15	6.98	6.85	6.76	6.68	6.62
	0.99	16.3	13.3	12.1	11.4	11.0	10.7	10.5	10.3	10.2	10.1
	0.995	22.8	18.3	16.5	15.6	14.9	14.5	14.2	14.0	13.8	13.6
6	0.005	0.00004	0.0050	0.022	0.045	0.069	0.090	0.109	0.126	0.140	0.153
	0.01	0.00017	0.010	0.036	0.066	0.094	0.118	0.139	0.157	0.172	0.186
	0.025	0.0011	0.025	0.068	0.109	0.143	0.172	0.195	0.215	0.231	0.246
	0.05	0.0043	0.052	0.112	0.162	0.202	0.233	0.259	0.279	0.296	0.311
	0.10	0.017	0.107	0.189	0.249	0.294	0.327	0.354	0.375	0.392	0.406
	0.90	3.78	3.46	3.29	3.18	3.11	3.05	3.01	2.98	2.96	2.94
	0.95	5.99	5.14	4.76	4.53	4.39	4.28	4.21	4.15	4.10	4.06
	0.975	8.81	7.26	6.60	6.23	5.99	5.82	5.70	5.60	5.52	5.46
	0.99	13.7	10.9	9.78	9.15	8.75	8.47	8.26	8.10	7.98	7.87
	0.995	18.6	14.5	12.9	12.0	11.5	11.1	10.8	10.6	10.4	10.2
7	0.005	0.00004	0.0050	0.023	0.046	0.070	0.093	0.113	0.130	0.145	0.159
	0.01	0.00017	0.010	0.036	0.067	0.096	0.121	0.143	0.162	0.178	0.192
	0.025	0.0010	0.025	0.068	0.110	0.146	0.176	0.200	0.221	0.238	0.253
	0.05	0.0042	0.052	0.113	0.164	0.205	0.238	0.264	0.286	0.304	0.319
	0.10	0.017	0.107	0.190	0.251	0.297	0.332	0.359	0.381	0.399	0.414
	0.90	3.59	3.26	3.07	2.96	2.88	2.83	2.78	2.75	2.72	2.70
	0.95	5.59	4.74	4.35	4.12	3.97	3.87	3.79	3.73	3.68	3.64
	0.975	8.07	6.54	5.89	5.52	5.29	5.12	4.99	4.90	4.82	4.76
	0.99	12.2	9.55	8.45	7.85	7.46	7.19	6.99	6.84	6.72	6.62
	0.995	16.2	12.4	10.9	10.0	9.52	9.16	8.89	8.68	8.51	8.38

TABLE D
(Continued)

v_2	p	\multicolumn{10}{c}{v_1}
		11	12	15	20	24	30	40	60	120	∞
4	0.005	0.145	0.153	0.172	0.193	0.204	0.216	0.229	0.242	0.255	0.269
	0.01	0.176	0.185	0.204	0.226	0.237	0.249	0.261	0.274	0.287	0.301
	0.025	0.234	0.243	0.263	0.284	0.296	0.308	0.320	0.332	0.346	0.359
	0.05	0.298	0.307	0.327	0.349	0.360	0.372	0.384	0.396	0.409	0.422
	0.10	0.394	0.403	0.424	0.445	0.456	0.467	0.478	0.490	0.502	0.514
	0.90	3.91	3.90	3.87	3.84	3.83	3.82	3.80	3.79	3.78	3.76
	0.95	5.94	5.91	5.86	5.80	5.77	5.75	5.72	5.69	5.66	5.63
	0.975	8.79	8.75	8.66	8.56	8.51	8.46	8.41	8.36	8.31	8.26
	0.99	14.4	14.4	14.2	14.0	13.9	13.8	13.7	13.7	13.6	13.5
	0.995	20.8	20.7	20.4	20.2	20.0	19.9	19.8	19.6	19.5	19.3
5	0.005	0.156	0.165	0.186	0.210	0.223	0.237	0.251	0.266	0.282	0.299
	0.01	0.188	0.197	0.219	0.244	0.257	0.270	0.285	0.299	0.315	0.331
	0.025	0.248	0.257	0.280	0.304	0.317	0.330	0.344	0.359	0.374	0.390
	0.05	0.313	0.322	0.345	0.369	0.382	0.395	0.408	0.422	0.437	0.452
	0.10	0.408	0.418	0.440	0.463	0.476	0.488	0.501	0.514	0.527	0.541
	0.90	3.28	3.27	3.24	3.21	3.19	3.17	3.16	3.14	3.12	3.10
	0.95	4.71	4.68	4.62	4.56	4.53	4.50	4.46	4.43	4.40	4.36
	0.975	6.57	6.52	6.43	6.33	6.28	6.23	6.18	6.12	6.07	6.02
	0.99	9.96	9.89	9.72	9.55	9.47	9.38	9.29	9.20	9.11	9.02
	0.995	13.5	13.4	13.1	12.9	12.8	12.7	12.5	12.4	12.3	12.1
6	0.005	0.164	0.174	0.197	0.224	0.238	0.253	0.269	0.286	0.304	0.324
	0.01	0.197	0.207	0.232	0.258	0.273	0.288	0.304	0.321	0.338	0.357
	0.025	0.258	0.268	0.293	0.320	0.334	0.349	0.364	0.381	0.398	0.415
	0.05	0.324	0.334	0.358	0.385	0.399	0.413	0.428	0.444	0.460	0.476
	0.10	0.418	0.429	0.453	0.478	0.491	0.505	0.519	0.533	0.548	0.564
	0.90	2.92	2.90	2.87	2.84	2.82	2.80	2.78	2.76	2.74	2.72
	0.95	4.03	4.00	3.94	3.87	3.84	3.81	3.77	3.74	3.70	3.67
	0.975	5.41	5.37	5.27	5.17	5.12	5.07	5.01	4.96	4.90	4.85
	0.99	7.79	7.72	7.56	7.40	7.31	7.23	7.14	7.06	6.97	6.88
	0.995	10.1	10.0	9.81	9.59	9.47	9.36	9.24	9.12	9.00	8.88
7	0.005	0.171	0.181	0.206	0.235	0.251	0.267	0.285	0.304	0.324	0.345
	0.01	0.205	0.216	0.241	0.270	0.286	0.303	0.320	0.339	0.358	0.379
	0.025	0.266	0.277	0.304	0.333	0.348	0.364	0.381	0.399	0.418	0.437
	0.05	0.332	0.343	0.369	0.398	0.413	0.428	0.445	0.461	0.479	0.498
	0.10	0.427	0.438	0.463	0.491	0.504	0.519	0.534	0.550	0.566	0.582
	0.90	2.68	2.67	2.63	2.59	2.58	2.56	2.54	2.51	2.49	2.47
	0.95	3.60	3.57	3.51	3.44	3.41	3.38	3.34	3.30	3.27	3.23
	0.975	4.71	4.67	4.57	4.47	4.42	4.36	4.31	4.25	4.20	4.14
	0.99	6.54	6.47	6.31	6.16	6.07	5.99	5.91	5.82	5.74	5.65
	0.995	8.27	8.18	7.97	7.75	7.65	7.53	7.42	7.31	7.19	7.08

TABLE D
(Continued)

v_2	p	\multicolumn{10}{c}{v_1}
		1	**2**	**3**	**4**	**5**	**6**	**7**	**8**	**9**	**10**
8	0.005	0.00004	0.0050	0.027	0.047	0.072	0.095	0.115	0.133	0.149	0.164
	0.01	0.00017	0.010	0.036	0.068	0.097	0.123	0.146	0.166	0.183	0.198
	0.025	0.0010	0.025	0.069	0.111	0.148	0.179	0.204	0.226	0.244	0.259
	0.05	0.0042	0.052	0.113	0.166	0.208	0.241	0.268	0.291	0.310	0.326
	0.10	0.017	0.107	0.190	0.253	0.299	0.335	0.363	0.386	0.405	0.421
	0.90	3.46	3.11	2.92	2.81	2.73	2.67	2.62	2.59	2.56	2.54
	0.95	5.32	4.46	4.07	3.84	3.69	3.58	3.50	3.44	3.39	3.35
	0.975	7.57	6.06	5.42	5.05	4.82	4.65	4.53	4.43	4.36	4.30
	0.99	11.3	8.65	7.59	7.01	6.63	6.37	6.18	6.03	5.91	5.81
	0.995	14.7	11.0	9.60	8.81	8.30	7.95	7.69	7.50	7.34	7.21
9	0.005	0.00004	0.0050	0.023	0.047	0.073	0.096	0.117	0.136	0.153	0.168
	0.01	0.00017	0.010	0.037	0.068	0.098	0.125	0.149	0.169	0.187	0.202
	0.025	0.0010	0.025	0.069	0.112	0.150	0.181	0.207	0.230	0.248	0.265
	0.05	0.0040	0.052	0.113	0.167	0.210	0.244	0.272	0.296	0.315	0.331
	0.10	0.017	0.107	0.191	0.254	0.302	0.338	0.367	0.390	0.410	0.426
	0.90	3.36	3.01	2.81	2.69	2.61	2.55	2.51	2.47	2.44	2.42
	0.95	5.12	4.26	3.86	3.63	3.48	3.37	3.29	3.23	3.18	3.14
	0.975	7.21	5.71	5.08	4.72	4.48	4.32	4.20	4.10	4.03	3.96
	0.99	10.6	8.02	6.99	6.42	6.06	5.80	5.61	5.47	5.35	5.26
	0.995	13.6	10.1	8.72	7.96	7.47	7.13	6.88	6.69	6.54	6.42
10	0.005	0.00004	0.0050	0.023	0.048	0.073	0.098	0.119	0.139	0.156	0.171
	0.01	0.00017	0.010	0.037	0.069	0.100	0.127	0.151	0.172	0.190	0.206
	0.025	0.0010	0.025	0.069	0.113	0.151	0.183	0.210	0.233	0.252	0.269
	0.05	0.0041	0.052	0.114	0.168	0.211	0.246	0.275	0.299	0.319	0.336
	0.10	0.017	0.106	0.191	0.255	0.303	0.340	0.370	0.394	0.414	0.430
	0.90	3.28	2.92	2.73	2.61	2.52	2.46	2.41	2.38	2.35	2.32
	0.95	4.96	4.10	3.71	3.48	3.33	3.22	3.14	3.07	3.02	2.98
	0.975	6.94	5.46	4.83	4.47	4.24	4.07	3.95	3.85	3.78	3.72
	0.99	10.0	7.56	6.55	5.99	5.64	5.39	5.20	5.06	4.94	4.85
	0.995	12.8	9.43	8.08	7.34	6.87	6.54	6.30	6.12	5.97	5.85
11	0.005	0.00004	0.0050	0.023	0.048	0.074	0.099	0.121	0.141	0.158	0.174
	0.01	0.00016	0.010	0.037	0.069	0.100	0.128	0.153	0.175	0.193	0.210
	0.025	0.0010	0.025	0.069	0.114	0.152	0.185	0.212	0.236	0.256	0.273
	0.05	0.0041	0.052	0.114	0.168	0.212	0.248	0.278	0.302	0.323	0.340
	0.10	0.017	0.106	0.192	0.256	0.305	0.342	0.373	0.397	0.417	0.435
	0.90	3.23	2.86	2.66	2.54	2.45	2.39	2.34	2.30	2.27	2.25
	0.95	4.84	3.98	3.59	3.36	3.20	3.09	3.01	2.95	2.90	2.85
	0.975	6.72	5.26	4.63	4.28	4.04	3.88	3.76	3.66	3.59	3.53
	0.99	9.65	7.21	6.22	5.67	5.32	5.07	4.89	4.74	4.63	4.54
	0.995	12.2	8.91	7.60	6.88	6.42	6.10	5.86	5.68	5.54	5.42

TABLE D
(Continued)

v_2	p	v_1 11	12	15	20	24	30	40	60	120	∞
8	0.005	0.176	0.187	0.214	0.244	0.261	0.279	0.299	0.319	0.341	0.364
	0.01	0.211	0.222	0.250	0.281	0.297	0.315	0.334	0.354	0.376	0.398
	0.025	0.273	0.285	0.313	0.343	0.360	0.377	0.395	0.415	0.435	0.456
	0.05	0.339	0.351	0.379	0.409	0.425	0.441	0.459	0.477	0.496	0.516
	0.10	0.435	0.445	0.472	0.500	0.515	0.531	0.547	0.563	0.581	0.599
	0.90	2.52	2.50	2.46	2.42	2.40	2.38	2.36	2.34	2.32	2.29
	0.95	3.31	3.28	3.22	3.15	3.12	3.08	3.04	3.01	2.97	2.93
	0.975	4.24	4.20	4.10	4.00	3.95	3.89	3.84	3.78	3.73	3.67
	0.99	5.73	5.67	5.52	5.36	5.28	5.20	5.12	5.03	4.95	4.86
	0.995	7.10	7.01	6.81	6.61	6.50	6.40	6.29	6.18	6.06	5.95
9	0.005	0.181	0.192	0.220	0.253	0.271	0.290	0.310	0.332	0.356	0.382
	0.01	0.216	0.228	0.257	0.289	0.307	0.326	0.346	0.368	0.391	0.415
	0.025	0.279	0.291	0.320	0.352	0.370	0.388	0.408	0.428	0.450	0.473
	0.05	0.345	0.358	0.386	0.418	0.435	0.452	0.471	0.490	0.510	0.532
	0.10	0.441	0.452	0.479	0.509	0.525	0.541	0.558	0.575	0.594	0.613
	0.90	2.40	2.38	2.34	2.30	2.28	2.25	2.23	2.21	2.18	2.16
	0.95	3.10	3.07	3.01	2.94	2.90	2.86	2.83	2.79	2.75	2.71
	0.975	3.91	3.87	3.77	3.67	3.61	3.56	3.51	3.45	3.39	3.33
	0.99	5.18	5.11	4.96	4.81	4.73	4.65	4.57	4.48	4.40	4.31
	0.995	6.31	6.23	6.03	5.83	5.73	5.62	5.52	5.41	5.30	5.19
10	0.005	0.185	0.197	0.226	0.260	0.279	0.299	0.321	0.344	0.370	0.397
	0.01	0.220	0.233	0.263	0.297	0.316	0.336	0.357	0.380	0.405	0.431
	0.025	0.283	0.296	0.327	0.360	0.379	0.398	0.419	0.441	0.464	0.488
	0.05	0.351	0.363	0.393	0.426	0.444	0.462	0.481	0.502	0.523	0.546
	0.10	0.444	0.457	0.486	0.516	0.532	0.549	0.567	0.586	0.605	0.625
	0.90	2.30	2.28	2.24	2.20	2.18	2.16	2.13	2.11	2.08	2.06
	0.95	2.94	2.91	2.85	2.77	2.74	2.70	2.66	2.62	2.58	2.54
	0.975	3.66	3.62	3.52	3.42	3.37	3.31	3.26	3.20	3.14	3.08
	0.99	4.77	4.71	4.56	4.41	4.33	4.25	4.17	4.08	4.00	3.91
	0.995	5.75	5.66	5.47	5.27	5.17	5.07	4.97	4.86	4.75	4.64
11	0.005	0.188	0.200	0.231	0.266	0.286	0.308	0.330	0.355	0.382	0.412
	0.01	0.224	0.237	0.268	0.304	0.324	0.344	0.366	0.391	0.417	0.444
	0.025	0.288	0.301	0.332	0.368	0.386	0.407	0.429	0.450	0.476	0.503
	0.05	0.355	0.368	0.398	0.433	0.452	0.469	0.490	0.513	0.535	0.559
	0.10	0.448	0.461	0.490	0.524	0.541	0.559	0.578	0.595	0.617	0.637
	0.90	2.23	2.21	2.17	2.12	2.10	2.08	2.05	2.03	2.00	1.97
	0.95	2.82	2.79	2.72	2.65	2.61	2.57	2.53	2.49	2.45	2.40
	0.975	3.47	3.43	3.33	3.23	3.17	3.12	3.06	3.00	2.94	2.88
	0.99	4.46	4.40	4.25	4.10	4.02	3.94	3.86	3.78	3.69	3.60
	0.995	5.32	5.24	5.05	4.86	4.76	4.65	4.55	4.45	4.34	4.23

TABLE D
(Continued)

v_2	p	\(v_1\) 1	2	3	4	5	6	7	8	9	10
12	0.005	0.00004	0.0050	0.023	0.048	0.075	0.100	0.122	0.143	0.161	0.177
	0.01	0.00016	0.010	0.037	0.070	0.101	0.130	0.155	0.176	0.196	0.212
	0.025	0.0010	0.025	0.070	0.114	0.153	0.186	0.214	0.238	0.259	0.276
	0.05	0.0041	0.052	0.114	0.169	0.214	0.250	0.280	0.305	0.325	0.343
	0.10	0.016	0.106	0.192	0.257	0.306	0.344	0.375	0.400	0.420	0.438
	0.90	3.18	2.81	2.61	2.48	2.39	2.33	2.28	2.24	2.21	2.19
	0.95	4.75	3.89	3.49	3.26	3.11	3.00	2.91	2.85	2.80	2.75
	0.975	6.55	5.10	4.47	4.12	3.89	3.73	3.61	3.51	3.44	3.37
	0.99	9.33	6.93	5.95	5.41	5.06	4.82	4.64	4.50	4.39	4.30
	0.995	11.8	8.51	7.23	6.52	6.07	5.76	5.52	5.35	5.20	5.09
15	0.005	0.00004	0.0050	0.023	0.049	0.076	0.102	0.125	0.147	0.166	0.183
	0.01	0.00016	0.010	0.037	0.070	0.103	0.132	0.158	0.181	0.202	0.219
	0.025	0.0010	0.025	0.070	0.116	0.156	0.190	0.219	0.244	0.265	0.284
	0.05	0.0041	0.051	0.115	0.170	0.216	0.254	0.285	0.311	0.333	0.351
	0.10	0.016	0.106	0.192	0.258	0.309	0.348	0.380	0.406	0.427	0.446
	0.90	3.07	2.70	2.49	2.36	2.27	2.21	2.16	2.12	2.09	2.06
	0.95	4.54	3.68	3.29	3.06	2.90	2.79	2.71	2.64	2.59	2.54
	0.975	6.20	4.76	4.15	3.80	3.58	3.41	3.29	3.20	3.12	3.06
	0.99	8.68	6.36	5.42	4.89	4.56	4.32	4.14	4.00	3.89	3.80
	0.995	10.8	7.70	6.48	5.80	5.37	5.07	4.85	4.67	4.54	4.42
20	0.005	0.00004	0.0050	0.023	0.050	0.077	0.104	0.129	0.151	0.171	0.190
	0.01	0.00016	0.010	0.037	0.071	0.105	0.135	0.162	0.187	0.208	0.227
	0.025	0.0010	0.025	0.071	0.117	0.158	0.193	0.224	0.250	0.273	0.292
	0.05	0.0040	0.051	0.115	0.172	0.219	0.258	0.290	0.318	0.340	0.360
	0.10	0.016	0.106	0.193	0.260	0.312	0.353	0.385	0.412	0.435	0.454
	0.90	2.97	2.59	2.38	2.25	2.16	2.09	2.04	2.00	1.96	1.94
	0.95	4.35	3.49	3.10	2.87	2.71	2.60	2.51	2.45	2.39	2.35
	0.975	5.87	4.46	3.86	3.51	3.29	3.13	3.01	2.91	2.81	2.77
	0.99	8.10	5.85	4.94	4.43	4.10	3.87	3.70	3.56	3.46	3.37
	0.995	9.94	6.99	5.82	5.17	4.76	4.47	4.26	4.09	3.96	3.85
24	0.005	0.00004	0.0050	0.023	0.050	0.078	0.106	0.131	0.154	0.175	0.193
	0.01	0.00016	0.010	0.038	0.072	0.106	0.137	0.165	0.189	0.211	0.231
	0.025	0.0010	0.025	0.071	0.117	0.159	0.195	0.227	0.253	0.277	0.297
	0.05	0.0040	0.051	0.116	0.173	0.221	0.260	0.293	0.321	0.345	0.365
	0.10	0.016	0.106	0.193	0.261	0.313	0.355	0.388	0.416	0.439	0.459
	0.90	2.93	2.54	2.33	2.19	2.10	2.04	1.98	1.94	1.91	1.88
	0.95	4.26	3.40	3.01	2.78	2.62	2.51	2.42	2.36	2.30	2.25
	0.975	5.72	4.32	3.72	3.38	3.15	2.99	2.87	2.78	2.70	2.64
	0.99	7.82	5.61	4.72	4.22	3.90	3.67	3.50	3.36	3.26	3.17
	0.995	9.55	6.66	5.52	4.89	4.49	4.20	3.99	3.83	3.69	3.59

TABLE D
(Continued)

v_2	p	v_1									
		11	12	15	20	24	30	40	60	120	∞
12	0.005	0.191	0.204	0.235	0.272	0.292	0.315	0.339	0.365	0.393	0.424
	0.01	0.227	0.241	0.273	0.310	0.330	0.352	0.375	0.401	0.428	0.458
	0.025	0.292	0.305	0.337	0.374	0.394	0.416	0.437	0.461	0.487	0.514
	0.05	0.358	0.372	0.404	0.439	0.458	0.478	0.499	0.522	0.545	0.571
	0.10	0.452	0.466	0.496	0.528	0.546	0.564	0.583	0.604	0.625	0.647
	0.90	2.17	2.15	2.11	2.06	2.04	2.01	1.99	1.96	1.93	1.90
	0.95	2.72	2.69	2.62	2.54	2.51	2.47	2.43	2.38	2.34	2.30
	0.975	3.32	3.28	3.18	3.07	3.02	2.96	2.91	2.85	2.79	2.72
	0.99	4.22	4.16	4.01	3.86	3.78	3.70	3.62	3.54	3.45	3.36
	0.995	4.99	4.91	4.72	4.53	4.43	4.33	4.23	4.12	4.01	3.90
15	0.005	0.198	0.212	0.246	0.286	0.308	0.333	0.360	0.389	0.422	0.457
	0.01	0.235	0.249	0.284	0.324	0.346	0.370	0.397	0.425	0.456	0.490
	0.025	0.300	0.315	0.349	0.389	0.410	0.433	0.458	0.485	0.514	0.546
	0.05	0.368	0.382	0.416	0.454	0.474	0.496	0.519	0.545	0.571	0.600
	0.10	0.461	0.475	0.507	0.542	0.561	0.581	0.602	0.624	0.647	0.672
	0.90	2.04	2.02	1.97	1.92	1.90	1.87	1.85	1.82	1.79	1.76
	0.95	2.51	2.48	2.40	2.33	2.39	2.25	2.20	2.16	2.11	2.07
	0.975	3.01	2.96	2.86	2.76	2.70	2.64	2.59	2.52	2.46	2.40
	0.99	3.73	3.67	3.52	3.37	3.29	3.21	3.13	3.05	2.96	2.87
	0.995	4.33	4.25	4.07	3.88	3.79	3.69	3.59	3.48	3.37	3.26
20	0.005	0.206	0.221	0.258	0.301	0.327	0.354	0.385	0.419	0.457	0.500
	0.01	0.244	0.259	0.297	0.340	0.365	0.392	0.422	0.455	0.491	0.532
	0.025	0.310	0.325	0.363	0.406	0.430	0.456	0.484	0.514	0.548	0.585
	0.05	0.377	0.393	0.430	0.471	0.493	0.518	0.544	0.572	0.603	0.637
	0.10	0.472	0.485	0.520	0.557	0.578	0.600	0.623	0.648	0.675	0.704
	0.90	1.91	1.89	1.84	1.79	1.77	1.74	1.71	1.68	1.64	1.61
	0.95	2.31	2.28	2.20	2.12	2.08	2.04	1.99	1.95	1.90	1.84
	0.975	2.72	2.68	2.57	2.46	2.41	2.35	2.29	2.22	2.16	2.09
	0.99	3.29	3.23	3.09	2.94	2.86	2.78	2.69	2.61	2.52	2.42
	0.995	3.76	3.68	3.50	3.32	3.22	3.12	3.02	2.92	2.81	2.69
24	0.005	0.210	0.226	0.264	0.310	0.337	0.367	0.400	0.437	0.479	0.527
	0.01	0.249	0.264	0.304	0.350	0.376	0.405	0.437	0.473	0.513	0.558
	0.025	0.315	0.331	0.370	0.415	0.441	0.468	0.498	0.531	0.568	0.610
	0.05	0.383	0.399	0.437	0.480	0.504	0.530	0.558	0.588	0.622	0.659
	0.10	0.476	0.491	0.527	0.566	0.588	0.611	0.635	0.662	0.691	0.723
	0.90	1.85	1.83	1.78	1.73	1.70	1.67	1.64	1.61	1.57	1.53
	0.95	2.21	2.18	2.11	2.03	1.98	1.94	1.89	1.84	1.79	1.73
	0.975	2.59	2.54	2.44	2.33	2.27	2.21	2.15	2.08	2.01	1.94
	0.99	3.09	3.03	2.89	2.74	2.66	2.58	2.49	2.40	2.31	2.21
	0.995	3.50	3.42	3.25	3.06	2.97	2.87	2.77	2.66	2.55	2.43

TABLE D
(Continued)

| v_2 | p | \multicolumn{10}{c}{v_1} |
		1	2	3	4	5	6	7	8	9	10
30	0.005	0.00004	0.0050	0.024	0.050	0.079	0.107	0.133	0.156	0.178	0.197
	0.01	0.00016	0.010	0.038	0.072	0.107	0.138	0.167	0.192	0.215	0.235
	0.025	0.0010	0.025	0.071	0.118	0.161	0.197	0.229	0.257	0.281	0.302
	0.05	0.0040	0.051	0.116	0.174	0.222	0.263	0.296	0.325	0.349	0.370
	0.10	0.016	0.106	0.193	0.262	0.315	0.357	0.391	0.420	0.443	0.464
	0.90	2.88	2.49	2.28	2.14	2.05	1.98	1.93	1.88	1.85	1.82
	0.95	4.17	3.32	2.92	2.69	2.53	2.42	2.33	2.27	2.21	2.16
	0.975	5.57	4.18	3.59	3.25	3.03	2.87	2.75	2.65	2.57	2.51
	0.99	7.56	5.39	4.51	4.02	3.70	3.47	3.30	3.17	3.07	2.98
	0.995	9.18	6.35	5.24	4.62	4.23	3.95	3.74	3.58	3.45	3.34
40	0.005	0.00004	0.0050	0.024	0.051	0.080	0.108	0.135	0.159	0.181	0.201
	0.01	0.00016	0.010	0.038	0.073	0.108	0.140	0.169	0.195	0.219	0.240
	0.025	0.00099	0.025	0.071	0.119	0.162	0.199	0.232	0.260	0.285	0.307
	0.05	0.0040	0.051	0.116	0.175	0.224	0.265	0.299	0.329	0.354	0.376
	0.10	0.016	0.106	0.194	0.263	0.317	0.360	0.394	0.424	0.448	0.469
	0.90	2.84	2.44	2.23	2.09	2.00	1.93	1.87	1.83	1.79	1.76
	0.95	4.08	3.23	2.84	2.61	2.45	2.34	2.25	21.8	2.12	2.08
	0.975	5.42	4.05	3.46	3.13	2.90	2.74	2.62	2.53	2.45	2.39
	0.99	7.31	5.18	4.31	3.83	3.51	3.29	3.12	2.99	2.89	2.80
	0.995	8.83	6.07	4.98	4.37	3.99	3.71	3.51	3.35	3.22	3.12
60	0.005	0.00004	0.0050	0.024	0.051	0.081	0.110	0.137	0.162	0.185	0.206
	0.01	0.00016	0.010	0.038	0.073	0.109	0.142	0.172	0.199	0.223	0.245
	0.025	0.00099	0.025	0.071	0.120	0.163	0.202	0.235	0.264	0.290	0.313
	0.05	0.0040	0.051	0.116	0.176	0.226	0.267	0.303	0.333	0.359	0.382
	0.10	0.016	0.106	0.194	0.264	0.318	0.362	0.398	0.428	0.453	0.475
	0.90	2.79	2.39	2.18	2.04	1.95	1.87	1.82	1.77	1.74	1.71
	0.95	4.00	3.15	2.76	2.53	2.37	2.25	2.17	2.10	2.04	1.99
	0.975	5.29	3.93	3.34	3.01	2.79	2.63	2.51	2.41	2.33	2.27
	0.99	7.08	4.98	4.13	3.65	3.34	3.12	2.95	2.82	2.72	2.63
	0.995	8.49	5.80	4.73	4.14	3.76	3.49	3.29	3.13	3.01	2.90
120	0.005	0.00004	0.0050	0.024	0.051	0.081	0.111	0.139	0.165	0.189	0.211
	0.01	0.00016	0.010	0.038	0.074	0.110	0.143	0.174	0.202	0.227	0.250
	0.025	0.00099	0.025	0.072	0.120	0.165	0.204	0.238	0.268	0.295	0.318
	0.05	0.0039	0.051	0.117	0.177	0.227	0.270	0.306	0.337	0.364	0.388
	0.10	0.016	0.105	0.194	0.265	0.320	0.365	0.401	0.432	0.458	0.480
	0.90	2.75	2.35	2.13	1.99	1.90	1.82	1.77	1.72	1.68	1.65
	0.95	3.92	3.07	2.68	2.45	2.29	2.18	2.09	2.02	1.96	1.91
	0.975	5.15	3.80	3.23	2.89	2.67	2.52	2.39	2.30	2.22	2.16
	0.99	6.85	4.79	3.95	3.48	3.17	2.96	2.79	2.66	2.56	2.47
	0.995	8.18	5.54	4.50	3.92	3.55	3.28	3.09	2.93	2.81	2.71

TABLE D
(Continued)

v_2	p	\multicolumn{10}{c}{v_1}									
		11	12	15	20	24	30	40	60	120	∞
30	0.005	0.215	0.231	0.271	0.320	0.349	0.381	0.416	0.457	0.504	0.559
	0.01	0.254	0.270	0.311	0.360	0.388	0.419	0.454	0.493	0.538	0.590
	0.025	0.321	0.337	0.378	0.426	0.453	0.482	0.515	0.551	0.592	0.639
	0.05	0.389	0.406	0.445	0.490	0.516	0.543	0.573	0.606	0.644	0.685
	0.10	0.481	0.497	0.534	0.575	0.598	0.623	0.649	0.678	0.710	0.746
	0.90	1.79	1.77	1.72	1.67	1.64	1.61	1.57	1.54	1.50	1.46
	0.95	2.13	2.09	2.01	1.93	1.89	1.84	1.79	1.74	1.68	1.62
	0.975	2.46	2.41	2.31	2.20	2.14	2.07	2.01	1.94	1.87	1.79
	0.99	2.91	2.84	2.70	2.55	2.47	2.39	2.30	2.21	2.11	2.01
	0.995	3.25	3.18	3.01	2.82	2.73	2.63	2.52	2.42	2.30	2.18
40	0.005	0.220	0.237	0.279	0.331	0.362	0.396	0.436	0.481	0.534	0.599
	0.01	0.259	0.276	0.319	0.371	0.401	0.435	0.473	0.516	0.567	0.628
	0.025	0.327	0.344	0.387	0.437	0.466	0.498	0.533	0.573	0.620	0.674
	0.05	0.395	0.412	0.454	0.502	0.529	0.558	0.591	0.627	0.669	0.717
	0.10	0.488	0.504	0.542	0.585	0.609	0.636	0.644	0.696	0.731	0.772
	0.90	1.73	1.71	1.66	1.61	1.57	1.54	1.51	1.47	1.42	1.38
	0.95	2.04	2.00	1.92	1.84	1.79	1.74	1.69	1.64	1.58	1.51
	0.975	2.33	2.29	2.18	2.07	2.01	1.94	1.88	1.80	1.72	1.64
	0.99	2.73	2.66	2.52	2.37	2.29	2.20	2.11	2.02	1.92	1.80
	0.995	3.03	2.95	2.78	2.60	2.50	2.40	2.30	2.18	2.06	1.93
60	0.005	0.225	0.243	0.287	0.343	0.376	0.414	0.458	0.510	0.572	0.652
	0.01	0.265	0.283	0.328	0.383	0.416	0.453	0.495	0.545	0.604	0.679
	0.025	0.333	0.351	0.396	0.450	0.481	0.515	0.555	0.600	0.654	0.720
	0.05	0.402	0.419	0.463	0.514	0.543	0.575	0.611	0.652	0.700	0.759
	0.10	0.493	0.510	0.550	0.596	0.622	0.650	0.682	0.717	0.758	0.806
	0.90	1.68	1.66	1.60	1.54	1.51	1.48	1.44	1.40	1.35	1.29
	0.95	1.95	1.92	1.84	1.75	1.70	1.65	1.59	1.53	1.47	1.39
	0.975	2.22	2.17	2.06	1.94	1.88	1.82	1.74	1.67	1.58	1.48
	0.99	2.56	2.50	2.35	2.20	2.12	2.03	1.94	1.84	1.73	1.60
	0.995	2.82	2.74	2.57	2.39	2.29	2.19	2.08	1.96	1.83	1.69
120	0.005	0.230	0.249	0.297	0.356	0.393	0.434	0.484	0.545	0.623	0.733
	0.01	0.271	0.290	0.338	0.397	0.433	0.474	0.522	0.579	0.652	0.755
	0.025	0.340	0.359	0.406	0.464	0.498	0.536	0.580	0.633	0.698	0.789
	0.05	0.408	0.427	0.473	0.527	0.559	0.594	0.634	0.682	0.740	0.819
	0.10	0.500	0.518	0.560	0.609	0.636	0.667	0.702	0.742	0.791	0.855
	0.90	1.62	1.60	1.55	1.48	1.45	1.41	1.37	1.32	1.26	1.19
	0.95	1.87	1.83	1.75	1.66	1.61	1.55	1.50	1.43	1.35	1.25
	0.975	2.10	2.05	1.95	1.82	1.76	1.69	1.61	1.53	1.43	1.31
	0.99	2.40	2.34	2.19	2.03	1.95	1.86	1.76	1.66	1.53	1.38
	0.995	2.62	2.54	2.37	2.19	2.09	1.98	1.87	1.75	1.61	1.43

TABLE E.1
Upper 5% of Studentized Range q

n_2	\\multicolumn{19}{c}{p^*}

n_2	2	3	4	5	6	7	8	9	10	11	12	13	14	15	16	17	18	19	20
1	18.0	26.7	32.8	37.2	40.5	43.1	45.4	47.3	49.1	50.6	51.9	53.2	54.3	55.4	56.3	57.2	58.0	58.8	59.6
2	6.09	8.28	9.80	10.89	11.73	12.43	13.08	13.54	13.99	14.39	14.75	15.08	15.38	15.65	15.91	16.14	16.36	16.57	16.77
3	4.50	5.88	6.83	7.51	8.04	8.47	8.85	9.18	9.46	9.72	9.95	10.16	10.35	10.52	10.69	10.84	10.98	11.12	11.24
4	3.93	5.00	5.76	6.31	6.73	7.06	7.35	7.60	7.83	8.03	8.21	8.37	8.52	8.67	8.80	8.92	9.03	9.14	9.24
5	3.61	4.54	5.18	5.64	5.99	6.28	6.52	6.74	6.93	7.10	7.25	7.39	7.52	7.64	7.75	7.86	7.95	8.04	8.13
6	3.46	4.34	4.90	5.31	5.63	5.89	6.12	6.32	6.49	6.65	6.79	6.92	7.04	7.14	7.24	7.34	7.43	7.51	7.59
7	3.34	4.16	4.68	5.06	5.35	5.59	5.80	5.99	6.15	6.29	6.42	6.54	6.65	6.75	6.84	6.93	7.01	7.08	7.16
8	3.26	4.04	4.53	4.89	5.17	5.40	5.60	5.77	5.92	6.05	6.18	6.29	6.39	6.48	6.57	6.65	6.73	6.80	6.87
9	3.20	3.95	4.42	4.76	5.02	5.24	5.43	5.60	5.74	5.87	5.98	6.09	6.19	6.28	6.36	6.44	6.51	6.58	6.65
10	3.15	3.88	4.33	4.66	4.91	5.12	5.30	5.46	5.60	5.72	5.83	5.93	6.03	6.12	6.20	6.27	6.34	6.41	6.47
11	3.11	3.82	4.26	4.58	4.82	5.03	5.20	5.35	5.49	5.61	5.71	5.81	5.90	5.98	6.06	6.14	6.20	6.27	6.33
12	3.08	3.77	4.20	4.51	4.75	4.95	5.12	5.27	5.40	5.51	5.61	5.71	5.80	5.88	5.95	6.02	6.09	6.15	6.21
13	3.06	3.73	4.15	4.46	4.69	4.88	5.05	5.19	5.32	5.43	5.53	5.63	5.71	5.79	5.86	5.93	6.00	6.06	6.11
14	3.03	3.70	4.11	4.41	4.64	4.83	4.99	5.13	5.25	5.36	5.46	5.56	5.64	5.72	5.79	5.86	5.92	5.98	6.03
15	3.01	3.67	4.08	4.37	4.59	4.78	4.94	5.08	5.20	5.31	5.40	5.49	5.57	5.65	5.72	5.79	5.85	5.91	5.96
16	3.00	3.65	4.05	4.34	4.56	4.74	4.90	5.03	5.15	5.26	5.35	5.44	5.52	5.59	5.66	5.73	5.79	5.84	5.90
17	2.98	3.62	4.02	4.31	4.52	4.70	4.86	4.99	5.11	5.21	5.31	5.39	5.47	5.55	5.61	5.68	5.74	5.79	5.84
18	2.97	3.61	4.00	4.28	4.49	4.67	4.83	4.96	5.07	5.17	5.27	5.35	5.43	5.50	5.57	5.63	5.69	5.74	5.79
19	2.96	3.59	3.98	4.26	4.47	4.64	4.79	4.92	5.04	5.14	5.23	5.32	5.39	5.46	5.53	5.59	5.65	5.70	5.75
20	2.95	3.58	3.96	4.24	4.45	4.62	4.77	4.90	5.01	5.11	5.20	5.28	5.36	5.43	5.50	5.56	5.61	5.66	5.71
24	2.92	3.53	3.90	4.17	4.37	4.54	4.68	4.81	4.92	5.01	5.10	5.18	5.25	5.32	5.38	5.44	5.50	5.55	5.59
30	2.89	3.48	3.84	4.11	4.30	4.46	4.60	4.72	4.83	4.92	5.00	5.08	5.15	5.21	5.27	5.33	5.38	5.43	5.48
40	2.86	3.44	3.79	4.04	4.23	4.39	4.52	4.63	4.74	4.82	4.90	4.98	5.05	5.11	5.17	5.22	5.27	5.32	5.36
60	2.83	3.40	3.74	3.98	4.16	4.31	4.44	4.55	4.65	4.73	4.81	4.88	4.94	5.00	5.06	5.11	5.15	5.20	5.24
120	2.80	3.36	3.69	3.92	4.10	4.24	4.36	4.47	4.56	4.64	4.71	4.78	4.84	4.90	4.95	5.00	5.04	5.09	5.13
∞	2.77	3.32	3.63	3.86	4.03	4.17	4.29	4.39	4.47	4.55	4.62	4.68	4.74	4.80	4.84	4.89	4.93	4.97	5.01

* p is the number of quantities (e.g., means) whose range is involved; n_2 is the degrees of freedom in the error estimate.

Source: J. M. May, "Extended and Corrected Tables of the Upper Percentage Points of the Studentized Range," *Biometrika*, Vol. 39, 1952, pp. 192–193.

TABLE E.2
Upper 1% Points of Studentized Range *q*

n_2	p^* 2	3	4	5	6	7	8	9	10	11	12	13	14	15	16	17	18	19	20
1	90.0	135	164	186	202	216	227	237	246	253	260	266	272	227	282	286	290	294	298
2	14.0	19.0	22.3	24.7	26.6	28.2	29.5	30.7	31.7	32.6	33.4	34.1	34.8	35.4	36.0	36.5	37.0	37.5	37.9
3	8.26	10.6	12.2	13.3	14.2	15.0	15.6	16.2	16.7	17.1	17.5	17.9	18.2	18.5	18.8	19.1	19.3	19.5	19.8
4	6.51	8.12	9.17	9.96	10.6	11.1	11.5	11.9	12.3	12.6	12.8	13.1	13.3	13.5	13.7	13.9	14.1	14.2	14.4
5	5.70	6.97	7.80	8.42	8.91	9.32	9.67	9.97	10.24	10.48	10.70	10.89	11.08	11.24	11.40	11.55	11.68	11.81	11.93
6	5.24	6.33	7.03	7.56	7.97	8.32	8.61	8.87	9.10	9.30	9.49	9.65	9.81	9.95	10.08	10.21	10.32	10.43	10.54
7	4.95	5.92	6.54	7.01	7.37	7.68	7.94	8.17	8.37	8.55	8.71	8.86	9.00	9.12	9.24	9.35	9.46	9.55	9.65
8	4.74	5.63	6.20	6.63	6.96	7.24	7.47	7.68	7.87	8.03	8.18	8.31	8.44	8.55	8.66	8.76	8.85	8.94	9.03
9	4.60	5.43	5.96	6.35	6.66	6.91	7.13	7.32	7.49	7.65	7.78	7.91	8.03	8.13	8.23	9.32	8.41	8.49	8.57
10	4.48	5.27	5.77	6.14	6.43	6.67	6.87	7.05	7.21	7.36	7.48	7.60	7.71	7.81	7.91	7.99	8.07	8.15	8.22
11	4.39	5.14	5.62	5.97	6.25	6.48	6.67	6.84	6.99	7.13	7.25	7.36	7.46	7.56	7.65	7.73	7.81	7.88	7.95
12	4.32	5.04	5.50	5.84	6.10	6.32	6.51	6.67	6.81	6.94	7.06	7.17	7.26	7.36	7.44	7.52	7.59	7.66	7.73
13	4.26	4.96	5.40	5.73	5.98	6.19	6.37	6.53	6.67	6.79	6.90	7.01	7.10	7.19	7.27	7.34	7.42	7.48	7.55
14	4.21	4.89	5.32	5.63	5.88	6.08	6.26	6.41	6.54	6.66	6.77	6.87	6.96	7.05	7.12	7.20	7.27	7.33	7.39
15	4.17	4.83	5.25	5.56	5.80	5.99	6.16	6.31	6.44	6.55	6.66	6.76	6.84	6.93	7.00	7.07	7.14	7.20	7.26
16	4.13	4.78	5.19	5.49	5.72	5.92	6.08	6.22	6.35	6.46	6.56	6.66	6.74	6.82	6.90	6.97	7.03	7.09	7.15
17	4.10	4.74	5.14	5.43	5.66	5.85	6.01	6.15	6.27	6.38	6.48	6.57	6.66	6.73	6.80	6.87	6.94	7.00	7.05
18	4.07	4.70	5.09	5.38	5.60	5.79	5.94	6.08	6.20	6.31	6.41	6.50	6.58	6.65	6.72	6.79	6.85	6.91	6.96
19	4.05	4.67	5.05	5.33	5.55	5.73	5.89	6.02	6.14	6.25	6.34	6.43	6.51	6.58	6.65	6.72	6.78	6.84	6.89
20	4.02	4.64	5.02	5.29	5.51	5.69	5.84	5.97	6.09	6.19	6.29	6.37	6.45	6.52	6.59	6.65	6.71	6.76	6.82
24	3.96	4.54	4.91	5.17	5.37	5.54	5.69	5.81	5.92	6.02	6.11	6.19	6.26	6.33	6.39	6.45	6.51	6.56	6.61
30	3.89	4.45	4.80	5.05	5.24	5.40	5.54	5.65	5.76	5.85	5.93	6.01	6.08	6.14	6.20	6.26	6.31	6.36	6.41
40	3.82	4.37	4.70	4.93	5.11	5.27	5.39	5.50	5.60	5.69	5.77	5.84	5.90	5.96	6.02	6.07	6.12	6.17	6.21
60	3.76	4.28	4.60	4.82	4.99	5.13	5.25	5.36	5.45	5.53	5.60	5.67	5.73	5.79	5.84	5.89	5.93	5.98	6.02
120	3.70	4.20	4.50	4.71	4.87	5.01	5.12	5.21	5.30	5.38	5.44	5.51	5.56	5.61	5.66	5.71	5.75	5.79	5.83
∞	3.64	4.12	4.40	4.60	4.76	4.88	4.99	5.08	5.16	5.23	5.29	5.35	5.40	5.45	5.49	5.54	5.57	5.61	5.65

* p is the number of quantities (e.g., means) whose range is involved; n_2 is the degrees of freedom in the error estimate.

Source: J. M. May, "Extended and Corrected Tables of the Upper Percentage Points of the Studentized Range," *Biometrika,* Vol. 39, 1952, pp. 192–193.

TABLE F
Coefficients of Orthogonal Polynomials

k	Polynomial	1	2	3	4	5	6	7	8	9	10	$\sum \xi_j^2$
3	Linear	−1	0	1								2
	Quadratic	1	−2	1								6
	Linear	−3	−1	1	3							20
4	Quadratic	1	−1	−1	1							4
	Cubic	−1	3	−3	1							20
	Linear	−2	−1	0	1	2						10
5	Quadratic	2	−1	−2	−1	2						14
	Cubic	−1	2	0	−2	1						10
	Quartic	1	−4	6	−4	1						70
	Linear	−5	−3	−1	1	3	5					70
6	Quadratic	5	−1	−4	−4	−1	5					84
	Cubic	−5	7	4	−4	−7	5					180
	Quartic	1	−3	2	2	−3	1					28
	Linear	−3	−2	−1	0	1	2	3				28
7	Quadratic	5	0	−3	−4	−3	0	5				84
	Cubic	−1	1	1	0	−1	−1	1				6
	Quartic	3	−7	1	6	1	−7	3				154
	Linear	−7	−5	−3	−1	1	3	5	7			168
	Quadratic	7	1	−3	−5	−5	−3	1	7			168
8	Cubic	−7	5	7	3	−3	−7	−5	7			264
	Quartic	7	−13	−3	9	9	−3	−13	7			616
	Quintic	−7	23	−17	−15	15	17	−23	7			2184
	Linear	−4	−3	−2	−1	0	1	2	3	4		60
	Quadratic	28	7	−8	−17	−20	−17	−8	7	28		2772
9	Cubic	−14	7	13	9	0	−9	−13	−7	14		990
	Quartic	14	−21	−11	9	18	9	−11	−21	14		2002
	Quintic	−4	11	−4	−9	0	9	4	−11	4		468
	Linear	−9	−7	−5	−3	−1	1	3	5	7	9	330
	Quadratic	6	2	−1	−3	−4	−4	−3	−1	2	6	132
10	Cubic	−42	14	35	31	12	−12	−31	−35	−14	42	8580
	Quartic	18	−22	−17	3	18	18	3	−17	−22	18	2860
	Quintic	−6	14	−1	−11	−6	6	11	1	−14	6	780

ANSWERS TO SELECTED PROBLEMS

CHAPTER 2

2.1 $\bar{y} = 18{,}472.9$, $s^2 = 41.72$.

2.3 For a two-sided alternative, some points are
$\mu = 18{,}473;\ 18{,}472;\ 18{,}471;\ 18{,}470;\ 18{,}469;\ 18{,}468;\ 18{,}467$
$P_a = \beta = 0.08, 0.39, 0.80, 0.95, 0.80, 0.39, 0.08$.

2.5 $n = 7$

2.7 Do not reject hypothesis, since $\chi^2 = 17.05$ with 11 df.

2.9 Do not reject hypothesis, since $f = 1.8$.

2.11 Reject hypothesis at 1% level, since $t = 3.56$.

2.13 **a.** Reject hypothesis, since $z = 4.0$.
b. Do not reject hypothesis, since $z = 1.71$.
c. Do not reject hypothesis, since $z < 1$.

2.15 For $\quad\quad \mu = 13 \quad 15 \quad 17 \quad 19$
Power $(1 - \beta) = 0.04 \quad 0.27 \quad 0.70 \quad 0.95$

2.17 Reject hypothesis at any reasonable α, since $t = 8.29$.

2.19 Reject hypothesis of equal variance at $\alpha = 0.05$, since $f = 17.4$. Do not reject hypothesis of equal means at $\alpha = 0.05$, since $t' \approx 0.6$.

2.21 Reject hypothesis at $\alpha = 0.05$, since $t = 2.68$.

2.23 Do not reject hypothesis at $\alpha = 0.05$, since $t = 1.62$.

2.25 Reject hypothesis at $\alpha = 0.05$, since $z = -1.69$.

2.27 No outliers; normality assumption seems reasonable.

2.29 a. Do not assume normality.
b. The sample distribution is skewed right. There are two outliers.

2.31 a. Yes, $t \approx 6.5$.
b. 4.73

CHAPTER 3

3.1

Source	DF	SS	MS	F	p
A	4	253.04	63.26	16.47	0.000
Error	20	76.80	3.84		
Total	24	329.84			

3.3 Two such contrasts might be

$$c_1 = T_A - T_C = 31 \qquad \frac{SS}{60.06}$$

$$c_2 = T_A - 2T_B + T_C = -79 \qquad \frac{130.02}{190.08}$$

Neither is significant at the 5% level.

3.5

	B	A	C
$\bar{y}_{.j} =$	25.4	22.4	18.5

None significantly different.

3.7 Two sets and their sums of squares might be

		SS
Set 1: $c_1 = 2T_1 \qquad\qquad\qquad\qquad - 2T_5 = -28$		49.0
$c_2 = \qquad\quad 4T_2 \qquad\quad - 6T_4 \qquad\qquad = -108$		48.6
$c_3 = 11T_1 + 11T_2 - 14T_3 + 11T_4 + 11T_5 = -71$		1.3
$c_4 = 10T_1 - 4T_2 \qquad\qquad - 4T_4 + 10T_5 = 8$		0.1
		99.0

SS

$$
\begin{aligned}
\text{Set 2: } c_1 &= 6T_1 - 2T_2 & = -18 & \quad 3.4 \\
c_2 &= 11T_1 + 11T_2 - 8T_3 & = -237 & \quad 33.6 \\
c_3 &= 4T_1 + 4T_2 + 4T_3 - 19T_4 & = -252 & \quad 36.3 \\
c_4 &= 2T_1 + 2T_2 + 2T_3 + 2T_4 - 23T_5 = -172 & & \quad 25.7
\end{aligned}
$$

99.0

3.9 75.6% due to the levels of A, 24.4% due to error.

3.11 Two might be $C_1 = T_{.1} - T_{.2}$ and $C_2 = 2T_{.1} - T_{.2} - T_{.3}$. Thus, $c_1 = -24$ and $c_2 = 7$. Neither is significant at 5%.

3.13

Source	DF	SS	MS	F	p
Temp	4	1268.53	317.13	70.27	0.000
Error	5	112.83	4.51		
Total	29	1381.37			

3.15

Source	DF	SS	MS	F	p
Coating	3	1135.0	378.3	29.79	0.000
Error	16	203.2	12.7		
Total	19	1338.2			

3.17 Newman-Keuls tests show:

Coating	IV	III	II	I
Means	42.0	43.6	57.2	58.4

3.19

Source	DF	SS	MS	F	p
Bonder	3	7,822	2607	1.95	0.139
Error	36	48,110	1336		
Total	39	55,932			

All bonders do about the same job.

3.21 One set of orthogonal contrasts might be

$$C_1 = 4T_{.1} - T_{.2} - T_{.3} - T_{.4} - T_{.5}$$
$$C_2 = \quad T_{.2} - T_{.3}$$
$$C_3 = \qquad\qquad T_{.4} - T_{.5}$$
$$C_4 = \quad T_{.2} + T_{.3} - T_{.4} - T_{.5}$$

3.23 For C_1 in Problem 3.21, $s_C = 24.5$.

3.25 Litter-to-litter differences accounts for 28.6% of the total variance.

3.27 Logical contrasts are $2T_A - T_B - T_C = -32.3$ and $T_B - T_C = -2.5$, whose SS are 28.98 and 0.52, respectively. The first is highly significant.

CHAPTER 4

4.1

Source	DF	SS	MS	F	p
Coater	3	1.53000	0.51000	5.67	0.035
Day	2	0.20667	0.10333	1.15	0.378
Error	6	0.54000	0.09000		
Total	11	2.27667			

4.3

	K	A	L	M
$\bar{y}_{.j}$:	5.27	4.87	4.73	4.27

K and M are significantly different at the 5% level.

4.5 Four contrasts might be

$C_1 = T_M - T_A$ Nonsignificant

$C_2 = T_M - T_K$ Significant at 5% by Scheffé

$C_3 = T_M - T_L$ Nonsignificant

$C_4 = T_A - T_K$ Nonsignificant

4.7

Source	DF	Seq SS	MS	F	P
Electrode	4	4.28960	1.07240	15.83	0.000
Position	4	0.56960	0.14240	2.10	0.143
Strip	4	0.31360	0.07840	1.16	0.377
Error	12	0.81280	0.06773		
Total	24	5.98560			

4.11 Error $df = 0$.

4.13 Three groups (treatments) and six lessons (blocks) ANOVA:

Source	df
Groups	2
Lessons	5
Error	10

4.15 Newman–Keuls on means:

Group:	C	B	A
Means:	12.33	19.17	22.17

4.17 a. $Y_{ijk} = \mu + S_i + T_j + D_k + \varepsilon_{ijk}$ and ANOVA:

Source	df
Scales	4
Times	4
Days	4
Error	12

b. ANOVA:

Source	df
Scales	4
Days	4
Error	16

4.19 $H_0 : F_j = 0$ for all j. Model: $Y_{ij} = \mu + F_j + \beta_i + \varepsilon_{ij}$. $F = 8.75$, which is significant at the 1% level.

4.21 From 38.1 to 44.4.

4.23 For example,

$$C_1 = T_{.1} - T_{.2}$$
$$C_2 = T_{.1} + T_{.2} - 2T_{.3}$$
$$C_3 = \qquad\qquad T_{.3} - T_{.4}$$

4.25

Source	DF	SS	MS	F	P
Vendor	2	10.8889	5.4444	49.00	0.020
Scale	2	32.88889	16.4444	148.00	0.007
Inspector	2	0.2222	0.1111	1.00	0.500
Error	2	0.2222	0.1111		
Total	8	44.2222			

4.27

Source	DF	SS	MS	F	P
Curr	2	6.2222	3.1111	28.00	0.034
Schools	2	3.5556	1.7778	16.00	0.059
Grade	2	0.8889	0.4444	4.00	0.200
Error	2	0.2222	0.1111		
Total	8	10.8889			

Only curricula significant at 5% level.
Newman–Keuls on curricula means gives

A	B	C
29.0	30.33	31.00

B and *C* better than *A*

4.29

Source	DF	SS	MS	F	P
Rep	2	3370.0	1685.0	4.71	0.045
Machine	4	3026.7	756.7	2.11	0.171
Error	8	2863.3	357.9		
Total	14	9260.0			

No significant differences between machines.

4.31 b. $Y_{ijk} = \mu + \beta_i + \tau_j + \gamma_k + \varepsilon_{ijk}$ with β_i the effect of the ith cavity ($i = 1, 2, 3, 4, 5$), τ_j the effect of the jth mold ($j = 1, 2, 3, 4, 5$), and γ_k the effect of the kth material ($k = 1, 2, 3, 4, 5$).

CHAPTER 5

5.1

Source	DF	SS	MS	F	P
Phosphor	2	1,244.4	622.2	8.96	0.004
Glass	1	13,338.9	13,338.9	192.08	0.000
Phosphor*Glass	2	44.4	22.2	0.32	0.732
Error	12	833.3	69.4		
Total	17	15,461.1			

5.3 G_2 better than G_1—lower current. For phosphors, Newman–Keuls:

Phosphor	C	A	B
Means:	253.33	260.00	273.33

so A or C is better than B.

5.5

Source	df	SS	MS
Humidity	2	9.07	4.53
Temperature	2	8.66	4.33
$T \times H$ interaction	4	6.07	1.52
Error	27	28.50	1.06
Totals	35	52.30	

5.7 $Y_{ijk} = \mu + H_i + T_j + (HT)_{ij} + \varepsilon_{k(ij)}$
Assumption: $\varepsilon_{k(ij)} \sim NID(0, \sigma^2)$
$H_1 : (HT)_{ij} = 0$ for $i = 1, 2, 3$ and $j = 1, 2, 3$
$H_2 : H_i = 0$ for $i = 1, 2, 3$
$H_3 : T_j = 0$ for $j = 1, 2, 3$

5.9 Plot of cell totals versus feed for the two material lines are not parallel (interaction), and both material and feed effect are obvious.

5.11

Source	DF	SS	MS	F	P
Temp	2	600.089	300.044	1.6E+04	0.000
Mix	2	1.585	0.793	42.27	0.000
Temp*Mix	4	0.979	0.245	13.05	0.000
Lab	3	3.844	1.281	68.33	0.000
Temp*Lab	6	1.546	0.258	13.74	0.000
Mix*Lab	6	0.706	0.118	6.27	0.000
Temp*Mix*Lab	12	0.953	0.079	4.24	0.000
Error	36	0.675	0.019		
Total	71	610.377			

5.13

Source	DF	SS	MS
Soil	4	41.467	10.367
Fertilizer	2	68.867	34.433
Soil*Fertilizer	8	16.133	2.017
Error	15	52.500	3.500
Total	29	178.967	

5.15 Newman–Keuls on fertilizers:

Types: 3   2   1
Means: 3.6   5.2   7.3

Hence type 1 gives best yield with any of the five soil types.

5.17 a. No.
   b. Yes.
   c. Yes.

5.19 Two-dimensional stimulus more variable than three dimensional.

J by A means:

1.04   1.14   1.17   1.20   1.26   1.42   1.58   1.59   1.70   1.71

5.21

Source	DF	SS	MS	F	P
Thickness A	1	846.81	846.81	310.72	0.000
Temperature B	1	5041.00	5041.00	1849.70	0.000
$A*B$	1	509.63	509.63	187.00	0.000
Drying condition C	1	5.88	5.88	2.16	0.152
$A*C$	1	1.44	1.44	0.53	0.473
$B*C$	1	15.21	15.21	5.58	0.024
$A*B*C$	1	0.14	0.14	0.05	0.822
Length of wash D	3	69.76	23.25	8.53	0.000
$A*D$	3	15.78	5.26	1.93	0.145
$B*D$	3	3.04	1.01	0.37	0.774
$A*B*D$	3	11.45	3.82	1.40	0.261
$C*D$	3	9.65	3.22	1.18	0.333
$A*C*D$	3	7.57	2.52	0.93	0.440
$B*C*D$	3	6.43	2.14	0.79	0.510
$A*B*C*D$	3	5.66	1.89	0.69	0.564
Error	32	87.21	2.73		
Total	63	6636.65			

5.23 a.

Source	df
Waxes	3
Times	2
$W \times T$ interaction	6
Error	12

b. Changes in gloss index due to the four waxes are different for the three polishing times.

5.25 a. Significants are A, B, C, and the ABC interaction.
b. Run Newman–Keuls on 24 cell means
d. 7.0 to 12.2

5.27

Source	DF	SS	MS	F	P
Cool	4	219.932	54.983	24.02	0.000
Preheat	1	0.520	0.520	0.23	0.639
Cool*Preheat	4	28.997	7.249	3.17	0.036
Error	20	45.777	2.289		
Total	29	295.227			

As interaction significant at 5% level, run N–K on 10 "cell" means and recommend $C = 10$, $P = 5$.

3.71   4.08   4.68   5.35   5.78   5.94   7.88   9.98   10.94   12.28

5.29 a. It seems that a post cure time of 120 minutes should be used with heat sag tests performed within 24 hours of molding.

b. Using Minitab, the following ANOVA summary is obtained. Since sample sizes differ, the sequential sums of squares (Seq SSs) should be ignored. The significant interaction effect tells us that further analysis should concentrate on the 9 cell means.

Source	DF	Seq SS	Adj SS	Adj MS	F	P
Post	2	0.73246	0.89867	0.44933	18.57	0.000
Time	2	0.12831	0.17252	0.08626	3.57	0.032
Post*Time	4	0.44185	0.44185	0.11046	4.57	0.002
Error	92	2.22575	2.22575	0.02419		
Total	100	3.52836				

c. It appears that homogeneity of variances should not be assumed. The normaility assumption is questionable, because there are several outliers among the residuals. This could make further analyses such as interval estimates and multiple comparison tests based on either assumption unreliable, especially since one cell has so few observations.

CHAPTER 6

6.1

Source	EMS
O_i	$\sigma_\varepsilon^2 + 10\sigma_O^2$
A_j	$\sigma_\varepsilon^2 + 2\sigma_{OA}^2 + 6\phi_A$
OA_{ij}	$\sigma_\varepsilon^2 + 2\sigma_{OA}^2$
$\varepsilon_{k(ij)}$	σ_ε^2

Tests are indicated by arrows. None significant at the 5% level.

6.3

Source	EMS
A_i	$\sigma_\varepsilon^2 + nc\sigma_{AB}^2 + nbc\sigma_A^2$
B_j	$\sigma_\varepsilon^2 + nc\sigma_{AB}^2 + nac\sigma_B^2$
AB_{ij}	$\sigma_\varepsilon^2 + nc\sigma_{AB}^2$
C_k	$\sigma_\varepsilon^2 + n\sigma_{ABC}^2 + na\sigma_{BC}^2 + nb\sigma_{AC}^2 + nab\phi_C$
AC_{ik}	$\sigma_\varepsilon^2 + n\sigma_{ABC}^2 + nb\sigma_{AC}^2$
BC_{jk}	$\sigma_\varepsilon^2 + n\sigma_{ABC}^2 + na\sigma_{BC}^2$
ABC_{ijk}	$\sigma_\varepsilon^2 + n\sigma_{ABC}^2$
$\varepsilon_{m(ijk)}$	σ_ε^2

Tests are obvious.
No direct test on C.

6.5

Source	EMS
A_i	$\sigma_\varepsilon^2 + nb\sigma_{ACD}^2 + nbc\sigma_{AD}^2 + nbd\sigma_{AC}^2 + nbcd\phi_A$
B_j	$\sigma_\varepsilon^2 + na\sigma_{BCD}^2 + nac\sigma_{BD}^2 + nad\sigma_{BC}^2 + nacd\phi_B$
AB_{ij}	$\sigma_\varepsilon^2 + n\sigma_{ABCD}^2 + nc\sigma_{ABD}^2 + nd\sigma_{ABC}^2 + ncd\phi_{AB}$
C_k	$\sigma_\varepsilon^2 + nab\sigma_{CD}^2 + nabd\sigma_C^2$
AC_{ik}	$\sigma_\varepsilon^2 + nb\sigma_{ACD}^2 + nbd\sigma_{AC}^2$
BC_{jk}	$\sigma_\varepsilon^2 + na\sigma_{BCD}^2 + nad\sigma_{BC}^2$
ABC_{ijk}	$\sigma_\varepsilon^2 + n\sigma_{ABCD}^2 + nd\sigma_{ABC}^2$
D_m	$\sigma_\varepsilon^2 + nab\sigma_{CD}^2 + nabc\sigma_D^2$
AD_{im}	$\sigma_\varepsilon^2 + nb\sigma_{ACD}^2 + nbc\sigma_{AD}^2$
BD_{jm}	$\sigma_\varepsilon^2 + na\sigma_{BCD}^2 + nac\sigma_{BD}^2$
ABD_{ijm}	$\sigma_\varepsilon^2 + n\sigma_{ABCD}^2 + nc\sigma_{ABD}^2$
CD_{km}	$\sigma_\varepsilon^2 + nab\sigma_{CD}^2$
ACD_{ikm}	$\sigma_\varepsilon^2 + nb\sigma_{ACD}^2$
BCD_{jkm}	$\sigma_\varepsilon^2 + na\sigma_{BCD}^2$
$ABCD_{ijkm}$	$\sigma_\varepsilon^2 + n\sigma_{ABCD}^2$
$\varepsilon_{q(ijkm)}$	σ_ε^2

No direct test on A, B, or AB.

6.7   **a.** $Y_{ijk} = \mu + T_i + S_j + TS_{ij} + \varepsilon_{k(ij)}$.

b.

Source	df	EMS
T_i	3	$\sigma_\varepsilon^2 + 2\sigma_{TS}^2 + 10\phi_T$
S_j	4	$\sigma_\varepsilon^2 + 8\sigma_S^2$
TS_{ij}	12	$\sigma_\varepsilon^2 + 2\sigma_{TS}^2$
$\varepsilon_{k(ij)}$	20	σ_ε^2

c. See arrows above.

6.9

Source	EMS
O_i	$\sigma_\varepsilon^2 + 2\sigma_{OI}^2 + 8\sigma_O^2$
I_j	$\sigma_\varepsilon^2 + 2\sigma_{OI}^2 + 8\sigma_I^2$
OI_{ij}	$\sigma_\varepsilon^2 + 2\sigma_{OI}^2$
$\varepsilon_{k(ij)}$	σ_ε^2

Instruments:	40%
Operators:	32%
Operator/ instrument interaction:	15%
Error:	13%

6.11 $n_o = 4.44$, $s_\varepsilon^2 = 1.16$, $s_A^2 = 5.32$.

6.15 To test A, $MS = MS_{AB} + MS_{AC} - MS_{ABC}$
B, $MS = MS_{AB} + MS_{BC} - MS_{ABC}$
C, $MS = MS_{AC} + MS_{BC} - MS_{ABC}$

6.17 Cards: 9.0%; Testers: 83.6%

CHAPTER 7

7.1

Source	DF	SS	MS	F	P	EMS
Lot	2	27.4200	13.7100	3.39	0.080	$\sigma_\varepsilon^2 + 3\sigma_R^2 + 12\phi_L$
Roll (Lot)	9	36.3800	4.0422	4.45	0.002	$\sigma_\varepsilon^2 + 3\sigma_R^2$
Error	24	21.8000	0.9083			σ_ε^2
Total	35	85.6000				

7.3

Source	EMS
A_i	$\sigma_\varepsilon^2 + 2\sigma_C^2 + 6\sigma_B^2 + 24\phi_A$
$B_{j(i)}$	$\sigma_\varepsilon^2 + 2\sigma_C^2 + 6\sigma_B^2$
$C_{k(ij)}$	$\sigma_\varepsilon^2 + 2\sigma_C^2$
$\varepsilon_{m(ijk)}$	σ_ε^2

Tests are obvious.

7.5 Let T, M, and S denote the type of machine, machine, and stock, respectively, with machine a random factor and machine nested in type.

Source	DF	SS	MS	F	P	EMS
T	1	5,489,354	5,489,354	26.89	0.035	$\sigma_\varepsilon^2 + 6\sigma_M^2 + 12\phi_T$
$M(T)$	2	408,250	204,125	10.31	0.001	$\sigma_\varepsilon^2 + 6\sigma_M^2$
S	1	8,971	8,971	0.57	0.529	$\sigma_\varepsilon^2 + 3\sigma_{MS}^2 + 12\phi_S$
$T * S$	1	37,446	37,446	2.38	0.263	$\sigma_\varepsilon^2 + 3\sigma_{MS}^2 + 6\phi_{TS}$
$S * M(T)$	2	31,521	15,760	0.80	0.467	$\sigma_\varepsilon^2 + 3\sigma_{MS}^2$
Error	16	316,929	19,808			σ_ε^2
Total	23	6,292,470				

7.7

Source	EMS
T_i	$\sigma_\varepsilon^2 + 18\sigma_m^2 + 36\phi_t$
$M_{j(i)}$	$\sigma_\varepsilon^2 + 18\sigma_m^2$
S_k	$\sigma_\varepsilon^2 + 9\sigma_{ms}^2 + 36\phi_s$
TS_{ik}	$\sigma_\varepsilon^2 + 9\sigma_{ms}^2 + 18\phi_{ts}$
$MS_{kj(i)}$	$\sigma_\varepsilon^2 + 9\sigma_{ms}^2$
P_m	$\sigma_\varepsilon^2 + 6\sigma_{mp}^2 + 24\phi_p$
TP_{im}	$\sigma_\varepsilon^2 + 6\sigma_{mp}^2 + 12\phi_{tp}$
$MP_{mj(i)}$	$\sigma_\varepsilon^2 + 6\sigma_{mp}^2$
SP_{km}	$\sigma_\varepsilon^2 + 3\sigma_{msp}^2 + 12\phi_{sp}$
TSP_{ikm}	$\sigma_\varepsilon^2 + 3\sigma_{msp}^2 + 6\phi_{tsp}$
$MSP_{mkj(i)}$	$\sigma_\varepsilon^2 + 3\sigma_{msp}^2$
$\varepsilon_{q(ijkm)}$	σ_ε^2

7.9   a, b.

Source	df	EMS
C_i	1	$\sigma_\varepsilon^2 + 2\sigma_F^2 + 8\sigma_C^2$
$F_{j(i)}$	6	$\sigma_\varepsilon^2 + 2\sigma_F^2$
V_k	1	$\sigma_\varepsilon^2 + \sigma_{FV}^2 + 4\sigma_{CV}^2 + 8\phi_V$
CV_{ik}	1	$\sigma_\varepsilon^2 + \sigma_{FV}^2 + 4\sigma_{CV}^2$
$FV_{kj(i)}$	6	$\sigma_\varepsilon^2 + \sigma_{FV}^2$

c. 75%.

7.11 Newman–Keuls shows experimental posttest group means better than control and one experimental group better than the other.

7.13 Newman-Keuls gives: <u>54.66 54.96 57.00</u> <u>65.66</u>
So posttest experimental is best.

7.15 Discuss based on EMS and results.

7.17 $Y_{ijkm} = \mu + G_i + C_j + GC_{ij} + S_{k(ij)} + T_m + GT_{im} + CT_{jm}$
$\qquad\qquad + GCT_{ijm} + TS_{mk(ij)}$

7.19

Source	df	EMS
D_i	2	$2\sigma_S^2 + 22\sigma_D^2$
G_j	1	$\sigma_{SG}^2 + 11\sigma_{DG}^2 + 33\phi_G$
DG_{ij}	2	$\sigma_{SG}^2 + 11\sigma_{DG}^2$
$S_{k(i)}$	30	$2\sigma_S^2$
$SG_{jk(i)}$	30	σ_{SG}^2

7.21

Source	df	EMS
R_i	1	$\sigma_\varepsilon^2 + 6\sigma_H^2 + 36\sigma_R^2$
$H_{j(i)}$	10	$\sigma_\varepsilon^2 + 6\sigma_H^2$
P_k	2	$\sigma_\varepsilon^2 + 2\sigma_{PH}^2 + 12\sigma_{RP}^2 + 24\phi_P$
RP_{ik}	2	$\sigma_\varepsilon^2 + 6\sigma_{PH}^2 + 12\sigma_{RP}^2$
$PH_{kj(i)}$	20	$\sigma_\varepsilon^2 + 2\sigma_{PH}^2$
$\varepsilon_{m(ijk)}$	36	σ_ε^2

ANSWERS TO SELECTED PROBLEMS 539

7.23

Source	df	EMS
F_i	4	$\sigma_\varepsilon^2 + 9\sigma_S^2 + 54\phi_F$
C_j	1	$\sigma_\varepsilon^2 + 9\sigma_S^2 + 135\phi_C$
FC_{ij}	4	$\sigma_\varepsilon^2 + 9\sigma_S^2 + 27\phi_{FC}$
$S_{k(ij)}$	20	$\sigma_\varepsilon^2 + 9\sigma_S^2$
P_m	2	$\sigma_\varepsilon^2 + 3\sigma_{SP}^2 + 90\phi_P$
FP_{im}	8	$\sigma_\varepsilon^2 + 3\sigma_{SP}^2 + 18\phi_{FP}$
CP_{jm}	2	$\sigma_\varepsilon^2 + 3\sigma_{SP}^2 + 45\phi_{CP}$
FCP_{ijm}	8	$\sigma_\varepsilon^2 + 3\sigma_{SP}^2 + 9\phi_{FCP}$
$SP_{mk(ij)}$	40	$\sigma_\varepsilon^2 + 3\sigma_{SP}^2$
$\varepsilon_{q(ijkm)}$	180	σ_ε^2

7.25 Recommend 1:1 concentration and any of the fillers—iron filings, iron oxide, or copper.

7.27

Source	df	EMS
S_i	9	$\sigma_\varepsilon^2 + 2\sigma_T^2 + 4\phi_S$
$T_{j(i)}$	10	$\sigma_\varepsilon^2 + 2\sigma_T^2$
P_k	1	$\sigma_\varepsilon^2 + \sigma_{PT}^2 + 20\phi_P$
SP_{ik}	9	$\sigma_\varepsilon^2 + \sigma_{PT}^2 + 2\phi_{SP}$
$PT_{kj(i)}$	10	$\sigma_\varepsilon^2 + \sigma_{PT}^2$
$\varepsilon_{m(ijk)}$	0	σ_ε^2

Test S vs. T and test P and SP vs. PT. No test on T.

7.29 b. $Y_{ijkmq} = \mu + D_i + B_j + (DB)_{ij} + M_k + (DM)_{ik} + (BM)_{jk}$
$+ (DBM)_{ijk} + W_{m(k)} + (DW)_{im(k)} + (BW)_{jm(k)}$
$+ (DBW)_{ijm(k)} + \varepsilon_{q(ijkm)}$

d. Tests are available on both die and bonder position, method, and the interactions among these factors. The method test has very low denominator degrees of freedom.

7.31 A good test exists for the factor of primary interest, supplier.

CHAPTER 8

8.1

Source	DF	SS	MS	f	P	EMS
R_i	2	337.15	168.57	30.34	0.000	$\sigma_\varepsilon^2 + 18\sigma_R^2$
S_j	1	16.67	16.67	0.26	0.659	$\sigma_\varepsilon^2 + 9\sigma_{RS}^2 + 27\phi_S$
RS_{ij}	2	126.78	63.39	11.41	0.000	$\sigma_\varepsilon^2 + 9\sigma_{RS}^2$
H_k	2	15.59	7.80	0.50	0.642	$\sigma_\varepsilon^2 + 6\sigma_{RH}^2 + 18\phi_H$
HS_{jk}	2	80.11	40.06	0.70	0.549	$\sigma_\varepsilon^2 + 3\sigma_{SRH}^2 + 9\phi_{HS}$
RH_{ik}	4	62.85	15.71	2.83	0.039	$\sigma_\varepsilon^2 + 6\sigma_{RH}^2$
SRH_{ijk}	4	229.44	57.36	10.33	0.000	$\sigma_\varepsilon^2 + 3\sigma_{SRH}^2$
$\varepsilon_{m(ijk)}$	36	200.00	5.56			σ_ε^2
Totals	53	1068.59				

Replications (blocks) and replications by all other factors are significant at the 5% level; however, these are not the factors of chief interest.

8.3

Source	df	SS	EMS
R_i	1	As before	$\sigma_\varepsilon^2 + 3\sigma_{RM}^2 + 12\sigma_R^2$
T_j	1	with R in	$\sigma_\varepsilon^2 + 3\sigma_{RM}^2 + 6\sigma_M^2 + 6\sigma_{RT}^2 + 12\phi_T$
RT_{ij}	1	place of S	$\sigma_\varepsilon^2 + 3\sigma_{RM}^2 + 6\sigma_{RT}^2$
$M_{k(j)}$	2		$\sigma_\varepsilon^2 + 3\sigma_{RM}^2 + 6\sigma_M^2$
$RM_{ik(j)}$	2		$\sigma_\varepsilon^2 + 3\sigma_{RM}^2$
$\varepsilon_{m(ijk)}$	16		σ_ε^2
Total	23		

No direct test on types, but results look highly significant. No other significant effects.

8.5

Source	df	EMS	
R_i	2	$10\sigma_R^2$	No test
M_j	4	$2\sigma_{RM}^2 + 6\phi_M$	
RM_{ij}	8	$2\sigma_{RM}^2$	No test
S_k	1	$5\sigma_{RS}^2 + 15\phi_S$	Poor test
RS_{ik}	2	$5\sigma_{RS}^2$	No test
MS_{jk}	4	$\sigma_{RMS}^2 + 3\phi_{MS}$	
RMS_{ijk}	8	σ_{RMS}^2	No test

8.7

Source	df	EMS
R_i	2	$30\sigma_R^2$
V_j	4	$6\sigma_{RV}^2 + 18\phi_V$
RV_{ij}	8	$6\sigma_{RV}^2$
S_k	5	$5\sigma_{RS}^2 + 15\phi_S$
RS_{ik}	10	$5\sigma_{RS}^2$
VS_{jk}	20	$\sigma_{RVS}^2 + 3\phi_{VS}$
RVS_{ijk}	40	σ_{RVS}^2

8.9

Source	df	EMS	
R_i	1	$\sigma_C^2 + 36\sigma_R^2$	
P_j	2	$\sigma_C^2 + 12\sigma_{RP}^2 + 24\phi_P$	Poor test
RP_{ij}	2	$\sigma_C^2 + 12\sigma_{RP}^2$	
H_k	5	$\sigma_C^2 + 6\sigma_{RH}^2 + 12\phi_H$	
RH_{ik}	5	$\sigma_C^2 + 6\sigma_{RH}^2$	
PH_{jk}	10	$\sigma_C^2 + 2\sigma_{RPH}^2 + 4\phi_{PH}$	
RPH_{ijk}	10	$\sigma_C^2 + \sigma_{RPH}^2$	
$C_{m(ijk)}$	36	$\sigma_C^2 + \sigma_C^2$	

8.13 Complete: $Y_{ijkm} = \mu + R_k + M_i + RM_{ik} + H_{j(i)} + RH_{kj(i)} + \varepsilon_{m(ijk)}$.
Reduced: $Y_{ijkm} = \mu + R_k + M_i + H_{j(i)} + \varepsilon_{m(ijk)}$.
Take fewer readings per head.

8.15 Complete: $Y_{ijkmq} = \mu + R_q + M_i + RM_{iq} + G_j + RG_{jq} + MG_{ij} + RMG_{ijq}$
$\qquad\qquad + T_{k(j)} + RT_{qk(j)} + MT_{ik(j)} + RMT_{iqk(j)} + \varepsilon_{m(ijkq)}$
Reduced: Omit all interactions with R.

8.17 a.

Source	df	EMS
R_i	1	$\sigma^2 + 2\sigma_S^2 + 20\sigma_R^2$
$S_{j(i)}$	18	$\sigma^2 + 2\sigma_S^2$
L_k	1	$\sigma^2 + 20\phi_L$
$\varepsilon_{m(ijk)}$	19	σ^2

b. For $\alpha < .098$, only S shows significance.

8.19 a. No significant results.
   b. Using $\alpha = 0.05$ both replicate and distance are significant.

CHAPTER 9

9.1

Source	DF	SS	MS	F	P
A	1	0.083	0.083	0.02	0.896
B	1	70.083	70.083	15.29	0.004
A*B	1	24.083	24.083	5.25	0.051
Error	8	36.667	4.583		
Total	11	130.917			

9.3

Source	DF	SS	MS	F	P
A	1	2,704	2,704	0.09	0.774
B	1	26,732	26,732	0.87	0.379
A*B	1	7,744	7,744	0.25	0.629
C	1	24,025	24,025	0.78	0.403
A*C	1	16,256	16,256	0.53	0.488
B*C	1	64,516	64,516	2.10	0.186
A*B*C	1	420	420	0.01	0.910
Error	8	246,284	30,786		
Total	15	388,682			

Nothing significant for any reasonable value of α.

9.5 No plots, since there are no significant effects at the 5% level.

9.7

Source	DF	SS	MS
A	1	13,736.5	
B	1	6,188.3	
AB	1	22,102.5	Same
C	1	22.8	as
AC	1	22,525.0	SS
BC	1	12,051.3	
ABC	1	20,757.0	

D	1	81,103.8	
AD	1	145,665.0	
BD	1	9,214.0	
ABD	1	126,630.3	
CD	1	148.8	
ACD	1	6757.0	
BCD	1	294.0	
ABCD	1	19,453.8	
Error	16	431,599.5	26,975.0
Totals	31	918,249.6	

9.9 a. (1), *a, b, ab, c, ac, bc, abc, d, ad, bd, abd, cd, acd, bcd, abcd.*
b.

Source	df	
Main effects	1 each for 4	
Two-way interaction	1 each for 6	
Three-way interaction	1 each for 4	Use as error
Four-way interaction	1 each for 1	
Total	15	

c. Find *A* effect by plus signs where a is present in treatment combination and minus signs were absent. Same for *D* and then multiply signs to get *AD* signs.

9.11 a. 32
b.

Source	df	
Main effects	1 each for 5	
Two-way interaction	1 each for 10	
Three-way interaction	1 cach for 10	
Four-way interaction	1 each for 5	Use as error
Five-way interaction	1 each for 1	
Total	31	

c. Get signs for *A, C,* and *E* and multiply.
d. True if no interactions are present. Otherwise use Newman-Keuls on treatment means.

9.13 Feed rate, 0.015; tool condition, new; tool type, precision.

9.15 2^6 factorial. Probably use four-, five-, and six-way interactions as error term with 22 df.

9.17

Source	df	SS and MS
Dropout temperature (A)	1	3.00
Back-zone temperature (B)	1	4.06*
$A \times B$ interaction	1	1.20
Atmosphere (C)	1	3.00
$A \times C$ interaction	1	1.71
$B \times C$ interaction	1	4.65*
$A \times B \times C$ interaction	1	15.96**

* Significant for $\alpha = 0.05$, based on MS for error of 0.37 with 3 df.
** Significant for $\alpha = 0.01$, based on MS for error of 0.37 with 3 df.

9.19

Source	df	SS and MS
A	1	45.60
B	1	1.36
AB	1	53.56
C	1	0.01
AC	1	0.00
BC	1	0.01
ABC	1	0.04

A and AB significant for $\alpha = 0.05$, based on MS for error of 4.41 with 3 df.
Newman–Keuls on A, B means: 0.00   3.95   4.35   9.95.
No significant gaps as error is too large.

9.21 Plot the following: $(1) = 0, a = -3, b = -2, ab = 5, c = -7, ac = 0, bc = 1$, and $abc = -2$.

9.23 No. Results of cold test are better on B.

9.25

Source	DF	SS	MS	F	P
Type, T	1	2,467.5	2,467.5	7.01	0.014
Position, P	1	14,921.3	14,921.3	42.38	0.000
$T*P$	1	1,582.0	1,582.0	4.49	0.045
Edge, E	1	4,255.0	4,255.0	12.08	0.002
$T*E$	1	195.0	195.0	0.55	0.464
$P*E$	1	5,644.5	5,644.5	16.03	0.001

$T*P*E$	1	569.5	569.5	1.62	0.216
Error	24	8,450.2	352.1		
Total	31	38,085.2			

Because the $P * E$ is significant at the 5% level, we examine the four means and find
0   3.5   20.12   69.75.

9.27 a. 16
b. $F = MS_D/MS_{error}$ with $v_1 = 1$ and $v_2 = 16$.

9.29 $A_{eff} = B_{eff} = 2.5$; $AB_{eff} = 7.5$

9.31 $A_{eff} = 0$, $B_{eff} = 2$, $C_{eff} = -6$

9.33 a. B is confounded with shift and C is confounded with day.

9.35 Paint the convector black, place the voltage regulator on the side of the unit, and mount the cooling fan on the side of the chassis.

CHAPTER 10

10.1

Source	DF	SS	MS	F	P
A	2	3.4844	1.7422	2.31	0.155
B	2	3.3378	1.6689	2.21	0.166
A*B	4	3.5822	0.8956	1.19	0.380
Error	9	6.8000	0.7556		
Total	17	17.2044			

10.3 $\left. \begin{array}{l} AB = 3.20 \\ AB^2 = 0.38 \end{array} \right\}$ 3.58

10.5

Source	DF	SS	MS	F	P
V	2	2.7900	1.3950	5.84	0.024
T	2	9.0133	4.5067	18.87	0.001
V*T	4	1.9867	0.4967	2.08	0.166
Error	9	2.1500	0.2389		
Total	17	15.9400			

10.7 $VT = 0.94$ and $VT^2 = 1.04$, each with 2 df.

10.9   Any combination except $T_H V_H$.

10.11

Source	DF	SS	MS	F	P
Surface A	2	25.4470	12.7235	480.47	0.000
Base B	2	47.8737	23.9369	903.91	0.000
$A*B$	4	1.8519	0.4630	17.48	0.000
Subbase C	2	41.6515	20.8257	786.43	0.000
$A*C$	4	1.8974	0.4744	17.91	0.000
$B*C$	4	0.2774	0.0694	2.62	0.057
$A*B*C$	8	1.0004	0.1250	4.72	0.001
Error	27	0.7150	0.0265		
Total	53	120.7143			

All except $B \times C$ are significant at the 5% level, but the main effects predominate.

10.13

Source		df		SS	
AB {	AB	4 {	2	1.85 {	0.67
	AB^2		2		1.18
AC {	AC	4 {	2	1.90 {	0.59
	AC^2		2		1.31
BC {	BC	4 {	2	0.28 {	0.06
	BC^2		2		0.22
	ABC		2		0.18
ABC {	ABC^2	8 {	2	1.00 {	0.08
	AB^2C		2		0.21
	AB^2C^2		2		0.53

10.15 Plots show strong linear effects of A, B, and C. They show AB and AC interaction— but it is slight compared with the main effects. A slight quadratic trend in A is also noted.

10.17 a. For $\alpha = 0.05$, we do not reject the hypothesis of no interaction (p value is 0.0743). Both current density and temperature are highly significant (p value ≈ 0.000). Using SNK on the main effects, conclude that the true average thickness is least when both temperature and current density are at their low levels and greatest when both are at their high levels.

b.

Source	df	SS	MS	f	p value
D	2	155378.13	77689.07		
D_L	1	155141.21	155141.21	1993.08	0.0000
D_Q	1	236.92	236.92	3.04	0.0983
T	2	3513.78	1756.89		
T_L	1	3438.14	3438.14	44.17	0.0000
T_Q	1	75.64	75.64	0.97	0.3377
DT	4	795.94	198.99		
$D_L T_L$	1	90.70	90.70	1.17	0.2937
$D_L T_Q$	1	126.23	126.23	1.62	0.2193
$D_Q T_L$	1	577.04	577.04	7.41	0.0140
$D_Q T_Q$	1	1.97	1.97	0.03	0.8644
Error	18	1401.07	77.84		
Total	26				

CHAPTER 11

11.1

	Source	df	EMS
Whole	R_i	1	$\sigma_\varepsilon^2 + 6\sigma_{RS}^2 + 18\sigma_R^2$
plot	H_j	2	$\sigma_\varepsilon^2 + 2\sigma_{RHS}^2 + 4\sigma_{HS}^2 + 6\sigma_{RH}^2 + 12\phi_H$
	RH_{ij}	2	$\sigma_\varepsilon^2 + 2\sigma_{RHS}^2 + 6\sigma_{RH}^2$
Split	S_k	2	$\sigma_\varepsilon^2 + 6\sigma_{RS}^2 + 12\sigma_S^2$
plot	RS_{ik}	2	$\sigma_\varepsilon^2 + 6\sigma_{RS}^2$
	HS_{jk}	4	$\sigma_\varepsilon^2 + 2\sigma_{RHS}^2 + 4\sigma_{HS}^2$
	RHS_{ijk}	4	$\sigma_\varepsilon^2 + 2\sigma_{RHS}^2$
	$\varepsilon_{m(ijk)}$	18	σ_ε^2
	Total	35	

11.3   **a.** Fill nine spaces in random order.

 b. Choose one of the orifice sizes at random, randomize flow rate through this orifice, then move to a second orifice size, and so on.

c.

Source	df
R_i	2
O_j	2
RO_{ij}	4
F_k	2
RF_{ik}	4
OF_{jk}	4
ROF_{ijk}	8

d. Take more replications.

11.5

	Source	df	EMS
Whole plot	R_i	3	$\sigma_\varepsilon^2 + 8\sigma_{RA}^2 + 40\sigma_R^2$
	S_j	1	$\sigma_\varepsilon^2 + 4\sigma_{RSA}^2 + 16\sigma_{SA}^2 + 20\sigma_{RS}^2 + 80\phi_S$
	RS_{ij}	3	$\sigma_\varepsilon^2 + 4\sigma_{RSA}^2 + 20\sigma_{RS}^2$
	J_k	1	$\sigma_\varepsilon^2 + 4\sigma_{RJA}^2 + 16\sigma_{JA}^2 + 20\sigma_{RJ}^2 + 80\phi_J$
	RJ_{ik}	3	$\sigma_\varepsilon^2 + 4\sigma_{RJA}^2 + 20\sigma_{RJ}^2$
	SJ_{jk}	1	$\sigma_\varepsilon^2 + 2\sigma_{RSJA}^2 + 10\sigma_{RSJ}^2 + 8\sigma_{SJA}^2 + 40\phi_{SJ}$
	RSJ_{ijk}	3	$\sigma_\varepsilon^2 + 2\sigma_{RSJA}^2 + 10\sigma_{RSJ}^2$
Split plot	A_m	4	$\sigma_\varepsilon^2 + 8\sigma_{RA}^2 + 32\sigma_A^2$
	RA_{im}	12	$\sigma_\varepsilon^2 + 8\sigma_{RA}^2$
	SA_{jm}	4	$\sigma_\varepsilon^2 + 4\sigma_{RSA}^2 + 16\sigma_{SA}^2$
	RSA_{ijm}	12	$\sigma_\varepsilon^2 + 4\sigma_{RSA}^2$
	JA_{km}	4	$\sigma_\varepsilon^2 + 4\sigma_{RJA}^2 + 16\sigma_{JA}^2$
	RJA_{ikm}	12	$\sigma_\varepsilon^2 + 4\sigma_{RJA}^2$
	SJA_{jkm}	4	$\sigma_\varepsilon^2 + 2\sigma_{RSJA}^2 + 8\sigma_{SJA}^2$
	$RSJA_{ijkm}$	12	$\sigma_\varepsilon^2 + 2\sigma_{RSJA}^2$
	$\varepsilon_{q(ijkm)}$	80	σ_ε^2

11.7

	Source	df	MS	EMS
Whole plot	R_i	2	60.46	$\sigma_\varepsilon^2 + 5\sigma_{RO}^2 + 20\sigma_R^2$
	D_j	4	276.19	$\sigma_\varepsilon^2 + \sigma_{RDO}^2 + 3\sigma_{DO}^2 + 4\sigma_{RD}^2 + 12\phi_D$
	RD_{ij}	8	40.63	$\sigma_\varepsilon^2 + \sigma_{RDO}^2 + 4\sigma_{RD}^2$

Split	O_k	3	16.73	$\sigma_\varepsilon^2 + 5\sigma_{RO}^2 + 15\sigma_O^2$
plot	RO_{ik}	6	57.44	$\sigma_\varepsilon^2 + 5\sigma_{RO}^2$
	DO_{jk}	12	12.78	$\sigma_\varepsilon^2 + 2\sigma_{RDO}^2 + 3\sigma_{DO}^2$
	RDO_{ijk}	24	46.83	$\sigma_\varepsilon^2 + \sigma_{RDO}^2$
	Total	59		

Days are significant by an F' test.

11.9 In a nested experiment the levels of B, say, are different within each level of A. In a split-plot experiment the same levels of B (in the split) are used at each level of A.

11.11

Laboratory means	1	3	2
	11.22	11.25	12.53

Mix means:	B	A	C
	10.29	11.58	13.13

Laboratory $\times$ mix means all different in decreasing order: 165, 155, 145 with B, A, and C.

11.13

To test T, $MS = 0.93$ and $F' = 2000$

To test M, $MS = 0.39$ and $F' = 189$

To test $T * M$, $MS = 0.11$ and $F' = 96$

All highly significant.

11.15

Source	df	EMS
R_i	$r - 1$	$\sigma_\varepsilon^2 + 15\sigma_R^2$
T_j	4	$\sigma_\varepsilon^2 + 3\sigma_{RT}^2 + 3r\phi_T$
RT_{ij}	$4(r - 1)$	$\sigma_\varepsilon^2 + 3\sigma_{RT}^2$
M_k	2	$\sigma_\varepsilon^2 + 5\sigma_{RM}^2 + 5r\phi_M$
RM_{ik}	$2(r - 1)$	$\sigma_\varepsilon^2 + 5\sigma_{RM}^2$
TM_{jk}	8	$\sigma_\varepsilon^2 + \sigma_{RMT}^2 + r\phi_{MT}$
RTM_{ijk}	$8(r - 1)$	$\sigma_\varepsilon^2 + \sigma_{RMT}^2$

Preferable design.

CHAPTER 13

13.1 Aliases:

Source	df	SS
$A = C$	1	30.25
$B = ABC$	1	2.25
$AB = BC$	1	0.25

No tests.

13.3 Aliases are
$$A = BC = AB^2C^2$$
$$B = AB^2C = AC$$
$$C = ABC^2 = AB$$
$$AB^2 = AC^2 = BC^2$$

13.5 $A = BD, B = AD, C = ABCD, D = AB, AC = BCD, BC = ACD, CD = ABC$.

13.7
$$A = AB^2C^2D = BCD^2$$
$$B = AB^2CD^2 = ACD^2$$
$$C = ABC^2D^2 = ABD^2$$
$$D = ABC = ABCD$$
$$AB = ABC^2D = CD^2$$
$$AB^2 = AC^2D = BC^2D$$
$$AC = AB^2CD = BD^2$$
$$AC^2 = AB^2D = BC^2D^2$$
$$AD = AB^2C^2 = BCD$$
$$BC = AB^2C^2D^2 = AD^2$$
$$BC^2 = AB^2D^2 = AC^2D^2$$
$$BD = AB^2C = ACD$$
$$CD = ABC^2 = ABD$$

13.9

Source	df
Main effects with 5- or 6-way	7
2-way interaction with 4- or 6-way	21
3-way interaction, and so on as error	35
Total	63

13.11

Source	df	SS	MS
Heats (or 2-way)	6	51.92	8.65*
Treatments (or part of H)	1	12.04	12.04*
Positions (or 3-way)	2	0.58	0.29
PH (or 3-way)	12	29.83	
PT (or 2-way)	2	3.59	33.42 2.39
Totals	23	97.96	

13.13 Combining three- and four-way interactions as error with 5 df gives

Source	MS
A	12.76
B	53.40*
AB	0.00
C	33.44
AC	7.06
BC	0.00
D	31.22
AD	42.48*
BD	1.62
CD	0.07
Error	5.11

B and AD significant at 5% level.

13.15 a. Replication I: $(1), ab, de, abde, bcd, acd, bce, ace.$
   b. $BC = DE = ABD = ACE$
   c. Replications $\times$ three-way $= 14$ df.

13.17 a. ACD
   b.

Source	df	MS
A (or CD)	1	5729.85*
B (or $ABCD$)	1	91.80
C (or AD)	1	4199.86*
D (or AC)	1	1217.71
AB, CD, BD, and their aliases	3	281.88

Some bad aliases, better to confound $ABCD$.